Critical issues in plant health: 50 years of research in African agriculture

It is widely recognised that agriculture is a significant contributor to global warming and climate change. Agriculture needs to reduce its environmental impact and adapt to current climate change whilst still feeding a growing population, i.e. become more 'climate-smart'. Burleigh Dodds Science Publishing is playing its part in achieving this by bringing together key research on making the production of the world's most important crops and livestock products more sustainable.

Based on extensive research, our publications specifically target the challenge of climate-smart agriculture. In this way we are using 'smart publishing' to help achieve climate-smart agriculture.

Burleigh Dodds Science Publishing is an independent and innovative publisher delivering high quality customer-focused agricultural science content in both print and online formats for the academic and research communities. Our aim is to build a foundation of knowledge on which researchers can build to meet the challenge of climate-smart agriculture.

For more information about Burleigh Dodds Science Publishing simply call us on +44 (0) 1223 839365, email info@bdspublishing.com or alternatively please visit our website at www.bdspublishing.com.

Related titles:

Achieving sustainable cultivation of cassava Volume 1: Cultivation techniques
Print (ISBN 978-1-78676-000-5) Online (ISBN 978-1-78676-002-9; 978-1-78676-003-6)

Achieving sustainable cultivation of cassava Volume 2:
Genetics, breeding, pests and diseases
Print (ISBN 978-1-78676-004-3) Online (ISBN 978-1-78676-006-7; 978-1-78676-007-4)

Achieving sustainable cultivation of grain legumes Volume 1:
Advances in breeding and cultivation techniques
Print (ISBN 978-1-78676-136-1) Online (ISBN 978-1-78676-138-5; 978-1-78676-139-2)

Achieving sustainable cultivation of grain legumes Volume 2:
Improving cultivation of particular grain legumes
Print (ISBN 978-1-78676-140-8) Online (ISBN 978-1-78676-142-2; 978-1-78676-143-9)

Achieving sustainable cultivation of maize Volume 1:
From improved varieties to local applications
Print (ISBN 978-1-78676-008-1) Online (ISBN 978-1-78676-010-4; 978-1-78676-011-1)

Achieving sustainable cultivation of maize Volume 2:
Cultivation techniques, pest and disease control
Print (ISBN 978-1-78676-012-8) Online (ISBN 978-1-78676-014-2; 978-1-78676-015-9)

Chapters are available individually from our online bookshop:
https://shop.bdspublishing.com

climate•SMART•publishing
IN AGRICULTURAL SCIENCE

BURLEIGH DODDS SERIES IN AGRICULTURAL SCIENCE
NUMBER 58

Critical issues in plant health: 50 years of research in African agriculture

Edited by Dr Peter Neuenschwander and Dr Manuele Tamò, International Institute of Tropical Agriculture (IITA), Benin

burleigh dodds
SCIENCE PUBLISHING

Published by Burleigh Dodds Science Publishing Limited
82 High Street, Sawston, Cambridge CB22 3HJ, UK
www.bdspublishing.com

Burleigh Dodds Science Publishing, 1518 Walnut Street, Suite 900, Philadelphia, PA 19102-3406, USA

First published 2019 by Burleigh Dodds Science Publishing Limited

Library of Congress Control Number: 2018967582

British Library Cataloguing in Publication Data
A catalogue record for this book is available from the British Library

ISBN 978-1-78676-232-0 (Print)
ISBN 978-1-78676-235-1 (PDF)
ISBN 978-1-78676-234-4 (ePub)
ISSN 2059-6936 (print)
ISSN 2059-6944 (online)

DOI 10.19103/AS.2018.0043

Typeset by Deanta Global Publishing Services, Chennai, India
Printed by CPI Group (UK) Ltd, Croydon CR0 4YY

Contents

Series list

Title	Series number
Achieving sustainable cultivation of maize - Vol 1 From improved varieties to local applications Edited by: Dr Dave Watson, CGIAR Maize Research Program Manager, CIMMYT, Mexico	001
Achieving sustainable cultivation of maize - Vol 2 Cultivation techniques, pest and disease control Edited by: Dr Dave Watson, CGIAR Maize Research Program Manager, CIMMYT, Mexico	002
Achieving sustainable cultivation of rice - Vol 1 Breeding for higher yield and quality Edited by: Prof. Takuji Sasaki, Tokyo University of Agriculture, Japan	003
Achieving sustainable cultivation of rice - Vol 2 Cultivation, pest and disease management Edited by: Prof. Takuji Sasaki, Tokyo University of Agriculture, Japan	004
Achieving sustainable cultivation of wheat - Vol 1 Breeding, quality traits, pests and diseases Edited by: Prof. Peter Langridge, The University of Adelaide, Australia	005
Achieving sustainable cultivation of wheat - Vol 2 Cultivation techniques Edited by: Prof. Peter Langridge, The University of Adelaide, Australia	006
Achieving sustainable cultivation of tomatoes Edited by: Dr Autar Mattoo, USDA-ARS, USA & Prof. Avtar Handa, Purdue University, USA	007
Achieving sustainable production of milk - Vol 1 Milk composition, genetics and breeding Edited by: Dr Nico van Belzen, International Dairy Federation (IDF), Belgium	008
Achieving sustainable production of milk - Vol 2 Safety, quality and sustainability Edited by: Dr Nico van Belzen, International Dairy Federation (IDF), Belgium	009
Achieving sustainable production of milk - Vol 3 Dairy herd management and welfare Edited by: Prof. John Webster, University of Bristol, UK	010
Ensuring safety and quality in the production of beef - Vol 1 Safety Edited by: Prof. Gary Acuff, Texas A&M University, USA & Prof. James Dickson, Iowa State University, USA	011
Ensuring safety and quality in the production of beef - Vol 2 Quality Edited by: Prof. Michael Dikeman, Kansas State University, USA	012
Achieving sustainable production of poultry meat - Vol 1 Safety, quality and sustainability Edited by: Prof. Steven C. Ricke, University of Arkansas, USA	013
Achieving sustainable production of poultry meat - Vol 2 Breeding and nutrition Edited by: Prof. Todd Applegate, University of Georgia, USA	014

The International Institute of Tropical Agriculture (IITA)

The International Institute of Tropical Agriculture (IITA) is a not-for-profit institution that generates agricultural innovations to meet Africa's most pressing challenges of hunger, malnutrition, poverty, and natural resource degradation. Working with various partners across sub-Saharan Africa, we improve livelihoods, enhance food and nutrition security, increase employment, and preserve natural resource integrity. IITA is a member of CGIAR, a global agriculture research partnership for a food secure future.

List of major IITA donors

- Administration générale de la coopération au développement, Belgium
- African Development Bank
- Bill & Melinda Gates Foundation
- Bundeskanzleramt, Austria
- Bundesministerium für Wirtschaftliche Zusammenarbeit und Entwicklung (BMZ), Germany
- Canadian International Development Agency (CIDA), Canada
- Danish International Development Agency (DANIDA), Denmark
- Department for development cooperation, Italy
- Department for International Development (DFID), United Kingdom
- Deutsche Gesellschaft für Internationale Zusammenarbeit (GIZ), Germany
- Directoraat Generaal Internationale Samenwerking (DGIS, now NEDA: Netherlands development assistance), The Netherlands
- European Union
- Food and Agriculture Organisation of the United Nations (FAO)
- Gatsby Trust
- Global Crop Diversity Trust
- International Development Research Center (IDRC), Canada
- International Fund for Agricultural Development (IFAD)
- King's College London Alumnae Association, UK
- Ministère de développement et de la coopération, France
- Overseas Development Administration (ODA), United Kingdom
- Robert Bosch Stiftung, Stuttgart, Germany
- Root, Tuber and Banana CRP of the CGIAR
- Rotary International
- Swiss Directorate for Development Cooperation and Humanitarian Aid (SDC), Switzerland
- Swiss National Science Foundation (SNSF) Marie Heim-Vögtlin research fellowship
- The Research Council of Norway
- The Rockefeller Foundation (RF)
- The transgenic research on control of banana Xanthomonas wilt was supported by USAID
- United Nations Development Program (UNDP)
- United States Agency for International Development (USAID), United States of America
- VVOB – education for development
- World Bank

Chapter lead author contact details

Lead authors for chapters	Email
Ch 1 Kenton Dashiell, IITA, Nigeria	k.dashiell@cgiar.org
Ch 2 Peter Neuenschwander, IITA, Benin	p.neuenschwander@cgiar.org
Ch 3 P. Lava Kumar, IITA, Nigeria	l.kumar@cgiar.org
Ch 4 Georg Goergen, IITA, Benin	g.goergen@cgiar.org
Ch 5 James P. Legg, IITA, Tanzania	j.legg@cgiar.org
Ch 6 Muaka Toko, IITA, Benin	m.toko@cgiar.org
Ch 7 Ranajit Bandyopadhyay, IITA, Nigeria	r.bandyopadhyay@cgiar.org
Ch 8 Djana Mignouna, IITA, Nigeria	d.mignouna@cgiar.org
Ch 9 Stefan Hauser, IITA, Nigeria	s.hauser@cgiar.org
Ch 10 Manuele Tamò, IITA, Benin	m.tamo@cgiar.org
Ch 11 Ignace Godonou, IITA, Benin	ignace.godonou@hotmail.fr
Ch 12 Peter Neuenschwander, IITA, Benin	p.neuenschwander@cgiar.org
Ch 13 Frederick Schreurs, IITA, Nigeria	f.schreurs@cgiar.org
Ch 14 David Chikoye, IITA, Zambia	d.chikoye@cgiar.org
Ch 15 Danny Coyne, IITA, Kenya	d.coyne@cgiar.org
Ch 16 Peter Neuenschwander, IITA, Benin	p.neuenschwander@cgiar.org

Foreword

Sanginga Nteranya

Director General, International Institute of Tropical Agriculture (IITA), Nigeria

As Director General of IITA I have the pleasure to present to you 'Critical issues in plant health: 50 years of research in African agriculture'. Fifty years is a long time in an institution that was created to assist newly independent African countries to develop their own research and outreach capacity for crops for their own populations. Together with similar institutions, IITA became a founding member of the Consultative Group on International Agricultural Research (CGIAR). With a global reach, its goal was to combat malnutrition and poverty through research for development (R4D). What started in 1967 with a research staff of a few dozen scientists in Ibadan, Nigeria, supported by the Ford and Rockefeller Foundations, has developed into an institute of over 200 scientists based at 22 stations in 18 African countries and supported by dozens of international and national donors.

From the beginning, IITA invested its resources in Resource and Crop Management and variety development of its 'mandate' crops: cassava, yam and other root crops, maize, cowpea and other legumes, banana and plantain. Through the years, the focus was widened to include all aspects of development of rural households, from achieving higher yields to sustainable farm production and income generation, from subsistence farming to commercialized farm enterprises. The aim was to lift the rural and urban poor out of poverty and drudgery to face the international developments of the 21st century. This was to be achieved in close collaboration with and according to the needs of national programs, subject to local political conditions. While it was recognized that basic studies on these mostly under-researched crops was essential for sustainable progress, the application of results in practice and adoption, even adaption, by the collaborating national programs was always essential. In the context of changing international pressures, and with a total of six Directors General, the relative importance and stress on research versus development has continued to change. The Institute still has constantly to decide where and to what degree to stress basic research, testing and implementation, education, and commercialization and at what point to pass on the baton to national research institutions, international and local non-government organizations that can better (i.e. more cheaply) distribute new research results and techniques.

Today, it is fascinating to observe what high percentage of all plant breeders, biocontrol specialists, economists, and other professionals and decision makers active in African governments, international organisations and NGOs received their education at IITA. To our great joy, IITA's achievements are regularly put into practice, sometimes without acknowledgement of the source – which is actually encouraged to improve adoption, though it may pose a handicap when we have to document impact for our donors. As a testimony to our outreach efforts, the percentage of African scientists among the family of international staff of IITA has constantly risen throughout the years, the collegial interaction with African universities has improved, and political acceptance at least in Africa must be considered excellent – perhaps better than on the larger world scene. This has only been possible thanks to the constant support by donors from Europe, America, but also from Africa, which is greatly appreciated. We can proudly tell them that their investment has paid of many times over and the benefits have gone directly to the poor as well as to upwardly mobile farmers and urban consumers.

Today, among all of IITA's R4D, plant protection occupies from one quarter to one third of all activities, next to crop improvement, soil and farm management and socio-economics. For this volume, two IITA staff who have worked here for 35 and 30 years, respectively, thereby covering a good part of IITA's history, have assembled all of our successes and failures in plant protection as editors of 16 chapters. First they asked 16 former IITA staff, former collaborators, and investors about their opinion on IITA's plant protection R4D. Then, 70 authors and co-authors describe their contributions to the common goal of protecting crops in a sustainable manner from pests like pathogens (including viruses), insects, mites, nematodes, and weeds. You will read many technical discussions; but throughout I hope you can feel the warm heart and determination of these scientists to improve the lot of African farmers. These include the richer segment of farmers who are able to profit from improved inputs created by our scientists, but also poor subsistence farmers, who can benefit from our research which they can use for free.

Some experts try to direct our research towards the mechanized, large-scale farming dependent on abundant fertilizer and pesticide use than can be seen in countries like the USA. Others want us to preserve small-scale family farms. We are well aware that the existing trend of increasing production by penetrating into forests, swamp-land and other habitats that are not particularly suited for agriculture or need to be protected as nature reserves is not sustainable. On the other hand, the ancient long-time fallow systems just do not suffice anymore. Relentless population increase creates an urgent need for new methods to produce more in a sustainable way on existing cultivated land. We therefore adopt pragmatic solutions, which in plant protection research mean the reduction of the horrendous crop losses and quality reduction due to pests in the field and in storage by all available means. These include developing resistant crop varieties, applying biological control in all its forms, developing so-called biorational pesticides to replace broad-spectrum synthetic pesticides that pollute the environment or at least to improve the application methods of existing pesticides, and finally - and perhaps most importantly - to develop cultural control methods, in other words, good farming practices. The present volume shows how far we have come in this quest, but also points out how far we still have to go in collaboration and with the support of, African governments, their extension services, universities across the globe, financial donors, whom we see as investors and, of course, African farmers.

Have a good read and judge for yourself! Nteranya Sanginga

Part 1

Managing threats to plant health

Chapter 1

Key challenges in plant health in sub-Saharan Africa: stakeholder priorities

Kenton Dashiell, *International Institute of Tropical Agriculture (IITA), Nigeria (corresponding author: k.dashiell@cgiar.org)*

1 Introduction

This chapter comprises short summaries from key stakeholders who have worked with the IITA on plant health. Each stakeholder was asked to:

- Introduce their organisation
- Comment on what they see as the main plant health issues affecting African agriculture
- Describe the nature of their organisation's collaboration with the IITA and the main achievements of that collaboration
- Reflect on future challenges

Stakeholders were categorised into the following groups:

- International organisations
- Universities
- Institutes working within national agricultural research systems (NARS)
- Donors/investors

Particular individuals who sent statements on behalf of their organisations were:
International organisations:

- Food and Agriculture Organisation (FAO) Dr Hans Dreyer, Director – Plant Production and Protection Division (AGP) (formerly IITA)

http://dx.doi.org/10.19103/AS.2018.0043.01
Published by Burleigh Dodds Science Publishing Limited, 2019.

- FAO: Professor Lukas Brader, former Director – Plant Production and Protection Division (AGP) and former Director General of IITA
- FAO: Dr Joyce Mulila-Mitti, Plant Production and Protection Officer for Southern Africa, Plant Production and Protection Division (AGP)
- International Rice Research Institute (IRRI): Dr Jacqueline Hughes, Deputy Director General for Research (formerly IITA)
- CABI: Dr George Oduor, Deputy Director of Research, CABI African Regional Centre
- Millennium Institute: Professor Hans Rudolf Herren (formerly IITA)

Universities:

- Professor Altus Viljoen, Stellenbosch University, South Africa
- Dr Olajumoke Alabi, University of Ibadan, Nigeria
- Professor Christian Borgemeister, University of Bonn, Germany (formerly IITA)
- Professor Barry Pittendrigh, Michigan State University, USA
- Professor Gilberto de Moraes, University of São Paulo, Brazil

Institutes working within NARS:

- Dr Willy Tata Hangy, National Institute for Agricultural Studies and Research (INERA), Democratic Republic of Congo
- Dr David Arodokoun, INRAB, Benin
- Dr Clémentine Dabiré/Binso, National Institute for Agricultural Studies and Research (INERA), Burkina Faso

Donors/investors:

- Dr Marlis Lindecke, Advisory Service on Agricultural Research for Development (BEAF), German Economic Corporation for International Cooperation (GIZ), Germany;
- Dr Stephan Krall, Director of the 'Forest, Biodiversity and Agriculture' section of GIZ

Statements are grouped by organisation type.

2 Priorities in plant health: international organisations

2.1 FAO: Hans Dreyer

Achieving food security for all is at the heart of FAO's efforts, while sustainable crop production and management of natural resources is at the core of FAO's mandate. FAO encourages sustainable crop production intensification, known as *'Save and Grow'*, which uses an ecosystems approach. It promotes Integrated Pest Management (IPM) as the preferred, ecologically based approach to growing healthy crops. FAO also provides the Secretariat for the International Plant Protection Convention (IPPC), the global agreement on plant health aiming at preventive measures against the introduction and spread of trans-boundary plant pests and diseases.

Trans-boundary plant pests and diseases are major plant health issues that increasingly challenge African agriculture. They affect many key crops and can thus have serious impacts on food security and livelihoods. Some of them have been – or may become – serious plant health threats to African agriculture: mealybug, mosaic virus and brown streak virus in cassava; *Xanthomonas* wilt, bunchy top virus and *Fusarium* wilt in banana and lethal necrosis and armyworms in maize.

IITA and FAO closely collaborate to tackle these challenges. In its support to member countries FAO draws on the scientific expertise and knowledge developed by IITA. Recent examples of collaboration include supporting seed systems and provision of farmers with disease-free planting material of cassava in the Great Lakes region and combatting the banana diseases bunchy top virus in Central Africa and *Fusarium* wilt globally. IITA has also been instrumental in the development of a specific and low-risk biopesticide in locust control.

Such collaboration is expected to become even more important in the future as plant health challenges related to trans-boundary pests and diseases will diversify and intensify due to climate change and intensified international trade. As a former research fellow at IITA, and in my present position at FAO, I am keen to further the good collaboration between FAO and this eminent centre of excellence, which is doing a marvellous job in developing innovative solutions to pressing problems that African small-scale farmers face these days.

2.2 FAO: Lukas Brader

Around the mid-1980s, as director of the Plant Production and Protection Division of the FAO, I found that plant protection did not describe adequately what we want to achieve after all the earlier efforts made in the development and introduction of IPM. I therefore suggested that plant health was a much better description of these various efforts to achieve sustainable results. However, this idea was too innovative to get support within the FAO at that time. After joining IITA in 1990, their inspiring biological control research programme offered an excellent opportunity to further evolve and apply the concept of plant health. The Plant Health Division was established, and this Division has been and still is one of the major driving forces behind this more holistic approach. The fact that it is now possible to publish a book on plant health shows that all these pioneering efforts have resulted in a definite change in attitudes on how to grow and protect crops more sustainably.

2.3 FAO: Joyce Mitti

Plant health is a priority for achieving sustainable crop production intensification, being a key component of one of the FAO's five Strategic Objectives. The major plant health interventions of the FAO to which IITA has significantly contributed include strengthening the capacity of national plant protection organisations (NPPOs) to implement the IPPC as well as management of trans-boundary plant pests.

IITA has played a key role as a strategic partner of the FAO by serving as member of the Steering Committee for the Inter-Africa Phytosanitary Council (IAPSC), the Regional Plant Protection Organization for sub-Saharan Africa. IITA has contributed significantly to the development of the IAPSC's strategic plan for establishing robust plant health systems for Africa. IITA also provides direct support to countries for management of trans-boundary plant pests. This support includes surveillance, pest identification and biological control. IITA has made significant contributions to the control of a number of major pests affecting some of Africa's most important crops, such as cassava mosaic virus, cassava brown streak disease (CBSD), maize lethal necrosis disease, banana bunchy top disease (BBTV), banana *Fusarium* wilt (TR4), banana *Xanthomonas* wilt and the fall armyworm which affects cereals such as maize and sorghum.

Capacity development, especially for NPPOs in ensuring IPPC compliance and effective management of trans-boundary plant pests, remain a key challenge. This requires continued collaboration between FAO and IITA, particularly to strengthen early detection and rapid response systems.

2.4 IRRI: Jacqueline Hughes

The IRRI, founded in 1960, is an autonomous, non-profit agricultural research and training organisation with offices in 17 rice-growing nations in Africa and Asia. IRRI aims to improve livelihoods and nutrition, abolishing poverty, hunger and malnutrition among those who depend on rice-based agri-food systems. In doing so, IRRI's work protects the health of rice farmers and consumers, and the environmental sustainability of rice farming in a world challenged by climate change. IRRI's work promotes the empowerment of women and supports opportunities for youth in an equitable agri-food system.

Historically, IITA paralleled IRRI's efforts on varietal improvement for both upland and irrigated environments for West and Central Africa with particular reference to greater yield potential and blast resistance until 1990 when lowland rice improvement activities were transferred to the West Africa Rice

Published by Burleigh Dodds Science Publishing Limited, 2019.

Development Association (WARDA, now the Africa Rice Center or AfricaRice) under the IRRI/IITA/WARDA Memorandum of Understanding signed in 1991. During this time, IRRI's collaboration with IITA and WARDA was facilitated by an IRRI Liaison Scientist and a coordinator for the International Network for Genetic Evaluation of Rice (INGER)-Africa based at IITA.

Through hosting IRRI scientists at IITA in Nigeria and Tanzania, IRRI and IITA, in collaboration with AfricaRice, have continued sharing and exchanging improved lines that served as a basis for rice breeding programmes in Africa, addressing specific constraints, such as low yield and, in particular, rice diseases and pests that further reduce yield. IRRI will continue its mandate of strong collaboration and partnership with AfricaRice, IITA and other strategic African partners through an alliance of RICE programme partners aimed at strengthening and extending partnership and building capacities in rice science.

My own career with the IITA, through subsequent jobs at AVRDC (the World Vegetable Center) and IRRI, has made me particularly aware of the importance of IITA in plant pathology and, more specifically, plant virology and seed health. Ten years of working as a plant virologist with IITA's mandate crops across sub-Saharan Africa, leading the Seed Health Unit, managing the biotechnology laboratories and supporting national partners and other CGIAR Centers enabled me to contribute to IITA's successes and minimise the risk of disease in both seed and vegetatively propagated crops. The strong institutional support for building capacity in virology at IITA and with partners, and for ensuring safe germplasm movement within Africa and beyond, was much appreciated within the region. The capacity of the staff and students, the laboratory facilities and linkages with partners were a firm base for subsequent scientists to build upon, and through which the reputation of IITA has been maintained and enhanced globally.

2.5 CABI: George Oduor

The mission of CAB International (CABI) is to 'improve people's lives worldwide by providing information and applying scientific expertise to solve problems in agriculture and the environment'. The principal beneficiaries of CABI's scientific research programmes are farmers (men, women and youth) better able to gain access to sufficient safe and nutritious food, as well as improved livelihoods from better market access through sustainable, climate-resilient agriculture in healthy ecosystems. In existence for more than 100 years, CABI's foundation is based on biological control of agricultural pests.

African agriculture faces several plant health challenges which include low capacity to conduct research and development (both in human resources and facilities); unfavourable policies leading to a lower priority given to agriculture;

weak regulatory systems; pre- and post-harvest pests (insects, diseases, weeds etc.); declining soil fertility and weak value chains. As a component of IPM, biocontrol is attractive as it is environmentally safe, more targeted than most pesticides, more affordable for the smallholder farmer and more sustainable.

CABI and IITA, as well as the Kenya Agricultural and Livestock Research Organisation, where I started my research career, have collaborated in a number of ways to address some of the above challenges. IITA has been highly instrumental in training African scientists, who are currently scattered in both national and international agricultural research centres (I am an example trained to MSc and PhD level). Examples of projects implemented jointly are the hugely successful biological control of the cassava mealybug (CM) and green mite. This brought a great relief to cassava farmers, and I'm proud to state that my villagers in Western Kenya still praise me for this achievement! We also collaborated in the biological control of locusts, identifying a strain of *Metarhizium anisopliae* and developing a commercial product (Green Muscle®). This still serves as an environmentally safe approach to manage desert locusts and other locusts and grasshoppers.

Despite this success in controlling major pests in African agriculture, there are still challenges for the future. These are mainly due to increased trade (movement of people and commodities), climate change and weak government regulatory services. These problems create a conducive environment for the spread of invasive pests, as shown by pests such as *Tuta absoluta*, Dodder (*Cuscuta* spp.) and, more recently, the fall armyworm (*Spodoptera frugiperda*). The IITA will continue to have a major role to play in addressing these evolving threats.

2.6 Millennium Institute: Hans R. Herren

Millennium Institute's (MI) vision is a world in which decision makers use system dynamics thinking and tools to analyse and understand the interconnectedness between economic, social, environmental factors, and issues of peace and security. MI seeks to play a catalytic role in creating a global network of system thinkers to solve critical twenty-first-century challenges. For this purpose, MI has developed the Integrated Sustainable Development Goals (iSDG) Model, a comprehensive policy simulation tool that generates country-specific development scenarios to show the implications of policy on a country's progress towards achieving the SDGs.

Plant health issues are multifaceted and therefore need a systemic approach to forestall problems. Climate change will add to the challenge of pest pressure. However, if the system is rebuilt according to the principle of regenerative, organic agriculture and agroecology, there is ample potential to overcome challenges in the next decades and beyond. There is also a need to take the longer view, something that IITA has been sacrificing for the short-term solutions, which would be more in the remit of the National Agricultural Research Organizations and perhaps regional research organisations too.

While at IITA from 1979 to 1994, I created the Africa-wide Cassava and Green Mite Biological Control Program, which eventually migrated to the IITA Cotonou research station for lack of space and government support for the special needs of the programme in Nigeria. With my colleagues from the Biocontrol Program, crop protection was transformed into the Plant Health Division with the aim to change the approach in plant protection from cure to prevention.

Since I left IITA to become first DG of icipe and then president of MI, collaboration between MI and IITA has unfortunately not worked out for the introduction of a systems' approach to research planning by using the tools developed at MI. Since the problems facing agriculture and also pest management/plant health are so complex and beyond the capacity of the human brain, we suggest using the system dynamics models developed by the MI and deployed in a number of countries.

In conclusion, we need a totally different approach to our food, fibre and feed production systems, which also includes the broader food system, as we cannot continue to treat production and consumption separately. The UN-commissioned International Assessment of Agricultural Knowledge, Science and Technology for Development (IAASTD) report 'Agriculture at a Crossroads' was already clear about that some 8 years ago; but, unfortunately, the CGIAR, donors and development partners have not followed up on this insight. Similarly, the more recent report from the International Panel of Experts on Sustainable Food Systems (IPES-Food), 'From Uniformity to Diversity', outlines the key blockages to the agricultural transformation needed. We therefore recommend that IITA moves more decisively towards a holistic agroecology framework and away from a reductionist approach prioritising short-term solutions such as GMOs, pesticides and synthetic fertilisers. If not, the true sustainable solutions tackling the root causes of the problem will not be found.

3 Priorities in plant health: universities

3.1 Stellenbosch University: Altus Viljoen

Stellenbosch University (SU) is home to an academic community of 29 000 students, including 4000 foreign students from 100 countries, and 3000 staff members (including 1000 academics). The university is recognised internationally as an academic institution of excellence. According to the Times Higher Education World University Rankings, SU is one of the top 300 universities in the world, and among the top 20 in BRICS countries. SU's Faculty of AgriSciences was established in 1918 and has made a major contribution to agricultural education and research in South Africa and on the African continent.

SU and IITA have been working together on plant health-associated topics for the past decade, a collaboration that was highlighted with the signing of a Memorandum of Understanding in September 2015 to feed a growing continental population through improved productivity and protection of African crops. Staple foods on the continent, such as maize and banana, are increasingly threatened by indigenous or introduced pathogens and pests, requiring environmentally sound and sustainable crop protection strategies. These include the use of microbes to protect plants against soil-borne pathogens and pests, and the conventional breeding of crops to enhance plant resistance.

As an example of sustainable crop protection, SU has worked with IITA to characterise and establish non-pathogenic host-specific strains of *Fusarium oxysporum* f. sp. *striga* as endophytes in banana plantlets to control nematodes and weevils. One of the most important collaborations between SU and IITA has been in combatting two destructive fungal diseases affecting African bananas, *Fusarium* wilt and black sigatoka, through breeding of resistant banana hybrids. More recently, the introduction of the Asian strain, *F. oxysporum* f. sp. *cubense* TR4, into Mozambique, has prompted SU and IITA to establish a consortium of researchers from 11 African countries to prevent its further spread and damage in the continent.

3.2 University of Ibadan: Olajumoke Alabi

Biotic stresses (insect pests, diseases and weeds) continue to threaten food security in Africa. Climate change has now contributed to the spread of invasive species. As an example, the spread of *Spodoptera frugiperda* (fall armyworm) and *Tuta absoluta* (tomato leaf miner) in Nigeria in 2016 resulted in significant shortages of maize grain and tomato fruits and the fall armyworm has continued to be a significant problem for smallholders in Nigeria. This highlights the importance of collaboration on plant health with organisations such as the IITA in such areas as monitoring and surveillance of insect pests to give early warning to farmers as well as the development of appropriate pest management techniques.

Founded in 1948, the University of Ibadan is the oldest university in Nigeria and is consistently ranked as one of the leading universities in Africa and around the world. The university has academic programmes in 16 faculties across both sciences and humanities. The Faculty of Agriculture (formerly the Faculty of Agriculture and Forestry) and IITA have worked closely together since the IITA was established in 1967, collaborating in such areas as breeding, crop physiology, soil fertility, plant pathology and entomology.

I joined IITA in 1995 as a doctoral student in the cowpea entomology unit and, since then, I have continued to collaborate with scientists on the IITA cowpea team, both in co-supervision of graduate students and in projects to deal with cowpea pests. As an example, my institution has worked in partnership with IITA, Wageningen University (The Netherlands) and the International Center of Insect Physiology and Ecology (icipe - Kenya), to develop traps that will attract and kill thrips (a pest of cowpea) using host plant volatiles. We have also worked together on the use of parasitoids to manage thrip pests of cowpea in a project funded by the Bill & Melinda Gates Foundation from a period of 2015 to 2018. As well as collaboration on particular projects, one of the many benefits that the IITA provides is a conducive environment for research and training for graduate students who lack such opportunities elsewhere. Several students in my institution have had the privilege of conducting their research in the IITA with great success.

3.3 University of Bonn: Christian Borgemeister

The Centre for Development Research (ZEF) at the University of Bonn in Germany deals with complex, development-related issues that often require inter- or transdisciplinary approaches. Sustainable intensification (SI) is a key concept to improve food security in Africa, and managing plant health issues in an efficient, environmentally friendly and sustainable manner is one of the major challenges in achieving SI.

As director of ZEF, it is a true pleasure being able to contribute to the 50 years anniversary compilation of IITA's plant health research and achievements. I started my career in international research as a postdoc at what was then the Africa-wide Biological Control Program of the IITA in Cotonou, Benin. The six years that I spent with IITA, working on the biological and integrated control of invasive post-harvest pests, were truly transformative for me and had a long-lasting effect on my professional career. At the time when I joined IITA's biocontrol team in 1992, the station in Cotonou was the place worldwide where the most exciting and groundbreaking biocontrol research and development took place; it was buzzing! Fuelled by the continent-wide success of the CM control programme, numerous important pest problems in African agriculture were being tackled, ranging from locusts, to cassava green mite, mango mealybug, water hyacinths to name a few. These were also 'golden years' in terms of the availability of resources for research.

Whilst not all projects came to fruition, the optimism and enthusiasm that biocontrol can really make a difference in pest control in Africa was palpable in the numerous African and international scientists that worked at IITA at the time, or in the numerous national research organisations across the continent with whom IITA developed close partnerships. What shaped my thinking most was the notion that through our work at IITA we were transforming biological control 'from an art into a science'. Sadly, biological control and IPM became less of a focus for the international donor community, and consequently levels of activity in the CGIAR in general and at IITA declined. Yet ironically, whenever the CGIAR recounts its great historic successes in food security and poverty alleviation through transforming agriculture, the story of the mealybug programme features very prominently.

So, where are we now and what can be learnt from these past successes and, in some cases, failures? In my personal experiences working in Africa, first at IITA and then later as the director general of the International Center of Insect Physiology and Ecology (icipe), if you are dealing with an invasive pest problem, try classical biological control first. In reality, if it works, nothing beats it in terms of costs, ease of implementation and sustainability. If it doesn't, try to intelligently combine interventions from the IPM toolbox. In addition, in my experience, the earlier you involve all stakeholders, and especially the farming community, the greater the probability of success. There is no one-size-fits-all solution. What works under drier conditions isn't necessarily easy to implement in humid lowlands. Likewise, tailor-made IPM for high-input agriculture will most likely flop with subsistence-level farmers. Talking to stakeholders from the start will help to identify these pitfalls. And last but not least, an efficient communication strategy, especially addressing women, is pivotal for success.

3.4 Michigan State University: Barry Pittendrigh

Michigan State University (MSU) was founded in 1855 and served as a prototype for the 69 other land-grant institutions. Located in East Lansing, Michigan, it is home to approximately 50 000 students, which includes approximately 39 000 undergraduates and 11 000 graduate and professional students. MSU employs over 5500 faculty and academic staff. It is ranked 75th within the world's top 100 universities, according to U.S. News and World Report. MSU has a long-standing commitment to international research and development, especially concerning grain legumes such as cowpea.

Cowpea is an important source of protein and income for tens of millions of people across Africa, especially for those that live on less than US$2 a day. The full yield potential of cowpeas is often not realised due to abiotic and biotic stresses. The most important biotic stress of cowpea crops is a complex of pest insects that can dramatically reduce yield. Although second-generation insecticide sprays are a short-term and localised solution for some farmers, development of environmentally friendly, scalable and sustainable pest control solutions is necessary for long-term and larger-scale pest population suppression of the cowpea pest complex.

MSU and IITA researchers, along with West African NARS programmes, have been working together for the past decade on IPM strategies to minimise pest problems in the cowpea cropping system. Two major strategies have emerged through IITA's leadership in this area. First, IITA has developed a biocontrol pipeline for the discovery, testing and deployment of biological control agents that can provide scalable and long-term suppression of pest populations. Second, IITA has provided global, regional and local leadership in the development of biopesticide control strategies. Its approach means that biocontrol products can be locally produced and used, providing economic opportunities for women and youth, as well as giving farmers access to affordable and effective pest control strategies that are both safe for human health and the environment.

Early in 2017, MSU participated in a workshop for the newly established International Research Center for Climate Change Resilient Agriculture (IRCCCA) in Cotonou, in which both MSU and IITA are partners. MSU is currently represented on the International Scientific Advisory Panel of IRCCCA, and looks forward to a fruitful collaboration with IITA and other partners in this new initiative.

3.5 University of São Paulo: Gilberto de Moraes

It has been my great pleasure to work with IITA since the mid-1980s, mainly in the biological control of the cassava green mite. Our main activities were centred along two main axes, one concerning the identification of mites in the different countries involved in the project, and the other concerning exploring prospective biological control agents of that pest in Brazil. Valuable research was undertaken on mite diversity in cassava ecosystems, especially in Africa and Latin America, with new descriptions and revised descriptions of dozens of species. Building on this was research which uncovered several promising predatory mite species, especially *Typhlodromalus aripo*. I also had the chance to collaborate with IITA in a project on the biological control of the coconut mite, again working on the identification of the mites found on this crop in Africa and searching for prospective natural enemies in Latin America. My collaboration with IITA has occurred throughout most of my professional career, and I have benefitted tremendously from the friendship and great expertise of my IITA counterparts. It will be my pleasure to continue that collaboration, as new pest mite species arrive in Africa, such as the tomato red spider mite.

4 Priorities in plant health: NARS

4.1 INERA, Democratic Republic of Congo: Willy Tata Hangy

The Institut National pour l'Etude et la Recherche Agronomique (INERA) is the main national institute responsible for agricultural research in the Democratic

Republic of Congo (DR Congo). Created in the 1930s as INEAC (Institut d'Etudes Agronomiques du Congo) during the colonial era, the institute was mostly interested in boosting the productivity of major industrial cash crops such as oil palm, cocoa, coffee, rubber and so on. Most research on plant health was done by European specialists, mainly in European laboratories. Less attention was given to food crops.

It is only after independence in the 1960s that INERA started research on food crops, mainly emphasising breeding. Because they were not seen as a significant threat, plant diseases and pests were of lesser concern. However, in the early 1970s, the main staple food crop in the country, cassava, was heavily infested with CM, an invasive pest accidently introduced from Latin America. This pest destroyed millions of hectares of cassava farms in the sub-Saharan region. DR Congo was the most affected country, as this crop covers more than half of cultivated area. Losses were evaluated at about US$2 billion. As if that were not enough, the crop was also attacked by another pest, the cassava green mite (CGM). Introduced from the neo tropics through East Africa, it quickly spread to all cassava-growing countries, causing yield loss of about 70%.

To fight the two pests, IITA initiated the Africa-wide biological control programme. Through the programme, IITA trained many technicians and scientists in biological control approaches, including myself. Through the same programme, the parasitic wasp *Anagyrus (Apoanagyrus,Epidinocarsi) lopezi* was identified in Latin America as a parasitoid of CM and introduced into Africa, including DR Congo, where it successfully established and controlled the pest within a few years. *Typhlodromalus aripo*, a predatory mite for CGM, was also identified and subsequently introduced into Africa and DR Congo to successfully fight the pest. After the successful biocontrol of these two cassava pests, IITA also helped INERA with the release of natural enemies against the mango mealybug *Rastrococcus invadens* in the 1990s. In the early 2000s, INERA and IITA worked together in the search for solutions to cassava mosaic disease. At present, IITA's expertise is still being called on to tackle CBSD in DR Congo.

4.2 INRAB, Benin: David Arodokoun

The National Agronomic Research Institute of Benin (INRAB) has been collaborating with IITA since 1967, the year of its foundation. I have been involved closely in the activities of IITA, firstly as a MSc and PhD student, then as director of the Plant Protection Laboratory of INRAB. I have also worked with IITA as the coordinator of the Sustainable Cowpea Protection and Cowpea Project for Africa, and later as scientific and then general director of INRAB.

African agriculture faces a range of abiotic and biotic constraints. INRAB's collaboration with IITA has aimed to improve plant health through research projects focusing on biological control of CM, CGM, mango mealybug and water hyacinth. INRAB and IITA have also been partners in projects such as Ecological Sustainable Cassava Production and Protection, Ecological Sustainable Cowpea Protection (PEDUNE) and the Cowpea Project for Africa (PRONAF).

The main achievements from INRAB's collaboration with IITA include improved quality of research in INRAB through scientific capacity building, and improved yields and quality of cowpea through better-performing varieties and advances in pest management. Future challenges we will need to address together include the threat of new pests and diseases related to climate change, including improving control methods against fruit flies and vegetable pests.

4.3 INERA, Burkina Faso: Clémentine Dabiré/Binso

I am director of the agricultural entomology laboratory at the Institut de l'Environnement et des Recherches Agricoles (INERA) in Kamboinsé, Burkina Faso. As national research coordinator for grain legumes and plant protection, I have developed collaborative links with the IITA cowpea entomology lab in Cotonou, Benin, particularly in the development and deployment of biological control and bio-pesticides. We have implemented joint projects and exchanged staff and students.

One of the main problems with grain legume cultivation and cowpea in particular is the intensive utilisation of chemical pesticides to exploit the maximum potential of improved varieties developed by the national programme breeders. The main collaboration with IITA has therefore been the search for more sustainable IPM techniques to reduce the reliance on pesticides.

Between 1994 and 2006, we collaborated on various phases of the SDC/IFAD-sponsored PEDUNE/PRONAF projects, led by IITA-Benin, to research biocontrol techniques for use in cowpea cropping. From 2007 to 2017, we collaborated with IITA on a USAID-sponsored Legume Innovation Lab project (IPM-OMICS) to develop scalable and sustainable biological solutions for pest management of insect pests of cowpea in Africa. These research projects included targeted studies on population dynamics and genetics of pests and natural enemies, the development of biocontrol agents and bio-pesticides, and novel approaches to deliver IPM advice to semi-literate farmers. The identification of insect pests and their natural enemies was carried out through the Biodiversity Center at IITA-Benin. The success of these projects has helped to significantly increase production over the past 10 years. Current challenges include new invasive species such as the fall armyworm, foliage thrips and spider mites, for which new biocontrol solutions need to be found.

5 Priorities in plant health: BEAF - GIZ

5.1 Marlis Lindecke, Advisory Service on Agricultural Research for Development (BEAF), German Economic Corporation for International Cooperation (GIZ), Germany

The Advisory Service on Agricultural Research for Development (BEAF) manages Germany's support of the 15 research centres of the CGIAR plus the World Vegetable Center and the International Center of Insect Physiology and Ecology (icipe) (known as CGIAR++) on behalf of the German Ministry of Cooperation and Development (BMZ). BEAF advises the BMZ on the strategy for international agricultural research for development (AR4D). It represents the BMZ in the European informal donor coordination platform, the European Initiative for Agricultural Research for Development and in the CGIAR System Council.

Germany's main instruments to support agricultural research is the funding for genebanks and two research grants that are offered each year (project funding worth €1.2 million for 3 years and small grants worth €100000 for 2 years). In addition, it supports the seconding of experts between Germany and Africa in cooperation with the Centrum for International Migration and Development (CIM).

Through project funding and small grants, BMZ/BEAF has supported at least seven IITA research projects directly related to plant health issues during the last 18 years. The topics supported covered a wide range of plant health issues. As an example, research towards the development of strategies to reduce aflatoxin contamination have provided highly promising development results. As a result initial funding in 1999 was followed by funding of other aflatoxin-related projects in 2003 and 2012. Translating research outputs into solutions that can be applied to farmer's fields remains one of the key challenges for AR4D.

5.2 Stephan Krall: GIZ - a personal view

I started employment with what was then GTZ (the forerunner of GIZ) in Togo in 1981, with the task of reducing post-harvest losses in Africa. Almost immediately I visited IITA in Ibadan and met Hans Herren, whose Africa-wide biological control project impressed me a lot. My collaboration with IITA and ICRISAT increased considerably in 1984, when I started work at the National Plant Protection Center in Porto-Novo, Benin. The links between my extension-oriented job and research were strengthened by sharing responsibility with the PhD students conducting research at the Center. At that time, GTZ published a large number of books and other materials, where we benefitted from the research of the IITA as well as of other CGIAR centres. A particular highlight was our participation in the LUBILOSA project from 1989 to 1999, which eventually produced the commercialised locust biopesticide Green Muscle®. After this time, I was responsible for overseeing the German contribution to the CGIAR research institutes including IITA. I was very conscious that the funds available to international research centres were only a fraction of the resources invested in agriculture by industry. I was saddened by the frequent reductions in funding which today have reached crisis level, though the German contribution over the years has shown a modest growth (to €20 million/year). Although some of my colleagues have been critical of more fundamental 'ivory tower' research, I was always aware of its importance, even when its potential practical applications were not immediately obvious. After all these years of collaboration with IITA, and as the director of the 'Forest, Biodiversity and Agriculture' section of GIZ, I can only stress how important the collaboration between research and extension still is, and my continuing friendship with Hans Herren and Peter Neuenschwander.

6 Conclusion

These statements from stakeholders who have worked with IITA highlight a number of key themes, including the following:

- The importance of collaboration both at an institutional and individual level; this was clearly the most frequently discussed topic that has contributed to successful outcomes.
- The role of IITA in capacity building for example in training of NARS staff; this was frequently highlighted and, while not discussed by this group, it is probably one of the most important contributions IITA has made.
- The value of a holistic concept of plant health which highlights the importance of IPM.
- The success of biocontrol techniques in IPM, many of them pioneered by IITA with its partners, particularly in managing major African crops such as cassava, maize and bananas.
- The importance of effective monitoring/surveillance and robust phytosanitary arrangements to limit the spread of new diseases and insect pests.
- The threat from invasive species related to the effects of climate change.

Some of the partners also commented on the decline in funding for plant health research after the great successes with the biological control of CM and green mite during the 1980s and 1990s. This caused a major decline in the number of plant health scientists working at IITA. In recent years there has been a gradual increase in funding and we expect this to continue given the expected increase in plant health problems related to climate change.

IITA has major research programmes on cassava, maize, cowpea, yam, banana and soybean. This group of stakeholders did not discuss two of IITA's mandate crops: soybean and yam. There was also limited reference to the many successes of breeding for host plant resistance, except for two diseases affecting African bananas, *Fusarium* wilt and black sigatoka.

One message is very clear: partnerships have always been, and will always be, required in all our plant health research efforts.

Published by Burleigh Dodds Science Publishing Limited, 2019.

Chapter 2

Introduction: meeting challenges in plant health in sub-Saharan Africa

Peter Neuenschwander *and* **Manuele Tamò**, *International Institute of Tropical Agriculture (IITA), Benin (corresponding author: p.neuenschwander@cgiar.org)*

1 The importance of plant protection in agriculture

Agricultural production is fundamentally determined by climate, soil fertility, crop adaptation and the actions of farmers and the societies in which they are embedded. The health of the growing plant, the resulting harvest and its marketing depend on how well these conditions are managed or mitigated. In other words, plant health management is not a stand-alone activity, but a holistic approach is quite a challenge, as this book clearly demonstrates.

1.1 Climate

This book only covers tropical agriculture in sub-Saharan Africa, that is south of the Sahara desert, north of the southern tip of South Africa (Fig. 1). The International Institute of Tropical Agriculture (IITA) mainly covers the humid tropics, so the northern Sahel region is only discussed here in relation to crops that are also planted further south.

Global climate change affects Africa, and disadvantaged people and their agriculture particularly badly, as the Intergovernmental Panel on Climate Change (IPCC, 2014) has stated with a 'high level' of confidence. The Sahel region has already seen a reduction in annual rainfall, and the more irregular distribution of rainfall observed in the humid forest zone is also likely to be a

http://dx.doi.org/10.19103/AS.2018.0043.02
Published by Burleigh Dodds Science Publishing Limited, 2019.

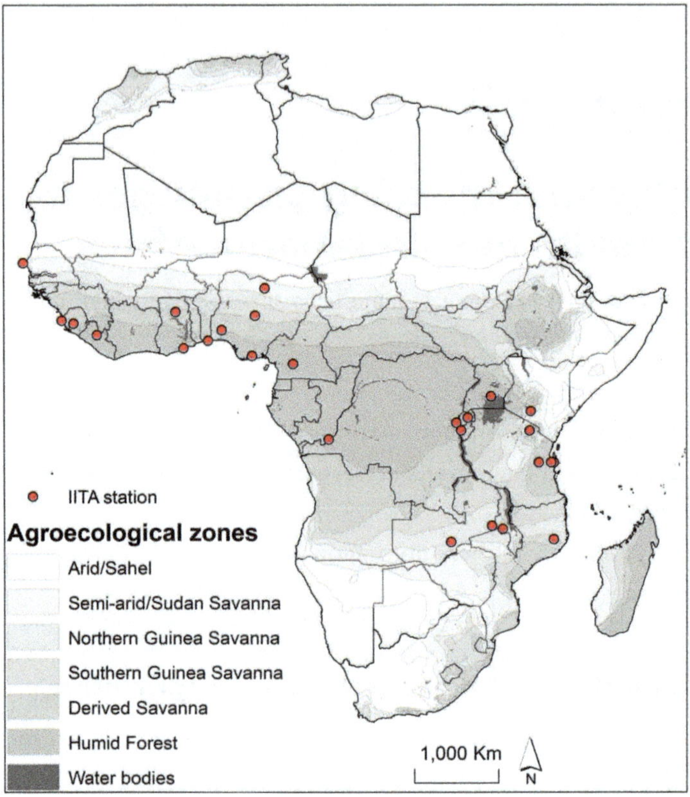

Figure 1 Ecological zones and IITA stations in Africa. *Source*: IITA-Ibadan, Nigeria.

sign of changing weather patterns. The IPCC report also states (again, with high confidence) that climate change already has a negative effect on crop yields, promoting heat waves, drought, floods and wildfires. Agricultural measures to mitigate and adapt to these issues include ecosystem management and conservation, afforestation as well as the use and ecologically sound management of crop varieties adapted to the local environment.

1.2 Soil fertility

African soils differ from those of northern climes as the bedrock is mostly basic granite/gneiss with few sedimentary rocks, covered locally by young sediments, forming fertile flood plains. This basaltic rock is covered by a thin layer of fragile top soil, made up of rapidly decaying organic material. When this topsoil is used for agriculture, its fertility is renewed by fallow and improved fallow periods, alley cropping, mulching and the application of inorganic fertilizers, as reviewed in detail and discussed by Mutsaers et al. (2017). Managing soil

Published by Burleigh Dodds Science Publishing Limited, 2019.

fertility is the basis of all agricultural activities, so much so that not even a high-yielding genotype can perform satisfactorily if management fails.

1.3 Adapted productive crop plant genotypes

Most crops grown in Africa today originate from tropical America (cassava, maize, cotton) and tropical Asia (banana, plantain, many legumes); few are native crops (oil palm, some legumes, some yams, some millets, numerous vegetables). Bananas, taro and some yams were introduced into Madagascar from Southeast Asia in the first millennium AD. Crops from America reached Africa's port areas in the sixteenth century, but penetration into the inner continent was sometimes as late as the early twentieth century. Many of these crops became more productive in their new environments than in their native origin. One of the reasons for this is the absence of adapted pests in the new location, which has important consequences for plant health management (Purslane, 1968, 1972; Diamond, 1997; Beaujard, 2011).

Gains from crop varieties created by the Consultative Group of International Agricultural Research (CGIAR) have improved agriculture throughout the tropics (Evenson and Gollin, 2003). Many of the plant protection measures discussed in this book would be ineffective without such crops being well-managed and grown using 'good farming practices' that preserve soil fertility. Weak plants are more susceptible to pests and diseases, and biological control measures cannot usually correct this completely.

1.4 Plant protection measures to minimize damage

Plant protection measures prevent crop damage. Losses start at planting, and continue in the field and through to final storage. Potential pre-harvest crop losses without crop protection measures differ by crop and region, ranging from 50% to more than 80%. Worldwide actual crop losses in maize and soybean, for instance, reach 31% and 26%, respectively, despite plant protection measures being used. Losses attributable to viruses are 3% for maize and 1% for soybean, pathogens 9% each, animal pests (mainly insects) 10% and 9%, respectively, and weeds 11% and 8%. In general, weeds cause the highest losses (34%), and losses due to animal pests and pathogens are less important (18% and 16%, respectively) (Oerke, 2006). A multidisciplinary approach is required to assess crop losses, particularly those caused by pathogens. Their harmful effects reduce food quality, so they not only reduce yields, but affect the economics of the food chain (Savary et al., 2012). Transport and storage are responsible for huge additional losses by damaging produce and seeds, and favouring fungal and insect infestation. They may also subject the product to high temperatures, depending on the transport/storage facility (Sallam, 2014). The huge scale of potential economic losses can be illustrated by the annual

Published by Burleigh Dodds Science Publishing Limited, 2019.

loss of US$17.7 billion in Brazil, representing a yield reduction of approximately 7.7% which is attributed to insects alone (Oliveira et al., 2014).

Unfortunately, despite continually increasing pesticide use worldwide, losses have remained rather constant since the 1960s (Oerke, 2006), so different and better plant protection measures are obviously needed. Since a large proportion of African farm land is still farmed using the so-called organic methods (although this epithet is not actually used in this context) in their strictest form, without the use of pesticides or inorganic fertilizers (Altieri and Toledo, 2011), the current aim is to ensure that modern plant health measures take into account these indigenous management strategies, which are suited to local conditions, including low soil fertility. While these farms are being nudged towards higher productivity (Mäder et al., 2002; Giller et al., 2015), the relatively few farms that employ the intensive farming practices used in the United States or Europe need to become more sustainable to survive the next 50 years.

To develop effective plant protection measures, you need to know your enemy. The pests targeted in this volume, along with many more pests, pathogens and weeds that affect African crops, are presented and illustrated by the following authors: insect pests by Prinsloo and Uys (2015), pathogens (fungi, bacteria, viruses) by Agrios (2005), nematodes by Sikora et al. (2018) and weeds by Akobundu et al. (2016). Winston et al. (2014) have also published a world catalogue of weeds and their antagonists.

These pests are either native, often polyphagous, African organisms that have adapted to the new crop, or invaders from the crop's native origin. Most of these pests are only severe under particular conditions, but some invaders like cassava mealybug (Chapter 6: Toko et al.) or fall armyworm (Chapter 7: Bandyopadhyay et al.) have been, or still are, so destructive that they have caused widespread total crop losses, even with good crop varieties and good soil management regimes.

2 Historical development of plant protection

It is difficult, and perhaps overambitious, to attempt to summarize agricultural development over vast time periods on such a diverse continent with little written history. A look back into the history of agriculture will help to clarify some concepts: new developments in agriculture reached only small niches at first, often starting with peri-urban agriculture or home gardens, before the new techniques became more widespread. Various partially overlapping stages can be distinguished, as discussed below.

2.1 Traditional agriculture in Africa and its limitations

In the 2000 years until the beginning of the twentieth century, human populations were low (1–10 persons per km^2) in most, but not all, areas. (The central parts

Published by Burleigh Dodds Science Publishing Limited, 2019.

of the Kingdom of Dahomey in Benin, Yoruba kingdoms in Nigeria and others were densely populated and farmed over centuries.) Agriculture was marked by long fallow periods with slash-and-burn practices which facilitated planting, maintained soil fertility and reduced pest pressure. In most parts, agriculture, local iron smelting and animal husbandry caused very little deforestation and no measurable negative effect on the environment (Giresse, 2008). Plant protection measures consisted of mechanical control, mainly weeding, and using other plants and their extracts to protect crop plants (Stoll, 2002).

However, this was no paradise, and the apparent harmony with nature was not that idyllic. The first agricultural revolution about 11 000 years ago set humankind on a path that eventually made it not only the top predator, but also the top herbivore, and led to nearly complete human domination of the earth's ecosystems (Vitousek et al., 1997). Agriculture set us on a collision course with our competitors, mostly arthropods, microorganisms, nematodes and similar, for food. The more we improved wild plant species through active, as well as inadvertent, selection and by planting them in monocultures, the more attractive they became to these competing organisms, which thereby became pests (the term is used here in reference to insects, nematodes, diseases and weeds). However, only about 350–500 out of a total of around ten million insect and slightly fewer nematode species, for instance, have ever become serious pests.

Today, farming conditions dating from before 1900 are found only in the remotest areas and are almost totally absent from West Africa. Even in these areas, solar panels and cell phones have changed communications and people's lives to an extent that would have seemed impossible 50 years ago, stimulating farmers' ambitions for a life of less drudgery and poverty. Populations have grown tenfold in many countries in the last 100 years and the consequences of this increase are dramatic. The total area under agriculture is still expanding to produce more food, and this expansion has encroached on forested or more marginal land in many areas. However, this is no longer feasible and more food must now be grown on already farmed land (Giller et al., 2015).

2.2 Pesticides and their limitations

In other continents, highly toxic products like cyanide were used as insecticides after the First World War. Africa was untouched by this development and subsistence agriculture, still mostly without the use of inputs (fertilizers, pesticides), remained widespread. Fallow periods became ever shorter; plant protection measures, if any, were mainly mechanical; and deterrents were employed during storage.

Since the Second World War, synthetic insecticides (first, chlorinated hydrocarbons, and later, phosphate esters) have been developed, accepted enthusiastically as 'silver bullet' solutions and then used abusively as routine

Published by Burleigh Dodds Science Publishing Limited, 2019.

calendar-based treatments. The huge detrimental effect of this 'pesticide treadmill' on the environment was first recognized and decried in the United States (Carson, 1962; Doutt and Smith, 1971; van den Bosch, 1978). Whiteflies, leaf miners, diamondback moth and tetranychid mites became the worst pests worldwide, but most do not deserve to be labelled as pests; they were made into pests by insecticide abuse. The same concerns arise again regarding neonicotinoids and their effects, for instance, on bees (Godfray et al., 2014). The first improvements were achieved in Europe and America by using damage thresholds to trigger treatments, leading to the development of Integrated Pest Management (IPM) techniques. These started as mere pesticide management programmes, beginning with cotton (van den Bosch et al., 1971), but subsequent consideration of existing biological control mechanisms in the environment (Hagen et al., 1971; Settle et al., 1996; van Driesche et al., 2010) and general concern for the health of the environment became the basis for more sustainable IPM schemes (Zadoks and Waibel, 2000).

In tropical Africa, the phase of (managed) pesticide abuse was bypassed in all field crops except vegetables, cowpeas and colonial export crops, including cotton and cacao. Today, insecticides based on plant extracts, *Bacillus thuringiensis* (*Bt*)-products and other biocides are available, as are systemic and non-systemic (specific) insecticides. However, the latter are often inaccessible to local farmers (Chapter 15: Coyne et al.). Even when the ultimate goal is a sustainable solution based on a combination of host plant resistance, biological control and biopesticides, as is the case for a long-term strategy against fall armyworm, preliminary measures still consist of using the least disruptive pesticides available locally as a stopgap (Chapter 4: Goergen et al.).

2.3 The Green Revolution and its limitations

In the 1920s, improved varieties of wheat, maize and Asian rice started to revolutionize agriculture in Europe, moving on to the United States and Mexico, and culminating in Southeast Asia in the 1960s, leading to Norman Borlaug's unique Nobel Peace Prize in 1970. The resulting huge increases in crop production exceeded the needs of the growing populations in these countries for the first time in their history. Africa was the only continent where this development did not occur, and human population increases consumed any increases in agricultural productivity that were made (Upton, 1993; Frison, 2016).

The Green Revolution depended heavily on synthetic pesticides, so its implementation was coupled with ever more serious damage to the environment, particularly in America and Asia. It has therefore come under close scrutiny (Gallagher et al., 1994; Pimentel, 1996) and the same approach will probably never be repeated in Africa (Chapter 15: Coyne et al.).

3 History of IITA's plant protection research and implementation

3.1 *Creation of the Consultative Group of International Agricultural Research (CGIAR)*

IITA, one of the CGIAR's founding institutions, was created in 1967 when revulsion at the abuse of synthetic insecticides reached its climax, and the Green Revolution claimed that genetic improvement of crop varieties was the pinnacle of agricultural development. Most scientific research in Africa before this had been directed towards colonial crops like cacao, coffee, tea and rubber, which were supported by dedicated agricultural research stations in many countries. Yet smallholder crops, which fed the local population and were not exported, had received little research assistance until the CGIAR was created.

3.2 *Plant protection as a service to breeding programmes*

In the sphere of plant protection, the CGIAR as a whole (including IITA) initially concentrated exclusively on breeding resistant cultivars of farmers' crops (see the IITA Annual Reports). The benefits of this approach have been reported across the entire CGIAR, and they are extremely impressive (Evenson and Gollin, 2003). IITA was particularly active in studying the mechanisms of virus transmission by insects and exploiting the results for its breeding research. Early successes against African cassava mosaic virus in cassava, maize streak in maize (honoured by the King Baudouin Award in 1986) and various viruses in legumes, and black sigatoka in *Musa* spp. (honoured by the King Baudouin Award in 1994) are detailed in Chapter 5 (Legg et al.). These studies show that breeding new resistant varieties is by no means at its end, but it needs to be developed further to incorporate novel molecular approaches (such as marker-assisted selection) and include robust management strategies to maintain resistance and overcome breakdown by genetic adaptation in the target pests.

In the sphere of safe handling and movement of germplasm and seed to prevent the transmission of diseases from one country to another, phytosanitary procedures and new diagnostic techniques have been developed and implemented in collaboration with the quarantine authorities in the countries involved (Hoeschle-Zeledon et al., 2013) (Chapter 3: Kumar et al.).

We will not discuss the breeding techniques that have led to the deployment of the resistant crop varieties examined in this book, as they will be dealt with in a planned companion fiftieth anniversary volume, but we shall look at their use in farming systems. In genetically modified organisms (GMOs), host plant resistance has been conferred by special techniques. Genetically modified crops, such as Golden Rice, are now available to small-scale farmers (Ye et al., 2000). Safety guidelines have been developed for deployment of GMOs

(Sharma et al., 2002), although biosafety legislation permitting distribution of GMOs to farmers is not yet in place in most countries, particularly in West Africa. IITA has recently developed its first transgenic plant; bananas transformed with a pepper gene that confers resistance to banana *Xanthomonas* wilt (BXW) (Chapter 9: Hauser et al.)

The IITA was originally divided into crop-based breeding programmes, and these were subsequently coordinated into the Crop Improvement Division. Soil-water issues and land management were joined with economic analysis into a vast resource and crop management program which was then renamed as a division. In the 1990s, crop improvement, soil and water management, and social sciences were integrated at benchmark sites, though these did not prioritize plant health (Mutsaers et al., 2017).

3.3 Biological control

The biological control approach to crop protection consists of introducing natural enemies from the native origin of the exotic pest into its new home. It has its modern roots in the late nineteenth century and interest was renewed in response to pesticide abuse after the Second World War. Biological control consists either of inoculative ('classical') or inundative introductions, or by encouraging and supporting native natural enemies (De Bach and Schlinger, 1964). Biological control is usually fully compatible with existing resistant plant varieties (Thomas and Waage, 1996), though there is still great potential for improvement by making crop varieties more attractive to the pest's natural enemies by exploiting chemical communication within the food web (Pickett et al., 2014).

The Africa-wide biological control of cassava mealybug and cassava green mite which started in 1979 was IITA's first project that did not primarily depend on resistant cultivars (Chapter 6: Toko et al.). Although biological control had been practised in Africa throughout the twentieth century, the new IITA project achieved a much wider reach and collaborated with national programmes more closely than before; it was soon recognized as a new phase in international efforts (Greathead, 2003). The project received the King Baudouin Award (IITA in collaboration with the International Center for Tropical Agriculture; CIAT) in 1990, and its original leader, Hans R. Herren, was awarded the World Food Prize in 1995.

Similar biological control projects have since been implemented by the same international collaborators, often with polyphagous pest species, in various crops, for some of which IITA did not have a breeding/selection programme (Chapter 12: Neuenschwander et al.). The gains from these large-scale biological control programmes alone equalled the crop improvement gains for the entire CGIAR breeding programme (Neuenschwander, 2004). Such programmes were often executed immediately after the new pest was introduced and, unlike cassava mealybug, the new pests were not allowed to cover the entire continent before effective biological control could be established.

Classical biological control has come under a great deal of criticism, mainly in the United States, Europe and Australia, because of unforeseen non-target effects (Howarth, 1991; Louda et al., 1997; Lynch and Thomas, 2000; Roy and Wajnberg, 2008) and the importation of exotic beneficial insects has become exceedingly difficult in some countries (Messing, 2000), although no cases of extermination of a non-target species by an introduced arthropod natural enemy have been documented. However, this is known to have occurred with introduced vertebrates and molluscs, which have since been banned as potential biological control agents (Strong and Pemberton, 2001). More recently, introduction of the efficient but polyphagous coccinellid *Harmonia axyridis* has demonstrated that polyphagous insect predators can affect local food webs (Roy et al., 2016) although no negative effects were reported for this insect in Africa (Mukwevho et al., 2017). It is now generally agreed that such natural enemies should not be introduced.

All IITA's projects in classical biological control have been executed in the spirit of the FAO code of conduct (FAO, 1996; ISPM-3, 2005; Bigler et al., 2006, which IITA helped formulate), even before these restrictions were formally imposed. A review of IITA's practices under new regulations revealed that no non-target effects were caused through their introductions of specific parasitoids or weevils (Neuenschwander and Markham, 2001). Where coccinellids had been introduced, they did not establish or remained inconspicuous (Chapter 6: Toko et al.). However, it must be mentioned that the taxonomic basis of African relatives of target species is sometimes so poor to the extent it can preclude thorough testing of indigenous sibling species. This knowledge base is currently being expanded by the activities of IITA's biodiversity collection in Benin (Chapter 4: Goergen et al.).

In Africa, augmentative or inundative releases of natural enemies (van Lenteren, 2011) have mainly employed pathogens: the entomopathogenic fungus *Metarhizium acridum* was developed in collaboration with other institutions and marketed as Green Muscle® in Africa, where it started being used against locusts and Sahelian grasshoppers, especially in Madagascar and Senegal (Lomer et al., 2001). A similar product was developed for Australia, using an Australian strain of the same fungus and marketed as Green Guard®. The current licence holder has lost interest in these products, so a third product, NOVACRID®, was developed in only 2 years, thanks to the valuable experience gained during the development of Green Muscle® (Chapter 13: Schreurs et al.). Entomopathogenic viruses are also being propagated in a rustic manner and applied to cowpea (Chapter 10: Tamò et al.).

Competitive displacement of indigenous organisms by released, better adapted and more efficient species has been observed in many biological control programmes. Maize and other crops can be protected from aflatoxin, a dangerous poison, for a year or more by releasing atoxigenic *Aspergillus flavus* which displaces the aflatoxin-producing strain of this fungus in the soil (Bandyopadhyay et al., 2016; Chapter 7: Bandyopadhyay et al.). This has led to the commercial development of Aflasafe®, which is currently being tested in 11 countries and will also soon be produced outside the IITA (Chapter 13: Schreurs et al.).

The effect of native natural enemies, such as coccinellids, will be discussed in many chapters of this book, but their importance is rarely quantified, as was the case in an IITA study on cassava (Gutierrez et al., 1988). Whether the performance of these natural enemies can be modified by manipulating the cropping system is reviewed by Wood and Lenné (1999). Vegetation diversity seems to be a mixed blessing for pest management, although varietal mixtures are definitely a viable strategy for sustainable productivity in subsistence agriculture (Zhu et al., 2014; see also Chapter 9: Hauser et al.; Chapter 14: Chikoye et al.). Improvements in landscape structure and natural environments affect agrobiodiversity, and increased biodiversity should be economically justified to become politically accepted (Landis et al., 2000; Balmford et al., 2002) (Chapter 4: Goergen et al.). Preserving biodiversity has become one of the most important issues facing us today and is an important criterion in all IPM interventions (Neuenschwander et al., 2003).

Considering the length of time that exotic crops have been established in Africa, it is not surprising that new food webs have been created, and associations formed between the exotic crops introduced into Africa and African insects, pathogens and their antagonists. As a corollary, attempts have been made to redistribute native natural enemies of some native and exotic lepidopteran stem borers within Africa (Chapter 7: Bandyopadhyay et al.).

Today, IITA pragmatically implements projects based on biological control across the continent. These solutions are usually more sustainable and less costly than other interventions (IAASTD, 2012), although, contrary to the recommendations of the IAASTD report, inorganic fertilizers and GMOs are not excluded *a priori*. To justify the initially high costs of biological control projects, quantitative impact assessments (Neuenschwander, 1996) and cost–benefit analysis (Zeddies et al., 2001) become absolutely essential and occupy a great deal of research time. In fact, the impact is often counter-intuitive because of the inconspicuous nature of the agents introduced. The classical biological control projects, which are often continent-wide, also become ideal experiments for scientific studies on insect ecology, advancing the science of biological control (Gutierrez et al., 1993).

3.4 Integrated Pest Management at IITA

The Africa-wide biological control programme developed into the Plant Health Management Division in 1991, and was dedicated to IPM in all its forms. This was the only administrative section for plant health research and implementation in the whole CGIAR. This approach was extended through the CGIAR Systemwide Program on IPM, which had been created and was led for most of its duration by IITA. SP-IPM facilitated cooperation on plant health research between the various centres and published a total of 28 briefs and guides to research methodology for extension services, promoting ecological approaches to IPM (www.spipm.cgiar.org; James et al., 2003; Hoeschle-Zeledon et al., 2013). SP-IPM could have served as a model for the current CGIAR Research Programs (CRPs), but it was discontinued. IITA's divisional organization was abandoned in 2001, and most activities were then grouped under Project B: developing plant health management options, although they were eventually scattered among the seven systematic projects formed in 2006, and since 2010–2011 superseded by the advent of the CRPs.

IPM based on strong biological control is presented for the following crops: maize and its stem borers (Chapter 7: Bandyopadhyay et al.), legumes with thrips and *Maruca* pod borer (Chapter 10: Tamò et al.) and yams (Chapter 8: Mignouna et al.). These chapters highlight the fact that a pragmatic approach involving all possible methods is probably the best way to achieve sustainable solutions to plant protection problems. Detailed studies sometimes reveal unexpected potential for classical biological control, also in native crops.

Many classical biological control interventions faced the problem that they had to be effective immediately because pesticides were already being used against diseases or other insects (Chapter 15: Coyne et al.), or when new invaders appeared and pesticide treatment was initially deemed necessary (Chapter 4: Goergen et al.; Chapter 12: Neuenschwander et al.). Similar integration of different IPM measures was required in projects that attempted to combat three very different categories of weeds: field weeds that were mainly suppressed by fallow techniques, parasitic weeds like *Striga* spp., and floating water weeds that blocked any local agricultural development. The floating weed projects included significant components of biological control; but combatting terrestrial weeds required the full panoply of techniques from herbicides to soil conservation (Chapter 14: Chikoye et al.). Totally new forms of control, combined with varietal development, were pioneered in bananas by integrating endophytes into banana roots to control banana weevil and sigatoka (Chapter 9: Hauser et al.). Similar new challenges arise today to develop previously untested biological controls as an important part of fall armyworm IPM (Chapter 4: Goergen et al.; Chapter 7: Bandyopadhyay et al.).

New products with fewer environmental effects have been developed and integrated into IPM programmes. Neem leaf or neem seed extracts (Schmutterer, 1990) and *Hyptis* have been promoted in research that involved farmer participation (Chapter 10: Tamò et al.; Chapter 15: Coyne et al.). The development of alternatives like Green Muscle® and Aflasafe® (Chapter 13:

Schreurs et al.) has further increased the tools available to IPM. Ecologically sound alternatives are particularly needed in IPM of vegetables, the crop that suffers most from insecticide abuse (James et al., 2010; Chapter 11: Godonou et al.). Despite these various new options, IPM is sometimes still used in the CGIAR as an excuse to just spray pesticides.

Farmer Field Schools, promoted by FAO, became a generator to implement biocontrol-based IPM (Ooi, 2000; Feder et al., 2004). However, they were often judged to be too expensive to fulfil their double function, namely to train farmers to apply new measures and to test farmers' own solutions. They were therefore adapted to become more practical (Chapter 15: Coyne et al.). With the advent of modern biotechnology in Africa (Pickett, 2016), the CGIAR (and the rest of the agricultural research world) sometimes imagined that IPM was obsolete and assumed that GMOs would solve all plant protection problems. However, the resistance to *Bt* maize that developed in Europe and America (Gassmann et al., 2014) demonstrates that this new 'silver bullet' is unlikely to remain useful unless deployed in a strict IPM context. GMOs - particularly those created by new techniques - can be cleverly integrated with biological control (Ryffel, 2017). This requires, however, a rethink about the so-called organic farming. Africa is also faced with an ever-increasing flow of potentially dangerous invaders from South America (Bellotti et al., 1999, 2012) and Asia, which will continue to challenge IITA's plant protection expertise and require a vigilant team to be prepared to tackle new challenges at any time.

3.5 Stronger focus on the environment

Widening the scope of IPM from mere insecticide management to complete landscape management means that agriculture and nature conservation need not be antagonistic (McNeely and Scherr, 2001), as is often the case in practice. A better understanding of food webs and the effect of biodiversity-rich habitats near agricultural fields is therefore essential to support sustainability. This task requires links to scientists and practitioners active in nature conservation, who are prepared to collaborate for their mutual benefit. Such research will benefit from complete habitat protection, on IITA-owned land (Neuenschwander et al., 2015; Chapter 4: Goergen et al.).

Biodiversity collections have proved indispensable to biosystematic studies and, more specifically, to studies that examine potential non-target effects of biological control agents. However, there were no adequate collections in West and Central Africa, so IITA has created its own first-class collection which offers open access and insect identifications, warns of new invaders and underpins biological control programmes throughout Africa. This collection also serves as a basis for studies on the pervasive effect of climate change (Chapter 4: Goergen et al.).

3.6 Stronger focus on the social and economic context

Understanding the changing role of the farmer has become more important with the advent of new products, approaches, and viewpoints. In particular, research into weed control strategies that are more sustainable and less laborious for farmers and their families requires integration across landscapes, and even across countries (Chapter 9: Hauser et al.; Chapter 14: Chikoye et al.).

Increasing urbanization has forced the focus to shift gradually from production in the field to sustainability from 'field to fork' and the transformation that adds value to farm products (Badiane et al., 2014). This has brought attention to a new type of participatory and gender-sensitive research centred around the 'farmer and her husband', and better integration of social sciences and economic studies at household, regional and country levels. Although this new approach is easily discussed far from the field by donors and scientists alike, it has often been painful to execute (Chambers et al., 1989; Bentley, 1994; Röling and Wagemakers, 1998; Mutsaers, 2007) (Chapter 15: Coyne et al.). As the importance of the farmers' well-being on productivity has become more apparent, the effect of different forms of pest management, particularly pesticide abuse, on human health had to be addressed.

4 Structure of this text

4.1 Coverage and demand

This volume, which mainly focuses on Africa, is a companion to Coll and Wajnberg (2017), with its more global focus. Research in plant health management by other international institutions active in Africa is only covered where IITA collaborated. Therefore, research by the International Center of Insect Physiology and Ecology (*icipe*) on tsetse fly (Mihok et al., 1992), the International Crops Research Institute for the Semi-Arid Tropics (ICRISAT) on millet head miner (Gahukar, 1984; Ba et al., 2013), various institutions on indigenous African armyworms (Rose et al., 2000) and some locusts (Krall et al., 1997), AfricaRice on rice pests (Wopereis et al., 2013; Heinrichs et al., 2017), the FAO on rodents and birds (Fiedler, 1994), and studies in RSA on Mediterranean crops (Prinsloo and Uys, 2015) is not included.

IITA's research is aligned with the agendas of regional and continental organizations, and is therefore demand-driven in every country. In order to listen to our colleagues active in plant protection, we asked the staff of national plant protection programmes, who know their farmers and their problems, as well as African university professors, and the staff and donors of international organizations for their opinions and viewpoints on how best to encourage agricultural development in Africa (Chapter 1: Dashiell et al.).

4.2 Authors and acknowledgements

This book was written by authors currently working at IITA, assisted by former IITA colleagues and colleagues at national and international institutions. As far as possible, statements are backed by publications in peer-reviewed journals, and particularly in the first half of the period, by unpublished yearly IITA annual reports available online from the IITA web portal.

The editors and authors would like to thank the thousands of students and collaborators who have not been cited, the unnamed laboratory and field assistants and, last but not least, all the farmers who offered assistance, counsel and encouragement. Finally, we are grateful to IITA management, the donors (who will be listed elsewhere) who made this research possible and the publisher Burleigh Dodds Scientific Publishing.

Enjoy the read!

5 Where to look for further information

Neuenschwander, P. and Tamò, M. (Eds), 2019, *Critical Issues in Plant Health: 50 Years of Research in African Agriculture*, Burleigh Dodds Scientific Publishing, Cambridge, UK (in press).

Chapter 1: Dashiell K. Key challenges in plant health in sub-Saharan Africa: stakeholder priorities.

Chapter 2: Neuenschwander P. and Tamò M. Introduction to critical issues in plant health: a 50-year retrospective on African agriculture.

Chapter 3: Kumar P. L., Legg J. P., Ayodele M., Mahuku G., Ortega-Beltran A., and Bandyopadhyay R. Disease surveillance, diagnostics and germplasm health in crop protection in sub-Saharan Africa.

Chapter 4: Goergen G., Neuenschwander P. and Coyne D. Conserving and exploiting biodiversity in crop cultivation in sub-Saharan Africa.

Chapter 5: Legg J., Kumar P. L., Mahuku G., Wosula E., Stavolone L., Terry E. and Bosque-Pérez N. Viruses affecting African crops and their vectors.

Chapter 6: Toko M., Neuenschwander P., Yaninek J. S., Ortega-Betran A., Fanou A., Zinsou V., Wydra K.. Hanna R., Fotso A., and Douro- Kpindou O. K. Identifying and managing plant health risks for key African crops: cassava.

Chapter 7: Bandyopadhyay R., Cardwell K. F., Ortega-Beltran A., Schulthess F., Meikle W., Sétamou M. and Cotty P. J. Identifying and managing plant health risks for key African crops: maize.

Chapter 8: Mignouna B. D., Kumar P. L., Coyne D., Bandyopadhyay R., Ortega-Beltran A., Bhattacharjee R. and De Koeyer D. Identifying and managing plant health risks for key African crops: yams, taro and cocoyam.

Chapter 9: Hauser S., Gold, C., Pasberg-Gauhl C., Gauhl F., Akello J., Jacobsen K., Norgrove L., Coyne D., Kumar L., Mahuku G., Kaushal M., Nakato V., Tripathi L.. and Tripathi J. Identifying and managing plant health risks for key African crops: banana and plantain.

Chapter 10: Tamò M., Afouda L., Bandyopadhyay R., Bottenberg H., Cortada-Gonzales L., Murithi, H., Ortega-Beltran A., Pittendrigh B., Sikirou R., Togola A. and Wydra K.D. Identifying and managing plant health risks for key African crops: legumes.

Chapter 11: Godonou I., Sæthre M-G., Tepa-Yotto G., Gnanvossou D., Douro Kpindou O. K. and Coyne D. Identifying and managing plant health risks for key African crops: vegetables.

Chapter 12: Neuenschwander P., Gnanvossou D., Hauser S., Goergen G., Hanna, R., Norgrove L., Negloh K. and Agboton C. Identifying and managing plant health risks for key African crops: fruit and other tree crops.

Chapter 13: Schreurs F., Bandyopadhyay R., Kooyman C., Ortega-Beltran A., Akande, A., Konlambigue M., and Van den Bosch N. Commercial products promoting plant health in African agriculture.

Chapter 14: Chikoye D., Ekeleme F., Hauser S., Menkir A., Kamara A. Y., Neuenschwander P., Ajuonu O. and Ajeigbe H. A. Weeds affecting field crops and water bodies in Africa.

Chapter 15: Coyne D., Abberton M., Adetonah S., Ayodele M., Cortada L., Gbaguidi B., Hauser S., Kumar L., Neuenschwander P., Schut M., Tamò M. and Togola A. What makes Integrated Pest Management (IPM) work in sub-Saharan Africa.

Chapter 16: Neuenschwander P., Tamò M. and Saethre M.-G. Improving plant health in sub-Saharan Africa: conclusions and future challenges.

6 References

Agrios, G. 2005. *Plant Pathology*, 5th ed. Elsevier Academic Press, Amsterdam, the Netherlands, pp. 26–7, 398–401.

Akobundu, I. O., Ekeleme, F., Agyakwa, C. W. and Ogazie, C. A. 2016. *A Handbook of West African Weeds*. IITA, Ibadan, Nigeria, 381pp.

Altieri, M. A. and Toledo, V. M. 2011. The agroecological revolution in Latin America: Rescuing nature, ensuring food sovereignty and empowering peasants. *The Journal of Peasant Studies* 38, 587–612.

Ba, M. N., Baoua, I. B., N'Diaye, M., Dabiré, C., Sanon, A. and Tamò, M. 2013. Biological control of the millet head miner *Heliocheilus albipunctella* in the Sahelian region by augmentative releases of the parasitoid wasp *Habrobracon hebetor:* Effectiveness and farmers' perceptions. *Phytoparasitica* 41, 569–76.

Badiane, O. S., Odjo, S. and Jemah, S. 2014. More resilient domestic food markets through regional trade. In: Badiane, O., Makombe, T. and Bahiigwa, G. (Eds), *Promoting Agricultural Trade to Enhance Resilience in Africa*. ReSAKSS Annual Trends and Outlook Report 2013, Washington DC, pp. 38–53. doi:10.2499/9780896298606.

Balmford, A., Bruner, A., Cooper, P., Costanza, R., Farber, S., Green, R. E., Jenkins, M., Jefferiss, P., Jessamy, V., Madden, J., Munro, K., Myers, N., Naeem, S., Paavola, J., Rayment, M., Rosendo, S., Roughgarden, J., Trumpern, K. and Turner, R. K. 2002. Economic reasons for conserving wild nature. *Science* 297, 950–3.

Bandyopadhyay, R., Ortega, A., Akande, A., Mutegi, C., Atehnkeng, J., Kaptoge, L., Senghor, A. L., Adhikari, B. N. and Cotty, P. J. 2016. Biological control of aflatoxins in Africa: Current status and potential challenges in the face of climate change. *World Mycotoxin Journal* 9, 771–89.

Beaujard, P. 2011. The first migrants to Madagascar and their introduction of plants: Linguistic and ethnological evidence. *Azania: The Journal of the British Institute of History and Archaeology in East Africa* 46, 169–89.

Bellotti, A. C., Smith, L. and Lapointe, S. L. 1999. Recent advances in cassava pest management. *Annual Review of Entomology* 44, 343–70.

Bellotti, A., Herrera Campo, B. V. and Hyman, G. 2012. Cassava production and pest management: Present and potential threats in a changing environment. *Tropical Plant Biology*. doi:10.1007/s12042-011-9091-4.

Bentley, J. W. 1994. Facts, fantasies, and failures of farmer participatory research. *Agriculture and Human Values* 11, 140–50.

Bigler, F., Babendreier, D. and Kuhlmann, U. 2006. *Environmental Impact of Invertebrates for Biological Control of Arthropods. Methods and Risk Assessment*. CABI Publishing, Wallingford, UK, 299pp.

Carson, R. 1962. *Silent Spring*. Houghton Mifflin Co., Boston, MA, 368pp.

Chambers, R., Pacey, A. and Thrupp, L. A. (Eds). 1989. *Farmer First. Farmer Innovation and Agricultural Research*. IT Publications, London, UK, 11pp.

Coll, M. and Wajnberg, E. 2017. *Environmental Pest Management. Challenges for Agronomists, Ecologists, Economists and Policy Makers*. John Wiley and Sons, New York, NY, 448pp.

De Bach, P., Schlinger, E. I. 1964. *Biological Control of Insect Pests and Weeds*. Chapman and Hall, London, 844pp.

Diamond, J. 1997. *Guns, Germs, and Steel: The Fates of Human Societies*. W.W. Norton & Company, New York, NY, 425pp.

Doutt, R. L. and Smith, R. F. 1971. The pesticide syndrome – diagnosis and suggested prophylaxis. In: Huffaker, C. B. (Ed.), *Biological Control*. Plenum Press, New York, NY, pp. 3–15.

Evenson, R. E. and Gollin, D. (Eds). 2003. *Crop Improvement and Its Effect on Productivity: The Impact of International Agricultural Research*. Chpt. 8 and 16. CABI, Wallingford, UK.

FAO (Food and Agriculture Organization of the United Nations). 1996. Code of conduct for the import and release of exotic biological control agents. International Standards for Phytosanitary Measures, Publication 3. FAO, Rome.

Feder, G., Murgai, R. and Quizon, J. B. 2004. Sending farmers back to school: The impact of farmer field schools in Indonesia. *Review of Agricultural Economics* 26, 45–62.

Fiedler, L. A. 1994. Rodent pest management in eastern Africa. FAO Plant Production and Protection Paper 123, 83pp. plus 7 Annexes.

Frison, E. A. 2016. *From Uniformity to Diversity: A Paradigm Shift from Industrial Agriculture to Diversified Agroecological Systems*. International Panel of Experts on Sustainable Food Systems. www.ipes-food.org. IPES-Food 2, 1–94.

Published by Burleigh Dodds Science Publishing Limited, 2019.

Gahukar, R. T. 1984. Insect pests of pearl millet in West Africa: A review. *Tropical Pest Management* 30, 142–7.

Gallagher, K. D., Kenmore, P. E. and Sogawa, K. 1994. Judicial use of insecticides deter planthopper outbreaks and extend the life of resistant varieties in southeast Asian rice. In: Denno, R. F. and Perfect, T. J. (Eds), *Planthoppers, Their Ecology and Management*. Chapman and Hall, New York, NY, pp. 599–614.

Gassmann, A. J., Petzold-Maxwell, J. L., Clifton, E. H., Dunbar, M. W., Hoffmann, A. M., Ingber, D. A. and Keweshan, R. S. 2014. Field-evolved resistance by western corn rootworm to multiple *Bacillus thuringiensis* toxins in transgenic maize. *Proceedings of the National Academy of Sciences USA* 111, 5141–6.

Giller, K., Andersson, J. A., Corbeels, M., Kirkegaard, J., Mortensen, D., Erenstein, O. and Vanlauwe, B. 2015. Beyond conservation agriculture. *Frontiers in Plant Science* 6, 1–16.

Giresse, P. 2008. Tropical and sub-tropical West Africa marine and continental changes during late quaternary. In: van der Meer, J. J. M. (Ed.), *Developments in Quaternary Science*, volume 10. Elsevier Academic Press, Amsterdam, the Netherlands, 395pp. ISSN: 1571-0866866, ISSN: 1571-0866.

Godfray, C. J., Blacquière, T., Field, L. M., Hails, R. S., Petrokofsky, G., Potts, S. G., Raine, N. E., Vanbergen, A. J. and McLean, A. R. 2014. A restatement of the natural science evidence base concerning neonicotinoid insecticides and insect pollinators. *Proceedings of the Royal Society B* 281: 20140558.

Greathead, D. J. 2003. Historical overview of biological control in Africa. In: Neuenschwander, P., Borgemeister, C. and Langewald, J. (Eds), *Biological Control in IPM Systems in Africa*. CABI Publishing, Wallingford, UK, pp. 1–26.

Gutierrez, A. P., Neuenschwander, P., Schulthess, F., Herren, H. R., Baumgaertner, J. U., Wermelinger, B., Löhr, B. and Ellis, C. K. 1988. Analysis of biological control of cassava pests in Africa: Cassava mealybug *Phenacoccus manihoti*. *Journal of Applied Ecology* 25, 921–40.

Gutierrez, A. P., Neuenschwander, P. and van Alphen, J. J. M. 1993. Factors affecting biological control of cassava mealybug by exotic parasitoids: A ratio-dependent supply-demand driven model. *Journal of Applied Ecology* 30, 706–21.

Hagen, K. S., van den Bosch, R. and Dahlsten, D. L. 1971. The importance of naturally-occurring biological control in the Western United States. In: Huffaker, C. B. (Ed.), *Biological Control*. Plenum Press, New York, pp. 253–93.

Heinrichs, E. A., Nwilene, F., Stout, M., Hadi, B. A. R. and Freitas, T. 2017. *Rice Insect Pests and Their Management*. Burleigh Dodds Science Publishing, Cambridge, UK, 292pp.

Hoeschle-Zeledon, I., Neuenschwander, P. and Kumar, L. 2013. Regulatory challenges for biological control. SP-IPM Secretariat, IITA, Ibadan, Nigeria, 43pp.

Howarth, F. G. 1991. Environmental impacts of classical biological control. *Annual Review of Entomology* 36, 485–509.

IAASTD. 2012. International assessment of agricultural knowledge, science and technology for development. Summary approved by governments attending the IAASTD Intergovernmental Plenary in Johannesburg, South Africa, 608pp.

IPCC. 2014. Synthesis Report. Summary for Policymakers. Working Groups of the Intergovernmental Panel on Climate Change, fifth Assessment Report, 31pp.

ISPM-3. 2005. Guidelines for the export, shipment, import and release of biological control agents and other beneficial organisms. International standards for phytosanitary measures No. 3. International Plant Protection Convention (IPPC) and FAO, Rome, Italy, 14pp.

James, B., Neuenschwander, P., Markham, R. H., Anderson, P., Braun, A., Overholt, W., Khan, K., Makkouk, K. and Emechebe, A. 2003. Bridging the gap with the CGIAR Systemwide Program on Integrated Management. In: Maredia, K., Dakouo, D. and Mota-Sanchez, D. (Eds), *Integrated Pest Management in the Global Arena*. CAB International, Wallingford, UK, pp. 419–34.

James, B., Atcha-Ahowé, C., Godonou, I., Baimey, H., Goergen, G., Sikirou, R. and Toko, M. 2010. Integrated pest management in vegetable production: A guide for extension workers in West Africa. International Institute of Tropical Agriculture, Ibadan, Nigeria, 120pp.

Krall, S., Peveling, R. and Ba Diallo, D. 1997. *New Strategies in Locust Control*. Birkhäuser Verlag, Basel, Switzerland, 522pp.

Landis, D. A., Wratten, S. D. and Gurr, G. M. 2000. Habitat management to conserve natural enemies of arthropod pests in agriculture. *Annual Review of Entomology* 45, 175–201.

Lomer, C., Bateman, R. P., Johnson, D. L., Langewald, J. and Thomas, M. B. 2001. Biological control of locusts and grasshoppers. *Annual Review of Entomology* 46, 667–702.

Louda, S. M., Kendall, D., Connor, J. and Simberloff, D. 1997. Ecological effects of an insect introduced for the biological control of weeds. *Science* 277, 1088–90.

Lynch, L. D. and Thomas, M. B. 2000. Nontarget effects in the biocontrol of insects with insects, nematodes and microbial agents: The evidence. *Biocontrol News and Information* 21, 117N–130N.

Mäder, P., Fliessbach, A., Dubois, D., Gunst, L., Fried, P. and Niggli, U. 2002. Soil fertility and biodiversity in organic farming. *Science* 296, 1694–7.

McNeely, J. A. and Scherr, S. J. 2001. Common ground, common future. How ecoagriculture can help feed the world and save wild biodiversity. IUCN, Gland, Switzerland and Future Harvest, Washington, DC, 24pp.

Messing, R. H. 2000. The impact of nontarget concerns on the practice of biological control. In: Follett, P. A. and Duan, J. J. (Eds), *Nontarget Effects of Biological Control*. Kluwer Academic, Boston, MA, pp. 45–55.

Mihok, S., Otieno, L. H. and Tarimo, C. S. 1992. Trypanosome infection rates in tsetse flies (Diptera: Glossinidae) and cattle during tsetse control operations in the Kagera River region of Rwanda. *Bulletin of Entomological Research* 82, 361–7.

Mukwevho, V. O., Pryke, J. S. and Roets, F. 2017. Habitat preferences of the invasive harlequin ladybeetle *Harmonia axyridis* (Coleoptera: Coccinellidae) in the Western Cape Province, South Africa. *African Entomology* 25, 86–97.

Mutsaers, H. J. W. 2007. *Peasants, Farmers and Scientists. A Chronicle of Tropical Agricultural Sciences in the Twentieth Century*. Springer, Dordrecht, the Netherlands, 612pp.

Mutsaers, H., Coyne, D., Hauser, S., Huising, J., Kamara, A., Nziguheba, G., Pipers, P., Taulya, G., van Asten, P. and Vanlauwe, B. 2017. *Soil and Fertility Management Research in sub-Saharan Africa. Fifty Years of Shifting Visions and Chequered Achievements*. Routledge and Taylor and Francis, London, UK, 361pp.

Neuenschwander, P. 1996. Evaluating the efficacy of biological control of three exotic Homopteran pests in tropical Africa. *Entomophaga* 41, 405–24.

Neuenschwander, P. 2004. Harnessing nature in Africa. Biological control can benefit the pocket, health and the environment. *Nature* 432, 801–2.

Published by Burleigh Dodds Science Publishing Limited, 2019.

Neuenschwander, P. and Markham, R. 2001. Biological control in Africa and its possible effects on biodiversity. In: Wajnberg, E., Scott, J. K. and Quimby, P. C. (Eds), *Evaluating Indirect Ecological Effects of Biological Control*. CAB International, Wallingford, UK, pp. 127–46.

Neuenschwander, P., Langewald, J., Borgemeister, C. and James, B. 2003. Biological control for increased agricultural productivity, poverty reduction and environmental protection in Africa. In: Neuenschwander, P., Borgemeister, C. and Langewald, J. (Eds), *Biological Control in IPM Systems in Africa*. CABI Publishing, Wallingford, UK, pp. 377–405.

Neuenschwander, P., Bown, D., Hèdégbètan, G. C. and Adomou, A. 2015. Long-term conservation and rehabilitation of threatened rainforest patches under different population pressures in West Africa. *Nature Conservation* 13, 21–46.

Oerke, E.-C. 2006. Centenary review: Crop losses to pests. *Journal of Agricultural Science* 144, 31–43.

Oliveira, C. M., Auad, A. M., Mendes, S. M. and Frizzas, M. R. 2014. Crop losses and the economic impact of insect pests on Brazilian agriculture. *Crop Protection* 56, 50–4.

Ooi, P. A. C. 2000. From passive observer to pest management expert: Science education and farmers. In: Guijt, I., Berdegue, J. A. and Loevinsohn, M. (Eds), *Deepening the Basis of Rural Resource Management*. ISNAR, The Hague, the Netherlands, pp. 167–77.

Pickett, J. A. 2016. The essential need for GM crops. *Nature Plants* 2, article 16078.

Pickett, J. A., Woodcock, C. M., Midega, C. A. O. and Khan, Z. R. 2014. Push–pull farming systems. *Current Opinion in Biotechnology* 26, 125–32.

Pimentel, D. 1996. Green revolution agriculture and chemical hazards. *The Science of the Total Environment* 188(Suppl.): S86–S98.

Prinsloo, G. L. and Uys, M. 2015. *Insects of Cultivated Plants and Natural Pastures in Southern Africa*. Entomological Society of Southern Africa, Hatfield, South Africa, 785pp.

Purslane, J. W. 1968. *Tropical Crops: Dicotyledons*. Longman Scientific & Technical, Harlow, England, 719pp.

Purslane, J. W. 1972. *Tropical Crops: Monocotyledons*. Longman Scientific & Technical, Harlow, England, 607pp.

Röling, N. and Wagemakers, M. A. E. (Eds). 1998. Facilitating sustainable agriculture. *Conservation Ecology* 6(2): 8.

Rose, D. J. W., Dewhurst, C. F. and Page, W. W. 2000. *The African Armyworm Handbook*, 2nd ed. NRI, Chatham, UK, 304pp.

Roy, H. E. and Wajnberg, E. (Eds). 2008. From biological control to invasion: The ladybird *Harmonia axyridis* as a model species. *BioControl* 53, 1–4.

Roy, H. E., Brown, P. M., Adriaens, T., Berkvens, N., Borges, I., Clusella-Trullas, S., Comont, R. F., De Clercq, P., Eschen, R., Estoup, A., et al. 2016. The harlequin ladybird, *Harmonia axyridis*: Global perspectives on invasion history and ecology. *Biological Invasions* 18, 997–1044.

Ryffel, G. U. 2017. Editorial. I have a dream: Organic movements include gene manipulation to improve sustainable farming. *Sustainability* 392, 1–9. doi:10.3390/su9030392

Sallam, M. N. 2014. Insect damage: Damage on Post-harvest. INPhO Post-harvest compendium, FAO, Rome, 37pp.

Savary, S., Ficke, A., Aubertot, J.-N. and Hollier, C. 2012. Crop losses due to diseases and their implications for global food production losses and food security. *Food Security*. doi:10.1007/s12571-012-0200-5.

Schmutterer, H. 1990. Properties and potential of natural pesticides from the neem tree *Azadirachta indica*. *Annual Review of Entomology* 35, 271–97.

Settle, W. H., Ariawan, H., Astuti, E. T., Cahyana, W., Hakim, A. L., Hindayana, D. and Lestari, A. S. 1996. Managing tropical rice pests through conservation of generalist natural enemies and alternative prey. *Ecology* 77, 1975-88.

Sharma, K. K., Sharma, H. C., Seetharama, N. and Ortiz, R. 2002. Development and deployment of transgenic plants: Biosafety considerations. *In Vitro Cell Development and Biology* 86, 1–11.

Sikora, R. A., Coyne, D. L., Hallman, J. and Timper, P. (Eds). 2018. *Plant Parasitic Nematodes in Subtropical and Tropical Agriculture*, 3rd ed. CAB International, Wallingford, UK, 876pp.

Stoll, G. 2002. *Protection naturelle des végétaux en zones tropicales*. Margraf, Weikersheim, Germany, 398pp.

Strong, D. R. and Pemberton, R. W. 2001. Food webs, risks of alien enemies and reform of biological control. In: Wajnberg, E., Scott, J. K. and Quimby, P. C. (Eds), *Evaluating Indirect Ecological Effects of Biological Control*. CABI Publishing, Wallingford, UK, pp. 57–79.

Thomas, M. and Waage, J. 1996. *Integration of Biological Control and Host-Plant Resistance Breeding*. CTA, Wageningen, the Netherlands, 99pp.

Upton, M. 1993. The economics of food production. In: Swaminathan, M. (Ed.), *Crop Protection and Sustainable Agriculture*. John Wiley & Sons, New York, pp. 61–75.

van den Bosch, R. 1978. *The Pesticide Conspiracy. An Alarming Look at Pest Control and the People Who Keep Us 'Hooked' on Deadly Chemicals*. Doubleday & Company Inc, New York, NY, 226pp.

van den Bosch, R., Leigh, T. F., Falcon, L. A., Stern, V. M., Gonzales, D. and Hagen, K. S. 1971. The developing program of integrated control of cotton pests in California. In: Huffaker, C. B. (Ed.), *Biological Control*. Plenum Press, New York, NY, pp. 377–94.

van Driesche, R. G., Carruthers, R. I., Center, T., Hoddle, M. S., Hough-Goldstein, J., Morin, L., Smith, L., Wagner, D. L., Blossey, B., Brancatini, V., et al. 2010. Review: Classical biological control for the protection of natural ecosystems. *Biological Control* 54, S2–S33.

van Lenteren, J. C. 2011. The state of commercial augmentative biological control: Plenty of natural enemies, but a frustrating lack of uptake. *BioControl* 57, 1–20.

Vitousek, P., Mooney, H. A., Lubchenco, J. and Melillo, J. M. 1997. Human domination of earth's ecosystems. *Science* 277, 494–9.

Winston, R. L., Schwarzlander, M., Hinz, H. L., Day, M. D., Cock, M. J. W., Julien, M. H. (Eds). 2014. *Biological Control of Weeds: A World Catalogue of Agents and Their Target Weeds*, 5th ed. USDA Forest Service, Forest Health Technology Enterprise Team, Morgantown, WV, 838pp.

Wood, D. and Lenné, J. M. (Eds). 1999. *Agrobiodiversity: Characterization, Utilization and Management*. CABI Publishing, Wallingford, UK, 490pp.

Wopereis, M. C. S., Johnson, D. E., Ahmadi, N., Tollens, E. and Jalloh, A. 2013. *Realizing Africa's Rice Promise*. CABI, Wallingford, UK, 451pp.

Published by Burleigh Dodds Science Publishing Limited, 2019.

Ye, X., Al-Babili, S., Klöti, A., Zhang, J., Lucca, P., Beyer, P. and Potrykus, I. 2000. Engineering the provitamin A (beta-carotene) biosynthetic pathway into (carotenoid-free) rice endosperm. *Science* 287, 303–5.

Zadoks, J. C. and Waibel, H. 2000. From pesticides to genetically modified plants: History, economics and politics. *Netherlands Journal of Agricultural Science* 48, 125–49.

Zeddies, J., Schaab, R. P., Neuenschwander, P. and Herren, H. R. 2001. Economics of biological control of cassava mealybug in Africa. *Agricultural Economics* 24, 209–19.

Zhu, P., Lu, Z., Heong, K., Chen, G., Zheng, X., Xu, H., Yang, Y., Nicol, H. I. and Gurr, G. M. 2014. Selection of nectar plants for use in ecological engineering to promote biological control of rice pests by the predatory bug, *Cyrtorhinus lividipennis* (Heteroptera: Miridae). *PLoS ONE* 9, 1–12.

Chapter 3

Disease surveillance, diagnostics and germplasm health in crop protection

P. Lava Kumar, *International Institute of Tropical Agriculture (IITA), Nigeria;* **James P. Legg**, *IITA, Tanzania;* **Maria Ayodele**, *IITA, Nigeria;* **George Mahuku**, *IITA, Tanzania; and* **Alejandro Ortega-Beltran** *and* **Ranajit Bandyopadhyay**, *IITA, Nigeria (corresponding author: l.kumar@cgiar.org)*

1 Introduction

Agricultural production in sub-Saharan Africa (SSA) is dominated by smallholder farmers (<5 ha landholdings) contributing 70–80% of the food production and an estimated 52% of total food calories consumed (Ricciardi et al., 2018). The bulk of the agricultural production occurs through manual labour, under low-input conditions or none, mainly using saved seeds. Often crop yields achieved are far below potential yields due to several abiotic and biotic threats. For over a century, diseases caused by a range of bacteria, fungi, oomycetes and viral pathogens have been documented to cause the most serious crop losses in SSA, contributing to a cumulative yield loss ranging from 20% to 40% (Savary et al., 2012) that could exceed 80% during epidemic situations. Outbreaks of pathogens coupled with human and livestock diseases and conflicts have often led to food scarcity at local and regional scales (Harsch, 2003). Pathogens also constrain seed production programs, owing to high incidence and prevalence of pathogens in the production environment and weak capacity in production of quality seeds in

http://dx.doi.org/10.19103/AS.2018.0043.03
Published by Burleigh Dodds Science Publishing Limited, 2019.

SSA. Often farmers plant contaminated seed sources due to unavailability of healthy planting material, especially those of vegetatively propagated crops, leading to situations of successive cultivation of infected seed sources, and high incidence and prevalence of pathogens in the farmers' fields (Bentley et al., 2018; Biemond et al., 2013). Pathogens as phytosanitary trade barriers are a significant check to regional and international trade of agricultural commodities from SSA. Estimated monetary value of losses due to pathogens is limited to a few scenarios (Oerke, 2006; Savary et al., 2012), but it is likely that cumulative losses from the direct and indirect impact of pathogens on production and trade in SSA could annually exceed US$50 billion.

From its inception in 1967, IITA has prioritized pathogen control as one of the main targets for plant health programmes to stabilize and enhance crop yields. Emphasis has been on breeding for resistance where possible and efforts were placed on integrated control measures in the absence of or in conjunction with durable host resistance. Success of these programmes hinged on the ability to accurately diagnose and detect pathogens at all stages of crop production and on phytosanitary surveillance programmes seeking pathogens for early detection, containment and elimination. This chapter summarizes the achievements and lessons from battling against important epidemics and emerging invasive pathogens of IITA mandate crops, emphasizing the contributions of diagnostics, surveillance and phytosanitary controls. We conclude that (1) surveillance and early detection, (2) containment action to prevent disease spread and (3) preventive management to safeguard disease-free areas are critical requirements to break the vicious cycle of recurrent disease epidemics and mitigate their deleterious impact on crop production. Detailed crop-specific case studies are covered in other chapters of this book; diseases caused by emerging insects and other invertebrate pests are presented in Chapter 4.

2 Disease surveillance

Important pathogens prevailing on the continent are generally categorized as (1) 'emerging pathogens' defined as agents that have newly appeared in a population or have existed but are rapidly increasing in incidence or geographic range (e.g. cassava brown streak ipomoviruses (CBSIs)); (2) 're-emerging pathogens' defined as agents that were problematic to crops, and then declined owing to interventions or other reasons, but are again becoming problematic in some areas (e.g. maize downy mildew); (3) 'invasive pathogens' defined as introduced or exotic agents that are not native to a specific location but on introduction cause significant problems (e.g. *Banana bunchy top virus* - BBTV, genus *Babuvirus*; *Maize chlorotic mottle virus* - MCMV, genus *Machlomovirus*); and (4) 'endemic pathogens' defined as agents that regularly occur in a particular area (e.g. *Colletotrichum gloeosporioides*). However, categorization of various pathogens occurring in SSA remains dynamic and changes depending on geography and time. For instance, CBSIs in Malawi can be termed as 'endemic' along the lakeshores, but these same pathogens in the Great Lakes region in East Africa are 'emerging', and for West Africa they are an 'invasive threat' because of their previous absence in the sub-region (Kamowa-Mbewe et al., 2015; Legg et al., 2015a). BBTV is another classic example of one pathogen with different categorization – 'endemic' in Central African countries such as the Democratic Republic of Congo (DRC) and Malawi, 'emerging' in West African countries of Benin and Nigeria, and 'invasive risk' to East African countries such as Uganda. Pathogen categorization by threat level is important to the adoption of appropriate control strategies, biosecurity adjustments and resource allocations. IITA plant health research specifically focuses on pathogens that are of economic significance and present a high risk of spread or high threat of invasion. It is noteworthy that most economically important diseases of the staple crops in Africa are caused by exotic introduced pathogens (EIPs) from other continents. The other category of priority for IITA are native pathogens termed new encounter pathogens (NEPs) that have shifted to non-indigenous crop species introduced from other continents (Table 1). Usually, native pathogens rarely cause severe epidemics on indigenous African crops. However, infection by such agents especially in susceptible cultivars results in significant yield reduction (e.g. *Yam mosaic virus* – YMV in white yam, *Dioscorea rotundata*).

2.1 New encounter pathogens

Almost all introduced food crops are affected by NEPs. IITA research on NEPs focuses on the characterization of pathogens and their diversity, tracking spread and developing control measures. The cassava mosaic disease (CMD) caused by cassava mosaic begomoviruses (CMBVs, genus *Begomovirus*) is a classic example of a pathogen that evolved on African native plants and infected cassava that was first introduced in the sixteenth century by the Portuguese (Geddes, 1990). The CMBVs are not known to occur in the centre of cassava domestication in Latin America. As an introduced plant species in Africa, the crop was vulnerable to NEPs due to a lack of any resistance genes, and this resulted in damaging outbreaks (Storey, 1936). Infection by CMBVs in most of the African native plant species causes only benign symptoms without any noteworthy damage to the host (Alabi et al., 2008b). Other economically important NEPs on introduced crop plants are CBSIs in cassava that result in cassava brown streak disease (CBSD) along the costal zones of Tanzania and Mozambique (Legg et al., 2015a); *Cocoa swollen shoot virus* (genus *Badnavirus*) responsible for swollen shoot disease in West Africa (Andres et al., 2017); *Maize*

Published by Burleigh Dodds Science Publishing Limited, 2019.

Table 1 Some economically important emerging infectious diseases (EID) and invasive threats (IT) of bacterial, fungal and viral origin to crops in sub-Saharan Africa

Disease	Pathogen	Hosts	EID/IT geographical location	Impact on African populations, economy and/or biodiversity	Reasons for emergence	IITA response to manage the EID/IT
Asian soybean rust	*Phakopsora pachyrhizi* (fungus)	Soybean, >150 legume species	Uganda, Kenya, Rwanda, Zimbabwe, Zambia, Nigeria, Mozambique, South Africa, Ghana, DRC, CAR, Tanzania and Malawi	Potential crop yield loss of up to 80%; continuous emergence of new races; loss of income	Introduction through urediniospores blowing from Western India to East Africa	Screening methods and evaluation of soybean resistance; deployment of resistant cultivars; virulence characterization, population genetic studies; epidemiological studies and disease management
Banana bunchy top disease	*Banana bunchy top virus* (virus)	Banana/ plantain	Angola, Benin, Burundi, Cameroon, CAR, Congo Brazzaville, DRC, Equatorial Guinea, Gabon, Malawi, Mozambique, Nigeria, Rwanda, South Africa, Zambia	Up to 90% loss of production; food insecurity; loss of income; loss of cultivar diversity	Introduction from South Pacific or Asia through infected suckers. Further spread through exchange of infected planting material and aphid (*Pentalonia nigronervosa*) vector	Epidemiology, ecology and mapping disease spread; development and implementation of clean disease eradication and replanting with seed systems; development of host resistance through conventional selection and transgenic approaches; quarantine controls; surveillance, early warning and pre-emptive controls; awareness and advocacy
Banana *Xanthomonas* wilt	*Xanthomonas campestris* pv. *musacearum* (bacterium)	Banana/ plantain	Burundi, Rwanda, DRC, Uganda, Kenya, Tanzania	Regional crop failure; loss of income; food insecurity; political instability; increased price	Introduction from Ethiopia to Uganda and then across the Great Lakes region	Development of detection tools; population genetic analysis; evaluation of cultivars' susceptibility; epidemiology; development of transgenic resistant germplasm; development of management strategies
Banana wilt tropical race IV	*Fusarium oxysporum* f. sp. *cubense* TR4 (fungus)	Banana/ plantain	Mozambique	Total crop loss and abandonment of the crop, food insecurity, political instability	Introduction from South East Asia	Member of the African Consortium for Foc TR4 (AC4FTR4) which focus on containing the disease and preventing its spread; awareness and advocacy

Cassava brown streak	*Cassava brown streak ipomoviruses* (virus)	Cassava	Burundi, Comoros, DRC, Kenya, Malawi, Mozambique, Rwanda, Tanzania, Uganda, Zambia	Unusable roots; up to 90% loss of production; food insecurity; loss of income; loss of landrace cultivars	Spread through exchange of infected stem cuttings and whitefly (*Bemisia tabaci*) vector	Epidemiology and mapping disease spread; development and implementation of clean seed systems; development of host resistance through conventional breeding and transgenic approaches; surveillance, early warning and pre-emptive controls; awareness and advocacy
Maize downy mildew	*Peronos-clerospora sorghi* (fungus)	Maize, sorghum	Nigeria, Botswana, Kenya, Malawi, Mozambique, Rwanda, Somalia, Sudan, Tanzania, Uganda, Zambia, DRC, Zimbabwe	Crop failure, up to 90% yield loss; food insecurity	Possible spread with seed exchange	Elucidation of disease cycle; population biology studies; development of screening methods and large-scale distribution of pathogen-free seeds; development of resistant germplasm for use in West and Central Africa
Maize lethal necrosis	*Maize chlorotic mottle virus* (virus)	Maize, sorghum, pearl millet	Burundi, DRC, Ethiopia, Kenya, Rwanda, Tanzania, Uganda	Crop failure, up to 100% loss of production; food insecurity; loss of income and seed trade	Possible spread from Southeast Asia with seed exchange	Virus characterization; epidemiology and mapping disease spread; development of diagnostics tools; strengthening quarantine controls; pre-emptive breeding for West Africa; early warning and pre-emptive controls; awareness and advocacy
Soybean witches' broom	16SrII-C subgroup phytoplasma (mollicutes)	Soybean and several wild legumes	Mozambique, Malawi, Tanzania	Up to 100% production loss	New encounter disease (pathogen suspected to have spread from native wild/weed species)	Pathogen characterization
Taro blight	*Phytophthora colocasiae* (oomycete)	Taro	Cameroon, Ghana, Nigeria and several other countries in SSA	Crop failure and up to 50–100% yield loss; destruction of local landraces	Possible spread from South Pacific islands through infected corms	Pathogen characterization and facilitation of improved taro germplasm with resistance to taro blight from South Pacific to West Africa

streak virus (MSV, genus *Mastrevirus*) responsible for streak disease in maize (Bosque-Pérez, 2000); maize downy mildew caused by *Peronosclerospora sorghi* (Cardwell, 1995); *C. gloeosporioides* on water yam (*D. alata*) (Thottappilly et al., 1999); and more recently banana *Xanthomonas* wilt (BXW; *Xanthomonas campestris* pv. *musacearum*) (Tripathi et al., 2009) and soybean witches' broom caused by the 16SrII subgroup phytoplasma (Kumar et al., 2011a; Murithi et al., 2015). Decades of interaction between the NEPs and their hosts have contributed to the evolution of NEP strains highly specific to the introduced host (e.g. *P. sorghi* of maize) and increased virulence (e.g. EACMV-Uganda). The spread of ACMV from cassava to soybean crops in the vicinity of infected cassava fields in Nigeria demonstrates the potential of a host-NEP system as a source for pathogens spreading to other crop species (Mgbechi-Ezeri et al., 2008a).

The geographic distribution of NEPs differs widely on the continent, with some NEPs (e.g. BXW, CBSIs) restricted to a few countries and some being widespread (e.g. ACMV, MSV). Variations in the genetic diversity of populations of widely distributed NEPs were detected, including cases of independent adaptation to introduced hosts in different sub-regions of Africa. For instance, *Africa cassava mosaic virus* (ACMV) in West Africa, a complex of East African cassava mosaic viruses (EACMVs) in East and Central Africa, and Southern African cassava mosaic (SACMV) in Southern Africa are dominant CMBVs involved in CMD aetiology (Alabi et al., 2011). Some NEPs with restricted spread, such as CBSIs known to be localized along the shores of Lake Malawi and coastal zones of Tanzania and Mozambique, have expanded their geographic ranges owing to the dispersal of infected planting material to East Africa, and subsequent spread by the omnipresent insect vectors (Kamowa-Mbewe et al., 2015; Legg et al., 2015a).

Discoveries of NEPs were driven by the need for identification because of their association with either a severe disease outbreak or the appearance of new symptoms in crops. It is plausible that several pathogens infecting introduced crop species have up to now evaded discovery through their obscurity. This was indeed proven in a number of recent studies using generic virus discovery tools (Ho and Zanetakis, 2014). Recent studies in Nigeria to determine the diversity of viruses infecting soybean resulted in the identification of two new 'legumoviruses' (genus *Begomovirus*), the *Soybean mild mottle virus* (SbMMV) and *Soybean chlorotic blotch virus* (SbCBV) (Alabi et al., 2010). One of these viruses, SbCBV, was later identified infecting cassava in Togo and Cameroon (Leke et al., 2016). Similar studies in maize resulted in the identification of *Maize yellow mosaic virus* (genus *Polerovirus*) in maize and neighbouring sugarcane in Nigeria (Yahaya et al., 2017b), and of MSV and a new strain of *Sugarcane streak virus* (genus *Mastrevirus*) in sugarcane plants adjacent to maize fields also in Nigeria (Yahaya et al., 2017a).

Available knowledge on NEPs on IITA mandate crops indicates the expansion of their host ranges (Alabi et al., 2008b) and the geographic spread of known NEPs (Kumar et al., 2009; Rabson et al., 2018) leading to 'new' new encounter diseases, changes in the NEPs' population structure from mutations and genetic reassortment leading to the genesis of new severe strains. Evidence also points to the occurrence of new pathogens awaiting discovery. This dynamic scenario warrants regular monitoring and preparedness to tackle new outbreaks. Excellent progress in breeding for host resistance to several NEPs (e.g. CMD, MSV, grey leaf spot – GLS, maize downy mildew) has offered sustainable and cost-effective control options (reviewed in crop-specific chapters in this book). However, for some NEPs, robust resistance is elusive. Efforts are ongoing at IITA and in partner organizations to develop resistance by conventional breeding (e.g. CBSIs) (Kayondo et al., 2018) and transgenic approaches (e.g. BXW) (Tripathi et al., 2014) whilst the interim measures based on integrated disease management (IDM) approaches are being used to mitigate disease impacts.

2.2 Exotic introduced pathogens

A number of diseases caused by EIPs have been reported since the beginning of the twentieth century in SSA. *Banana bunchy top virus* (BBTV, genus *Babuvirus*) in the 1960s in DRC (Kumar et al., 2015), cassava bacterial blight (CBB; *Xanthomonas axonopodis* pv. *manihotis*) in the 1970s in East Africa (Bart et al., 2012), black Sigatoka (*Mycosphaerella fijiensis*) in Zambia in 1973 (Zandjanakou-Tachin et al., 2009) and Asian soybean rust (*Phakopsora pachyrhizi*) may have been present in Zambia since 1978 (Javid and Ashraf, 1978; Twizeyimana et al., 2009). All these pathogens have since expanded their geographic range from the point of introduction. Some EIPs have spread to all parts of SSA (e.g. CBB, black Sigatoka) while the spread of some pathogens is restricted to a few countries (e.g. BBTV). In addition to the EIPs introduced in the twentieth century, the occurrence of new EIPs has been recorded in the twenty-first century. Most notorious EIPs of the present decade are as follows: *Maize chlorotic mottle virus* (MCMV, genus *Machlomovirus*) first detected in 2011 in Kenya (Wangai et al., 2012) that is responsible for the outbreak of maize lethal necrosis (MLN) in East Africa (Mahuku et al., 2015), banana fungal wilt caused by the *Fusarium oxysporum* f. sp. *cubense* Tropical Race IV (FoC TR4) in Mozambique (Fusarium wilt of banana, 2015) and taro blight of taro caused by *Phytophthora colocasiae* in West and Central Africa (Bandyopadhyay et al., 2011; Omane et al., 2012).

The first detection of almost all these EIPs was made at the peak of an epidemic, and too late in many cases for containment. For instance, reports of the spread of MLN to countries neighbouring Kenya appeared before the identification of the causal pathogen (MCMV) associated with MLN. By 2014, MCMV was recorded in Burundi, DRC, Ethiopia, Rwanda, Uganda and Tanzania (Lukanda et al., 2014; Mahuku et al., 2015). This case and most of the scenarios of first reports of EIPs reveal an unfortunate trend of EIP detection followed by the public call for attention to a damaging outbreak. This exposes serious weaknesses in the system's ability to recognize EIPs that may be present but do not attract attention owing to their as yet insignificant effect on the host.

The introduction pathways of EIPs in many cases remain undetermined, but we often presume an introduction through the inadvertent transfer of infected planting material or contaminated plant parts with trade and human migration, or an aerial dispersal of fungal spores. Once introduced into the continent, each point of infection serves as an inoculum source for further spread through planting material, insect vectors or aerial dispersal. Many predisposing factors are suspected in the rapid spread of EIPs in SSA, including a lack of sufficient knowledge on pathogen risks and pathways of entry and spread, traditional markets and cultural linkages, and porous borders which heightens risk of pathogen spread from country to country.

Ongoing efforts to contain EIPs are focused on mitigating disease impact in the EIP-invaded countries. For some EIPs, breeding for host resistance (e.g. Asian soybean rust and CBB) have resulted in the successful deployment of resistant cultivars (Twizeyimana et al., 2008; Hartman et al., 2011), and some EIPs are best managed by developing disease-free planting material as for BBTV (Ajambo et al., 2018) and CBSIs (Tumwegamire et al., 2018). Research is in progress to develop resistant sources to important EIPs such as BBTV (Ngatat et al., 2017) and MCMV (Beyene et al., 2017). Increased efforts are also devoted to raise awareness on the risks of EIPs, to improve surveillance at the expanding front and to increase training and capacity development for early detection and eradication in risk-prone countries, and breeding for augmenting host resistance. Some of the ongoing transcontinental initiatives on such a mission in SSA include the West African Virus Epidemiology project focused on CBSD control; the ALLIANCE for BBTV control in Africa; the CGIAR Research Program (CRP) on Roots, Tubers and Bananas work to contain FoC TR4; the CRP MAIZE efforts to contain MLN; and the Borlaug Global Rust Initiative to contain the spread of UG99 strain of wheat stem and leaf rusts.

3 Factors for emergence and re-emergence of epidemics in SSA

Since the 1990s, a dramatic surge in epidemics caused by NEPs and EIPs has been observed in SSA. These were caused by pathogens already established in the continent (e.g. CMD, CBSD, BBTV, BXW) and also those newly introduced in the twenty-first century (e.g. taro blight, MLN, FoC TR4). Several human-made and natural factors are responsible for these new outbreaks. These are grouped into two categories: predisposing factors and causal factors.

3.1 Predisposing factors for new outbreaks

3.1.1 Omnipresent host, pathogen and vectors

Year-round production of short duration crops (maize), continuous availability of long duration crops (cassava, yam) or perennial/semi-perennial plantations (banana) in the field provide a perfect situation for the persistence of pathogens and their vectors in the field throughout the year. Pathogens readily propagated in the seeds have spread along with the dispersal of planting material, and the spread of pathogens not seed-borne is aided by the insect vectors or wind dispersal of spores. Many pathogens and vectors have wide host range and can survive on alternate hosts in the absence of their main hosts during off-seasons and initiate new disease cycles with the onset of a new cropping season. In general, the abundance of pathogen inoculum and insect vectors has contributed to major outbreaks, as realized in the outbreaks of CMD and CBSD around the Great Lakes region in East Africa (Alicai et al., 2007; Legg et al., 2015a,b).

3.1.2 Agricultural intensification

Increase in agricultural production in SSA was mainly achieved by an increase in the production area. For instance, from 1966 to 2016 the total area under cassava in Africa almost tripled, from <6 million ha in 1966 to >17 million ha in 2016 (FAOSTAT, 2018). Production intensification has dramatically increased cropping density and reduced the separation distance between fields. External sources of infection could thereby spread readily to newly established fields in the vicinity, assisted by the wind-borne dispersal of pathogens and spores as well as by insect vectors (e.g. CBSIs, MCMV, Asian soybean rust). Any incidence of EIP introduction into one farm can quickly evolve into an epidemic because of the close proximity of the susceptible hosts (Xing et al., 2017).

3.1.3 Smallholder farming practices

About 80% of the agricultural production in SSA is achieved by smallholder farmers with landholdings of <2 ha (Wiggins, 2009). A large percentage of these farmers grow crops under low or no-input conditions often using saved seeds very likely to be contaminated with pathogens. Planting dates of smallholder farmers stretch several weeks, leading to a mix of young and older crops at any given time. Young plantations serve as green bridges for pathogens and vector spread from older fields thereby helping survival and an increase in inoculum pressure over successive crop cycles. This factor indeed has played a major role in several epidemics, including in the recent epidemic of MLN in East Africa supported by the migration of insect vectors from necrotized diseased maize to young healthy plants sown later (Mahuku and Kumar, 2017).

3.1.4 Increase in human migration, trade and transportation

Movement, whether of live plants, contaminated plant parts or planting material through trading commodities and of migrants in search of opportunities or people displaced by conflict represents the most important pathways of pathogen spread into new areas. Many cases of pathogen outbreaks could be linked to this factor alone. The most recent EIPs, MCMV and FoC TR4, appear to have spread from Southeast Asia to Africa through trade and human travel (Mahuku and Kumar, 2017) and CBSD from DRC to Zambia (Rabson et al., 2018).

3.1.5 Weak surveillance and quarantine controls

Many countries in SSA lack adequate capacity to enforce phytosanitary regulations and implement quarantine checks along the trade and transportation routes. This situation, coupled with porous borders, facilitates trade through non-border posts and limits possible checks on the movement of material in and out of a country. National or regional surveillance systems that can facilitate early detection of outbreaks and rapid containment action are non-existent in most countries and, where available, systems are weak with insufficient resources and lack of sufficient coordination among implementing agencies. This situation delays disease recognition and subsequent action, making the region vulnerable to new outbreaks.

3.2 Causal factors for new outbreaks

The components of the disease triangle, the pathogen, its host and the vector/environment, form the important causal factors for disease outbreaks. The pathogen is the important causal factor in all epidemics recorded to date in SSA. A susceptible host exposed to a native pathogen, an exotic pathogen or a new strain rapidly evolve into a local and regional epidemic because of favourable predisposing factors. The insect vector is the second most important causal factor influencing an outbreak and the magnitude of an epidemic especially of virus diseases. Detailed studies on the influence of vectors are limited to CMBVs and whitefly (*Bemisia tabaci*) vectors in cassava systems in SSA. Changes in host suitability of vectors (Legg and Ogwal, 1998), genetic changes in whitefly vector populations (Legg et al., 2014) and the loss of whitefly endosymbiont infections (Ghosh et al., 2018) have been proposed as likely causes for the increases in whitefly populations, making them a key driver of CMD and CBSD epidemics in East Africa (Legg et al., 2011). Host susceptibility to a pathogen is the third most important factor in propelling the outbreaks. Susceptible hosts provide a perfect recipe for an epidemic provided the pathogen and vector find a suitable environment. Environmental conditions have a major influence on epidemics caused by the fungi, oomycetes and vector-transmitted viruses.

Although it is hard to measure the quantitative effects of each of these factors, their cumulative effect is significant in epidemics of NEPs and EIPs recorded to date in SSA. Studies based on ecological approaches are required to understand the influence of each of the predisposing and causal factors for improving control measures and developing risk prediction models necessary for forecasting. Recent studies on banana, cassava and maize are exploring this approach (Hilker et al., 2017; Szyniszewska et al., 2017; Xing et al., 2017).

4 Tools for disease diagnostics

Great progress has been achieved in the development and application of a wide diversity of laboratory-based techniques for the detection and diagnosis of pests and diseases affecting the crops IITA is working with in its 50+ years of operation. The recognition of a disease based on the symptoms expressed by affected plants remains the primary diagnostic approach for the vast majority of diagnostic events. These comprise diagnoses undertaken during pest/disease surveillance exercises, during the screening of landraces and new varieties produced by conventional and transgenic methods. Almost all major diseases of IITA mandate crops in SSA can be identified in most circumstances in this way, although there are some important exceptions. Symptoms of CMD, maize streak, banana bunchy top, banana bacterial wilt and few others can be diagnosed with a high certainty based on the symptoms. A few other diseases such as leaf symptoms of CBSD, soybean rust, yam anthracnose and GLS of maize requires a higher level of expertise for symptom-based diagnosis. In addition, symptom expression is variable and depends on the stage of infection, plant age, co-infection with other pathogens and environmental conditions, thus complicating symptom-based diagnoses. Some plant stages, especially *in vitro* plantlets seldom express symptoms. Consequently, the limitations of symptom-based diagnoses became clear at an early stage to IITA researchers who then started adopting highly sensitive diagnostic tests for the identification and confirmation of pathogens.

4.1 Conventional methods

Early diagnostics at IITA used classical microscopy and microbiological methods for the identification of fungal and bacterial pathogens. Fungal

Published by Burleigh Dodds Science Publishing Limited, 2019.

agents are identified based on growth and colony morphology, spore and/or mycelial structure and ornamentation, whereas bacteria are identified based on the colony colour, morphology, biochemical tests, hypersensitive test in tobacco and pathogenicity tests on host plants. For virus detection, three methods were used: bioassays by grafting or mechanical inoculation to test plants, enzyme-linked immunosorbent assay (ELISA) using antibodies specific to target viruses and transmission electron microscopy (TEM) for identification based on the particle morphology. These methods, except TEM-based virus detection, are still in use in plant health research programmes at IITA. Of late, modern methods based on the nucleic acid and protein sequence properties have been used more frequently for pathogen detection.

4.2 Antibody-based methods

Virus detection by bioassays and TEM is a tedious process, expensive to perform and includes high maintenance costs for TEM. Antibody-based methods developed in the 1960s have fast emerged as reliable methods for the detection of viruses. Agar-gel double diffusion immuno assays performed on glass plates offers a simple approach for virus detection. However, this method consumes large amounts of antibody and is less sensitive and this method was abandoned for routine diagnostics use. Advent of the 96-well format ELISA replaced the gel-based immuno assays. Although ELISA was relatively efficient, especially for testing large numbers of samples, it was constrained by a limited availability of antibodies. In the 1970s and 1980s, commercial antibody production for plant viruses was non-existent and researcher-produced antibodies were often exchanged as goodwill. Antibody supplies were limited and not available for a number of target viruses of interest to IITA mandate crops. Therefore, in the 1980s, greater emphasis was placed on developing ELISA-based methods for the detection of viruses infecting its mandate crops. Facilities for virus purification and production of poly- and monoclonal antibodies were established at the IITA Ibadan station in Nigeria. Early efforts focused almost exclusively on developing polyclonal antibodies to viruses infecting cowpea. Antibodies were produced for six viruses infecting cowpea in West Africa (see Chapter 5 for details on cowpea viruses). These antibodies enabled the conduct of surveys concerning the distribution and diversity of viruses of cowpea (Thottappilly and Rossel, 1992; Shoyinka et al., 1997) and soybean (Time et al., 2010), germplasm screening for resistance (Atiri et al., 1984) and many other applications. Production of poly- and monoclonal antibodies remained one of the core functions of the Virology Unit at IITA Ibadan. This task was however used preferentially for viruses that could be purified relatively easily, mainly including mechanically transmissible viruses produced on herbaceous hosts. Thus, antibodies were produced for major viruses of cowpea and soybean (*Bean common mosaic virus-black eye cowpea strain, Southern bean mosaic virus, Cowpea mild mosaic virus, Cowpea chlorotic mottle virus, Cucumber mosaic virus*), yam (YMV), maize (MSV and MCMV) and banana (*Banana streak virus,* BSV, genus *Badnavirus*) (Agindotan et al., 2003, 2006; Akinbade et al., 2007; Eni et al., 2012; Mgbechi-Ezeri et al., 2008b; Thottappilly et al., 1998). However, the production of antibodies to some viruses such as BBTV and several that infect yam proved to be difficult because of problems with virus purification from host plants. Despite that, the successful purification and production of poly- and monoclonal antibodies to BSV are a noteworthy achievement (Agindotan et al., 2003, 2006; Thottappilly et al., 1998) and helped in differentiating episomal infections from endogenous badnavirus sequences (see Chapter 9 for more details on BSV).

Efforts to produce antibodies against the recombinant virus coat protein expressed in vitro resulted in the production of high-quality polyclonal antibodies to MCMV, but the same approach was not successful for YMV and *Yam mild mosaic virus* (genus *Potyvirus*) due to the insoluble nature of the bacterial-expressed coat protein. Antibody production and the development of ELISA-based assays were not limited to viruses, but were also used for the production of antibodies to the bacterial agents of CBB and BXW (Nakato et al., 2013) and the mycotoxin aflatoxin B_1 produced by *Aspergillus flavus* and closely related fungi. Since many viruses infecting IITA mandate crops have broad host ranges these antibodies have also become handy for virus detection in vegetable (Arogundade et al., 2018a) and fruit crops (Arogundade et al., 2018b) in other organizations in SSA. This in-house capacity to produce antibodies, and their standardization and validation for use in diagnostic applications has also benefited surveillance surveys, virus characterization, germplasm screening and seed health testing. IITA has supplied antibodies to partners' programmes and used them in several training courses aimed at developing diagnostic capacity among the national programmes in SSA.

4.3 Nucleic acid-based methods

Early stage nucleic acid (NA-based) diagnostics depended on hybridization methods involving lengthy and cumbersome procedures, including the preparation of reporter-labelled probes, which often included nucleotides labelled with radioactive dye such as ^{32}P, hybridization and exposure to detect positive reactions. Several restrictions with importing radioactive dyes have constrained adoption of such methods for diagnostics at IITA. However, the advent of the polymerase chain reaction (PCR) for nucleic acid amplification dramatically shifted the pathogen diagnostics arena. PCR technology simplified the development of NA-based tests for all types of pathogens. PCR diagnostics were extensively used during the late 1990s for disease surveys, tracking the

Published by Burleigh Dodds Science Publishing Limited, 2019.

proportion of EACMV-UG, ACMV and EACMV infections in cassava fields in the areas of East and Central Africa affected by the CMD pandemic (Legg, 1999). Initial methods used for differentiating the three viruses involved amplification of viruses using generic CMBV primers, and digestion of amplified products with restriction enzymes and identification of viruses based on the fragment patterns. The increase of sequences of CMBVs in the public database later facilitated the design of genus- and virus-specific oligonucleotide primers for direct amplification of the target viruses. Although the use of generic primers offered ready detection of CMBVs, high variation in CMBV populations and the involvement of nine species of begomoviruses in CMD aetiology across SSA created a challenge for specific identification by differential PCR. Since epidemiological studies required the identification of AMCV, EACMV and EACMV-UG, multiplex PCR methods were developed for simultaneous detection and differentiation of these three viruses in a single reaction (Alabi et al., 2008a; Otti et al., 2016). This approach offered excellent advantages by dramatically reducing the number of reactions to be performed to identify key CMBVs and lowered the cost of testing. It is anticipated that the diagnostics will continue to improve with increasing knowledge about CMBVs and the continued evolution of new diagnostic technologies.

A similar type of evolutionary approach involved in the development and standardization of PCR-based diagnostics for several pathogens at IITA. Each newly established method is extensively validated for sensitivity, specificity and robustness for application in pathogen detection in a variety of samples and working conditions. This process involves extensive checking to ensure a high level of confidence for the generated results to avoid false positives and false negatives. Procedures of validated methods have been disseminated through training courses, protocol manuals and formal publications. Presently, PCR, reverse transcription-PCR (RT-PCR) and variations of these methods, including real-time versions using TaqMan probes and SYBR Green dye, have become mainstay methods for pathogen diagnostics at IITA. Amplified products are often sequenced for accurate diagnoses especially in the identification of a new pathogen (Mgbechi-Ezeri et al., 2008a) and diversity studies (Kamowa-Mbewe et al., 2015). They are used routinely for virus surveillance and diagnostics for almost all important pathogens: BBTV (Kumar et al., 2011b), CBSIs (Legg et al., 2015b), potyviruses and cucumoviruses of yam (Eni et al., 2008; Nkere et al., 2013), viruses infecting maize (Yahaya et al., 2017a), BXW (Odipio et al., 2009), GLS (Sharma et al., 2010), black Sigatoka (Zandjanakou-Tachin et al., 2009) and several other targets.

Various innovations were made in pathogen detection including the use of multiple PCR for simultaneous detection of viruses (Alabi et al., 2008a; Nkere et al., 2013; Otti et al., 2016) and real-time PCR assays for the quantitative detection of MSV, BBTV, YMV and CBSIs (Shirima et al., 2017). However, the detection of badnaviruses in banana and yam pose challenges because of the presence of integrated viral sequences in the host genome (endogenous viral sequences) (Kumar et al., 2015). The conventional PCR method amplifies both endogenous and episomal virus sequences. For diagnostic testing, differentiating one from the other is important as plants with episomal viruses can pose a quarantine risk. To overcome this problem, immuno-capture PCR was established to specifically detect episomal viruses (Agindotan et al., 2006). Although this procedure was effective for the detection of BSV, the unavailability of antibodies for yam badnavirus limited its amplification. To overcome this challenge, a rolling circle amplification method was developed for the differential detection of episomal badnaviruses (Bömer et al., 2016).

More recently, isothermal NA amplification methods such as loop-mediated isothermal amplification (LAMP) (Nkere et al., 2018) and recombinase polymerase amplification (RPA) (Silva et al., 2015) were developed for the detection of yam viruses. As the name implies, the isothermal methods amplify RNA or DNA templates at a single incubation temperature and thereby eliminates the need for expensive thermal cyclers and can be completed in less than 40 min compared to 3–4 h for conventional PCR methods. The isothermal assays retain all the qualities of diagnostics such as specificity, sensitivity, and robustness and offer flexibility for use in the lab and field. Both LAMP and RPA assays established for viruses infecting banana (BBTV), maize (MSV and MCMV), yam (YMV and YMMV) and cassava (ACMV and EACMVs). These assays were optimized to be used directly with leaf sap in a closed tube format. Amplification of target molecules are visualized by chromogenic detection using SYBR Green dye (Nkere et al., 2018) or fluorophore-labelled probes which can be visualized using probe detectors (Silva et al., 2018). Both RPA and LAMP offer good advantages for on-farm diagnostics using pre-mixed ready-made reagents using mobile detection devices. Efforts are ongoing to convert isothermal assays into ready-to-use diagnostic devices for simultaneous detection of multiple viruses in a single reaction (Ellis et al., 2017).

New virus detection approaches based on the high-throughput next-generation sequencing (NGS) technologies (Kreuze et al., 2009) have also been adopted for virus detection in IITA programmes. This method, which is based on the extraction of 21- to 24-nucleotide small RNAs, sequencing and in silico reassembly and bioinformatics-based virus identification, provides unparalleled advantages for detecting known as well as previously undetected viruses in samples. The NGS method is shown to be equally or more sensitive than real-time PCR and sequence-independent, and is fast emerging for use in surveys and for screening germplasm collections as an efficient method for determining virus diversity and pre-emptive detection of latent viruses. The NGS method has been used for detecting viruses in maize samples collected during field surveys in Nigeria (Yahaya et al., 2017b) and to verify the virus-free status of planting materials generated in vitro as a confirmatory test (Bömer et al., 2018). As part of the ongoing 'NextGen Phytosanitation' initiative, work is in progress to identify virus diversity in clonal crop germplasm collections held in the IITA Genebank to improve phytosanitary therapy and virus indexing procedures. At this stage the NGS method is being used as a tool for virus discovery than as a routine diagnostic tool because of its high cost compared to other conventional diagnostics and length of time to obtain results (at least 2–3 weeks). However, rapid advancements in sequencing and bioinformatics technologies are poised to make NGS as a regular virus diagnostic tool in the near future (Greninger, 2018).

4.4 ICT tools for diagnostics

Most established diagnostic tools require a certain degree of expertise, some equipment and training. The frequent emergence of new diseases and changes in technology demand continuous training and retraining of personnel using the technology. While this could be easy for few lab-based staff, it is difficult to cover a large number of extension officers working remotely from the main stations. The consequence of this is that only a small fraction of extension officers can access diagnostic support for pests and diseases affecting their crops. Most extension officers base their decisions on symptoms and relay information to the next level for appropriate action or else collect and send samples to referral labs for diagnostic testing. Strengthening field-based diagnostics requires new approaches for diagnosis and communication, communication updates and decision support. Mobile phone-based diagnostic systems offer great potential for addressing this gap. Africa has rapidly caught up with the rest of the world by offering stable access to the internet through smartphones; low-priced smartphones are achieving rapid penetration throughout African markets, even in rural areas. Recognizing this potential, researchers in the United States and at IITA in Tanzania have developed a phone application (app) for cassava disease/pest recognition which is based on artificial intelligence (AI). As a first step, large numbers of images of five major disease/pest conditions affecting cassava were used to train deep convolutional neural networks to recognize the symptoms through a process of transfer-learning (Ramcharan et al., 2017). Using this AI approach achieved accuracies of >90% in identifying damage from CMD, CBSD, brown leaf spot, red mite and cassava green mite. This AI was subsequently incorporated into a smartphone app which used an instance of Google's open source image-recognition software – TensorFlow – to deliver the offline identification of four cassava leaf symptom conditions: CMD, CBSD, cassava mites and healthy (i.e. disease and pest-free). Accuracy for pronounced CMD symptoms was >83% when either image capture or video was used. In comparison with assessment by an expert, accuracies for CBSD (76%) and mites (82%) were lower for pronounced symptoms using image capture. Mild symptoms gave lower accuracies, although the most common alternative results were either a failure to detect anything or the identification of the leaf as healthy. The app has been specifically designed to minimize false positives by taking a 'cautious' approach to the detection of mild symptoms. Future development of this app and related apps will almost certainly result in increased levels of accuracy and offer the potential to put powerful diagnostic tools into the hands of tens of millions of farmers. Further advancement of this app will cover other crop diseases; it is expected to integrate low-cost portable chemistry-based diagnostics and communicate directly with smartphones.

Although diagnostics of pathogens have been explained in this chapter, the same capacity is available at IITA for detecting arthropod (Goergen et al., 2016) and nematode (Kolombia et al., 2017) pests (see Chapter 4).

5 Early warning and response systems

An early warning and response system (EWRS) is an approach based on a precautionary principle to prepare, predict and respond to emergencies. It is considered the most important part of national biosecurity plans to protect agriculture, biodiversity and the environment. Setting up EWRS requires knowledge on pest risk and spread pathways, perceived economic damage to cropping systems, technologies for risk prediction and threat detection, supportive policies and procedures, and a strategy for effective implementation backed by appropriate resources (Miller et al., 2009). The national plant protection organization (NPPO) strategies usually encompass a generic EWRS-like plan with the objective of preventing pest and pathogen incursions and implementing emergency containment actions.

However, the establishment of effective EWRS has suffered in SSA owing to limited capacity, lack of a clear implementation plan, outdated policies and insufficient fund allocation. Consequently, Africa's NPPO capacity has remained very low in preventing the introduction of invasive pathogens or appropriately countering introduced invasive pathogens (Early et al., 2016). Resource limitation has also affected the continental risk mitigation plans of the African Union Inter-African Phytosanitary Council (AU-IAPSC) in delivering its mandatory strategic support to countries in emergency situation and in adopting a 'One Health' framework for the detection, identification and monitoring of infectious diseases of humans, animals and plants in Africa by the African Union Science and Technology Research Commission (STRC, 2011). However, a spate of damaging epidemics during the last decade and the quest by many countries to capture international trade is reviving the national interest in strengthening EWRS plans (IAPPS, 2018).

The IITA team has been proactively supporting national and regional plans to develop EWRS with (1) the contribution of knowledge and information on the status of important pathogens at country, regional, continental and global scale, (2) pest risk assessment under different scenarios and use of modelling for predictions about future changes including effects of a warming climate and increased human activities, (3) provision of technologies for diagnostics and surveillance, (4) advocacy for appropriate policies and investments and (5) development of physical and technical capacities for staff responsible for EWRS.

IITA has also pioneered surveillance systems for cassava viruses. An early warning system was established to tackle the spread of severe CMD and CBSD epidemics in northwestern Tanzania (IITA, 2011). The aim of this initiative was the provision of training on CMD and CBSD symptom recognition to farmers living in areas immediately ahead of regions recently affected by the diseases, and for farmer groups to send regular SMS-based reports on the presence or absence of these diseases to a central research station 'hub'. In ten participating districts, 1281 farmers were trained in 60 farmer groups, and reports were sent by groups on a monthly basis. The established network was referred to as the Digital Early Warning Network (DEWN). Researchers provided the necessary training at the outset and re-visited all participating villages after 6 months. During these visits, researchers assessed farmers' fields to confirm the presence/absence of diseases reported by farmers. Farmers' reports of CMD were shown to closely match the observations of the researchers. In contrast, there was much less agreement in CBSD reporting, as there were frequent reports from farmers' groups of CBSD that could not be substantiated during the visits by researchers. However, two farmers' groups reported on CBSD in districts where the disease had not previously been recorded. At the completion of the initiative, it was concluded that the approach provided a useful means of highlighting pest/disease threats within farming communities, raising awareness and increasing the engagement of farmers in monitoring pests/diseases in their own fields. Weaknesses of the system, however, included the insufficient phone network coverage of the region which made the 'tech' unreliable; the high cost of travel for researchers to provide training and confirm farmers' reports; and the relative difficulty for farmers in identifying CBSD, which often has cryptic symptoms. The need remains as great as ever for the effective early warning of expanding epidemics, such as severe CMD/CBSD, MLN, BBTV, BXW, FoC TR4 as well as outbreaks of invasive pests, such as the fall and southern Armyworms and potato cyst nematode.

New approaches are currently being developed to enable farmers to do their own diagnoses using AI-equipped smartphone apps (see the previous section), with smartphones geo-tagging to produce maps of pest/disease occurrence in real-time. A WebApp for cassava disease surveillance (CDS) was developed and piloted in Nigeria (Kumar et al., unpublished). The CDS system operated by the Nigerian Agricultural Quarantine Service (NAQS) facilitates early diagnosis through digital imaging, communication among the quarantine officials, notification of pest risk and facilitation of emergency response. A digital surveillance lab has been established at the NAQS post-entry quarantine station in Ibadan, Nigeria, and NAQS staff were trained in operating the system. With a crowdsourcing model, this app uses a web interface for information gathering from quarantine officials posted in the field and international border posts, researchers, extension agents and farmers. The CDS facilitates confirmatory testing of samples for accurate diagnosis and alerts appropriate authorities for a mitigation response. As smartphone technology continues its rapid penetration of African markets, early warning and rapid response strategies look set to be revolutionized with concomitant improvements in pest/disease management.

6 Germplasm safety and preventing the spread of seed-borne pathogens

Exchange (import and export) of genetic resources through formal and informal procedures within and between countries over several centuries has contributed to the global distribution and adoption of economically important crops for agriculture and food systems diversification which has led to the present scenario of exotic crops contributing up to 68.7% of the national food production around the world (Khoury et al., 2016). For more than 50 years IITA has been at the forefront of the international exchange of food crop germplasm to tropical parts of the world (wild and landraces from the Genebank, breeding lines and varieties from the crop improvement programmes, international trials and seed stock for commercial seed production). Each year IITA exchanges genetic resources of its mandate crops with 30–50 countries, including those for its own research programmes across SSA, the national and international partners from the public and private sectors, and the farmers (Kumar and Abberton, 2016). IITA as the global custodian of genetic resources conserved 33 713 accessions of its six mandate crops (banana/plantain, cassava, cowpea and other *Vigna* species, maize, soybean and yam) in its ex situ Genebank collection located in Ibadan. Success of IITA's international crop improvement programmes depends on the unrestricted flow of germplasm between IITA and its national and international partners spread around the world. However, international distribution of germplasm is a risky event due to the threat of movement by seed-borne pathogens with contaminated germplasm.

6.1 Preventing pathogen spread with crop germplasm

The spread of pathogens and pests through the movement of contaminated planting materials (seeds and vegetative propagules) and plant products is one of the major routes of pathogens' introduction into new territories. Several cases for this type of spread have been recorded from the Irish famine in the 1840s caused by the outbreak of *Phytophthora infestans* introduced from South America (Klinkowski, 1970), to the recent outbreaks of BBTV, MLN and FoC-TR4 in SSA in the 2010s (explained in Section 2). Spread of the seed-borne pathogens through vegetative planting material is recognized as the main cause for the rapid within-country and regional spread of pathogens in SSA (Bentley et al., 2018). Seed-borne pathogens affect the seed viability, and the germination and performance of the established plants. National and international policies and legislative procedures were established to prevent the risk of seed-borne pathogens spread (IPPC, 2012). The first set of plant quarantine measures were enacted in 1873 in Germany as a legal measure for inspection for Colorado potato tuber beetles in imported potatoes from the Americas. Many countries subsequently followed suit and by 1951 quarantine measures were adapted as an international treaty under the auspices of the FAO-International Plant Protection Convention (IPPC) (IPPC, 2012). However, imbalance in implementation capacity due to poor resources and lack of adequate physical and technical capacity in developing and underdeveloped countries have affected implementation; as a consequence, pathogen continues to find their way into new territories resulting in serious economic damage (Waage et al., 2006). An analysis of the risk of 1300 known invasive pests and pathogens to 124 countries recognized SSA as the region most vulnerable to invasive risks due to weak phytosanitary capacities in preventing entry or controlling spread after the entry of biological threats (Paini et al., 2016). Not surprisingly, several new pest and pathogen outbreaks continue to occur on an annual basis in one or other of the sub-regions.

6.2 Phytosanitary procedures for safe distribution of germplasm

The establishment of IITA coincided with the first post-independence decade for a number of countries in SSA. Many institutions were under development, including NPPOs and the IAPSC was established in 1956 as the continental phytosanitary body of the Organization of African Unity that has evolved into the present-day 'African Union'. IITA's germplasm exchange program was constrained during that stage by multiple challenges because of a lack of adequate baseline knowledge on pests and pathogens and the weak quarantine infrastructure. For the IITA plant health group it became a simultaneous task of discovery and research to identify pathogens and safeguard germplasm health. Since most of the germplasm was exchanged from IITA Ibadan station in Nigeria, the IITA team has worked following the guidelines of the Nigeria Plant Quarantine Service (NPQS; renamed in 2010 as Nigeria Agricultural Quarantine Service) which in turn followed the IAPSC guidelines.

Germplasm health testing was performed in the 'Seed Health Unit (SHU)' which acted as liaison with NPQS and IITA's Plant Health group facilitating inspection and testing by an interdisciplinary team comprising pathologist, entomologist and virologist. Conditions for the importation of germplasm ranged from conditional approval to holding germplasm in a contained facility for testing and validation of health status prior to release or complete prohibition. This decision was influenced by categorization of the suspected pathogen to one of the three groups (Aluko, 1991): (1) pathogens that are not present in Nigeria or any country in West Africa are classified in 'Category A' and treated as high-level threat and regulated through strict quarantine; (2) pathogens of restricted distribution in Nigeria and other countries in West Africa are classified in 'Category B' and controlled through seed health testing, seed treatment and the selection of pathogen-free seeds; and (3) the pathogens that are widely distributed in Nigeria and other countries are classified in 'Category C' and, depending on the certification standards, seed lots with certain levels of infection (tolerance level) are approved for use.

Specific seed health testing protocols were established over a period of time for IITA's mandate crops including those for rice and sweet potato which were then part of the IITA mandate (Schilder et al., 1999), together with adoption of the 'Technical Guidelines for Safe Movement of Germplasm' developed by an international expert group coordinated by FAO and the International Bureau of Plant Genetic Resources (IBPGR) (FAO-IPGRI, 1996). Virus-free tissue culture plants were generated for clonally propagated crops such as cassava for export from Nigeria. IITA depended on an intermediary quarantine station in the Netherlands for the importation of cassava from outside Africa. Intensive germplasm exchange activities of IITA in the 1970s and 1980s were estimated to have contributed to over two-thirds of the total quarantine activities of NPQS (Aluko, 1982) and were instrumental in making NPQS the only functional quarantine station in West and Central Africa (Chiarappa, 1985).

6.3 Establishment of CGIAR Germplasm Health Units

Phytosanitary constraints similar to those faced by IITA were experienced by all CGIAR centres located in Africa Asia, and Latin America. A CGIAR commissioned study in the 1980s evaluated quarantine services and challenges to the exchange of germplasm and made several recommendations for improvements (Plucknett and Smith, 1988). These included the development of diagnostics and therapy procedures, and a call for the global community to improve national quarantine capacity. The international agricultural research centres also adopted the IPPC recommendation made in the Sixth International Plant Protection Congress, August 1993, in Montreal, to establish 'a Germplasm Health Unit (GHU)' as an independent unit in all the CGIAR centres with a main responsibility of ensuring full compliance with IPPC, regional and national plant protection regulations.

Following the 1993 Montreal recommendation, SHU at IITA was replaced with GHU as a stand-alone unit to serve as the gateway for the safe exchange of germplasm and biological resources (plant parts, DNA, biocontrol agents and others). Core functions of the GHU include the following: (1) produce healthy germplasm of IITA crops, (2) ensure full compliance with IPPC, IAPSC and national regulations applicable to germplasm and bioresource exchange, (3) ensure biological material imported and exported is free of regulated pests and pathogens, (4) inspect and provide quality certification to all germplasm exported from IITA, (5) issue a health statement for each consignment and (6) serve as a liaison between IITA and national quarantine authorities in organizing phytosanitary inspections and permits. The main base of GHU is in IITA Ibadan station in Nigeria, but the Unit oversees and supports germplasm exchange activities in all IITA stations, and that of other CGIAR centres housed within its campuses such as ILRI and AfricaRice stations. IITA stations housed in the other CGIAR centres such as IITA station housed in ILRI campus in Nairobi or other international agricultural research organizations such as ICIPE in Nairobi derive phytosanitary services from the hosting centre.

6.4 Essential functions of IITA GHU

Details of GHU activities and crop-specific phytosanitary procedures are beyond the scope of this chapter. Some of the key functions of GHU involve the following:

- Checks import and export requirements for each material meant for exchange, and provides consultation and advice to scientists on phytosanitary matters,
- Maintenance of records of biological materials distributed for traceability,
- Arranging inspections at active growth stage-, pre- and post-entry by the national quarantine authorities,
- Conducts seed health testing for pests and pathogens; seed treatment to eliminate seed-borne pathogens; grow-out tests as a check for seed-transmitted pathogens, production of healthy seed stocks for conservation and distribution,
- Periodic updates of GHU procedures to stay in line with the changing policies and phytosanitary threats,
- Develops/adapts/adopts procedures and protocols, diagnostics and therapy procedures to eliminate pathogens from infected/infested germplasm,

- Ensures the availability of adequate infrastructure needed for handling germplasm exchanges,
- Updates knowledge on pest and disease distribution relevant to IITA mandate crops and geographies; and develops standard operating procedures for germplasm health testing and healthy seed production,
- Maintenance of the phytosanitary health status of germplasm conserved in vitro and ex situ collections of IITA Genebank and breeders' germplasm and breeding lines, and
- Conducts pest risk assessment of research stations and seed production sites to mitigate biotic risks; and organizes training workshops to transfer know-how and knowledge to the partners.

GHU is also responsible for the import and export of genetically modified organisms (GMO). In this case, permits for GMO importation or export are obtained from the national biosafety management authorities prior to an application for phytosanitary/import permits from the national quarantine organizations. The Biological Control Program at the IITA Station in Cotonou, Republic of Benin, includes a special facility for the rearing and testing of exotic biocontrol agents for biocontrol research program. IITA GHU is represented as the international observer to the IAPSC Steering Committee, national germplasm committee and a number of technical and policy meetings offering scope for interaction with international stakeholders and policy makers. Such platforms have offered IITA opportunities to share latest knowledge and technologies and provide feedback on the phytosanitary decisions and policies. Advocacy through such platforms and forums has led to the development of IAPSC guidelines for production and distribution of virus-free cassava stem cuttings as an alternative to tissue culture plants for intracontinental distribution; and a decision model for distributing *Musa* germplasm with integrated BSV (see Chapter 9).

6.5 Preparedness for future phytosanitary challenges

From the former SHU to the present day, GHU has ensured institutional compliance with national and global phytosanitary regulations and provided effective phytosanitary services to IITA research programmes serving as conduits for the export and import of all types of bioresources (Kumar and Abberton, 2016). Despite this progress, new challenges continue to arise owing to changes in phytosanitary policies or new pest outbreaks or reports of the occurrence of pathogens new to IITA mandate crops and geographies prompting an additional set of monitoring procedures, thus increasing effort and costs. Such an activity sometimes involves a phytosanitary sweep of an entire germplasm collection to be reassured of germplasm safety, contributing to a temporary cessation of germplasm distribution until the evaluation has been completed. This uncompromising approach to uphold phytosanitary safety of the import and export of bioresources and the unswerving compliance with national and global procedures have earned trust among the partners and national quarantine authorities, enabling a smooth flow of the bioresources vital for the international mandate of IITA. Formal incorporation of 'GHU Component' within the 'Conservation Module' of the CGIAR Genebank Platform Program 2017–21 boosted funds for improving GHU capacity and procedures to face emerging phytosanitary challenges and improve operational efficiency to fast-track germplasm exchanges.

7 Conclusion and future trends

Since its inception, IITA together with a range of national and international collaborators has played a pivotal role in combating almost all the major pathogen outbreaks on its mandate crops in SSA. Several lessons can be drawn from five decades of IITA's experience in combating pathogens in SSA. Effective disease control can be achieved using resistant varieties and IDM, including the use of clean planting material. However, successes can be compromised by the emergence of new outbreaks caused by novel or latent pathogens. For instance, benefits achieved through a decade of efforts to contain the CMD pandemic in the Great Lakes region of East Africa in the 1990s were compromised owing to the emergence of the CBSD epidemic in the same region in the 2000s. Disease outbreaks by unsuspected pathogens, such as MCMV in maize in East Africa, can cripple crop production and gains achieved over several decades of R4D efforts.

Major pathogen outbreak events enforce redesigning of crop improvement programmes and considerably slowdown the progress. It is therefore imperative to break the cyclic nature of epidemics caused by established, emerging and invasive pathogens in SSA. Events and experiences from the past three decades suggest this task is anything but easy owing to the shifts in agricultural systems influenced by a variety of factors: (1) rising temperatures and climate change influencing the distribution of pathogens and their vectors; (2) changes in cropping cycles due to new agronomic practices, and the introduction of new crop species and varieties; (3) increase in inter-continental and within continental transport; and (4) increase in the frequency of regional and global agricultural trade of a variety of goods, including seeds produced in different countries by public and private organizations. The latter two factors are likely to increase the risk of new invasive because of new trade connections and high volumes of commodity exchanges between the countries and regions.

These scenarios have been creating a complex dynamic making prediction of pathogen outbreaks a difficult task in SSA. General preparedness for early detection of outbreaks and mitigation plans are therefore critical to stop outbreaks from becoming epidemics and pandemics. This is easier said than done under the present status quo regarding funding and capacity to perform regular monitoring. However, these limitations can be avoided by adopting new technologies that are less reliant on infrastructure, specialized skills or operational costs. For instance, adoption of citizen science/crowdsourcing model, using image-recognition smartphone applications to identify and report pathogens, coupled with on-farm quick diagnostics for confirmatory testing and better information communication among the key stakeholders and quarantine authorities, can make early detection and responsive action affordable and effective. In addition, there is a pressing need to strengthen import and export quarantine monitoring in all countries in SSA to prevent the spread of pathogens. This is becoming extremely important with increases in the volume of agricultural commodity trade between and within the regions and across continents.

The pathogen diagnostics capability in IITA has evolved and stayed contemporary to meet present and future needs. The Institute has the capacity to detect all known pathogens of its mandate crops prevailing in SSA and is also equipped to detect pathogens of relevance to crops in Africa not yet present for use in surveillance and quarantine indexing programmes. Such preparedness indeed enables IITA and partners to track pathogens associated with new outbreaks. Diagnostics remains one of the core strengths of IITA and gives unparalleled advantage to the multifaceted plant health research programmes operating in all the sub-regions in SSA. On numerous occasions, IITA diagnostics have contributed to the first diagnoses of pathogens associated with disease outbreaks. Current developments at the Institute are gearing towards the increased use of field-ready tools and ICTs for crowdsourcing to improve surveillance and the diagnostic capacity in SSA. The fast-paced improvement in diagnostic technology is opening-up a plethora of opportunities to develop fit-for-purpose tools for research and development in SSA. It is likely that the range of methods used for diagnostics in IITA plant health programmes will continue to evolve.

In conclusion, emerging and re-emerging pathogens are likely to remain as a major threat to agricultural production in SSA. Recent advances in technology for rapid identification, prevention and scouting are offering cutting-edge tools to control these agents. Better awareness about the pathogen risks, likely routes of entry, knowledge of technology and funding, especially during non-epidemic scenarios, can equip nations to face these ever-increasing challenges.

8 References

Agindotan, B. O., Thottappilly, G., Uwaifo, A. and Winter, S. 2003. Production of monoclonal and polyclonal antibodies against a Nigerian isolate of banana streak virus. *African Journal of Biotechnology* 2: 171–8.

Agindotan, B., Winter, S., Lesemann, D., Uwaifo, A., Mignouna, J., Hughes, J. and Thottappilly, G. 2006. Diversity of banana streak-inducing viruses in Nigeria and Ghana: Twice as many sources detected by immunoelectron microscopy (IEM) than by TAS-ELISA or IC-PCR. *African Journal of Biotechnology* 5: 1194–203.

Ajambo, S., Rietveld, A., Nkengla, L. W., Niyongere, C., Dhed'a, D. B., Olaosebikan, D. O., Nitunga, E., Toengaho, J., Kumar, P. L., Hanna, R., Sufo Kankeu, R. and Omondi, A. 2018. Recovering banana production in bunchy top-affected areas in Sub-Saharan Africa: Developing gender-responsive approaches. *Acta Horticulture* 1196: 219–28.

Akinbade, S. A., Mgbechi-Ezeri, J. U., Oben, T. T. and Kumar, P. L. 2007. Production of monoclonal antibodies against Cowpea mottle virus. In: *Abstracts of the 10th International Plant Virus Epidemiology Symposium*, 15–19 October 2007, ICRISAT, Hyderabad, India, p. 187.

Alabi, O. J., Kumar, P. L. and Naidu, R. A. 2008a. Multiplex PCR method for the detection of African cassava mosaic virus and East African cassava mosaic Cameroon virus in cassava. *Journal of Virological Methods* 154: 111–20.

Alabi, O. J., Ogbe, F. O., Bandyopadhyay, R., Kumar, P. L., Dixon, A. G. O, Hughes, J. d'A. and Naidu, R. A. 2008b. Alternative hosts of African cassava mosaic virus and East African cassava mosaic Cameroon virus in Nigeria. *Archives of Virology* 153: 1743–7.

Alabi, O. J., Kumar, P. L., Mgbechi-Ezeri, J. U. and Naidu, R. A. 2010. Two new 'legumoviruses' (genus *Begomovirus*) naturally infecting soybean in Nigeria. *Archives of Virology* 155: 643–56.

Alabi, O. J., Kumar, P. L. and Naidu, R. A. 2011. Cassava mosaic disease: A curse to food security in sub-Saharan Africa. *APSnet Features* (ISSN: 2153-0297). doi:10.1094/APSnetFeature-2011-0701. http://www.apsnet.org/publications/apsnetfeatures/Pages/cassava.aspx

Alicai, T., Omongo, C. A., Maruthi, M. N., Hillocks, R. J., Baguma, Y., Kawuki, R., Bua, A., Otim-Nape, G. W. and Colvin, J. 2007. Re-emergence of cassava brown streak disease in Uganda. *Plant Disease* 91: 24–9.

Aluko, M. D. 1982. *International System for Safe and Efficient Movement of Materials in Global Germplasm Exchange Networks*, 15–17 June 1982, CIAT, Cali, Colombia.

Aluko, M. O. 1991. The role of plant quarantine in Nigerian agriculture development. In: *Crop Genetic Resources of Africa, Volume II* (N. Q. Ng, P. Perrino, A. Attere and H. Zedan, Eds). IITA and International Board of Plant Genetic Resources, Rome and International Environment Program, Nairobi, pp. 241–9.

Andres, C., Gattinger, A., Dzahini-Obiatey, H. K., Blaser, W. J., Offei, S. K. and Six, J. 2017. Combatting Cocoa Swollen Shoot Virus Disease: What do we know? *Crop Protection* 98: 76–84.

Arogundade, O., Balogun, O. S. and Kumar, P. L. 2018a. Seed transmissibility of Cucumber mosaic virus in *Capsicum* species. *International Journal of Vegetable Science* 1–8 (in press). doi:10.1080/19315260.2018.1487498.

Arogundade, O., Oyekanmi, J., Oresanya, A., Ogunsanya, P., Akinyemi, S. O. S. and Kumar, P. L. 2018b. First report of Passion fruit woodiness virus associated with passion fruit woodiness disease of passion fruit in Nigeria. *Plant Disease* 102: 1181.

Published by Burleigh Dodds Science Publishing Limited, 2019.

Atiri, G. I., Ekpo, E. J. A. and Thottappilly, G. 1984. The effect of aphid resistance in cowpea on infestation and development of *Aphis craccivora* and the transmission of cowpea aphid-borne mosaic virus. *Annals of Applied Biology* 104: 339–46.

Bandyopadhyay, R., Sharma, K., Onyeka, T. J., Aregbesola, A. and Kumar, P. L. 2011. First report of taro leaf blight caused by *Phytophthora colocasiae* in Nigeria. *Plant Disease* 95: 618.

Bart, R., Cohn, M., Kassen, A., McCallum, E. J., Shybut, M., Petriello, A., Krasileva, K., Dahlbeck, D., Medina, C., Alicai, T., Kumar, L., Moreira, L. M., Rodrigues Neto, J., Verdier, V., Santana, M. A., Kositcharoenkul, N., Vanderschuren, H., Gruissem, W., Bernal, A. and Staskawicz, B. J. 2012. High-throughput genomic sequencing of cassava bacterial blight strains identifies conserved effectors to target for durable resistance. *Proceedings of the National Academy of Sciences of the United States of America* 109: 1972–9.

Bentley, J. W., Andrade-Piedra, J., Demo, P., Dzomeku, B., Jacobsen, K., Kikulwe, E., Kromann, P., Kumar, P. L., McEwan, M., Mudege, N., Ogero, K., Okechukwu, R., Orrego, R., Ospina, B., Sperling, L., Walsh, S. and Thiele, G. 2018. Understanding root, tuber and banana seed systems and coordination breakdown: A multi-stakeholder framework. *Journal of Crop Improvement* 32: 599–621.

Beyene, Y., Gowda, M., Suresh, L. M., Mugo, S., Olsen, M., Oikeh, S. O., Juma, C., Tarekegne, A. and Prasanna, B. M. 2017. Genetic analysis of tropical maize inbred lines for resistance to maize lethal necrosis disease. *Euphytica* 213: 224.

Biemond, P. C., Oguntade, O., Kumar, P. L., Stomph, T. J., Termorshuizen, A. and Struik, P. 2013. Does the informal seed system threaten cowpea seed health? *Crop Protection* 43: 166–74.

Bömer, M., Turaki, A., Silva, G., Kumar, P. L. and Seal, S. E. 2016. A sequence-independent strategy for amplification and characterization of episomal badnavirus sequences reveals three previously uncharacterized yam badnaviruses. *Viruses* 8: 188.

Bömer, M., Rathnayake, A. I., Visendi, P., Sewe, S. O., Paolo, J., Sicat, A., Silva, G., Kumar, P. L. and Seal, S. E. 2018. Tissue culture and next-generation sequencing: A combined approach for detecting yam (*Dioscorea* spp.) viruses. *Physiological and Molecular Plant Pathology* (in press). doi:10.1016/j.pmpp.2018.06.003.

Bosque-Pérez, N. A. 2000. Eight decades of Maize streak virus research. *Virus Research* 71: 107–21.

Cardwell, K. F. (Eds). 1995. *Proceedings of the Maize Downy Mildew Eradication Campaign*, 22–24 February 1994, IITA, Ibadan, Nigeria, 135pp.

Chiarappa, L. 1985. Plant protection at FAO-40 years of service international plant pest management. *FAO Plant Protection Bulletin* 33: 131–8.

Early, R., Bradley, B. A., Dukes, J. S., Lawler, J. L., Olden, J. O., Blumenthal, D. M., Gonzalez, P., Grosholz, E. D., Ibañez, I., Miller, L. P., Sorte, C. J. and Tatem, A. J. 2016. Global threats from invasive alien species in the twenty-first century and national response capacities. *Nature Communications* 7: 12485.

Ellis, D., Gueye, B. J., Kreuze, J., Valkonen, J., Weibel, D. and Kumar, P. L. 2017. NextGen phytosanitation: Rapid elimination of viruses from root and tuber plants for crop improvement and seed systems. In: *Abstracts, ISTRC-AB Symposium*, 5–10 March 2017, Whitesands Hotel, Dar es Salaam, Tanzania.

Eni, A. O., Hughes J. d'A. and Rey, M. E. C. 2008. Survey of the incidence and distribution of five viruses infecting yam in the major yam producing zones in Benin. *Annals of Applied Biology* 153: 223–32.

Eni, A., Hughes, J., Asiedu, R. and Rey, M. 2012. Re-evaluation of Yam mosaic virus (YMV) detection methods. *Academic Journal of Plant Sciences* 5: 18–22.

FAO-IPGRI. 1996. FAO/IPGRI technical guidelines for the safe movement of germplasm. FAO, Rome.

FAOSTAT. 2018. Crop production data for the year 2016. Accessed on 12 July 2018 at: http://www.fao.org/faostat/en/#data/QC

Fusarium wilt of banana. 2015. Accessed on 12 June 2018 at: http://www.promusa.org/Fusarium+wilt

Geddes, A. M. 1990. The relative importance of crop pests in sub-Saharan Africa. Natural Resources Institute Bulletin, No. 36. NRI, Chatham, 78pp.

Ghosh, S., Bouvaine, S., Richardson, S. C. W., Ghanim, M. and Maruthi, M. N. 2018. Fitness costs associated with infections of secondary endosymbionts in the cassava whitefly species *Bemisia tabaci*. *Journal of Pest Science* 91: 17–28.

Goergen, G., Kumar, P. L., Sankung, S. B., Togola, A. and Tamò, M. 2016. First report of outbreaks of the fall armyworm *Spodoptera frugiperda* (J. E. Smith) (Lepidoptera: Noctuidae), a new alien invasive pest in West and Central Africa. *PLoS ONE* 11(10): e0165632.

Greninger, R. 2018. A decade of RNA virus metagenomics is (not) enough. *Virus Research* 244: 218–29.

Harsch, E. 2003. Africa beyond famine – new strategies needed to combat hunger, disease and rural poverty in Africa Recovery 17. Accessed on 10 July 2018 at: http://www.un.org/en/africarenewal/vol17no1/171food1.htm

Hartman, G., Hill, C., Twizeyimana, M., Miles, M. R. and Bandyopadhyay, R. 2011. Interaction of soybean and *Phakopsora pachyrhizi*, the cause of soybean rust. *CAB Reviews: Perspectives in Agriculture, Veterinary Science, Nutrition and Natural Resources* 6: 1–13.

Hilker, F. M., Allen, L. J. S., Bokil, V. A., Briggs, C. J., Feng, Z., Garrett, K. A., Gross, L. J., Hamelin, F. M., Jeger, M. J., Manore, C. A., Power, A. G., Redinbaugh, M. G., Rúa, M. A. and Cunniffe, N. J. 2017. Modelling virus coinfection to inform management of maize lethal necrosis in Kenya. *Phytopathology* 107: 1095–108.

Ho, T. and Zanetakis, I. E. 2014. Development of a virus detection and discovery pipeline using next generation sequencing. *Virology* 471–473: 54–60.

IAPPS. 2018. Tackling invasive species in Africa: A stakeholder workshop. *Crop Protection* 109: 142–4.

IITA. 2011. DEWN: A novel surveillance system. Accessed on 25 May 2018 at: http://r4dreview.iita.org/index.php/2011/04/14/dewn-a-novel-surveillance-system/

IPPC. 2012. Protecting the world's plant resources from pests – an international framework for cooperation, 30pp. Accessed on 2 July 2018 at: https://www.ippc.int/static/media/files/mediakit/IPPCOverviewBrochure2012-03-en.pdf

Javid, I. and Ashraf, M. 1978. Some observations on soybean diseases in Zambia and occurrence of *Pyrenochaeta glycines* on certain varieties. *Plant Disease Reporter* 62: 46–7

Kamowa-Mbewe, W., Kumar, P. L., Changadeya, W., Ntawuruhunga, P. and Legg, J. P. 2015. Diversity, distribution and effects on cassava cultivars of cassava brown streak viruses in Malawi. *Journal of Phytopathology* 163: 433–43.

Kayondo, S. I., Del Carpio, D. P., Lozano, R., Ozimati, A., Wolfe, M., Baguma, Y., Gracen, V., Offei, S., Ferguson, M., Kawuki, K. and Jannink, J.-L. 2018. Genome-wide association mapping and genomic prediction for CBSD resistance in *Manihot esculenta*. *Scientific Reports* 8: 1549.

Published by Burleigh Dodds Science Publishing Limited, 2019.

Khoury, C. K., Achicanoy, H. A., Bjorkman, A. D., Navarro-Racines, C., Guarino, L., Flores-Palacios, X., Engels, J. M. M., Wiersema, J. H., Dempewolf, H., Sotelo, S., Ramírez-Villegas, J., Castañeda-Álvarez, N. P., Fowler, C., Jarvis, A., Rieseberg, L. H. and Struik, P. C. 2016. Origins of food crops connect countries worldwide. *Proceedings of the Royal Society of London: Biological Sciences* 283: 20160792. doi:10.1098/rspb.2016.0792.

Klinkowski, M. 1970. Catastrophic plant diseases. *Annual Review of Phytopathology* 8: 37–60.

Kolombia, Y. A., Karssen, G., Viaene, N., Kumar, P. L., De Sutter, N., Joos, L., Coyne, D. L. and Bert, W. 2017. Diversity of root-knot nematodes associated with tubers of yam (*Dioscorea* spp.) established using isozyme analysis and mitochondrial DNA-based identification. *Journal of Nematology* 49: 177–88.

Kreuze, J., Perez, A., Untiveros, M., Quispe, D., Fuentes, S., Barker, I. and Simon, R. 2009. Complete viral genome sequence and discovery of novel viruses by deep-sequencing of small RNAs: A generic method for diagnosis, discovery and sequence of viruses. *Virology* 388: 1–7.

Kumar, P. L. and Abberton, M. 2016. Navigating the international exchange of plant genetic resources amidst biosecurity challenges: Experiences of IITA in Africa. *Indian Journal of Plant Genetic Resources* 29: 404–6.

Kumar, P. L., Akinbade, S. A., Dixon, A. G. O., Mahungu, N. M., Londa, L., Mutunda, M. P., Kiala, D. and Legg, J. P. 2009. First report of the occurrence of East African cassava mosaic virus-Uganda (EACMV-UG) in Angola. *Plant Pathology* 58: 402.

Kumar, P. L., Sharma, K., Boahen, S., Tefera, H. and Tamo, M. 2011a. First report of soybean witches'-broom disease caused by group 16SrII phytoplasma in soybean in Malawi and Mozambique. *Plant Disease* 95: 492.

Kumar, P. L., Hanna, R., Alabi, O. J., Soko, M. M., Oben, T. T., Vangu, G. H. P. and Naidu, R. A. 2011b. Banana bunchy top virus in sub-Saharan Africa: Investigations on virus distribution and diversity. *Virus Research* 159: 171–82.

Kumar, P. L., Selvarajan, R., Iskra-Caruana, M.-L., Chabannes, M. and Hanna, R. 2015. Biology, etiology and control of virus diseases of banana and plantain. *Advances in Virus Research* 91: 229–69.

Legg, J. P. 1999. Emergence, spread and strategies for controlling the pandemic of cassava mosaic virus disease in east and central Africa. *Crop Protection* 18: 627–37.

Legg, J. P. and Ogwal, S. 1998. Changes in the incidence of African cassava mosaic geminivirus and the abundance of its whitefly vector along south-north transects in Uganda. *Journal of Applied Entomology* 122: 169–78.

Legg, J. P., Jeremiah, S. C., Obiero, H. M., Maruthi, M. N., Ndyetabula, I., Okao-Okuja, G., Bouwmeester, H., Bigirimana, S., Tata-Hangy, W., Gashaka, G., Mkamilo, G., Alicai, T. and Lava Kumar, P. 2011. Comparing the regional epidemiology of cassava mosaic and cassava brown streak virus pandemics in Africa. *Virus Research* 159: 161–70.

Legg, J. P., Sseruwagi, P., Boniface, S., Okao-Okuja, G., Shirima, R., Bigirimana, S., Gashaka, G., Herrmann, H. W., Jeremiah, S., Obiero, H., Ndyetabula, I., Tata-Hangy, W., Masembe, C. and Brown, J. K. 2014. Spatio-temporal patterns of genetic change amongst populations of cassava *Bemisia tabaci* whiteflies driving virus pandemics in East and Central Africa. *Virus Research* 186: 61–75.

Legg, J. P., Kumar, P. L. and Kanju, E. 2015a. Cassava brown streak. In: *Viral Diseases of Tropical and Subtropical Crops* (P. Tennant and G. Fermin, Eds). CABI, Wallingford, UK, pp. 42–55.

Legg, J. P., Kumar, P. L., Makeshkumar, T., Tripathi, L., Ferguson, M., Kanju, E., Ntawuruhunga, P. and Cuellar, W. 2015b. Cassava virus diseases: Biology, epidemiology, and management. *Advances in Virus Research* 91: 85–142.

Leke, W. N., Mignouna, D. B., Brown, J. K., and Fondong, V. N. 2016. First report of *Soybean chlorotic blotch virus* and *West African Asystasia virus 1* infecting cassava and a wild cassava relative in Cameroon and Togo. *New Disease Reports* 33: 24.

Lukanda, M., Owati, A., Ogunsanya, P., Valimunzigha, K., Katsonga, K., Ndemere, H. and Kumar P. L. 2014. First report of *Maize chlorotic mottle virus* infecting maize in the Democratic Republic of the Congo. *Plant Disease* 98: 1448.

Mahuku, G. and Kumar, P. L. 2017. Rapid response to disease outbreaks in maize cultivation: A case study of maize lethal necrosis. In: *Achieving Sustainable Cultivation of Maize*, Vol. 2 (D. Watson, Ed.). Cambridge Burleigh Dodds Series in Agricultural Science 2, Burleigh Dodds Science Publishing, Cambridge, UK, pp. 339–70.

Mahuku, G., Lockhart, B. E. L., Wanjala, B., Jones, M. W., Kimunye, N. J., Stewart, L. R., Cassone, B. J., Sevgan, S., Nyasani, J. O., Kusia, E., Kumar, P. L., Niblett, C. L., Kiggundu, A., Asea, G., Pappu, H. R., Wangai, A., Prasanna, B. M. and Redinbaugh, M. G. 2015. Maize lethal necrosis, an emerging threat to maize-based food security in sub-Saharan Africa. *Phytopathology* 105: 956–65.

Mgbechi-Ezeri, J. U., Alabi, O. J., Naidu, R. A. and Kumar, P. L. 2008a. First report of the occurrence of African cassava mosaic virus in soybean in Nigeria. *Plant Disease* 92: 1709.

Mgbechi-Ezeri, J. U., Ogunsanya, P. A., Akinbade, S. A. and Kumar, P. L. 2008b. Production and characterization of polyclonal antibodies against Cowpea mild mottle virus (CPMMV) isolated from mosaic disease-affected soybean in Nigeria. In: *Edited Abstracts* (J. Osuji, Ed.), 35th Annual Conference of Nigerian Society for Plant Protection (NSPP), Nigeria, pp. 29–30.

Miller, S. A., Beed, F. D. and Harmon, C. L. 2009. Plant disease diagnostic capabilities and networks. *Annual Review of Phytopathology* 47: 15–38.

Murithi, H., Owati, A., Madata, C. S., Joosten, M., Beed, F. and Kumar, P. L. 2015. First report of 16SrII-C subgroup phytoplasma causing phyllody and witches'-broom disease in soybean in Tanzania. *Plant Disease* 99: 886.

Nakato, G. V., Akinbade, S. A., Kumar, P. L., Bandyopadyhay, R. and Beed, F. 2013. Development of ELISA for the detection of *Xanthomonas campestris* pv. *musacearum*, the causal agent of BXW: Banana Xanthomonas Wilt. In: *Banana Systems in the Humid Highlands of Sub-Saharan Africa: Enhancing Resilience and Productivity* (G. Blomme, B. Vanlauwe and P. van Asten, Eds). CABI, Wallingford, UK, pp. 93–100.

Ngatat, S., Hanna, R., Kumar, P. L., Gray, S. M., Cilia, M., Ghogomu, R. T. and Fontem, D. A. 2017. Relative susceptibility of *Musa* genotypes to banana bunchy top disease in Cameroon and implication for disease management. *Crop Protection* 101: 116–22.

Nkere, C. K., Seal, S. E., Atiri, G. I., Onyeka, J. T. and Kumar, P. L. 2013. Relative incidence and distribution of viruses in plants grown from different portions of seed yams (*Dioscorea* spp.). In: *Program and Book of Abstracts of the 12th International Plant Virus Epidemiology Symposium*, 28 January – 1 February 2013, Arusha, Tanzania. IITA Publication, pp. 67.

Nkere, C. K., Oyekanmi, J., Silva, G., Bömer, M., Atiri, G. I., Onyeka, J., Maroya, N., Seal, S. E. and Kumar, P. L. 2018. Chromogenic detection of Yam mosaic virus by closed tube reverse transcription loop-mediated isothermal amplification (CT-RT-LAMP). *Archives of Virology* 163: 1057–61.

Odipio, J., Tusiime, G., Tripathi, L. and Aritua, V. 2009. Genetic homogeneity among Ugandan isolates of *Xanthomonas campestris* pv. *musacearum* revealed by Randomly Amplified Polymorphic DNA Analysis. *African Journal of Biotechnology* 8: 5652–60.

Oerke, E.-C. 2006. Crop losses to pests. *Journal of Agricultural Science* 144: 31–43.

Omane, E., Oduro, K. A., Cornelius, E. W., Opoku, I. Y., Akrofi, A. Y., Sharma, K., Kumar, P. L. and Bandyopadhyay, R. 2012. First report of leaf blight of taro (*Colocasia esculenta*) caused by *Phytophthora colocasiae* in Ghana. *Plant Disease* 96: 292.

Otti, G., Bouvaine, S., Kimata, B., Mkamillo, G., Kumar, P. L., Tomlins, K. and Maruthi, M. N. 2016. High throughput multiplex real time PCR assay method for the simultaneous quantification of DNA and RNA viruses infecting cassava plants. *Journal of Applied Microbiology* 120: 1346–56.

Paini, D. R., Shappard, A. W., Cook, A. W., De Barro, P. J., Woomer, S. P. and Thomas, M. B. 2016. Global threat to agriculture from invasive species. *Proceedings of the National Academy of Sciences of the United States of America* 113: 7575–9.

Plucknett, D. L. and Smith, J. H. 1988. Plant quarantine and the international transfer. CGIAR Study Paper Number 25. World Bank, Washington DC, 65pp.

Rabson, M., Mulenga, R. M., Boykin, L. M., Chikoti, P. C., Sichilima, S., Ng'uni, D. and Alabi, O. J. 2018. Cassava brown streak disease and Ugandan cassava brown streak virus reported for the first time in Zambia. *Plant Disease* 102: 1410–18.

Ramcharan, A., Baranowski, K., McCloskey, P., Ahmed, B., Legg, J. P. and Hughes, D. P. 2017. Deep learning for image-based cassava disease detection. *Frontiers in Plant Science* 8: 1852. doi:10.3389/fpls.2017.01852.

Ricciardi, V., Ramankutty, N., Mehrabia, Z., Jarvisa, L. and Chookolingo, B. 2018. How much of the world's food do smallholders produce? *Global Food Security* 17: 64–72.

Savary, S., Ficke, A., Aubertot, J. and Hollier, C. 2012. Crop losses due to diseases and their implications for global food production losses and food security. *Food Security* 4: 519–37.

Schilder, A. M. C., Florini, D. A., Berner, D. A., Hughes, J. d'A., Ng, S. Y. C., Ng, N. Q., Thottappilly, G. and Haug, T. M. 1999. Containment facilities and safeguards at the International Institute of Tropical Agriculture. In: R. P. Kahn and S. B. Mathur (Eds), *Exclusion of Pests and Pathogens: Containment Facilities and Safeguards*. APS Press, pp. 66–78.

Sharma, K., Ayodele, M., Bandyopadhyay, R., Aregbesola, E. A., Menkir, A. and Kumar, P. L. 2010. Genetically distinct *Cercospora* species cause grey leaf spot of maize (*Zea mays* L.) in Nigeria. *Phytopathology* 100: S117.

Shirima, R. R., Maeda, D. G., Kanju, E., Ceasar, G., Tibazarwa, F. I. and Legg, J. P. 2017. Absolute quantification of cassava brown streak virus mRNA by real-time qPCR. *Journal of Virological Methods* 245: 5–13.

Shoyinka, S. A., Thottappilly, G., Adebayo, G. G. and Anno-Nyako, F. O. 1997. Survey on cowpea virus incidence and distribution in Nigeria. *International Journal of Pest Management* 43: 127–32.

Silva, G., Bömer, M., Nkere, C., Kumar, P. L. and Seal, S. E. 2015. Rapid and specific detection of Yam mosaic virus by reverse-transcription recombinase polymerase amplification. *Journal of Virological Methods* 222: 138–44.

Silva, M., Oyekanmi, J., Nkere, C. K., Bömer, M., Kumar, P. L. and Seal, S. E. 2018. Rapid detection of potyviruses from crude plant extracts. *Analytical Biochemistry* 546: 17–22.

Storey, H. H. 1936. Virus diseases on East African plants. VI. A progress report on the studies of the diseases of cassava. *East African Agricultural Journal* 2: 34–9.

STRC. 2011. African Union Science and Technology Framework for the detection, identification and monitoring of infectious diseases of humans, animals and plants in Africa. Viewed on 15 June 2018 at: https://au.int/sites/default/files/newsevents/workingdocuments/27580-wd-ids_framework_english.pdf

Szyniszewska, A. M., Busungu, C., Boni, S. B., Shirima, R., Bouwmeester, H. and Legg, J. P. 2017. Spatial analysis of temporal changes in the pandemic of severe cassava mosaic disease in northwestern Tanzania. *Phytopathology* 107: 1229–42.

Thottappilly, G. and Rossel, H. W. 1992. Virus diseases of cowpea in Tropical Africa. *Tropical Pest Management* 38: 337–48.

Thottappilly, G., Dahal, G. and Lockhart, B. E. L. 1998. Studies on a Nigerian isolate of banana streak badnavirus: I. Purification and enzyme linked immunosorbent assay. *Annals of Applied Biology* 132: 253–61.

Thottappilly, G., Mignouna, H .D., Onasanya, A., Abang, M., Oyelakin, O. and Singh, N. K. 1999. Identification and differentiation of isolates of *Colletotrichum gloeosporioides* from yam by random amplified polymorphic DNA markers. *African Crop Science Journal* 7: 195–205.

Time, I., Atiri, G. I. and Kumar, P. L. 2010. Viruses infecting soybean (*Glycine max* L. *Merill) in Nigeria. Phytopathology* 100: S126.

Tripathi, L., Mwangi, M., Abele, S., Aritua, V., Tushemereirwe, W. K. and Bandyopadhyay, R. 2009. Xanthomonas Wilt: A Threat to Banana Production in East and Central Africa. *Plant Disease* 93: 440–51.

Tripathi, L., Kumar, P. L., Patil, B. L. and Fatokun, C. 2014. Pathogen-resistant genetically engineered crops in Africa. In: *Genetically Engineered Crops in Developing Countries* (D. V. R. Reddy, A. P. Kumar, P. L. Kumar, G. Loebenstein and C. K. Rao, Eds). Studium Press LLC, Huston, USA, pp. 217–40.

Tumwegamire, S., Kanju, E., Legg, J., Shirima, R., Kombo, S., Mkamilo, G., Mtunda, K., Sichalwe, K., Kulembeka, H., Ndyetabura, I., Saleh, H., Kawuki, R., Alicai, T., Adiga, G., Benesi, I., Mhone, A., Zacarias, A., Sofrimento, F. M., Munga, T., Ateka, E., Navangi, L., Maruthi, M. N., Mwatuni, F., Ngundo, G., Mwangangi, M., Mbugua, E., Ndunguru, J., Rajabu, C. and Mark, D. 2018. Exchanging and managing in-vitro elite germplasm to combat cassava brown streak disease (CBSD) and cassava mosaic disease (CMD) in Eastern and Southern Africa. *Food Security* 10: 351–68.

Twizeyimana, M., Ojiambo, P. S., Ikotun, T., Ladipo, J. L., Hartman, G. L. and Bandyopadhyay, R. 2008. Evaluation of soybean germplasm for resistance to soybean rust (*Phakopsora pachyrhizi*) in Nigeria. *Plant Disease* 92: 947–52.

Twizeyimana, M., Ojiambo, P. S., Sonder, K., Ikotun, T., Hartman, G. L. and Bandyopadhyay, R. 2009. Pathogenic variation of *Phakopsora pachyrhizi* infecting soybean in Nigeria. *Phytopathology* 99: 353–61.

Waage, J. K., Woodhall, J. W., Bishop, S. J., Jones, D. R., Smith, J. J. and Spence, N. J. 2006. Patterns of new disease spread: A plant pathogen database analysis. In: *Foresight 2006, UK Government Office of Science and Innovation, Infectious Diseases: Preparing for the Future*, Study T15.pdf.

Wangai, A. W., Redinbaugh, M. G., Kinyua, Z. M., Mahuku, G., Sheets, K. and Jeffers, D. 2012. First report of Maize chlorotic mottle virus and maize lethal necrosis in Kenya. *Plant Disease* 96: 1582.

Wiggins, S. 2009. Can the smallholder model deliver poverty reduction and food security for a rapidly growing population in Africa? Paper for the Expert Meeting on How to feed the World in 2050, Rome, 20pp. Accessed on 9 May 2018 at: http://www.fao.org/tempref/docrep/fao/012/ak982e/ak982e00.pdf

Xing, Y, Hernandez Nopsa, J., Andrade-Piedra, J., Beed, F., Blomme, G., Carvajal-Yepes, M., Coyne, D., Forbes, G., Kreuze, J., Jürgen, K., Kumar, L., Legg, J., Parker, M., Schulte-Geldermann, E. and Garrett, K. 2017. Global cropland connectivity: A risk factor for invasion and saturation by emerging pathogens and pests. *BioRxiv* 1–29. doi:10.1101/106542.

Yahaya, A., Dangora, D. B., Alegbejo, M. D., Kumar, P. L. and Alabi, O. J. 2017a. Identification and molecular characterization of a novel sugarcane streak mastrevirus and an isolate of the A-strain of maize streak virus from sugarcane in Nigeria. *Archives of Virology* 162: 597–602.

Yahaya, A., Al Rwahnih, M., Dangora, D. B., Gregg, L., Alegbejo, M. D., Kumar, P. L. and Alabi, O. J. 2017b. First report of Maize yellow mosaic virus infecting sugarcane (*Saccharum* spp.) and itch grass (*Rottboellia cochinchinensis*) in Nigeria. *Plant Disease* 101: 1355.

Zandjanakou-Tachin, M., Vroh-Bi, I., Ojiambo, P. S., Tenkouano, A., Gumedzoe, Y. M. and Bandyopadhyay, R. 2009. Identification and genetic diversity of *Mycosphaerella* species on banana and plantain in Nigeria. *Plant Pathology* 58: 536–46.

Chapter 4

Conserving and exploiting biodiversity in crop cultivation in sub-Saharan Africa

Georg Goergen *and* **Peter Neuenschwander**, *International Institute of Tropical Agriculture (IITA), Benin; and* **Danny Coyne**, *International Institute of Tropical Agriculture (IITA), Kenya (corresponding author: g.goergen@cgiar.org)*

1 Introduction

Recognizing the importance of naturally occurring biological control agents in or around cultivated crops was one of the pillars of the Integrated Pest Management (IPM) concept since its inception (Hagen et al., 1971). The transition from exploiting this agrobiodiversity to the conservation of natural fauna and flora was formulated when the CGIAR institutes and the World Wide Fund (WWF) for Conservation of Nature framed a common platform, arguing that agriculture and nature conservation need not be antagonistic as is most often still the case (McNeely and Scherr, 2001). Today, biodiversity in general, and that of tropical forests and their inhabitants in particular, need to be preserved for ethical reasons, for assuring sustainability of agriculture under the protection by biological control (van Driesche et al., 2010), for use in varietal selection (Bhattacharjee et al., 2011) and for the preservation of water regimes, mitigation of climate change (Corlett, 2014) and other ecosystem services (Kuyah et al., 2016; Rowland et al., 2017).

In order to appreciate and protect biodiversity one needs to be able to accurately identify and understand the target organisms, which is a particular challenge for certain taxonomic groups, particularly for insects, mites and

http://dx.doi.org/10.19103/AS.2018.0043.04
Published by Burleigh Dodds Science Publishing Limited, 2019.

nematodes in the under-researched tropics. For example, whole important functional insect groups such as ants remain relatively poorly assessed in the Afrotropical region, with an estimated 4000 species of which about 60% remain to be described (Robertson, 2000). IITA therefore brought together and expanded its modest collections on the IITA-Benin campus once the biological control group had moved from Ibadan, Nigeria, to Benin in 1988. Today, this collection is the largest and, most importantly, best curated insect collection in West and Central Africa (Kirk-Spriggs, 2017), bringing together IITA specialists and taxonomic authorities from across the globe. It offers open access and diagnostic services, underpinning biological control programs.

Before natural enemies can be imported into Africa for classical biological control their taxonomic status, biology and ecological impact needs to be precisely ascertained, in order to exclude possible ill-effects on useful insects, exclude hyperparasitoids and, controversially today, make sure that the introduced organism does not attack indigenous non-target species (Bigler et al., 2006). The taxonomic status of African relatives of target species is, however, sometimes so poor as to exclude a thorough testing of indigenous sibling species.

A striking example for the importance of such a collection led by a dedicated curator is given by the taxonomic problems surrounding the recent accidental introduction of a noctuid moth onto the African continent. Control of the insect that has now invaded most of the Afrotropics causing yield losses on maize in the range of several billion US$ annually (Abrahams et al., 2017) could only be researched once the moth had been accurately identified as the neotropical fall armyworm (*Spodoptera frugiperda* (J. E. Smith); Lepidoptera: Noctuidae) using both classical morphological taxonomy and DNA analyses (Goergen et al., 2016).

Since most specimens in IITA's biodiversity collection have been collected over the last 20 years and are therefore relatively young by comparison to well-established insect collections, their DNA can mostly be extracted with sequences in full length and identified, which becomes an important contribution to the international barcode of life initiative (www.boldsystems.org) (Jordaens et al., 2015a,b). In addition, with its insect specimens identified by location and date, this collection also provides an important resource to follow and document the impact of climate change on biodiversity.

Based on the inventory of predatory mites researched in the framework of biological control against the cassava green mite as presented in Chapter 6, the collection now also comprises a unique repository of phytoseiid mites. Similarly, collections of beneficial microorganisms have been initiated including, for example, from soil microbiodiversity studies targeting rhizobial diversity, microbial antagonists of banana pests and diseases, pathogens of aquatic weeds or arbuscular mycorrhizal fungi (AMF), which on yams are particularly diverse (Tchabi et al., 2009, 2010). The studies on yam in West Africa led to the description of several new species of AMF and demonstrated the negative impact of agricultural intensification on the diversity of beneficial AMF (Tchabi et al., 2008), emphasizing the importance of cataloguing and preserving populations of soil microbiology, especially those sensitive to ecological disturbances and changes in agricultural practices.

Although there has been much progress towards improved diagnostics of nematodes, their taxonomy continues to pose a challenge, in part due to the limited expertise in Africa. IITA has been among few CGIAR centres to develop nematological expertise. Research concerned banana and plantain systems, diagnosing and documenting parasitic species, and developing management options according to accurate species diagnosis in the different banana and plantain systems across the continent as described in more detail in Chapter 9. Over the past 15 years, IITA collaborated with numerous partners to develop robust diagnostic protocols and establish a more accurate documentation of *Meloidogyne* species, the most damaging and probably the best known genus, which tended to be referred to as *Meloidogyne* spp. only (Jones et al., 2013; Pagan et al., 2015; Janssen et al., 2016; dos Santos et al., 2018; Coyne et al., 2018).

Finally, we shall describe efforts not only to exploit biodiversity but also to protect it. Biodiversity loss through forest destruction across Africa and particularly in West Africa is huge (Poorter et al., 2004), which places additional importance on regenerating regrowth (Chazdon, 2014). IITA has therefore become active in preserving and sometimes rehabilitating at least the remaining forest fragments on their experimental stations in view of integrating them in various research programmes.

2 Developing the insect collection at IITA

One important foundation that contributed to the success of the Africa-wide biological control project against the cassava mealybug, *Phenacoccus manihoti* Matile-Ferrero (Hemiptera: Pseudococcidae), that stands to this date for one of the strongest impacts of the CGIAR on agriculture in sub-Saharan Africa (SSA) as detailed in Chapter 6, was a thorough study of all biodiversity components at species level to understand functional linkages established in its food web. Thus, with the collection of roughly 1000 voucher specimens and the identification of about 300 species collected in many of the 31 affected African countries, the cassava mealybug food web has probably become one of the most comprehensively studied ecosystems within the framework of a biological control programme (Neuenschwander et al., 1987; Yaninek et al., 1989). This research involving meticulous taxonomic work led to a detailed understanding of interactions between the cassava mealybug, its specific natural enemy *Anagyrus lopezi* (De Santis) (Hymenoptera; Encyrtidae) and a rich guild of native insects across several trophic levels and within different ecological contexts. One key lesson of this exercise was that the introduced parasitoid *A. lopezi* proved to be strictly host-specific and showed no non-target effect, that is, attacked no indigenous African insects (Neuenschwander and Markham, 2001).

Responding to the need for expertise and reference specimens to conduct such ecosystem analysis, an entomological collection was initiated in 1994. Its establishment and steady growth was fuelled by regular taxonomic back-stopping at the IITA research station in Benin that became in 1997 the seat of the 'Biological Control Centre for Africa', where several projects developed management options against introduced or indigenous pests on various crops. Thus, thorough assessments of insect guilds within projects targeting African maize stem borers, thrips and pod borers of legume crops, whiteflies on cassava and horticultural crops, West African migratory locusts and grasshoppers, floating waterweeds, as well as mango mealybug and other Homoptera as described in the present crop chapters became a good starting point to enrich the collection with reference material.

The core collection was rapidly enlarged thanks to field inventories made in a biodiversity research project running in successive phases over 14 years since 1995. Insects sampled under agricultural and natural conditions in various ecological regions mainly of Benin, Togo, Nigeria and Cameroon allowed to maintain acquisition rates of 10 000–15 000 specimens per year. Today the collection encompasses more than 365 000 specimens and has become the most important of its kind in West and Central Africa and at the same time the largest non-plant biodiversity repository within the whole CGIAR. More than 6000 species from 335 insect families have been identified from the sampled material, of which the Coleoptera and Lepidoptera constitute the most biodiverse orders with 2300 and 1200 identified species, respectively. Presently the collection provides the basis for a regional identification service by IITA over two decades. Several alien invasive species new to the African continent were discovered, among them the whitefly *Paraleyrodes minei* Iaccarino (Hemiptera: Aleyrodidae), the Oriental fruit fly *Bactrocera dorsalis* (Hendel) (Diptera: Tephritidae) for West Africa (Drew et al., 2005), the papaya mealybug *Paracoccus marginatus* Williams and Granara de Willink (Hemiptera: Pseudococcidae) (Goergen et al., 2011a), the fall armyworm *Spodoptera frugiperda* (J.E. Smith) (Lepidoptera: Noctuidae) (Goergen et al., 2016) and lately the southern armyworm *Spodoptera eridania* (Stoll) (Goergen, 2018).

Published by Burleigh Dodds Science Publishing Limited, 2019.

3 Specific examples demonstrating the importance of an accurate diagnosis

Accurate diagnosis is critical as it represents the unique entry point to access existing information about any organism. The primary benefits of such taxonomic services are the initiation of adequate measures for the conservation of potentially threatened species, the prevention of pest introduction, early detection and suitable management decisions for established pests, for biological control exploiting native natural antagonists or by introducing specific natural enemies. Much of the information needed to conduct accurate identifications is found in well-preserved and -curated biological collections. They contain voucher specimens labelled with all relevant data that are precious references for species identification, comparison and verification especially in the context of new technologies allowing genetic analysis also of older material (Gaskin et al., 2011). Selected examples of how taxonomic accuracy has been critical to successfully resolve problems of identifying and combatting introduced invasive pests in tropical Africa are presented.

3.1 Cassava mealybug

Cassava mealybug, *Phenacoccus manihoti*, was an unknown species when it was first detected in Africa following devastating outbreaks on cassava in the surroundings of Brazzaville and Kinshasa in 1973. In the absence of adequate control options, the pest rapidly spread over the entire cassava growing area threatening the food security of some 200 million people in tropical Africa (Herren and Neuenschwander, 1991). Exploration for natural enemies in Latin America led to the discovery of a morphologically similar mealybug species, *Phenacoccus herreni* Cox and Williams (1981), which also attacked cassava. Initial attempts to control cassava mealybug in Africa with parasitoids reared on *P. herreni* failed because these parasitoids did not accept *P. manihoti* as their host. Later, subtle, yet significant, differences in the morphology and reproductive biology of the two mealybugs were detected enabling the two species to be separated. Subsequently, foreign exploration efforts were focused specifically on cassava mealybug, which had meanwhile been detected in Paraguay and which resulted in the discovery of the host-specific encyrtid parasitoid *Anagyrus lopezi* with the results described in Chapter 6.

3.2 Floating fern

Floating fern, *Salvinia molesta* D. S. Mitchell (Salviniaceae), an aquatic plant native to south-eastern Brazil has spread throughout the world and is currently listed among the 100 most invasive species globally (Luque et al., 2014). In Africa, it occurs in 26 countries. Attempts to control this weed by the introduction of three natural herbivores, that is, the weevil *Cyrtobagous singularis* Hustache (Coleoptera: Curculionidae), the moth *Samea multiplicalis* (Guenée) (Lepidoptera: Crambidae) and the grasshopper *Paulinia acuminata* (De Geer) (Orthoptera: Acrididae) was based on the initial identification of the aquatic weed as *Salvinia auriculata* Aubl. However, the poor control performance in Florida and South Africa by these agents prompted a closer examination of herbarium specimens in Brazil that then revealed the presence of an unknown species, which was subsequently described as *Salvinia molesta*. A new assessment of natural antagonists of the weed led to the discovery of a sibling species of *C. singularis*, namely the semi-aquatic weevil *C. salviniae* Calder and Sands, which proved to be highly effective in suppressing *S. molesta* in Southern Africa (Cilliers, 1991) and with IITA's involvement in Central Africa as detailed in Chapter 14.

3.3 Oriental fruit fly

By 2003, an unknown fruit fly belonging to the highly damaging *Bactrocera dorsalis* (Hendel) species complex was discovered for the first time in Kenya and shortly thereafter in Benin. The species, believed to originate from Sri Lanka, was considered as new to science and described as *Bactrocera invadens* (Drew et al., 2005) (Fig. 1). The new pest proved to be highly invasive. By 2017, it had spread to 42 countries in SSA attacking fruit of at least 46 plant species belonging to 23 families (Goergen et al., 2011b). The concern raised by this new threat to several important horticultural crops, including mango and citrus, strong competitive displacement with native fruit fly species and economic consequences prompted much research to establish its geographic distribution, assess its biological potential, economic impact and to develop control options. Despite a disparate geographic origin of *B. invadens* and *B. dorsalis*, both species are morphologically and genetically very similar and the taxonomic delimitation of both species became increasingly questioned. Therefore, the variation range of morphological characters, molecular analyses involving several types of markers, phylogenetic analyses and results on mating compatibility were brought together in an integrative taxonomic approach. Results provided sufficient evidence to consider both species as synonyms (Schutze et al., 2014). Such validation exerted profound consequences, especially on quarantine regulation of affected countries and fruit export to international markets as discussed in Chapters 11 and 12.

3.4 Siam weed

Siam weed, *Chromolaena odorata* (L) RM King and H Rob (Asteraceae), is a perennial shrub native to South and Central America that became an invasive

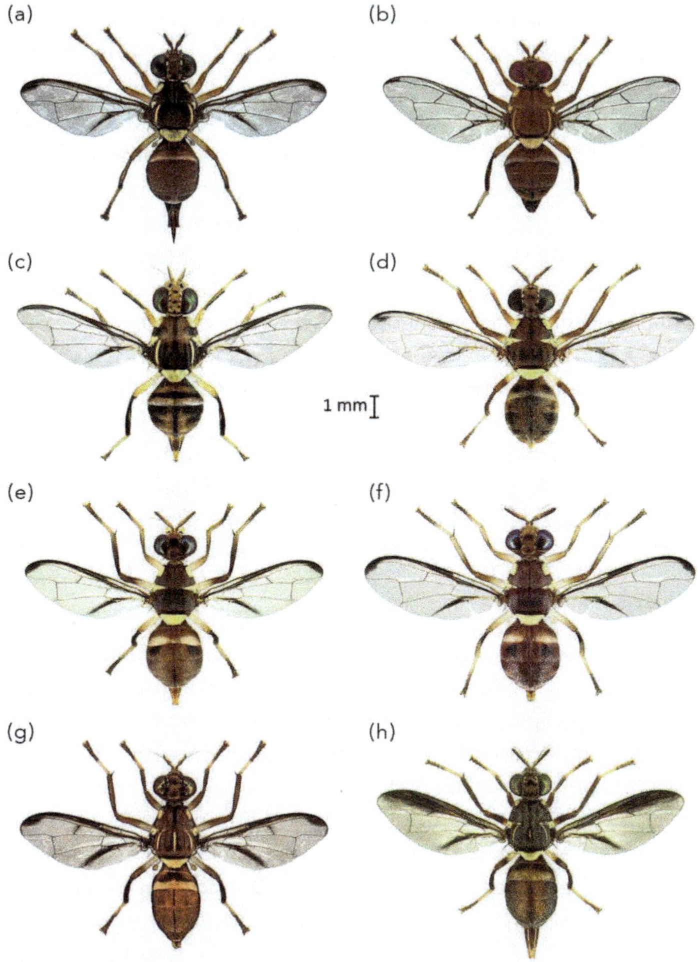

Figure 1 *Bactrocera dorsalis* belongs to an economically important species complex of fruit flies with more than 500 described species. The present plate illustrates the difficulty for non-specialists to differentiate between the introduced pest and species native to Africa from the same or closely related genus. (a) *Bactrocera latifrons* (Hendel); (b) *B. zonata* (Saunders); (c) *B. dorsalis* (Hendel); (d) *Dacus vertebratus* Bezzi; (e) *D. frontalis* Becker; (f) *D. ciliatus* Loew; (g) *D. punctatifrons* Karsch; (h) *D. bivitatus* (Bigot).

weed in West Africa since the 1930s. Its ability to rapidly proliferate and produce large amounts of wind-dispersed seeds facilitated its swift spread from its initial foothold in West Africa to most of tropical Africa within just six decades. Though its pest status remains controversial the plant is generally perceived as a noxious weed and has a severe impact on native African agriculture as discussed in Chapters 9 and 14. Whereas introduction of the leaf-feeding moth

Published by Burleigh Dodds Science Publishing Limited, 2019.

Pareuchaetes pseudoinsulata Rego Barros (Lepidoptera: Erebidae) succeeded in West Africa following repeated releases during the 1990s (Timbilla and Braimah, 2000; Uyi et al., 2009), the same species could not establish in South Africa. Though sustained efforts were made in South Africa for over a decade to introduce a sibling species, *P. insulata* (Walker), the mass release of strains from Florida (USA), Jamaica and Cuba resulted in localized establishment of only one moth population. However, weed control levels remained highly variable (Zachariades et al., 2011). Apart from the climatic factors, the overall weak performance of the introduced biological control agents could be explained by substantial morphological differences in the pubescence, odour and flower colour between *C. odorata* strains occurring in South Africa and elsewhere in tropical Africa (Von Senger et al., 2002). This discovery of biotypes of the host plant from different origins, namely southern Caribbean and northern Caribbean, revitalized efforts to control this invasive weed and recently resulted in the introduction of four new potential biocontrol agents with prospects for further releases of two promising candidates (Zachariades et al., 2016).

3.5 Root knot nematode

Root knot nematode, *Meloidogyne africana* (Whitehead): With the development and improvement of molecular techniques, the separation and identification of species of *Meloidogyne* has improved over time and recently provided clarity on a taxonomic conundrum of root knot nematodes causing damage to coffee in Tanzania (Janssen et al., 2017). Specimens recovered from galled coffee roots were morphologically identified as three different species, depending on whether females, males or juveniles were used. Molecular analyses, however, indicated that just one species was present. During the initial descriptions of *M. africana, Meloidogyne oteifae* (Elmiligy) and *Meloidogyne decalineata* (Whitehead), the different life stages were apparently mixed up and confused with specimens of other species. All three species are now synonymized with the re-described *M. africana* (Talwana et al., 2016). Current advances in molecular diagnostics are enabling us to better document the species present and responsible for damage symptoms in Africa. This has great relevance in defining management options, especially in relation to selecting and breeding for host resistance as discussed in Chapters 8, 9 and 11.

3.6 Potato cyst nematodes

Potato cyst nematodes, *Globodera* spp.: In 2014, this globally important quarantine pest was detected in Kenya, effectively it first established foothold in SSA (Mwangi et al., 2015). Belonging to the most important bioconstraints of potato cultivation, it was important to determine which of the two morphologically similar species of potato cyst nematodes were present in Kenya. Previously, the inability to detect the presence of both species following introductions of these invasive pests proved critical, and extremely costly, to their management. During a nation-wide survey of potato fields in Kenya, the employment of molecular diagnostic techniques enabled the detection of both *Globodera rostochiensis* Wollenweber and a very small incidence of *Globodera pallida* Stone (Mburu et al., 2018). The known presence of both species had major implications for the deployment of host resistance as the two species are affected by different genetic sources of resistance.

3.7 Fall armyworm

Fall armyworm, *Spodoptera frugiperda*: This pest has now further extended its range to the Indian subcontinent (Sharanabasappa et al., 2018) and assistance provided for its initial identification has since proved crucial for a rapid response, partly building on experience gained to manage this new pest on the African continent, as described in Chapter 7. The reason why this exotic invader was only identified after it had spread across Africa lies in the inconspicuous wing patterns of the adult moths occurring on a crop that already is a host plant for numerous moth species lacking strong distinctive characters for the non-specialist. Larvae show morphological differences, but interspecies variation is substantial. In tropical Africa there already exist eight indigenous species of *Spodoptera* (Fig. 2), which are best identified by the male genitalia. DNA-based taxonomy is therefore important and has, in fact, led to the identification of yet another American invader, namely *Spodoptera eridania* (Stoll) (Goergen, 2018).

3.8 The legume pod borer

The legume pod borer *Maruca vitrata* Fabricius (Lepidoptera: Crambidae) has been tacitly considered an indigenous pest in Africa for many years, mainly because no one ever looked at its taxonomy in detail. Observations by Kalshoven and van der Vecht (1950) from Indonesia showed that this insect was not considered a major problem there, because of the activity of natural enemies. At the same time, taxonomists had pointed out that tropical Asia might be the centre of origin of the genus *Maruca*, which comprises three additional sympatric species. Subsequently, a biodiversity assessment in Taiwan allowed the identification of a first set of hymenopteran parasitoids unknown from Africa, of which *Apanteles taragamae* Viereck (Hymenoptera: Braconidae) looked to be the most promising candidate (Huang et al., 2003). Hypotheses about the probable origin of the pest triggered extensive collections in ten countries within Africa, Asia, Oceania and Latin America from 22 different host plants in order to conduct genetic studies. The analysis of 77 different populations

Figure 2 *Spodoptera frugiperda* (a) and *S. eridania* (e) are two species of noctuid moths that recently became invasive on the African continent. The fact that both species bear relatively inconspicuous differentiation characters compared to relatives of the same genus known to occur in tropical Africa has contributed to their late detection. Most females of the group cannot be reliably identified by external features only. (b) *S. littoralis*; (c) *S. exigua*; (d) *S. triturata*.

using the cytochrome c oxidase subunit 1 gene sequences allowed to confirm South-East Asia as the putative area of origin of *M. vitrata* and suggested the presence of two subspecies in Asia and SSA (Periasamy et al., 2015). These findings renewed efforts in the search for natural enemies which soon led to the recovery of two braconid parasitoids from continental Asia known to effectively

Published by Burleigh Dodds Science Publishing Limited, 2019.

control *M. vitrata* (Kalshoven and van der Vecht, 1950) and initially recorded under the taxon *Phanerotoma philippinensis* (Ashmead) and *Bassus javanus* (Bhat & Gupta). However, following taxonomic revisions, the two latter were recently renamed *Phanerotoma syleptae* and *Therophilus javanus,* respectively (Srinivasan et al., 2015). This discovery was critical for the biological control program at IITA, and both parasitoids have since been introduced into Benin and Burkina Faso for experimental releases as described in Chapter 10.

3.9 Mites

Despite their microscopic size, plant parasitic mites are important pests of many agricultural crops worldwide. About 6000 species have been described within this group that encompasses five families, namely, Eriophyidae, Tarsonemidae, Tenuipalpidae, Tetranychidae and Tuckerellidae. The cassava green mite (CGM) *Mononychellus tanajoa* (Bondar) (Acari: Tetranychidae) is one among several tetranychid species within the complex of phytophagous mites attacking cassava in South America. Its accidental introduction in Uganda in 1971 (Nyiira, 1972) soon resulted in significant root yield losses ranging from 13% to 80% depending on cassava varietal susceptibility and cropping conditions. The spread of this mite in most of the African cassava belt prompted new taxonomic challenges as the search for specific natural enemies required a thorough assessment of the natural enemy fauna in South America. Hence, in the framework of a collaborative effort between CIAT and EMBRAPA in South America and IITA in Africa, more than ten neotropical phytoseiid mite species were identified to prey upon CGM. Natural enemies that were found under climatic conditions matching those prevailing in tropical Africa were imported, laboratory tested and released in pilot sites before control efforts on the African continent were scaled up using the most promising species, that is *Amblydromalus manihoti* (Moraes), *Typhlodromalus aripo* De Leon and *Neoseiulus idaeus* Denmark and Muma, as described in Chapter 6.

To assess the predatory performance of indigenous mites and possible non-target effects of introduced species, the phytoseiid mite fauna native to Africa was also studied in more detail. Following extensive field surveys, mite samples collected over several decades throughout tropical Africa served to review a total of 216 species distributed among five subtribes and eight tribes of African phytoseiid mites. Results of these studies were published between 2006 and 2008 in a series of eight successive revisions in which roughly 60 predatory mite species new to science were described (Moraes et al., 2006, 2007a,b, 2008; Ueckermann et al., 2008; Zannou et al., 2006, 2007).

It is important to note that satisfactory levels of CGM control achieved under field conditions were partly obtained thanks to sustained research in host plant resistance that led to the selection of cassava varieties expressing not only strong antixenosis and antibiosis against the mite pest, but also of cultivars producing hairiness at apexes serving as shelters ('domatia') for some phytoseiid natural antagonists like *T. aripo*, which eventually became the most important biological control agent in Africa. This synergetic effect is an impressive example of how collaboration between plant protectionists and plant breeders can be complementary to enhance pest control levels.

Another mite pest that required in-depth taxonomic assessment of associated natural enemies is the eriophyid coconut mite *Aceria guerreronis* Keifer (Acari: Eriophyidae), the most economically important coconut pest in the Americas and Africa, as presented in Chapter 12. Damage attributable to the pest such as premature coconut fruit drop and quality reduction due to necrosis of the fruit meristem can cause yield reduction up to 80% in coconut plantations (Haq et al., 2002). Explorations to find natural enemies started in the late 2000s in West and East Africa and were extended to Brazil, the presumed area of origin of the coconut mite. Comparison of the four most common natural enemies at morphological and molecular levels from various geographical origins revealed the presence of an efficient phytoseiid species from Brazil previously attributed to *Neoseiulus baraki* Athias-Henriot. Related biological studies showed that the new species is a promising candidate for classical biological control of the coconut mite.

All studies of the various acarofaunas were based on extensive exploration work conducted in many countries in western, eastern and Southern Africa. Voucher specimens were slide mounted, labelled, curated and preserved in the arthropod collection housed at the IITA station in Benin. Presently the mite collection amounts to more than 65 000 slides representing roughly 350 identified species distributed in 33 families. With its digitalized database the mite collection is a unique scientific repository of an arthropod subclass sorely lacking active taxonomic experts within tropical Africa.

4 Exploiting scientific collections

Specimens in scientific collections can have unanticipated relevance to research in fields other than the one for which they were originally collected. An example of how faunal activities have contributed to formulate recommendations for biodiversity conservation is the exploitation of long-term collection data for the constitution of a Red List of arthropods for Benin (Neuenschwander et al., 2011).

Numerous specialist taxonomists visited this collection to study selected groups of relevance of their own research field. New distribution data for many family representatives and newly detected species led to the publication of several monographs with IITA participation, in particular darkling beetles (Robiche et al., 2002) and sawflies (Koch et al., 2015). In the course of such studies a total of 58 insect species were described new to science, 10 of them with the species name *goergeni*.

Similar efforts were recently made to compile distribution records from butterflies sampled during the last two decades in Benin. This group is particularly suited to serve as environmental indicators because butterflies have a relatively short life cycle, adults generally display a limited migratory ability and larvae develop on a restricted host plant range. Consequently, their response to environmental changes is comparatively rapid. Moreover, the butterfly fauna is particularly well studied and known because of the interest by numerous amateur lepidopterologists. The contribution of data from IITA's collection for the production of an atlas of the butterflies of Benin revealed unexpected results (Coache et al., 2017): Although butterflies belong to the most extensively studied insect taxa with relative recent data on the West African fauna (Larsen, 2005), a total of 310 species could be newly added to the previously 160 known records for Benin. In addition, a species belonging to the biodiverse genus *Euphaedra* could be described as new to science. These results contradict the general belief that Benin, which is situated in a biogeographic area with comparatively low annual rainfall and sparse forest cover (Dahomey Gap), should be consequently less biodiverse.

5 New challenges and opportunities

Traditionally, the identification of specimens has relied on morphological characters, host associations and life history data. This process can become challenging especially when taxa are microscopic, cryptic, poorly known, available specimens are damaged or immatures or lacking stringent characters. To date molecular techniques have become refined and are widely used with major advantages: identifications are reliable, and accuracy can be obtained with the same degree of reliability from well-preserved specimens as from different life stages, sexually dimorphic species, body fragments and damaged specimens. The comparatively young age of IITA's collection makes it particularly well suited for the application of molecular techniques such as DNA barcoding.

Besides agricultural pests and their natural enemies, attention is currently given to crop pollinators because of their vital services to ecosystems and the concern raised by their global decline. So far, the DNA sequences of representative species among the 1500 hoverfly specimens from the IITA collection have been obtained and will be made publicly accessible thus advancing the goals set by the Consortium for the Barcode of Life (CBOL). At the same time this technology has enabled the detection of a hoverfly species newly introduced from South America into Africa (Jordaens et al., 2015a,b) and revealed up to 20 cryptic species new to science.

Moreover, sending colour photographs with total depth of focus obtained from top photographic equipment allows remote identifications by specialists in other museums without the cumbersome and risky (for the specimens) transport of insects. This has profoundly facilitated collaboration among the few specialists of taxonomic groups across the globe.

The utility of such technology in terms of public accessibility and sharing of reference specimens with their taxonomic information has led to the creation of free accessible picture databases such as MorphoBank (http://www.morphobank.org/) that may well become the future standard when species are newly described.

6 From using to protecting biodiversity

IITA maintains 24 stations in 16 African countries, covering all major ecological zones of the continent. On many stations, forest pockets have been left, often just as a reserve for future expansions for experimental fields or buildings. These stations were sometimes founded far from urban centres surrounded by forests or fallow fields. Nowadays, some like the IITA-Ibadan campus have been engulfed by urban sprawl. While forest destruction across the continent continues practically unabated (Poorter et al., 2004), these forest fragments suddenly have a much more important role. They can serve as endpoints in studies with gradients from relatively untouched natural conditions, through fallow of different ages, to crop fields with different soil qualities, to totally abused fields covered by Imperata grass and unfit for further cultivation. This allows the quantification of changes in physical soil constituents, soil organisms, vegetation and animals, changes that take tens of years, but can now be observed at the same time.

To prepare this unique instrument for ecological and agronomic research, the forest fragments and methods for conserving them need to be described in detail. Such a study of sites with different distances to urban centres is available for the IITA-Ibadan and IITA-Cotonou sites, with comparisons to local forests managed by NGOs or village elders (Neuenschwander et al., 2015). In addition, a detailed botanical study in a forest that was rehabilitated during the last 20 years and belongs now to IITA-Benin reveals that such a botanically enriched site, which also harbours a critically endangered endemic monkey species, can become a protected reference forest and make an important contribution to nature conservation. The site now shelters about 70% of all endangered plant species of Benin that could possibly survive there (Neuenschwander and Adomou, 2017).

Through the resulting close collaboration with local scientists active in land management and nature conservation, a Red List for Benin on the basis of IUCN criteria was established (Neuenschwander et al., 2011). It now forms the basis for national actions and contributes to McNeely and Scherr's (2001) call for bringing agriculture and nature conservation together to preserve biodiversity and strengthen sustainability of agricultural activities.

7 Conclusion and future trends

Studies that document the outcomes of several IITA-led projects against introduced pests have demonstrated that the economic benefits of biological control programmes rival the benefits of much better supported breeding efforts in the CGIAR (Neuenschwander, 2004). However, though identifications and taxonomic research are recognized to be critical for the success of such biological control attempts, related costs are most often still not included in the budgets of these projects and remain a core contribution by the institute. In order for costs for biodiversity research to become politically acceptable further research on economic benefits accruing from protected biodiversity is essential (Balmford et al., 2002; Juffe-Bignoli et al., 2016).

The use of molecular tools has become the standard in many collections for specimen identification through comparison of DNA sequence information from freely accessible large databanks such as GenBank (www.ncbi.nlm.nih.gov/genbank/) or Barcode of Life Data System (v4.boldsystems.org/). Moreover, the development of non-destructive extraction methods has made the need to damage valuable specimens by removing a leg from them obsolete. Expanding IITA's molecular diagnostic capabilities to the insect collection has therefore become an urgent need. This is all the more justified as the technology has become widely affordable and the Afrotropical insect fauna is yet poorly represented in available databanks and whole insect orders are only weakly covered. Of the roughly 5 million Barcodes and 210 000 with species names available (BOLD, 2018) only 5.6% are now from Africa. Thus, a considerable contribution could be made by IITA in the form of pioneering work for future generations.

This chapter amply demonstrates why IITA and the CGIAR should give similar attention to non-plant genetic resources as they have given routinely to plant genetic resources, since their creation.

8 References

Abrahams P., Bateman M., Beale T., Clottey V., Cock M., Colmenarez Y., Corniani N., Day R., Early R., Godwin J. and Gomez J. 2017. Fall Armyworm: Impacts and Implications for Africa. Evidence Note (2). Report to DFID, 144pp.

Balmford A., Bruner A., Cooper P., Costanza R., Farber S., Green R. E., Jenkins M., Jefferiss P., Jessamy V., Madden J., Munro K., Myers N., Naeem S., Paavola J., Rayment M., Rosendo S., Roughgarden J., Trumpern K. and Turner R. K. 2002. Economic reasons for conserving wild nature. *Science* 297, 950–3.

Bhattacharjee R., Gedil M., Sartie A. M., Otoo E., Dume D., Kikuo H., Kumar P. L. and Asiedu R. 2011. Chapter 4: Dioscorea. In: Kole C. (Ed.), *Wild Crop Relatives: Genomic and Breeding Resources, Industrial Crops*. Springer, Berlin Heidelberg, pp. 71–96.

Bigler F., Babendreier D. and Kuhlmann U. 2006. *Environmental Impact of Invertebrates for Biological Control of Arthropods. Methods and Risk Assessment*. CABI Publishing, Wallingford, UK, 299pp.

BOLD. 2018. v4.boldsystems.org/ (consulted on 7 September 2018).

Cilliers C. J. 1991. Biological control of water fern, *Salvinia molesta* (Salviniaceae), in South Africa. *Agriculture, Ecosystems and Environment* 37, 219–24.

Chazdon R. L. 2014. *Second Growth: The Promise of Tropical Forest Regeneration in an Age of Deforestation*. University of Chicago Press, Chicago and London, 449pp.

Coache A., Rainon B. and Sinzogan A. 2017. *Atlas illustré des Rhopalocères du Bénin*. Centre d'Etudes et de Recherches Entomologiques Béninois, Université d'Abomey-Calavi, 732pp.

Corlett R. T. 2014. Forest fragmentation and climate change. In: Kettle C. and Koh L. P. (Eds), *Global Forest Fragmentation*. CAB International, Wallingford, pp. 69–78.

Cox J. M. and Williams D. J. 1981. An account of cassava mealybugs (Hemiptera: Pseudococcidae) with a description of a new species. *Bulletin of Entomological Research* 71, 247–58.

Coyne D. L., Cortada L., Dalzell J. J., Claudius-Cole A. O., Haukeland S., Luambano N. and Talwana H. 2018. Plant parasitic nematodes and food security in sub-Saharan Africa. *Annual Review of Phytopathology*, 56, 381–403.

dos Santos M. F. A., Monteiro J. S. M., da Silva Mattos V., Almeida M. R. A., Castagnone Sereno P., Coyne D. and Carneiro R. M. D. G. 2018. Diversity of *Meloidogyne* spp. from periurban areas of sub-Saharan Africa and their genetic similarity with American populations. *Physiological and Molecular Plant Pathology* (in press).

Drew R. A. I., Tsuruta K. and White I. M. 2005. A new species of pest fruit fly (Diptera: Tephritidae: Dacinae) from Sri Lanka and Africa. *African Entomology* 13, 149–54.

Gaskin J. F., Bon M.-C., Cock M. J. W., Cristofaro M., De Biase A., De Clerck-Floate R., Ellison C. A., Hinz H. L., Hufbauer R. A., Julien M. H. and Sforza R. 2011. Applying molecular-based approaches to classical biological control of weeds. *Biological Control* 58, 1–21.

Goergen G. 2018. Southern armyworm, a new alien invasive pest identified in West and Central Africa. *Crop Protection* 112, 371–2.

Goergen G., Tamò M., Kyofa-Boamah M. E., Bokonon-Ganta A. and Neuenschwander P. 2011a. Papaya mealybug, *Paracoccus marginatus* (Hemiptera: Pseudococcidae), a new invading pest in West Africa. *Biocontrol News and Information* 32, 9N–10N.

Goergen G., Vayssières J.-F., Gnanvossou D. and Tindo M. 2011b. *Bactrocera invadens* (Diptera: Tephritidae) a new invasive fruit fly pest for the Afrotropical region: Host-plant range and distribution in West and Central Africa. *Environmental Entomology* 40, 844–54.

Goergen G., Kumar P. L., Sankung S. B., Togola A. and Tamò M. 2016. First report of outbreaks of the fall armyworm *Spodoptera frugiperda* (J.E. Smith) (Lepidoptera: Noctuidae), a new alien invasive pest in West and Central Africa. *PLoS ONE* 11(10), e0165632.

Hagen K. S., van den Bosch R. and Dahlsten D. L. 1971. The importance of naturally-occurring biological control in the Western United States. In: Huffaker C. B. (Ed.), *Biological Control*. Plenum Press, New York, NY, pp. 253–93.

Haq M. A., Sumangala K. and Ramani N. 2002. Coconut mite invasion, injury and distribution, pp. 41-9. In: Fernando L. C. P., Moraes G. J. and Wickramananda I. R. (Eds), *Proceedings of the International Workshop on Coconut Mite (Aceria guerreronis)*. Coconut Research Institute, Sri Lanka, 117pp.

Herren H. R. and Neuenschwander P. 1991. Biological control of cassava pests in Africa. *Annual Review of Entomology* 36, 257–83.

Huang C. C., Peng W. K. and Talekar N. S. 2003. Characteristics of infestation by the bean pod borer, *Maruca vitrata* (Lepidoptera: Pyralidae) on *Sesbania cannabina*. *Formosan Entomology* 23, 1–11.

Janssen T., Karssen G., Verhaeven M., Coyne D. and Bert W. 2016. Mitochondrial coding genome analysis of tropical root-knot nematodes (*Meloidogyne*) supports haplotype based diagnostics and reveals evidence of recent reticulate evolution. *Scientific Reports* 6, 22591.

Janssen T., Karssen G., Topalović O., Coyne D. and Bert W. 2017. Integrative taxonomy of root-knot nematodes reveals multiple independent origins of mitotic parthenogenesis. *PLoS ONE* 12, e0172190.

Jones J. T., Haegeman A., Danchin E. G. J., Gaur H. S., Helder J., Jones M. G., Kikuchi T., Manzanilla-López R., Palomares-Rius J. E., Wesemael W. M. and Perry R. N. 2013. Top 10 plant-parasitic nematodes in molecular plant pathology. *Molecular Plant Pathology* 14, 946–61.

Jordaens K., Goergen G., Virgilio M., Backeljau T., Vokaer A. and De Meyer M. 2015a. DNA barcoding to improve the taxonomy of the afrotropical hoverflies (Insecta: Diptera: Syrphidae). *PLoS ONE* 10(10), e0140264.

Jordaens K., Goergen G., Kirk-Spriggs A. H., Vokaer A., Backeljau T. and De Meyer M. 2015b. A second New World hoverfly, *Toxomerus floralis* (Fabricius) (Diptera: Syrphidae), recorded from the Old World, with description of larval pollen- feeding ecology. *Zootaxa* 4044, 567–76.

Juffe-Bignoli D., Brooks T. M., Butchart S. H., Jenkins R. B., Boe K., Hoffmann M., Angulo A., Bachman S., Böhm M., Brummitt N. and Carpenter K. E. 2016. Assessing the cost of global biodiversity and conservation knowledge. *PLoS ONE*, 11(8), e0160640.

Kalshoven L. G. E. and van der Vecht J. 1950. *De plagen van de culturgewassen in Indonesië*. Van Hoeve, La Haye, 512pp.

Kirk-Spriggs A. H. 2017. Chapter 1: Introduction and brief history of Afrotropical dipterology. In: Kirk-Spriggs A. H. and Sinclair B. J. (Eds), *Manual of Afrotropical Diptera. Volume 1. Introductory Chapters and Keys to Diptera families*. Suricata 4. South African National Biodiversity Institute, Pretoria, pp. 1–67.

Koch F., Goergen G. and van Noort S. 2015. The sawflies of Namibia and the western South Africa (Symphyta, Hymenoptera). *ABC Taxa* 15, 262. Museum für Naturkunde der Humboldt-Universität zu Berlin.

Kuyah S., Öborn I., Jonsson M., Dahlin A. S., Barrios E., Muthuri C., Malmer A., Nyaga J., Magaju C., Namirembe S., Nyberg Y. and Sinclair F. L. 2016. Trees in agricultural landscapes enhance provision of ecosystem services in Sub-Saharan Africa. *International Journal of Biodiversity Science, Ecosystem Services and Management* 12, 255-73.

Larsen T. B. 2005. *Butterflies of West Africa*. Apollo Books, Svendborg, Denmark. vol. 1: 595pp; vol. 2: 270pp.

Luque G. M., Bellard C., Bertelsmeier C., Bonnaud E., Genovesi P., Simberloff D. and Courchamp F. 2014. The 100th of the world's worst invasive alien species. *Biological Invasions* 16, 981–5.

Mburu H., Cortada L., Mwangi G., Gitau K., Kiriga A., Kinyua Z., Ngundo G., Ronno W., Coyne D., Holgado R. and Haukeland S. 2018. First report of potato cyst nematode *Globodera pallida* infecting potato (*Solanum tuberosum* L.) in Kenya. *Plant Disease* 102, 1671.

McNeely J. A. and Scherr S. J. 2001. *Common Ground, Common Future: How Ecoagriculture Can Help Feed the World and Save Wild Biodiversity*. IUCN, Gland; Future Harvest, Washington DC., 24pp.

Moraes G. D. de, Zannou I. D., Oliveira A. R., Yaninek J. S. and Hanna R. 2006. Phytoseiid mites of the subtribes Typhlodromalina and Euseiina (Acari: Phytoseiidae: Euseiini) from sub-Saharan Africa. *Zootaxa* 1114, 1–52.

Moraes G. D. de, Zannou I. D., Ueckermann E. A., Oliveira A. R., Hanna R. and Yaninek J. S. 2007a. Species of the subtribes Arrenoseiina and Proprioseiopsina (tribe Amblyseiini) and the tribe Typhlodromipsini (Acari: Phytoseiidae) from sub-Saharan Africa. *Zootaxa* 1448, 1–39.

Moraes G. D. de, Zannou D. I., Ueckermann E. A., Oliveira A. R., Yaninek J. S. and Hanna R. 2007b. Phytoseiid mites of the tribes Afroseiulini, Kampimodromini and Phytoseiulini, and complementary notes on mites of the tribes Euseiini and Neoseiulini (Acari: Phytoseiidae) from sub-Saharan Africa. *Zootaxa* 1628, 1–22.

Moraes G. D. de, Zannou I. D., Ueckermann E. A., Oliveira A. R., Hanna R. and Yaninek J. S. 2008. Phytoseiid mites of the tribe Paraseiulini Wainstein (Acari: Phytoseiidae) from sub-Saharan Africa. *Zootaxa* 1687, 1–34.

Mwangi J. M., Kariuki G. M., Waceke J. W. and Grundler F. M. 2015. First report of *Globodera rostochiensis* infesting potatoes in Kenya. *New Disease Reports* 31, 18.

Neuenschwander P. 2004. Harnessing nature in Africa: Biological control can benefit the pocket, health and the environment. *Nature* 432, 801–2.

Neuenschwander P. and Adomou A. C. 2017. Reconstituting a rainforest patch in southern Benin for the protection of threatened plants. *Nature Conservation* 21, 57–82.

Neuenschwander P. and Markham R. 2001. Biological control in Africa and its possible effects on biodiversity. In: Wainberg, E., Scott, J. S. and Quimby, P. C. (Eds), *Evaluating Indirect Ecological Effects of Biological Control*. CAB International, Wallingford, UK, pp. 127–46.

Neuenschwander P., Hennessey R. D. and Herren H. R. 1987. Food web of insects associated with the cassava mealybug, *Phenacoccus manihoti* Matile-Ferrero (Hemiptera: Pseudococcidae), and its introduced parasitoid, *Epidinocarsis lopezi* (De Santis) (Hymenoptera: Encyrtidae), in Africa. *Bulletin of Entomological Research* 77, 177–89.

Neuenschwander P., Sinsin B. and Goergen G. (Eds). 2011. *Protection de la nature en Afrique de l'Ouest: Une Liste Rouge pour le Bénin (Nature Protection in West Africa: A Red List for Benin)*. IITA, Ibadan, Nigeria, 388pp.

Neuenschwander P., Bown D., Hèdégbètan G. C. and Adomou A. 2015. Protecting rain forest patches in West Africa. *Nature Conservation* 13, 21–46.

Nyiira Z. M. 1972. *Report of Investigation of Cassava Mite,* Mononychellus tanajoa *Bondar*. Kawanda Research Station, Kampala, Uganda, 14pp.

Pagan C., Coyne D., Carneiro R., Kariuki G., Luambano N., Affokpon A. and Williamson V. M. 2015. Mitochondrial haplotype-based identification of ethanol-preserved root-knot nematodes from Africa. *Phytopathology* 105, 350–7.

Periasamy M., Schafleitner R., Muthukalingan K. and Ramasamy S. 2015. Phylogeographical structure in mitochondrial DNA of legume pod borer

Published by Burleigh Dodds Science Publishing Limited, 2019.

(*Maruca vitrata*) population in tropical Asia and sub-Saharan Africa. *PLoS ONE* 10, e0124057

Poorter L., Bongers F., Kouamé F. N. and Hawthorne W. D. (Eds). 2004. *Biodiversity of West African Forests. An Ecological Atlas of Woody Plants Species*. CABI Publishing, Cambridge, UK, 521pp.

Robertson H. G. 2000. Afrotropical ants (Hymenoptera: Formicidae): Taxonomic progress and estimation of species richness. *Journal of Hymenoptera Research* 9, 71–84.

Robiche G., Le Gall P. and Goergen G. 2002. Contribution à l'étude de la biodiversité des coléoptères Tenebrionidae de la République du Bénin: Premier inventaire. *Lambillionea* CII, 381–431.

Rowland D., Ickowitz A., Powell B, Nasi R. and Sunderland T. 2017. Forest foods and healthy diets: Quantifying the contributions. *Environmental Conservation* 44, 102–14.

Schutze M. K., Mahmood K., Pavasovic A., Bo W., Newman J., Clarke A. R., Krosch M. N. and Cameron S. L. 2014. One and the same: Integrative taxonomic evidence that *Bactrocera invadens* (Diptera: Tephritidae) is the same species as the Oriental fruit fly *Bactrocera dorsalis. Systematic Entomology* 40, 472–86.

Sharanabasappa, Kalleshwaraswamy C. M., Asokan R., Mahadeva Swamy H. M., Marutid M. S., Pavithra H. B., Hegde K., Navi S., Prabhu S. T. and Goergen G. 2018. First report of the Fall armyworm, *Spodoptera frugiperda* (J E Smith) (Lepidoptera: Noctuidae), an alien invasive pest on maize in India. *Pest Management in Horticultural Ecosystems* 24, 23–9.

Srinivasan R., Yule S., Lin M. Y. and Khumsuwan C. 2015. Recent developments in the biological control of legume pod borer (*Maruca vitrata*) on yard-long bean. *Acta Horticulturae* 1102, 143–50.

Talwana H., Sibanda Z., Wanjohi W., Kimenju W., Luambano-Nyoni N., Massawe C., Manzanilla-López R. H., Davies K. G., Hunt D. J., Sikora R. A., Coyne D. L., Gowen S. R. and Kerry B. R. 2016. Agricultural nematology in East and Southern Africa: Problems, management strategies and stakeholder linkages. *Pest Management Science* 72, 226–45.

Tchabi A., Coyne D., Hountondji F., Lawouin L., Wiemken A. and Oehl F. 2008. Arbuscular mycorrhizal fungal communities in sub-Saharan Savannahs of Benin, West Africa, as affected by agricultural land use intensity and ecological zone. *Mycorrhiza* 18, 181–95.

Tchabi A., Burger S., Coyne D., Hountondji F., Lawouin L., Wiemken A. and Oehl F. 2009. Promiscuous arbuscular mycorrhizal symbiosis of yam (*Dioscorea* spp.), a key staple crop in West Africa. *Mycorrhiza* 19, 375–92.

Tchabi A., Coyne D., Hountondji F., Lawouin L., Wiemken A. and Oehl F. 2010. Efficacy of indigenous arbuscular mycorrhizal fungi for promoting white yam (*Dioscorea rotundata*) growth in West Africa. *Applied Soil Ecology* 45, 92–100.

Timbilla J. A. and Braimah H. 2000. Establishment, spread and impact of *Pareuchaetes pseudoinsulata* (Lepidoptera: Arctiidae) an exotic predator of the Siam Weed, *Chromolaena Odorata* (Asteraceae: Eupatoriae) in Ghana. In: Spencer N. R. (Ed.), *Proceedings of the X International Symposium on Biological Control of Weeds*, Montana State University, Bozeman, Montana, USA, pp. 105–11.

Ueckermann E. A., Zannou I. D., De Moraes G. J., Oliveira A. R., Hanna R. and Yaninek S. J. 2008. Phytoseiid mites of the tribe Typhlodromini (Acari: Phytoseiidae) from sub-Saharan Africa. *Zootaxa* 1901, 1–122.

Uyi O., Wilson D. D. and Zachariades C. 2009. The impact of *Pareuchaetes pseudoinsulata* (Lepidoptera: Arctiidae), on the growth rate of *Chromolaena odorata* (L.) in Southern Ghana. *Journal of Applied Sciences and Environmental Management* 13, 59–65.

van Driesche R. G., Carruthers R. I., Center T., Hoddle M. S., Hough-Goldstein J., Morin L., Smith L., Wagner D.L., Blossey B., Brancatini V., Casagrande R., Causton C.E., Coetzee J.A., Cuda J., Ding J., Fowler S.V., Frank J.H., Fuester R., Goolsby J., Grodowitz M., Heard T.A., Hill M.P., Hoffmann J.H., Huber J., Julien M., Kairo M.T.K., Kenis M., Mason P., Medal J., Messing R., Miller R., Moore A., Neuenschwander P., Newman R., Norambuena H., Palmer W.A., Pemberton R., Perez-Panduro A., Pratt P.D., Rayamajhi M., Salom S., Sands D., Schooler S., Schwarzländer M., Sheppard A., Shaw R., Tipping P.W. and Klinken van R.D. 2010. Review: Classical biological control for the protection of natural ecosystems. *Biological Control* 54, S2–S33.

Von Senger I., Barker N. and Zachariades C. 2002. Preliminary phylogeography of *Chromolaena odorata*: Finding the origin of a South African weed. In: *Proceedings of the 5th International Workshop on Biological Control and Management of Chromolaena odorata*. ARCPPRI, Pretoria, South Africa, pp. 90–9.

Yaninek J. S., Moraes G. J. and Markham R. H. 1989. *Handbook on the Cassava Green Mite (Mononcychellus tanajoa) in Africa: A Guide to Their Biology and Procedures for Implementing Classical Biological Control*. IITA, Ibadan, 140pp.

Zachariades C., Strathie L. W., Retief E. and Dube N. 2011. Progress towards the biological control of *Chromolaena odorata* (L) RM King and H Rob (Asteraceae) in South Africa. *African Entomology* 19, 282–302.

Zachariades C., Uyi O. O., Dube N., Strathie L. W., Muir D., Conlong D. E. and Assefa Y. 2016. Biological control of *Chromolaena odorata: Pareuchaetes insulata* spreads its wings. In: *Proceedings of the Annual Congress-South African Sugar Technologists' Association* 89, pp. 291–306.

Zannou I. D., de Moraes G. J., Ueckermann E. A., Oliveira A. R., Yaninek J. S. and Hanna R. 2006. Phytoseiid mites of the genus *Neoseiulus* Hughes (Acari: Phytoseiidae) from sub-Saharan Africa. *International Journal of Acarology* 32, 241–76.

Zannou I. D., De Moraes G. J., Ueckermann E. A., Oliveira A. R., Yaninek J. S. and Hanna R. 2007. Phytoseiid mites of the subtribe Amblyseiina (Acari: Phytoseiidae: Amblyseiini) from sub-Saharan Africa. *Zootaxa* 1550, 1–47.

Published by Burleigh Dodds Science Publishing Limited, 2019.

Chapter 5

Viruses affecting African crops and their vectors

James P. Legg, *IITA, Tanzania;* **P. Lava Kumar**, *IITA, Nigeria;* **George Mahuku**, *IITA, Tanzania;* **Everlyne Wosula**, *IITA, Tanzania;* **Livia Stavolone**, *IITA, Nigeria;* **Eugene Terry**, *New Markets Lab, USA; and* **Nilsa Bosque-Pérez**, *University of Idaho, USA (corresponding author: j.legg@cgiar.org)*

1 Introduction

1.1 The importance of plant virus disease in Africa and elsewhere

Viruses and the diseases that they cause are one of the major constraints to global agricultural production. In Africa, problems are particularly acute since control methods are either not available or not readily accessible to farmers. Surprisingly, the first steps towards researching crop virus diseases in Africa were taken more than a century ago. Towards the end of the nineteenth century, the disease condition 'Kräuselkrankheit' was recorded, which was later shown to be caused by the viruses causing cassava mosaic disease (CMD). Around the same time, in South Africa, Claude Fuller made reference in his 'First Report of the Government Entomologist. 1899–1900. Natal' to a variegated 'mealie' (= maize) condition. Although at the time he misinterpreted the cause of this, it was subsequently recognized as the first published observation of maize

http://dx.doi.org/10.19103/AS.2018.0043.05
Published by Burleigh Dodds Science Publishing, 2019.

streak disease (MSD). Although the first crop virus was described as early as 1898 (Beijerinck, 1898), it was not until the 1920s/30s that the first reports were made of viruses as the causal agents of plant disease in Africa (Storey, 1936a - MSD; Storey, 1939b - CMD; Jones and Mason, 1926 - Cotton leaf curl disease, CLCuD). Also, at around this time, researchers began to conduct the earliest experiments to confirm that the plant viruses causing these diseases were transmitted by insect vectors (Kufferath and Ghesquière, 1932 - CMD; Kirkpatrick, 1931 - CLCuD; Storey, 1939a - MSD).

In more recent years, the magnitude of importance of many of the viruses affecting crops grown in Africa has been demonstrated. In a recent review of the 'top ten viruses' with respect to their global impact (Rybicki, 2015), it can be noted that many on the list either only occur in Africa or have their major impacts in Africa. This list includes the cassava mosaic begomoviruses (CMBs), maize streak virus (MSV), banana bunchy top virus (BBTV), banana streak virus (BSV) and sugarcane mosaic virus (SCMV, which affects maize). In addition, the major impact of cassava brown streak disease (CBSD) caused by cassava brown streak ipomoviruses (CBSIs), particularly since 2004, was recognized as Pennisi (2010) included this in a report in *Science* as one of the seven greatest plant disease threats to food security in the world.

1.2 Progress in research into major African crop viruses

IITA has since its inception in 1967 aimed to contribute to the reduction of poverty in Africa through research for development targeting several major crop staples. These 'mandate' crops comprise banana/plantain, cassava, cowpea, maize (in the lowlands of West Africa), soybean and yam. All of these crops are affected by several viruses, although the effects range from major losses with occasional localized devastation, to symptomless infections that have no detectable effect on yield. In this chapter we will discuss primarily the most important of the viruses and the diseases that they cause. Our discussion is not intended to be a technical examination of the viruses affecting IITA's mandate crops, but rather a review of the progress that IITA and its partners have made in enhancing understanding of these virus diseases and delivering effective control measures to farmers. The pathosystems that have been selected include:

- Banana/plantain: banana streak virus - mealybug vector; banana bunchy top virus - aphid vector
- Cassava: cassava mosaic and cassava brown streak viruses - whitefly vector
- Legumes: cowpea aphid-borne mosaic virus and soybean mosaic virus - aphid vectors
- Yam: yam mosaic virus - aphid vector

During the 50 years that IITA has been working towards its mission, there has been significant progress made in improving knowledge of these important viral pathogens. This is illustrated with the results of searches on Elsevier's Scopus database, which is one of the largest repositories for published articles (Fig. 1). The largest number of published documents available for five viruses/virus groups are: BBTV, BSV, CMBs, MSV and SMV (Fig. 1a), whilst those for which there are relatively less published articles are CBSIs, CABMV, the viruses causing MLN and YMV (Fig 1b). What is evident in all cases, however, is the

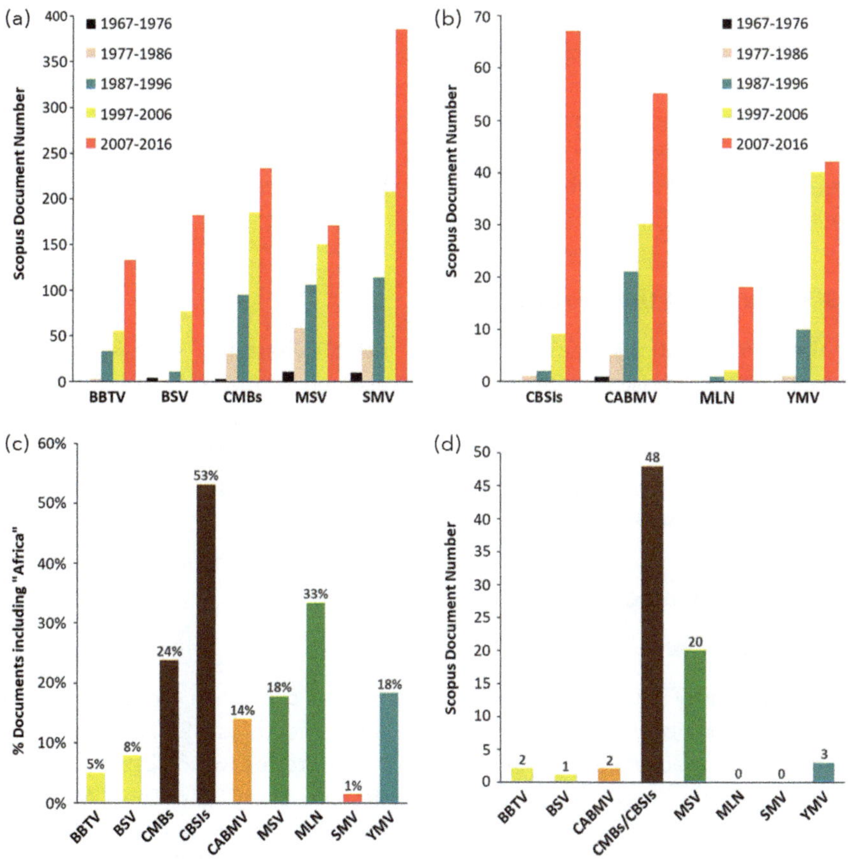

Figure 1 Number of publication references in the Scopus (Elsevier) database for selected important plant viruses. Searches queried for title, abstract and keywords. (a) Five viruses with most published documents. (b) Four viruses/virus diseases with least published documents. (c) Proportion of documents referencing each of the nine viruses/virus diseases in which the term 'Africa' was included for the period 1960–2017. (d) Number of published documents for each of the viruses/virus disease where the terms 'Africa' and 'Vector' were both included: a measure of the amount of vector research on each of the pathosystems.

large increase in the volume of published research, decade-on-decade, from 1967 to 2016. The most rapid increases in publishing frequency over the last decade (2007–16) have been recorded for MLN and CBSIs, both of which are notable in having experienced rapid epidemiological expansions in recent years. Since the documentary records of publications are not restricted to Africa, an additional set of searches was run incorporating 'Africa' in the search query. The results generated through this (Fig 1c) highlight a strong Africa focus for publications on CBSIs, MLN, CMBs, MSV and YMV. All of these viruses/virus disease conditions either occur exclusively in Africa, or are of much greater importance in Africa than elsewhere. BSV and BBTD occur almost worldwide where bananas are grown, and SMV is of major importance in the Americas where most soybean production is concentrated. Finally, database searches using terms for the vector groups responsible for transmitting each of these virus groups reveal a heavy emphasis on research into the whitefly *Bemisia tabaci* (Genn.) (Hemiptera; Aleyrodidae), which transmits the two groups of cassava viruses, as well as the *Cicadulina* leafhoppers (Hemiptera; Cicadellidae) that transmit MSV (Fig 1d).

1.3 Structure of the chapter

As these observations demonstrate, the last 50 years has seen substantial change in the 'landscape' of plant virus research in Africa. In the following sections, which focus on specific pathosystems, we will further highlight the nature of these changes, and will emphasize the ways in which research methods and control measures have been adapted to cope with the strongly dynamic plant virus situation.

2 Cassava virus diseases

2.1 Introduction

Several groups of viruses have been recorded from cassava in Africa, and these have been described in detail in reviews (Calvert and Thresh, 2002; Legg et al., 2015). In this chapter we will consider only two groups, which are by far the most important. These groups are the CMBs and the CBSIs. The CMBs are from the genus *Begomovirus* and family *Geminiviridae*, while CBSIs are from the genus *Ipomovirus* and family *Potyviridae*. Nine species of CMBs have been recorded from Africa (Legg et al., 2015) but all are closely related and elicit the same mosaic symptoms in infected plants, which is why the disease caused is known as CMD. CMD has been reported from all territories of continental Africa where cassava is cultivated, and in part due to the extent of this distribution, is responsible for losses which have been estimated at more than US$1 billion annually (Legg et al., 2006). There are two species of CBSIs: *Cassava brown*

Published by Burleigh Dodds Science Publishing, 2019.

streak virus and *Ugandan cassava brown streak virus* (Mbanzibwa et al., 2011). Both species produce similar symptoms, which comprise a blotchy vein-associated chlorosis on lower leaves, occasional stem lesions, as well as areas of brown, dry, necrotic rot in mature tuberous roots (Nichols, 1950). CBSD was for many years confined to coastal East Africa and the shores of Lake Malawi, but since 2004 has spread more widely through East and Central Africa (Alicai et al., 2007; Legg et al., 2011). Losses due to CBSD have been estimated at US$ 75 million annually (Manyong et al., unpublished).

As is apparent from these figures, and the background provided in the introduction to this chapter, these are two of the most important groups of viral pathogens, not just in Africa, but in the global context. In the following three sections we discuss some of the key stages of research on cassava viruses in Africa.

2.2 'The resistance-breeding years'

Following the first reports of cassava viruses from East Africa at the outset of the twentieth century, there was a period during the 1920s and 1930s during which there was a 'flurry' of new disease reports for CMD in particular, throughout Africa. It was at this time that the first organized research was initiated into CMD in several parts of Africa, and CBSD in East Africa. First reports were made of the viral cause of CMD and CBSD (Storey, 1936b) and of the whitefly transmission of pathogens causing CMD (Kufferath and Ghesquière, 1932). A concerted breeding effort was launched by Storey and colleagues in Tanzania in 1937 to identify sources of resistance to cassava viruses (Nichols, 1947), and this gave rise to the 'Amani breeding programme'. This initiative, directed from the Amani Research Station, in Tanga Region, north-eastern Tanzania, aimed to introgress resistance into cultivated cassava (*Manihot esculenta* Crantz) from wild relatives. The most important of these turned out to be ceará rubber (*M. glaziovii* Müll.-Arg.). Third backcrosses to cultivated cassava from *M. esculenta* × *M. glaziovii* hybrids demonstrated high levels of resistance to both CMD and CBSD. Seed taken from resistant genotypes produced in this way was introduced to Nigeria where seedling and clonal evaluations were conducted which eventually resulted in the identification of the CMD-resistant clone TMS 58308 – selected in 1958.

It was this clone that provided the basis for the first significant research programme launched by IITA with a major focus on CMD, which was led by S. K. Hahn. Since TMS 58308 had poor quality roots and low yield, an initial task was to cross TMS 58308 with other genotypes with high yield potential and/or improved quality characteristics. In order to do this, Hahn and colleagues used a recurrent selection approach (Hahn et al., 1980). Although CMD work at IITA during this time was primarily focused on breeding for resistance, important supporting research was conducted on other aspects of the pathosystem. Vector entomological studies demonstrated that the abundance of the whitefly vector – *B. tabaci* – was a key factor driving seasonal patterns of CMD spread (Leuschner, unpublished), and pathology research described the yield impacts of CMD and related the severity of CMD symptoms to components of yield (Terry and Hahn, 1980). One of the most straightforward but useful outputs of this period of research was the development of the 1–5 severity scoring system for CMD (Hahn et al., 1980), which continues to be universally used to this day. Some of the CMD-resistant cassava genotypes generated during this period continue to be the most widely disseminated and grown varieties. One important example is TMS 30572; first cloned in 1972, but still widely cultivated and continuing to retain an effective level of resistance to CMD. The proportion of IITA germplasm contributing to variety release in the national research programmes of African countries increased from 55% during the 1970 to 1979 period to almost 80% between 1980 and 1989, illustrating the scale of the Africa-wide impact of this resistance breeding effort.

2.3 'The surveillance and epidemiology years'

The second major 'era' of IITA's research on cassava viruses began following the shocking stories of cassava production collapse that emerged from Uganda during the early 1990s. Reports were received of CMD so severe, that plants became stalks, yields were reduced to almost nothing and people in northern Uganda were suffering from food shortages and localized famine (Otim-Nape et al., 1997; Thresh et al., 1994). At this time, only two species of CMB were known – based on serological studies – and these were *African cassava mosaic virus* (ACMV) and *East African cassava mosaic virus* (EACMV) (Swanson and Harrison, 1994). However, the arrival of new nucleic acid-based diagnostics using the polymerase chain reaction (PCR), together with affordable sequencing meant that it became possible to 'look inside' viruses and make comparisons between their genetic material. Although serological methods suggested that the usual ACMV was associated with the severe CMD occurring in northern Uganda, when PCR and sequencing methods were employed, it was demonstrated that an EACMV-ACMV recombinant hybrid virus was always present in areas affected by severe CMD (Zhou et al., 1997). Furthermore, the severest symptoms were the result of mixed infections of both ACMV and the new recombinant, which became known as EACMV-UG, or the 'Uganda Variant' (Harrison et al., 1997).

In spite of initial scepticism within the IITA breeding community about this new form of CMD, the institute initiated a new programme of research in partnership with Uganda's National Agricultural Research Organisation (NARO) to characterize, monitor and mitigate the impacts of this rapidly expanding epidemic of severe CMD. The programme was housed at IITA's Eastern and Southern Africa Regional Centre (IITA-ESARC), which was based at NARO's

Namulonge Research Station, about 25 km north of Kampala. It soon became apparent that the severe CMD, which was associated with both EACMV-UG and unusually high populations of *B. tabaci*, was spreading through Uganda and into the neighbouring countries of Kenya to the east and Tanzania and Rwanda to the west.

In response, IITA launched a regional CMD pandemic mitigation effort, in partnership with NARS in each of the affected countries. This had the aims of monitoring the regional spread of severe CMD, forecasting future patterns of spread, identifying worst-affected regions, evaluating the best available CMD-resistant germplasm and working with NARS and local NGOs to multiply and disseminate planting material of CMD-resistant varieties to farmers. During this period, an extensive set of published data was compiled, including important contributions from Advanced Research Institutes in Europe and the United States, NARS researchers in affected countries, as well as IITA scientists. The extent and speed of the pandemic spread were characterized (Otim-Nape et al., 1997; Legg and Ogwal, 1998), regional risk maps were developed (Legg, 1999), the nature of virus–virus interactions causing severe disease were described (Pita et al., 2001), mild strain protection was demonstrated (Owor et al., 2004) and new studies characterizing CMBs were initiated in neighbouring countries, such as Tanzania (Ndunguru et al., 2005) and Kenya (Were et al., 2004).

As the pandemic spread through East and Central Africa, and the monitoring activities of IITA and partners tracked it, a series of 'first reports' were generated for countries newly affected (Legg et al., 2001 – Rwanda; Neuenschwander et al., 2002 – Republic of Congo; Legg et al., 2004 – Gabon; Bigirimana et al., 2004 – Burundi; Lava Kumar et al., 2009 – Angola; and Akinbade et al., 2010 – Cameroon). Major steps were made in pandemic mitigation during this period, and progress was facilitated by carefully managed germplasm introduction programmes to countries in the region using 'open quarantine' procedures. These sources of CMD-resistant varieties would provide the basis for massive multiplication and dissemination programmes which gradually increased in scale from the mid-1990s through to the end of the 2000s. A further crucial feature of the CMD research work of IITA during this period was capacity building at postgraduate level. Between 2000 and 2005, nine MSc and five PhD students from within the affected region were supervised specifically on cassava viruses and their whitefly vector. Several of these trained researchers have gone on to take up positions of considerable responsibility within their national research systems, which means that their training is having a transformative impact on agricultural research for development in this part of Africa.

2.4 'The building quality seed system years'

For seven decades, CBSD had been a problem of minor importance in Africa, being confined to the eastern coastal areas and the shores of Lake Malawi.

The situation changed dramatically, however, following reports of a new outbreak in 2004 in southern Uganda, at altitudes (>1000 masl) (Alicai et al., 2007) hitherto considered to be inimical to the survival and spread of CBSD (Nichols, 1950). In the years immediately following, the emerging pandemic of CBSD spread to Kenya, Tanzania, Rwanda, Burundi and eastern Democratic Republic of Congo (Bigirimana et al., 2011; Legg et al., 2011; Mulimbi et al., 2012). This spread was associated with large-scale losses in newly affected regions, an effect that was exacerbated by the fact that most of the locally available varieties were highly susceptible to this disease that had never previously been encountered.

In response, there was a rapid increase in the level of human and financial investment directed towards improving understanding of the disease and controlling the pandemic. For the first time this had become a truly African problem, as westwards spread threatened the major cassava production centres of Central and West Africa. Several of the young researchers who had been trained during the period of the CMD pandemic were able to transfer their newly acquired knowledge and skills to tackling this new challenge. The consequence was an unprecedented level of research output. This is clearly demonstrated by the Scopus publication data (Fig. 1b) which reveal that whilst there were only two publications on CBSD in the period 1987–96 (0.02/month), this number increased to 9 between 1997 and 2006 (0.08/month) and reached as high as 67 for 2007 and 2016 (0.6/month). Moreover, in the first ten months of 2017, there were 19 new CBSD publications in the Scopus database (1.9/month). It is significant to note that of these 19 papers, roughly half include authors who are students and half include either IITA scientists or NARS researchers who did their postgraduate training with IITA. These points highlight the research impact that is being realized through previous and ongoing capacity development efforts within the region.

CBSD produces symptoms that are not easily recognized, and infected plants may sometimes be completely asymptomatic. The practical consequence of this is that this disease is much more readily spread through infected planting material than CMD, which usually presents clear symptoms. In view of the potential for rapid hidden spread of the viruses causing CBSD through infected planting material, a key focus of CBSD pandemic management efforts has been to strengthen systems for the production, exchange and dissemination of near virus-free planting material. A 'Quality Management Protocol' was developed that prescribed tolerance levels for the major pests and diseases of cassava, as well as other quality characteristics for planting material. This was extensively applied through the USAID-funded Crop Crisis Control Project (C3P) in the six target countries of Burundi, Democratic Republic of Congo, Kenya, Rwanda, Tanzania and Uganda. Application of the protocol ensured that cassava multiplication and dissemination work carried out through the project from 2006 to 2007 conformed to high quality standards and minimized the chance of inadvertent virus spread.

The need for these types of approaches became much more acute, however, during the period of operation of the larger Great Lakes Cassava Initiative (GLCI), which ran from 2007 to 2012. It was during this period that rapid new spread of CBSD was taking place through the Great Lakes region of East and Central Africa, which largely matched the target zone of the GLCI project. New virus testing protocols were introduced based on real-time PCR (Adams et al., 2013) and a zero-tolerance plan for CBSIs was initially recommended for movements of planting material from one multiplication site to other multiplication sites or farmers' fields. Eventually the impracticality of this extreme approach was recognized as large-scale virus testing was simply not feasible nor affordable and farming areas being served were in most cases so much worse affected than multiplication sites that it made no practical sense to ban movements of planting material from multiplication sites where there were very low levels of CBSIs.

A new model was therefore developed through the New Cassava Varieties and Clean Seed to Combat CMD and CBSD (5CP) project in which stringent yet pragmatic tolerance levels were set for each of the major cassava pests and diseases, including the CMBs and CBSIs. Most importantly, these levels were set in a consultative way with the involvement of the full range of cassava seed sector stakeholders, including farmers. Additionally, the process was managed by the national agency responsible for seed certification, which in Tanzania was the Tanzania Official Seed Certification Institute (TOSCI). With the twin aims of enhancing the quality of cassava planting material production and facilitating the emergence of commercially viable cassava 'seed' business within Tanzania, TOSCI succeeded in officially appending the new certification guidelines to the Government of Tanzania Seed Act on 17 January 2017. This provides specified tolerance levels for disease and other quality parameters for pre-basic, basic and certified levels of seed production, as well as specifying virus testing requirements and service charges payable by seed producers. Significantly, virus testing is only required for CBSIs at the pre-basic level. In view of the success of this programme, as well as parallel efforts implemented by NARO in Uganda, concerted efforts are currently being made by IITA, NARS and seed certification agencies in several other countries (Burundi, DRC and Rwanda) to deliver similar cassava virus management outcomes through high quality seed systems. An approach that combines the deployment of cassava virus-resistant varieties with effective systems for the management of seed health offers great promise for the sustainable management of cassava viruses in East/Central Africa and beyond.

3 Maize virus diseases

3.1 Introduction

Numerous viral diseases of maize occur in Africa, but that caused by *Maize streak virus* (MSV; genus *Mastrevirus*; family *Geminiviridae*) has been by far the most important and widespread (Thottappilly et al., 1993) since it was first reported more than a century ago. In more recent years, however, Maize lethal necrosis (MLN), also known as Corn lethal necrosis, has spread through East and Southern Africa following the introduction through infected seed of *Maize chlorotic mottle virus* (MCMV; genus *Machlomovirus*; family *Tombusviridae*). The MLN condition results from the synergistic interaction of MCMV with *Sugarcane mosaic virus* (SCMV; genus *Potyvirus*; family *Potyviridae*). SCMV is endemic in Africa, but MCMV seems to have been recently introduced, most probably from the Far East.

3.2 The period of research into MSV (1980s–1990s)

Early years of research at IITA on MSV examined the epidemiology of the virus, interactions with its leafhopper vectors (*Ciccadulina* spp.) and managing the disease through resistance breeding. Vector aspects are examined in detail in the subsequent section on virus vectors. In this section we discuss epidemiology and control through host plant resistance.

MSV is widely distributed in sub-Saharan Africa (SSA) and adjacent islands (Thottappilly et al., 1993). Symptoms of MSV in maize first appear as small, spherical, chlorotic spots. As leaves expand streaks develop along the veins on most of the leaf but since the virus is systemic, symptoms appear only on the inoculated and subsequent leaves (Thottappilly et al., 1993). Symptom severity depends on genotype susceptibility and plant age at the time of infection (Bosque-Pérez et al., 1998). In highly sensitive varieties, chlorosis of the entire leaf lamina might be observed, followed by progressive necrosis and plant death, particularly if infection has occurred at an early plant growth stage. MSV reduces yield and growth of maize and in susceptible varieties yield reductions often exceed 70% (Bosque-Pérez et al., 1998).

MSV is a new encounter disease, which emerged after maize was introduced to Africa from the Americas over 400 years ago. It is believed that only some of the original isolates of this African virus were capable of naturally infecting maize plants (Bosque-Pérez, 2000). A shift from native grasses to maize would have required that both virus and leafhopper vectors expand their host ranges. Possibly, infection events occurred independently in various locations in Africa, from where further virus spread took place (Bosque-Pérez and Buddenhagen, 1999). With the expansion of maize cultivation across SSA, outbreaks then ensued and continue to occur sporadically.

Epidemiological and ecological studies of MSV and its vectors were done by IITA scientists in the 1980s and 1990s. Characterization of 24 isolates originating from grasses and cereal crops in Nigeria revealed differences in virulence among the MSV isolates tested. Some of them caused mild symptoms similar to the original symptoms in grasses and were transmissible to maize only at low rates (Mesfin et al., 1992). Isolates from maize, *Brachiaria lata* (Schumach.), *Axonopus compressus* (Sw.) P. Beauv., *Dactyloctenium aegyptium* (L.) Willd., *Digitaria horizontalis* Willd. and *Setaria barbata* (Lam. Kunth) (all Poaceae) were identified as the most virulent in maize (Mesfin et al., 1992). Of the isolates studied by Mesfin et al. (1992), nine could not infect the maize cultivar Pool 16, which was known for its susceptibility to MSV in farmers' fields in West Africa. The authors concluded that some MSV isolates appear to be adapted to their grass hosts and that many grasses in West Africa would not be a source of maize-infecting isolates in the field (Mesfin et al., 1992).

Understanding the survival of the grasses that harbour vectors and MSV isolates capable of infecting maize during the dry season (November to March), particularly in low-lying hydromorphic soils (i.e. fadamas) was deemed critical for understanding MSV ecology and epidemiology (Bosque-Pérez and Buddenhagen, 1999). Perpetuation of MSV and its vectors during the dry season in West Africa appears to occur mostly in riverine areas and low-lying fields where there is residual moisture, and both maize and grass weeds are present. MSV outbreaks appear to occur only when favourable weather conditions allow vector survival and population build-up, and where MSV isolates capable of causing disease in maize occur in grass hosts (Bosque-Pérez, 2000). Based on host range, serological tests and symptom severity of isolates tested, Mesfin et al. (1992) concluded that weeds of importance in perpetuating epidemiologically competent MSV in Nigeria include the native annuals *S. barbata* and *B. lata* and the introduced perennial *A. compressus* (Asanzi et al., 1994). These weeds harbour MSV under field conditions, are prevalent in many areas of the country and in the humid forest zone are frequently found in hydromorphic areas (Mesfin et al., 1992). Additionally, leafhoppers were commonly found in farmers' fields on *S. barbata* and *B. lata*, but no indigenous perennial grass has been found in West Africa harbouring a maize-competent MSV strain (Bosque-Pérez, 2000).

In addition to ecological zone, incidence of MSV has been found to vary according to variety and planting date (Asanzi et al., 1994; Bosque-Pérez et al., 1998). Late season plantings (July to October) had higher virus incidence than early season ones (March to May) (Fajemisin et al., 1976). Additionally, maize cultivars differ in their ability to sustain virus disease. For example, the susceptible cultivar TZB-Gusau had a higher incidence of MSV in the field than the resistant hybrid 8321-21, which showed a relatively low incidence across locations, planting dates and years (Asanzi et al., 1994; Bosque-Pérez et al., 1998). Low field incidence in 8321-21 is in part due to non-preference for feeding by vectors (Mesfin and Bosque-Pérez, 1998).

Sporadic epidemics of MSV still occur in parts of Africa and the resulting losses can be devastating. Resistance in maize was noted as early as 1931 in South Africa, in the variety 'Peruvian Yellow'. Later, tolerance was found in other materials and utilized to develop a hybrid population, PxH (Soto et al., 1982). Resistance to MSV was also identified in East Africa (Storey and Howland, 1967). At IITA in Nigeria, resistance was found within IITA's maize population TZ-Yellow and in the local variety Revolution in Réunion (Soto et al., 1982; Buddenhagen and Bosque-Pérez, 1999). Improved challenge methods using artificial infestations with infective leafhopper vectors both in screenhouses and in the field were developed and further perfected at IITA over the years (Soto et al., 1982).

Resistance to MSV in many of the IITA open-pollinated varieties and hybrids manifests itself as reduced disease severity combined with low virus incidence in the field (Soto et al., 1982; Asanzi et al., 1994; Bosque-Pérez et al., 1998). Highly resistant varieties have been shown to have low virus titre (Asanzi, unpublished). Yield comparisons between infected and non-infected plots of resistant and susceptible varieties demonstrated the superiority of resistant ones. A 71% yield reduction was observed in the susceptible open-pollinated variety TZB Gusao when infected with MSV compared with the protected control, while the highly resistant hybrid 8321-21 averaged a 10% reduction when infected compared with the protected control (Bosque-Pérez et al., 1998). Reduction in yield due to MSV is a function of the age of plants at the time of infection, with younger plants sustaining greater yield reduction than mature ones (Vogel et al. 1993; Bosque-Pérez et al., 1998).

Maize varieties that combine resistance to MSV with other desirable characteristics were developed at IITA during the 1980s and 1990s and more recently at the Harare station of the International Maize and Wheat Improvement Center (CIMMYT). Both high-yielding varieties and varieties traditionally grown in various countries in Africa were converted into MSV-resistant ones at IITA in cooperation with CIMMYT and National Programmes in Africa. Numerous breeding programmes in Africa used resistance sources developed at IITA to incorporate resistance into their varieties (Efron et al., 1989). By the 1990s streak-resistant maize germplasm had been released and was being grown by farmers, was recommended for use or was being considered for release in 20 countries in SSA (Bosque-Pérez, 2000).

3.3 The period of research into MLN (2011-the present)

The spread of MLN has intensified since the first outbreak was reported in September 2011 in Kenya (Wangai et al., 2012a). Currently, MLN has been reported in Tanzania, Uganda, Rwanda, DRC, Ethiopia and South Sudan (Mahuku et al., 2015a). In areas of Kenya where MLN was very serious, farmers experienced extensive or complete crop loss (Wangai et al., 2012b). Infected plants are frequently barren and the ears formed are small, deformed and set little or no seeds, drastically reducing the yield. Maize grain yield losses ranging from 23% to 100% have been reported in Kenya (De Groote et al., 2016). The monetary value lost to MLN annually has been estimated at US$ 261 million in Ethiopia and US$ 198 million in Kenya (Marenya et al., 2018).

Each of the viruses associated with MLN can cause symptoms when present in a single infection. Symptoms depend upon genotype, age of infection and environmental conditions (Mahuku et al., 2015a). Symptoms are much more severe where there are mixed infections of MCMV together with a *Potyvirus*, the most commonly occurring of which is SCMV. This synergistic condition is what characterizes MLN and what makes it so damaging to the plant. The maize plant is susceptible to MLN infection at all stages of plant growth; the earlier the plant becomes infected, the greater the damage and the entire crop can frequently be killed before tasselling (Niblett and Claflin, 1978; Uyemoto et al., 1980, 1981; Wangai et al., 2012b; Makone et al., 2014).

MLN was first identified in the United States in 1976 (Niblett and Claflin, 1978). In Africa, the first outbreak of MLN was reported in 2011 in Bomet County in the Southern Rift Valley Region in Kenya (Wangai et al., 2012a). The disease subsequently spread to several other countries in East and Central Africa (Mahuku et al., 2015a,2015b; Lukanda et al., 2014; Adams et al., 2014). Some projections based on the current situation indicate that MLN could spread and intensify across Africa by 2020 which would result in a more significant impact on food security in the region (Isabirye and Rwomushana, 2016).

MCMV is transmitted mechanically and is also spread by several insect vectors including maize thrips (*Frankliniella williamsi* Hood) (Thysanoptera; Thripidae) (Cabanas et al., 2013; Jiang et al., 1992), maize rootworms (*Diabrotica undecimpunctata* (L.), *Diabrotica longicornis* (Say) and *Diabrotica virgifera* LeConte) (Coleoptera; Chrysomelidae), cereal leaf beetles (*Oulema melanopus* (L.)) (Coleoptera; Chrysomelidae), corn flea beetle (*Systena frontalis* (Fabricius)) (Coleoptera; Chrysomelidae) and *Chaetocnema pulicaria* F.E. Melsheimer (Coleoptera; Chrysomelidae) (Nault et al., 1978; Jensen, 1985; Reyes and Castillo, 1988). SCMV is spread by maize aphids. Also, seed transmission of MCMV has been reported (Jensen et al., 1991).

The best approach for the management of MLN is to employ integrated pest management practices encompassing the implementation of sound quarantine and phytosanitary measures, cultural control (such as a closed season, crop rotation and crop diversification), vector control using seed treatment followed by foliar sprays with contact insecticides, and host plant resistance, all supported by appropriate policy measures.

The use of seed that is free of the viruses that cause MLN is an effective strategy for managing MLN. All seed coming into a country should be certified as MCMV-free, and there should be no movement of commercial seed from MLN endemic to unaffected countries. National plant protection organizations may also need to apply internal quarantine restrictions. Combining the enforcement of such regulations with measures to raise awareness amongst the public and within farming communities can help to reduce the spread of the disease.

Crop rotation and maize-free windows can be used to contain MLN (Uyemoto, 1983; Wangai et al., 2012b; Kiruwa et al., 2016). Rotating maize crops with a non-grass crop or non-cereals reduces the virus inoculum of both MCMV and SCMV. It is necessary to use good field sanitation methods, including weed control and burning grasses around the field to eliminate alternative hosts for potential vectors (Wangai et al., 2012b). Seeds should not be recycled; farmers should use clean and certified seeds. Furthermore, crop debris should not be moved around, as it represents a potential source of inoculum (Samson et al., 2014).

Reduction of vector populations through insecticide application is unlikely to have any impact on the disease unless applied consistently and over large areas. These conditions are difficult to achieve in the dense patchworks of smallholder plantings found in Africa. In Hawaii, chemical spraying at early establishment was deemed successful in mitigating MLN spread (Nelson et al., 2011). However, seed dressing can be effective, as this prevents secondary transmission of the viruses, but this is only practical in seed production fields.

The use of tolerant or resistant varieties is probably the most effective means of managing MLN. Superior resistance to MCMV is widely available in tropical maize seed stocks and provides the best control for this disease (Nelson et al., 2011). Ultimately, resistant varieties need to be developed for all agro-ecological zones because most maize producing areas are either affected or are vulnerable (Isabirye and Rwomushana, 2016). Although resistance to both MCMV and SCMV is available, no complete immunity has been observed. Preliminary studies on the inheritance of MCMV resistance suggest a polygenic control of the disease, with resistance being partially dominant (Gowda et al., 2015). A similar situation has been observed for SCMV and other potyviruses (Jones et al., 2011). This encourages the commercial production of hybrids only if both parents are resistant to the pathogen. As the symptoms of MLN are due to co-infection with two viruses, resistance against any one of the viruses can substantially reduce the damage due to the disease. In the long run, deployment of varieties that are resistant to both MCMV and SCMV will be the best means of managing MLN. Through breeding, resistance to MCMV can be incorporated into susceptible maize varieties within a 4-year period. National and international programmes should intensify efforts to introgress MLN resistance into the commercial genetic pools in all countries including those that are not yet affected by MLN.

Published by Burleigh Dodds Science Publishing, 2019.

MLN continues to be a major threat to maize production in Africa. Effective management of the disease will require a rigorous surveillance system for MCMV, awareness creation of farmers and seed producers of the impending threat and policy regulations to minimize movement of seed from MCMV endemic areas. Research is needed on vectors of MCMV, vector biology, epidemiology, ecology, molecular biology and effective management strategies. Levels of resistance to the viruses among all varieties of maize grown in the region and the available germplasm are also low. Continued efforts are therefore required to understand the mechanisms of resistance to MCMV and to develop MLN-resistant maize varieties, and to have these deployed in both affected and threatened regions.

4 Banana and plantain virus diseases

4.1 Introduction

Banana virology at IITA commenced in the 1990s at around the time of reports of the integration of *Banana streak virus* (BSV; genus *Badnavirus*; family *Caulimoviridae*) genomic sequences into the genome of *Musa* and its implications on international germplasm distribution and banana breeding programmes (Harper et al., 1999). Of about 20 different viruses reported to infect *Musa* worldwide, the bulk of the IITA virology work was focused on BSV, responsible for 'leaf streak', and *Banana bunchy top virus* (BBTV; genus *Babuvirus*; family *Nanoviridae*) responsible for bunchy top (Kumar et al., 2015). Details of achievements on BSV and BBTV are presented in Chapter 10. In this section, a synthesis of the IITA banana virology programme is presented to highlight the focus based on the emerging priorities and future directions.

The objectives of the banana virology programme have changed over the course of time. Initial research was focused on the diagnosis of banana viruses to support the production of virus-free tissue culture clones for international distribution. The discovery of BSV genomic sequence integration into the *Musa* genome led to an increase in the scale and coverage of banana virus research at IITA from the mid-1990s to 2005. This involved a focus on several key elements of BSV research. These included understanding the biological significance of BSV integration on plant performance, breeding and banana micropropagation, as well as the development of reliable tools for detection of episomal (replication form) BSV infections in banana. The discovery of BSV genome integration halted the distribution of *Musa* germplasm in the 2000s, especially newly bred synthetic banana and plantain hybrids, as well as the BITA and PITA series clones developed by the IITA *Musa* breeding programme in West Africa. This situation restricted the progress of *Musa* improvement for several years. After 2008, however, IITA adopted a new strategy to distribute *Musa* clones within Africa based on the African Union

Inter-African Phytosanitary Council (AU-IAPSC) guidelines that allowed the distribution of clones with integrated BSV, with informed consent. This step provided momentum and reinvigorated banana improvement efforts. The IITA protocol was later adopted at global level in 2015 for the international distribution of *Musa* clones (Thomas et al., 2015). Subsequent to this work on BSV, the research focus shifted to BBTV, which emerged as a serious threat to *Musa* production in Southern and Central Africa. The research work on BBTV since 2008 has represented the core of banana virology research at IITA. Banana virus diagnostics and the production of virus-free planting material continue to be key components of the virology programme, and these also enable IITA to fulfil its role as an internationally recognized source of *Musa* germplasm.

Objectives of the current programme are to map the distribution and diversity of BBTV in Africa, to understand virus-host-vector interactions, to characterize the diversity of the banana aphid (*Pentalonia nigronervosa* (Coquerel) (Hemiptera; Aphididae)) that vectors BBTV, to elucidate the role of endosymbionts in banana aphid biology, to design biocontrol approaches for banana aphid management, to develop versatile diagnostic tools for application under lab and field conditions, to establish improved procedures for the production of virus-free *Musa* planting material and to develop BBTV-resistant clones using conventional and transgenic approaches. The virology programme is well equipped to diagnose all viruses reported to infect banana around the world and endowed with excellent capacity to develop new diagnostics and conduct virus discovery research using next-generation sequencing-based approaches.

4.2 Future prospects

Of about 20 different viruses reported to infect *Musa* worldwide, BBTV, BSV and *Cucumber mosaic virus* (CMV; genus *Cucumovirus*; family *Bromoviridae*) are known to occur in all or several *Musa*-producing countries in Africa, whereas *Banana bract mosaic virus* (genus *Potyvirus*; family *Potyviridae*) is confined to a few countries in Asia and has not been reported from Africa (Kumar et al., 2015). IITA research work has mainly focused on BSV and BBTV owing to their importance to banana and plantain production systems in Africa. An important achievement has been to break the decade-old impasse on *Musa* germplasm distribution through advocating for a pragmatic approach, which was supported by the AU-IAPSC guidelines for *Musa* germplasm distribution and which was eventually accepted as the global standard. This work has had the important spin-off benefit of reviving the *Musa* improvement programme at IITA, whose success and relevance hinged on having a safe but effective system for the international distribution of clones.

Published by Burleigh Dodds Science Publishing, 2019.

While BSV posed technical challenges, the real impact on *Musa* production has been caused by BBTV, an aphid-transmitted virus that has now spread to 16 countries in Central, Southern and West Africa. IITA initiatives have contributed to the establishment of the status of the disease and factors contributing to its spread (Kumar et al., 2011), which eventually led to the development of a continental strategy to contain BBTV spread and restore banana production. This approach termed the 'ALLIANCE against BBTV in Africa', which is co-led by IITA, Bioversity International and CIRAD, is an important priority R4D theme for the Roots, Tubers and Bananas Programme of the CGIAR (CRP) (www.bbtvalliance.org). A multi-pronged approach is being used to tackle BBTV together with a range of national and international partners in SSA.

An important component of this approach is the search for durable host resistance. Recent advances in *Musa* genomics and new phenotyping methods are ushering in new hope for sustainable alternatives to current management approaches which involve cultural methods such as the eradication and replanting of infected mats with clean planting material obtained from virus-indexed source material. IITA and the ALLIANCE team are poised to achieve these milestones in the near future. Capacity development, especially training in the application of diagnostics, clean planting material production and surveillance procedures, remains a key requirement for the strengthening of future banana virus management efforts. Postgraduate students and national research programmes remain key partners with the banana virology programme for collaborating in research as well as disseminating knowledge and technologies to partners. Finally, in order to achieve the greatest improvements to banana productivity and the optimum use of limited resources, IITA needs to develop an integrated banana disease management strategy. In addition to tackling viral pathogens and the vectors that spread them, this should also address the major bacterial (banana *Xanthomonas* wilt) and fungal (*Fusarium oxysporum* f. sp. *cubense* Tropical Race 4) diseases.

5 Yam viruses

5.1 Introduction

Several species of yams are cultivated by farmers for their edible starchy tubers. The white yam (*Dioscorea rotundata* Poir.) dominates production in the West African yam belt, followed by water yam (*D. alata* L.) and several other less important species such as *D. cayenensis*, *D. esculenta* (Lour.) Burkill, *D. dumetorum* (Kunth) Pax and *D. bulbifera* L. Yams are propagated vegetatively using tubers. Mosaic disease caused by virus infection of yam is one of the leading contributors to low tuber yields in the yam belt which is the main target area for IITA's yam improvement programme. The yam belt, stretching from the west of Cameroon to Ivory Coast is the world's largest food yam production zone, accounting for about 95% of the global production (FAO, 2016). Research into yam viruses began at IITA in the 1980s and focused on characterization of viruses, development of diagnostic tools for virus detection and breeding for *Yam mosaic virus* (YMV) resistance in *D. rotundata*.

5.2 Yam virus disease aetiology, epidemiology and effects on yield

Various survey studies since the 2000s have revealed that the viruses causing disease in yams in the West African yam belt belong to three genera. The virus species and their genera are YMV and *Yam mild mosaic virus* (YMMV) (previously known as *Dioscorea alata virus*) (genus *Potyvirus*); CMV; and a complex of several yam badnaviruses (YBVs) belonging to the genus *Badnavirus* (Eni et al., 2008; Turaki et al., 2017). All of these viruses have insect vectors. The vectors are aphids for members of the genus *Potyvirus* and *Cucumovirus* and mealybugs for members of the genus *Badnavirus*. The significance of vector transmission is yet to be elucidated.

YMV is most prevalent and known to reduce yields by up to 60% in most of the yam varieties. Vegetative propagation, sequential re-use of farmer-saved seed yams, farmer-to-farmer seed exchange and the informal seed trade have contributed to the wide dissemination of virus-infected seed yams and consequent spread of yam viruses in West Africa. Surveys conducted in the 2000s confirmed a virus incidence of >80% in white yam and an overall prevalence of >90% in the yam belt (Eni et al., 2008). Virus infection therefore is a threat to both plant and seed health. Durable virus resistance was not available in cultivated yam landraces. The production and supply of virus-free seed yams was prioritized as a tactic to control virus diseases in West Africa. In support of this effort, a major thrust was placed on the development of reliable diagnostic tools for yam viruses necessary for seed health indexing and certification. This work was implemented through the Yam Improvement for Income and Food Security in West Africa (YIIFSWA) Project. Since 2012 this work has resulted in the development of a range of diagnostic tools for the detection of viruses in the field and lab, including conventional PCR-based assays as well as isothermal methods such as Loop-mediated Isothermal Amplification (LAMP) (Nkere et al., 2018) and Recombinase Polymerase Amplification (RPA) (Silva et al., 2018) methods for the detection of YMV, YMMV and CMV. The rolling circle amplification method has been used for the detection of badnaviruses whose genomes are known to be integrated into the yam genome (Bömer et al., 2016). Methods for the production of virus-free plants were also established to support the production of clean seed (Balogun et al., 2017). More details about clean seed systems are given in Chapter 9.

Published by Burleigh Dodds Science Publishing, 2019.

5.3 Seed degeneration

Seed degeneration is one of the major challenges to clean seed systems. Virus-free seed introduced to the field gets re-infected by viruses as all the released cultivars are susceptible to viruses. To help improve the situation various on-farm management approaches such as 'positive selection' are being implemented, by selectively harvesting seed from healthy plants for replanting. This approach is effective but laborious and not readily practised by farmers. Therefore, increasing efforts are being made to augment host resistance through breeding. Systematic screening of yam landraces conserved in the IITA Genebank identified promising YMV-resistant sources which are being exploited as parents for breeding (Mignouna et al., 2002). This work is being continued as part of the AfricaYam Project.

5.4 Future prospects

There has been significantly less donor investment in yam research and development than for other African food staples such as bananas, cassava and maize. This has resulted in relatively less progress in virus research on this crop. Core support in the early years focused on breeding initiatives, although interest in screening germplasm for the major pests and diseases did generate valuable knowledge on the relative importance of viruses and their impact on yield reduction, prompting a stakeholder consultation in the West African sub-region and subsequent baseline studies in the 2000s. These efforts have demonstrated the high value and importance of yams to West African livelihoods and led to major new project investments such as the YIIFSWA and AfricaYam Projects, which have in turn led to strong progress in the development and application of state-of-the-art diagnostics and new breeding approaches. However, all ongoing efforts on yam viruses are focused on West Africa, and Nigeria in particular. Although, yam production is less significant in other parts of the world, a basic understanding of virus threats is necessary for surveillance and to support clean seed production programmes.

6 Cowpea viruses

6.1 Introduction

Cowpea (*Vigna unguiculata* (L) Wallp.) is an annual legume that was first domesticated in Africa and is widely cultivated in semi-arid tropical West Africa for its protein-enriched seed for human consumption and the haulms that are used as fodder for livestock (Boukar et al., 2013). Although the crop is grown in all parts of the tropics and sub-tropics, the most intensive cultivation is confined to Nigeria, Niger and Burkina Faso, which collectively represent 81% of the total global production of 12.3 million ha (FAO, 2016). Viral diseases in cowpea have been recorded wherever it is cultivated (Hema et al., 2014). IITA initiated virus research on cowpea in the late 1970s with a focus on (i) distribution and diversity of viruses infecting cowpea in West Africa, (ii) breeding for virus disease resistance and (iii) development of diagnostic tools, especially for virus indexing and production of virus-free seed stocks for conservation and international distribution of cowpea genetic resources.

6.2 Understanding the diversity of viruses infecting cowpea in the field

Surveys of cowpea virus diversity confirmed the occurrence of at least ten virus species belonging to seven genera in West Africa among the more than twenty viruses recorded in cowpea around the world (Atiri et al., 1984; Biemond et al., 2013; Shoyinka et al., 1997; Thottappilly and Rossel, 1992; Hema et al., 2014). All the cowpea-infecting viruses detected in West Africa are known to have broad host ranges and infect other cultivated legumes, weeds and wild plants. Natural spread of these viruses in the field is through insect vectors and seed. However, exceptions are CGMV and SHMV, which are not seed-transmitted and SHMV which is not known to have any insect vector. The same insect vectors are involved in the transmission of diverse cowpea viruses albeit with different transmission mechanisms. CMMV, CYMV (genus *Comovirus*; family *Secoviridae*), CpMoV (genus *Carmovirus*; family *Tombusviridae*) and SBMV (genus *Sobemovirus*; family unassigned) are transmitted by leaf beetles (*Chrysomelidae* family), whereas the whitefly (*Bemisia tabaci*) transmits CGMV (genus *Begomovirus*; family *Geminiviridae*) and CPMMV (genus *Carlavirus*; family *Betaflexiviridae*), and aphids (*Aphis craccivora* Koch) are involved in the transmission of BCMV-BlCMV and CABMV (genus *Potyvirus*; family *Potyviridae*) and CMV.

The involvement of similar vectors in transmission often results in multiple infections in cowpea as revealed in several survey studies. Of the ten viruses, the most economically important ones are BCMV-BlCMV and CABMV which were found to occur in every season and which cause significant yield loss in either single or mixed infection with any other cowpea virus, especially in the presence of CMV. These infections can sometimes result in the premature death of the plant. The incidence of other viruses is known to be sporadic, although infections of these viruses, especially at early stages of growth, can result in significant reduction in yield. A peculiar case is that of CMV and CMMV which in single infection are either asymptomatic or produce only mild mottling symptoms, in spite of high virus titres in the plants. The knowledge of virus diversity has been used for planning strategies to breed for resistance to BCMV-BlCMV and CABMV, and for the development of diagnostic tools for all of the

Published by Burleigh Dodds Science Publishing, 2019.

viruses, regardless of their economic significance. The effective application of diagnostics supports germplasm exchange as well as quality seed production (as described in detail in Chapter 3).

6.3 Several viruses, one disease and the unpredictability of virus diversity in cowpea fields

When cowpea is infected by different viruses, the result is a 'mosaic disease' typified by mild to severe mosaic and yellowing symptoms on leaves. There may also be mild to severe stunting and 20–80% reduction in grain yield. Virus disease symptoms are severe in plants infected at an early stage and result in the most serious losses. Symptom type and severity is influenced by cowpea genotype and time of infection. Considerable variation in incidence and distribution of cowpea viruses were observed within and between seasons and in different cowpea growing areas (Shoyinka et al., 1997). For instance, SBMV incidence and prevalence was highest in Nigeria in surveys conducted in 1991, although this virus was not detected in surveys conducted in 1992, and its incidence was very low in 1993 (Thottappilly and Rossel, 1992). Similar situations were observed for other viruses, complicating the prediction of field virus incidence. Although extensive epidemiological investigations still need to be carried out, the type of virus and incidence observed in cowpea fields seems to depend on the seed-borne virus inoculum, vector spread and alternative hosts or volunteer plants from the previous crop (Biemond et al., 2013; Odedara and Kumar, 2016). Understanding the factors driving virus incidence and diversity is essential for enhancing preparedness to tackle virus disease outbreaks in cowpea.

6.4 Overcoming challenges to cowpea virus diagnosis

Symptoms induced by the different virus infections in cowpea lack distinctiveness to use as virus-specific diagnostic markers. This means that diagnostic tools have to be used for virus identification. As a consequence of this, the knowledge on virus diversity generated is skewed by the availability of diagnostic tests and the ability of those tests to distinguish related viruses. Surveys without the use of diagnostics at best provide incidence and severity of mosaic disease and miss out viruses that do not cause clear symptoms, such as single infections of CPMMV and CMV. These factors have affected the virus reports from various surveys, particularly those during the early years.

Transmission electron microscopy has been used for virus identification based on particle morphology. Another commonly used diagnostic method is host range studies using sap inoculation on a range of herbaceous hosts. Based on host range and particle morphology knowledge, IITA virologists procured antibodies from other research groups for confirmation of virus identity or embarked on virus purification procedures to produce antibodies which were subsequently used for diagnostics. This approach was reliable but also led to the misidentification of certain viruses. For instance, BlCMV and CABMV were initially considered as distinct serotypes of the same species. However, it was later clarified that they were separate species in the genus *Potyvirus* (Huguenot et al., 1993). Another example is the case of *Blackeye cowpea mosaic virus* (BlCMV), initially considered to be a separate species. Later work, however, revealed that it is identical to the previously described *Bean common mosaic virus* (BCMV), leading to its renaming as a cowpea strain of BCMV (BCMV-BlCMV) (Khan et al., 1993). This taxonomic confusion was effectively addressed with the better understanding of virus properties made possible by advances in virus purification, gene cloning and PCR. Current diagnosis at IITA is based on serological methods such as enzyme-linked immunosorbent assay and reverse transcription-PCR. All the important viruses have been characterized by genomic sequencing to assign them to exact taxonomic groups. This improved understanding of diversity has led directly to improved diagnostics.

6.5 Cowpea virus control through resistant varieties

Breeding for virus resistance has been the main approach used by IITA for cowpea virus disease control. Although virus resistance breeding was not the primary target, the evaluation of germplasm and breeding lines for virus response was considered as part of the process of variety development and any promising lines selected were advanced. Considering their widespread occurrence, BCMV-BlCMV and CABMV were identified as priority targets (Boukar et al., 2013). Inconsistencies in infection rate encountered during field screening experiments led to IITA establishing protocols for screenhouse-based screening for targeted viruses. This improvement led to the identification of several highly resistant cowpea lines that were used in breeding programmes to incorporate resistance to potyviruses (e.g. IT98K-1092-1). However, recent studies have shown that resistance selected against potyviruses is compromised in co-infection with other viruses (e.g. CMV) due to the phenomenon known as synergistic interaction. This finding led to the modification of the screening procedure to include simultaneous testing of genotypes under single and mixed infections for selected lines with the most promising levels of resistance.

6.6 Future prospects

Despite excellent progress, cowpea remains vulnerable to virus epidemics as most of the popular landraces and improved varieties lack durable resistance to even the most commonly occurring viruses. It is fortunate that there are no records of cowpea virus disease epidemics in West Africa or elsewhere in Africa. However, it is unlikely that this situation will remain the same given the rapid changes occurring within farming systems including crop intensification. These changes contribute to breaking the natural seasonal barriers that check the perpetuation of disease cycles. Additionally, it seems likely that anthropogenic climate change will make conditions increasingly conducive for vectors and disease establishment. In order to counter these threats, it is important for IITA to perform scenario analysis by considering various risk situations. This should improve understanding of the likely impact of virus diseases on cowpea production and enable the preparation of informed mitigation strategies. Also, efforts should be made to conduct comprehensive baseline studies on the status of viruses in Southern Africa, where cowpea is fast emerging as a dual-purpose legume for food and fodder use. Cowpea collections along with wild relatives need to be phenotyped in order to select robust resistance to multiple virus infections and investments should be made to tap advanced technologies for cowpea virus disease surveillance, diagnostics, phenotyping and the development of genomic markers for accelerated breeding for virus resistance.

7 Soybean viruses

7.1 Introduction

Soybean (*Glycine max* L. Merr.) is widely grown in Nigeria and several countries in Southern Africa. The industrial potential of this crop is making it attractive for farmers and as a consequence the soybean production area in many countries is rapidly increasing. From 2006 to 2016 land area under soybean increased 32.5% in Africa (FAO, 2016). Soybean virology has not been strongly prioritized at IITA, and research activities have been limited primarily to seed health testing.

7.2 Occurrence, diversity and incidence of soybean viruses

Although virus symptomatic plants have been noted in many soybean fields, the incidence and severity of such plants has often been regarded as being below the economic threshold. The first comprehensive surveys for soybean viruses were conducted in Nigeria between 2008 and 2010 (Time et al., 2010). This study revealed a very low incidence of viruses in most soybean production areas in Nigeria. However, in Benue State, which was the first state to commercialize soybean cultivation over 45 years ago, and where soybean cultivation is most intense, a high incidence and diversity of viruses was recorded. This suggests that endemic viruses have adapted to African soybean production systems, perhaps due to the movement of vectors between crops. Over 15 viruses were detected in soybean in Nigeria, including two novel begomoviruses,

Soybean mild mottle virus (SbMMV) (genus *Begomovirus*; family *Geminiviridae*) and *Soybean chlorotic blotch virus* (SbCBV) (genus *Begomovirus*; family *Geminiviridae*) (Alabi et al., 2010), and the first report of ACMV on soybean (Mgbechi-Ezeri et al., 2008).

Several viruses detected in soybean are known to infect cowpea as well, and they are transmitted by aphids, beetles and whiteflies. Virus infection results in foliar symptoms such as mosaic and mottling, leaf distortion, severe reduction in leaf size and stunting of plants. Mixed infection with more than one virus is common under field conditions. *Cowpea mild mottle virus* (CPMMV) was the most frequently detected virus in soybean. Unlike in cowpea, CPMMV causes severe mosaic symptoms similar to *Soybean mosaic virus* (SMV). Evaluation of over 30 soybean cultivars, released or part of advanced yield evaluation trials for release, revealed high susceptibility to CPMMV (Time et al., 2010).

7.3 Future prospects

Future efforts should focus on breeding for CPMMV resistance because of its widespread occurrence and the ability of this virus to cause serious yield reduction. Future priorities for soybean virology are similar to those indicated for cowpea. Although virus epidemics in soybean have not yet been recorded anywhere in Africa, studies have shown the widespread susceptibility of soybean germplasm to endemic virus infections in Africa. Changes in soybean farming systems associated with agricultural intensification and climate change could alter the balance, exposing the crop to a greater risk of damage from viruses. Currently, there are no measures in place to tackle such an eventuality. This should be addressed by performing scenario analyses considering various levels of risks with the aim of developing soybean virus disease risk control strategies.

8 Virus vectors

8.1 Introduction

Most plant virus species use an arthropod vector as a means of transportation from one host to another. The insect vectors of plant viruses mostly belong to three orders: Hemiptera, Coleoptera and Thysanoptera. Hemiptera comprise the majority of the virus vectors in the families Aphididae (aphids), Aleyrodidae (whiteflies), Cicadellidae (leafhoppers), Fulgoridae (planthoppers), Psyllidae (psyllids), Pseudococcidae (mealybugs) and Coccidae (soft scales). Aphids, whiteflies and leafhoppers transmit more than 80% of plant viruses which represents close to 400 virus species within 39 different genera (Fereres and Moreno, 2009). In this section we provide a short summary of some of the key research findings on two of the most important vector groups affecting the IITA mandate crops, namely the whitefly vector of cassava viruses and the leafhopper vectors of MSV.

8.2 The whitefly vector of cassava viruses

Both the CMBs causing CMD and the CBSIs causing CBSD are transmitted by the whitefly *Bemisia tabaci* (Storey and Nichols 1938; Dubern 1994; Maruthi et al. 2005; 2017). *Bemisia tabaci* is a cryptic species complex that is currently reputed to have at least 34 morphologically indistinguishable species (De Barro 2012; Tay et al. 2012). The cassava-colonizing species have five major haplogroups SSA1 to SSA5 with SSA1 further divided into five subgroups SSA1-SG1 to SG5 based on partial sequences of the mitochondrial cytochrome oxidase I (mtCOI) gene (Ghosh et al. 2015; Legg et al., 2014a). Recently, these were reclassified into six major haplogroups (SSA-ECA, SSA-ESA, SSA-CA, SSA-WA, SSA2 and SSA4) based on SNP genotyping (Wosula et al. 2017).

Bemisia tabaci has six developmental stages: egg, four nymphal instars and the adult. All but the adult stage and brief, early first instar crawler are functionally immobile (Byrne and Bellows 1991). Reproduction is arrhenotokous; unmated females produce males and mated females produce both males and females (Byrne and Bellows 1991). *B. tabaci* is a multivoltine insect that has no diapause or quiescent stage. As a result, populations are sustained through the continual exploitation of multiple host resources, both wild and cultivated, over the annual cycle (Byrne and Bellows 1991). Mean development duration varies depending on the region, host plant and prevailing conditions (Byrne and Bellows 1991). On cassava, *B. tabaci* completes its life cycle in about 21 days with prevailing temperatures of 26.6°C to 31.5°C (Fishpool et al. 1995).

Bemisia tabaci transmits viruses mostly in the genus *Begomovirus* in a persistent, circulative and non-propagative manner (Rosen et al. 2015). These viruses are acquired through phloem feeding and require a latent period (time interval between acquisition and transmission) to translocate from the insect midgut epithelial cells into the haemolymph, and to the salivary glands from which they are inoculated into the phloem with saliva (Brown and Czosnek 2002). Begomoviruses such as CMBs are thought to be retained in the whitefly vector for its entire life, and whiteflies are capable of inoculating plants in as little as 10 min (Dubern, 1994). *Bemisia tabaci* transmits viruses in the genus *Ipomovirus* such as the CBSIs in a semi-persistent manner (Ng and Falk, 2006; Maruthi et al., 2017).

Severe CMD pandemics in the 1990s led to the inclusion of cassava whiteflies in a major project, The Tropical Whitefly IPM Project, an initiative of the System-wide IPM Programme of the Consultative Group on International Agricultural Research (CGIAR). Under this project IITA was mandated to carry out baseline surveys in SSA to identify whiteflies colonizing cassava and their parasitoids with the aim of using this information to understand the dynamics of CMD and development of control tactics (Legg 2005; Sseruwagi et al. 2005). In nine major cassava-growing countries that were sampled, the most dominant whiteflies on cassava were *B. tabaci* and *B. afer* (Priesner and Hosny), with the latter not known to transmit cassava viruses. Two major parasitoid groups *Encarsia* spp. and *Eretmocerus* spp. (both Hymenoptera Aphelinidae) were also reported to parasitize *B. tabaci*. The information generated from this project laid the foundation for the numerous studies on genetic diversity of cassava-colonizing *B. tabaci* and its parasitoids, and its role in the spread of cassava virus pandemics.

There are two major groups of *B. tabaci* in Africa that are referred to as cassava-colonizing and non-cassava-colonizing. Studies in East and West Africa in the 1990s showed that *B. tabaci* found colonizing cassava were largely restricted to this plant and its close relatives (Burban et al 1992; Legg 1996). The severe CMD pandemic in the 1990s was driven by super-abundant *B. tabaci* populations (>100 adults per top five leaves) that contributed to rapid disease spread (20–30 km/year) (Legg and Ogwal 1998). MtCOI sequencing of *B. tabaci* specimens collected along transects from CMD pandemic to non-pandemic locations in 1997 revealed the occurrence of two major haplogroups designated Ug1 and Ug2, with the latter found to be abundant in locations with severe disease (Legg et al. 2002), although several years later, Ug2 was not detected, having apparently been displaced by Ug1 (Sseruwagi et al. (2006).

A study on genetic diversity of cassava *B. tabaci* collected from six countries in East and Central Africa from 1997 to 2010 reported the existence of six haplogroups under the SSA group, namely SSA1-SG1 (formerly Ug1), SSA1-SG2, SSA1-IntSG1/2, SSA1-SG4, SSA2 (formerly Ug2) and SSA3 (Legg et al. 2014a). This study demonstrated that although there was an initial association between the CMD pandemic and the SSA2 haplotype (1997–9), the SSA1-SG1 haplogroup eventually replaced SSA2, becoming the primary haplogroup associated with the spread of the pandemic, and expanding its geographical range coincident with the CMD pandemic from 2000 to 2010 (Legg et al. 2014a). Recently, using SNP genotyping, Wosula et al. (2017) reported that mtCOI is not reliable at discriminating between cassava *B. tabaci* genotypes. Based on this finding, cassava *B. tabaci* were reclassified into six major haplogroups (SSA-ECA, SSA-ESA, SSA-CA, SSA-WA, SSA2 and SSA4), and these results will be used to develop robust SNP-based diagnostic tools. There remains a strong geographical association of one of the haplogroups (SSA-ECA) with regions affected by the severe CMD and CBSD pandemics, although it is likely that there are other non-genetic factors associated with the super-abundance phenomenon. Tajebe et al. (2015) noted a low frequency of the bacterial endosymbiont *Arsenophonus* in *B. tabaci* populations occurring in areas of Tanzania affected by the severe CMD pandemic, and Ghosh et al. (2018) have demonstrated the deleterious effects of *Arsenophonus* infection on both fitness parameters and virus transmission of cassava-colonizing *B. tabaci*.

8.3 Control of Bemisia tabaci

Although insecticides are widely used to control *B. tabaci* in commercial agriculture, their use in cassava systems is confined to experimental trials and large-scale 'clean seed' production sites, but rarely used amongst smallholder farms in Africa (Legg et al. 2014b). At least eleven species of parasitoids, mainly in the *Encarsia* and *Eretmocerus* genera, have been used in attempts to control *B. tabaci* through augmentation (Gerling et al., 2001). Two detailed studies on parasitoids of cassava-colonizing *B. tabaci* in Uganda and Tanzania reported the prevalence of *Encarsia* spp. and *Eretmocerus* spp., with *Encarsia sophia* (Girault and Dodd) and *Eretmocerus mundus* Mercet the most dominant species (Guastella et al. 2015; Otim et al. 2006). Entomopathogen/mycoinsecticide products based on *Lecanicillium lecanii* R. Zare & W. Gams (order Hypocreales; family *Cordycipitaceae*), *Isaria fumosorosea* Wize (order Hypocreales; family *Clavicipitaceae*) and *Beauveria bassiana* Bals.-Criv. (order Hypocreales; family *Cordycipitaceae*) have been explored and used under greenhouse and field conditions to control whiteflies including *B. tabaci* (Faria and Wraight 2001). So far, no literature is available on identification of entomopathogens and their efficacy against *cassava B. tabaci* in Africa. Host plant resistance against *B. tabaci* has been aggressively pursued in crops such as tomato, soybean, cotton, alfalfa, cucurbits and pepper (Nombela and Muñiz 2010), although there has been little significant progress in identifying sources of resistance to *B. tabaci* on cassava in Africa. Although there are several potential control options for whiteflies on cassava in Africa, none are currently being regularly used. Insecticides are too expensive, whilst there are currently still no practical applications for the use of parasitoids, entomopathogens or host plant resistance. Further research is therefore required to identify solutions that can be effective and practical in the context of SSA, where most cassava producers are smallholder farmers.

8.4 The leafhopper vectors of MSV

Twenty-two species of *Cicadulina* are recognized, of which 18 occur in Africa (Webb, 1987), which is presumed to be the centre of origin of this genus (Rose, 1978). Nine of the species in Africa have been reported as vectors of MSV. Studies in Nigeria demonstrated that *C. mbila* (Naudé) is the most abundant vector across major ecological zones (forest, southern Guinea savanna and northern Guinea savanna) and seasons (Asanzi et al., 1994).

MSV is transmitted in a persistent circulative manner. The behaviour of *Cicadulina* on maize varieties exhibiting resistance to the vector in addition to the virus may help reduce MSV spread in the field. For viruses transmitted by aphids in a non-persistent manner, frequent probing is likely to increase virus transmission. However, for a persistent virus such as MSV, short-duration probes that do not reach the phloem would not enhance virus transmission (Bosque-Pérez, 2000).

Cicadulina leafhoppers are generally regarded as grassland species. In West Africa, the greatest numbers of *Cicadulina* leafhoppers were detected on the annuals *Brachiaria deflexa* (Schumach.) (Poaceae), *B. lata*, *Scutellaria barbata* D. Don (Lamiaceae), *D. horizontalis* and *Eleusine indica* (L.) (Poaceae) (Mesfin et al., 1992; Asanzi et al., 1994). Leafhoppers were found also on *Pennisetum* spp. and *Rottboellia cochinchinensis* (Lour.) (Poaceae), but only towards the end of the rainy season in Guinea savanna locations.

Regular surveys of MSV and leafhopper vectors in maize and grasses in farmers' fields and research stations across ecological zones in Nigeria revealed differences in leafhopper species composition (Okoth et al., 1988; Dabrowski et al., 1991). Five *Cicadulina* species, all known to be vectors of MSV were found, although *C. mbila* was the most abundant vector across ecological zones and seasons (Asanzi et al., 1994). In the southern and northern Guinea savannas, leafhopper populations appear to build-up with the rains as grass hosts suitable for feeding and reproduction become available. Compared to northern locations, the humid forest locations in the south often have consistently low leafhopper populations during most of the year (Bosque-Pérez, 2000).

Asanzi et al. (1994) showed that the percentages of viruliferous insects were low at the beginning of the rainy period (0–4%), and higher (15–22%) as the season progressed. As the rainy season progressed and disease incidence increased, the probability of insects feeding on an infected plant and becoming infective increased. In these surveys, however, high percentages of infective leafhoppers did not correspond with high leafhopper densities, but rather occurred immediately following leafhopper population density peaks.

Systemic insecticides have been shown to control leafhopper vectors of MSV, but ongoing influxes of the insects into treated fields from surrounding vegetation mean that repeated applications are required for sustained control (Martin and Shepherd, 2009). Since the majority of affected farmers are smallholders, very few have the means to pay for the expense of chemical vector control. Management of MSV therefore focuses almost entirely on control of the virus through the development and deployment of resistant varieties.

9 Emerging techniques and future perspectives for plant virus management in Africa

In the past 50 years, conventional approaches have been successfully applied to tackle viral diseases in Africa. In particular, selection of resistant genotypes and improved understanding of the pathosystems have significantly contributed to controlling viral diseases. However, in the last two decades, important technological advances have opened up new possibilities for gaining a deeper understanding of viral diseases at the molecular level. High-throughput sequencing, such as next-generation sequencing (NGS), has become a key technique, essential for innovative research with impact. Nowadays genotyping of hundreds of plants, viruses and their vectors is routine work that can be executed, with the appropriate infrastructure, in as little as two weeks at an affordable cost (Gore et al., 2009). Furthermore, NGS allows highly accurate detection of any known virus and also facilitates the discovery of unknown or unexpected viruses (Carvajal-Yepes et al., 2014). Importantly, NGS technology can be applied reliably to any organism or virus of interest.

The availability of whole-genome sequences coupled with NGS technologies has increased the volume and speed of genetic marker discovery. Functional genomic studies have enabled the identification of loci and sometimes genes responsible for resistance to some of the most severe African viral diseases described above, including the important CMD2 and MSV1 QTLs for resistance to CMD and MSV, respectively (Rabbi et al., 2014; Wolfe et al., 2016; Masumba et al., 2017; Umber et al., 2016; Nair et al., 2015; Welz et al., 1998). This has allowed the development and use of molecular markers leading to the acceleration of breeding programmes as well as the production of resistant transgenic lines (Zhang et al., 2005; Vanderschuren et al., 2012; Shepherd et al., 2007).

Although genetic differences have until now been considered to be the main source of susceptibility/tolerance to variations in both virus and host crop populations, evidence is emerging to support the existence of additional components of heritability linked to epigenetic modification of DNA and chromatin, which operate in parallel with genetics to influence heritable variation, adaptation and defence in plants. There are several examples of this, and one which has been confirmed in cassava is where a loss of resistance to CMD has been shown to be associated with changes in the genomic methylation profile. Other examples involve gene regulation and RNA function.

The involvement of RNA has proved to be central to the potential for heritable genetic and epigenetic variation because, unlike DNA sequences that do not vary, the expression of RNA can change under the influence of biotic and abiotic stresses and explain some environmental traits in crop selection. Comparative transcriptomics analysis of healthy and virus-infected genotypes or genotypes with different levels of susceptibility to virus infections is proving to be a straightforward strategy to identify genes regulating plant tolerance to viral diseases (Amuge et al., 2017; Anjanappa et al., 2017; Allie et al., 2014). Finally, the discovery of the mechanism of post-transcriptional gene silencing that regulates gene expression by means of small RNA molecules (namely microRNAs and siRNAs) and the elucidation of the central role of such a mechanism in plant–virus–vector interaction during virus infection has opened up new perspectives to addressing viral disease control (Bond and Baulcombe, 2014). An example of this is the siRNA profiling of virus-infected cassava plants that has contributed to clarifying differences in host and viral behaviours during infection (Ogwok et al., 2016; Chen et al., 2015; Wagaba et al., 2017; Xia et al., 2016).

The deeper application of such new knowledge and modern technologies combined with conventional strategies promises to uncover critical aspects of plant–virus–vector pathosystems and to provide new tools to face present and future viral outbreaks.

10 Future trends and conclusion

During the first two decades of virus research at IITA, most of the focus was on providing support to breeders in identifying and incorporating sources of virus resistance into IITA's mandate crops. Great successes were achieved, and many of the virus-resistant varieties developed during this period are still being actively disseminated and grown by farmers today. The development and application of diverse virus diagnostic techniques has provided vital tools to help researchers conduct monitoring and surveillance studies, as well as to assure the production of healthy seeds and vegetatively propagated planting material. During the second part of IITA's 50+ years of virus research, the institute dedicated substantial resources towards the 'fight' against a series of continentally important virus disease pandemics, including severe CMD, CBSD, BBTV and MLN. A key element of several of these efforts has been the establishment of procedures for the production and dissemination of high quality certified seed/planting material of varieties with resistance to each of these major diseases. This work has had a truly transformational effect on the production of cassava, bananas and maize – some of Africa's most important food staple crops.

Key challenges for the future are likely to be the development and application of next-gen technologies for virus characterization and diagnostics development; maximizing the potential for using IT and AI systems for point-of-use disease diagnosis, area-wide surveillance and early warning systems;

Published by Burleigh Dodds Science Publishing, 2019.

releasing the potential for transgenics and gene editing technologies to supplement existing sources of virus resistance obtained through conventional breeding; and putting in place pre-emptive crop virus management strategies to replace the current reactive approaches. Finally, IITA will need to play a lead role within Africa in establishing global networks to develop solutions to virus problems that show an increasing propensity to move between continents. The BBTV Global Alliance is an excellent example of such a network that should serve as a paradigm for similar networks focusing on other crop/virus disease pathosystems.

11 Where to look for further information

There are several books which provide good general coverage of the topic of crop viruses in Africa, and which also feature significant contributions from scientists working at IITA.

Whitefly and Whitefly-borne Viruses in the Tropics: Building a Knowledge Base for Global Action, edited by Anderson, P. K. and Morales, F. J.; with the collaboration of Jones, A. L. and Markham, R. H., Centro Internacional de Agricultura Tropical (CIAT), Cali, Colombia, 2005. 351p. - (CIAT publication no. 341) ISBN 958-694-074-8.

Virus and Virus-like Diseases of Major Crops in Developing Countries, Thottappilly, G. and Loebenstein, G. (Gad), Kluwer Academic, Dordrecht and London, 2003.

Plant Virology in sub-Saharan Africa: Proceedings of a Conference Organized by IITA, edited by Hughes, J. d'A. and Odu, B. O., 4–8 June 2001, International Institute of Tropical Agriculture, Ibadan, Nigeria. https://books.google.com.ng/books?id=NbLlvMhVxxsC&pg=PA549&lpg=PA549&dq=plant+virology+in+sub-saharan+africa&source=bl&ots=xoOHvTZ65q&sig=hiR8yLBvgJeyxxlVqEfkRsvPS6s&hl=en&sa=X&ved=2ahUKEwjRyZSAxuDeAhVFsaQKHTa4BVIQ6AEwCHoECAgQAQ#v=onepage&q=plant%20virology%20in%20sub-saharan%20africa&f=false

Integrated Pest Management in Tropical Regions, edited by Rapisarda, C. and Massimino Cocuzza, G. E., CAB International, Wallingford, Oxford, UK.

Achieving Sustainable Cultivation of Maize, Volume 2: Cultivation Techniques, Pest and Disease Control, edited by Watson, D., Burleigh Dodds Science Publishing, Sawston, Cambridge, UK, 2017.

Achieving Sustainable Cultivation of Cassava, Volume 2: Genetics, Pests and Diseases, edited by Hershey, C., Burleigh Dodds Science Publishing, Sawston, Cambridge, UK, 2017.

Olmos, A., Boonham, N., Candresse, T., Gentit, P., Giovani, B., Kutnjak, D., Liefting, L., Maree, H. J., Minafra, A., Moreira, A., Nakhla, M. K., Petter, F., Ravnikar, M., Rodoni, B., Roenhorst, J. W., Rott, M., Ruiz-García, A. B., Santala, J., Stancanelli, G., van der Vlugt, R., Varveri, C., Westenberg, M., Wetzel, T., Ziebell, H. and Massart, S., 'High-throughput sequencing technologies for plant pest diagnosis: Challenges and opportunities', *EPPO Bulletin* 48, no. 2(2018), 219–24. https://doi.org/10.1111/epp.12472

Thomas, J. H., 'High-throughput sequencing of cytosine methylation in plant DNA', *Plant Methods* 9, no. 1(2013), 16. doi:10.1186/1746-4811-9-16.

General websites of relevance to the topic of this chapter:

https://www.iita.org
https://www.plantwise.org/#
https://talk.ictvonline.org/

12 References

Adams, I. P., Abidrabo, P., Miano, D. W., Alicai, T., Kinyua, Z. M., Clarke, J., Macarthur, R., Weekes, R., Laurenson, L., Hany, U., Peters, D., Potts, M., Glover, R., Boonham, N. and Smith, J. (2013), 'High throughput real-time RT-PCR assays for specific detection of cassava brown streak disease causal viruses, and their application to testing of planting material', *Plant Pathology* 62, 233–42.

Adams, I. P., Harju, V. A., Hodges, T., Hany, U., Skelton, A., Rai, S., Deka, M. K., Smith, J., Fox, A., Uzayisenga, B., Ngaboyisonga, C., Uwumukiza, B., Rutikanga, A., Rutherford, M., Ricthis, B., Phiri, N. and Boonham, N. (2014), 'First report of maize lethal necrosis disease in Rwanda', *New Disease Reports* 29, 22.

Akinbade, S. A., Hanna, R., Nguenkam, A., Njukwe, E., Fotso, A., Doumtsop, A., Ngeve, J., Tenku, S. T. N. and Lava Kumar, P. (2010), 'First report of the *East African cassava mosaic virus*-Uganda (EACMV-UG) infecting cassava (*Manihot esculenta*) in Cameroon', *New Disease Reports* 21, 22.

Alabi, O. J., Kumar, P. L., Mgbechi-Ezeri, J. U. and Naidu, R. A. (2010), 'Two new "legumoviruses" (genus *Begomovirus*) naturally infecting soybean in Nigeria', *Archives of Virology* 155, 643–56.

Alicai, T., Omongo, C. A., Maruthi, M. N., Hillocks, R. J., Baguma, Y., Kawuki, R., Bua, A., Otim-Nape, G. W. and Colvin, J. (2007), 'Re-emergence of cassava brown streak disease in Uganda', *Plant Disease* 91, 24–9.

Allie, F., Pierce, E. J., Okoniewski, M. J. and Rey, C. (2014), 'Transcriptional analysis of South African cassava mosaic virus-infected susceptible and tolerant landraces of cassava highlights differences in resistance, basal defense and cell wall associated genes during infection', *BMC Genomics* 15, 1006.

Amuge, T., Berger, D. K., Katari, M. S., Myburg, A. A., Goldman, S. L. and Ferguson, M. E. (2017), 'A time series transcriptome analysis of cassava (*Manihot esculenta* Crantz) varieties challenged with *Ugandan cassava brown streak virus*', *Scientific Reports* 7, 9747.

Anjanappa, R. B., Mehta, D., Okoniewski, M. J., Szabelska-Beręsewicz, A., Gruissem, W. and Vanderschuren, H. (2017), 'Molecular insights into *Cassava brown streak virus* susceptibility and resistance by profiling of the early host response', *Molecular Plant Pathology* 19, 476–89.

Asanzi, C. M., Bosque-Pérez, N. A., Buddenhagen, I. W., Gordon, D. T. and Nault, L. R. (1994), 'Interactions among maize streak virus disease, leafhopper vector populations and maize cultivars in forest and savanna zones of Nigeria', *Plant Pathology* 43, 145–57.

Atiri, G. I., Ekpo, E. J. A. and Thottappilly, G. (1984), 'The effect of aphid resistance in cowpea on infestation and development of *Aphis craccivora* and the transmission of cowpea aphid-borne mosaic virus', *Annals of Applied Biology* 104, 339–46.

Balogun, M., Maroya, N., Augusto, J., Ajayi, A., Kumar, P. L., Aighewi, B. and Asiedu, R. (2017), 'Relative efficiency of positive selection and tissue culture for generating pathogen-free planting materials of yam (*Dioscorea* spp.)', *Czech Journal of Genetics and Plant Breeding* 53, 9–16.

Beijerinck, M. W. (1898), 'Über ein Contagium vivum fluidum als Ursache der Fleckenkrankheit der Tabaksblätter', *Verhandelingen der Koninklijke Akademie van Wetenschappen te Amsterdam* (in German) 65, 1–22. Translated into English in Johnson, J. (Ed.) (1942), *Phytopathological Classics*, American Phytopathological Society, St. Paul, MN, No. 7, pp. 33–52.

Biemond, P. C., Oguntade, O., Kumar, P. L., Stomph, T. J., Termorshuizen, A. and Struik, P. (2013), 'Does the informal seed system threaten cowpea seed health?', *Crop Protection* 43, 166–74.

Bigirimana, S., Barumbanze, P., Obonyo, R. and Legg, J. P. (2004), 'First evidence for the spread of *East African cassava mosaic virus*– Uganda (EACMV-UG) and the pandemic of severe cassava mosaic disease to Burundi', *Plant Pathology* 53, 231.

Bigirimana, S., Barumbanze, P., Ndayihanzamaso, P., Shirima, R. and Legg, J. P. (2011), 'First report of cassava brown streak disease and associated *Ugandan cassava brown streak virus* in Burundi', *New Disease Reports* 24, 26.

Bömer, M., Turaki, A., Silva, G., Kumar, P. L. and Seal, S. E. (2016), 'A sequence-independent strategy for amplification and characterization of episomal badnavirus sequences reveals three previously uncharacterized yam badnaviruses', *Viruses* 8, 188.

Bond, D. M. and Baulcombe, D. C. (2014), 'Small RNAs and heritable epigenetic variation in plants', *Trends in Cell Biology* 24, 100–7.

Bosque-Pérez, N. A. (2000), 'Eight decades of *Maize streak virus* research', *Virus Research* 71, 107–21.

Bosque-Pérez, N. A. and Buddenhagen, I. W. (1999), 'Biology of *Cicadulina* leafhoppers and epidemiology of maize streak virus disease in West Africa', *South African Journal of Plant Soil* 16, 50–5.

Bosque-Pérez, N. A., Olojede, S. O. and Buddenhagen, I. W. (1998), 'Effect of maize streak virus disease on the growth and yield of maize as influenced by varietal resistance levels and plant stage at time of challenge', *Euphytica* 101, 307–17.

Boukar, O., Bhattacharjee, R., Fatokun, C., Kumar, P. L. and Gueye, B. (2013), 'Cowpea', in *Genetic and Genomic Resources of Grain Legume Improvement*, Singh, M., Upadhyaya, H. D. and Bisht, I. S. (Eds.), Elsevier, London, UK, pp. 137–56.

Brown, J. K. and Czosnek, H. (2002), 'Whitefly transmission of plant viruses', *Advances in Botanical Research* 36, 65–100.

Buddenhagen, I. W. and Bosque-Pérez, N. A. (1999), 'Historical overview of breeding for durable resistance to *Maize streak virus* for tropical Africa', *South African Journal of Plant Soil* 16, 106–11.

Burban, C. L. D. C., Fishpool, L. D. C., Fauquet, C., Fargette, D. and Thouvenel, J.-C. (1992), 'Host-associated biotypes within West African populations of the whitefly *Bemisia tabaci* (Genn.), (Hom., Aleyrodidae)', *Journal of Applied Entomology* 113, 416–23.

Byrne, D. N. and Bellows, T. S. (1991), 'Whitefly biology', *Annual Reviews of Entomology* 36, 431–57.

Cabanas, D., Watanabe, S., Higashi, C. H. and Bressan, A. (2013), 'Dissecting the mode of *Maize chlorotic mottle virus* transmission (Tombusviridae: Machlomovirus) by *Frankliniella williamsi* (*Thysanoptera: Thripidae*)', *Journal of Economic Entomology* 106, 16–24.

Calvert, L. A. and Thresh, J. M. (2002), 'The viruses and virus diseases of cassava', in *Cassava: Biology, Production and Utilization*, Bellotti, A. C., Hillocks, R. J. and Thresh, J. M. (Eds), CABI, Wallingford, UK, pp. 237–60.

Carvajal-Yepes, M., Olayaa, C., Lozanoa, I., Cuervob, M., Castañoa, M. and Cuellar, W. J. (2014), 'Unraveling complex viral infections in cassava (*Manihot esculenta* Crantz) from Colombia', *Virus Research* 186, 76–86.

Chen, S., Huang, Q., Wu, L. and Qian Y. (2015), 'Identification and characterization of a maize-associated mastrevirus in China by deep sequencing small RNA populations', *Virology Journal* 12, 156.

Dabrowski, Z. T., Nwilene, F. and Kumar, R. (1991), 'First regular observations on leafhoppers, *Cicadulina* spp., vectors of *Maize streak virus* in southeastern Nigeria', *Insect Science and its Application* 12, 249–61.

De Barro, P. J. (2012), 'The *Bemisia* species complex: Questions to guide future research', *Journal of Integrative Agriculture* 11, 187–96.

De Groote, H., Oloo, F., Tongruksawattana, S. and Biswanath, D. (2016), 'Community-survey based assessment of the geographic distribution and impact of maize lethal necrosis (MLN) disease in Kenya', *Crop Protection* 82, 30–5.

Dubern J. (1994), 'Transmission of African cassava mosaic geminivirus by the whitefly (*Bemisia tabaci*)', *Tropical Science* 34, 82–91.

Efron, Y., Kim, S., Fajemisin, J., et al. (1989), 'Breeding for resistance to *Maize streak virus*: A multi-disciplinary team approach', *Plant Breeding*, 103: 1–36.

Eni, A. O., Hughes, J. d'A. and Rey, M. E. C. (2008), 'Survey of the incidence and distribution of five viruses infecting yam in the major yam producing zones in Benin', *Annals of Applied Biology* 153, 223–32.

Fajemisin, J. M., Cook, G. E., Okusanya, F. and Shoyinka, S. A. (1976), '*Maize streak virus* epiphytotic in Nigeria', *Plant Disease Reporter* 60, 443–7.

FAO. (2016), '*FAO Cowpea and Soybean Production Statistics for 2016*', http://www.fao.org/faostat/en/#data/QC (accessed 26 May 2018).

Faria, M. and Wraight, S. P. (2001), 'Biological control of *Bemisia tabaci* with fungi', *Crop Protection* 20, 767–78.

Fereres, A. and Moreno, A. (2009), 'Behavioural aspects influencing plant virus transmission by homopteran insects', *Virus Research*, 141, 158–68.

Fishpool, L. D. C., Fauquet, C., Fargette, D., Thouvenel, J.-C., Burban, C. and Colvin, J. (1995), 'The phenology of *Bemisia tabaci* (Homoptera: Aleyrodidae) populations on cassava in southern Côte d'Ivoire', *Bulletin of Entomological Research* 85, 197–207.

Gerling, D., Alomar, O. and Arnò, J. (2001), 'Biological control of *Bemisia tabaci* using predators and parasitoids', *Crop Protection* 20, 779–99.

Ghosh, S., Bouvaine, S. and Maruthi, M. N. (2015), 'Prevalence and genetic diversity of endosymbiotic bacteria infecting cassava whiteflies in Africa', *BMC Microbiology* 15, 93.

Ghosh, S., Bouvaine, S., Richardson, S. C. W., Ghanim, M. and Maruthi, M. N. (2018), 'Fitness costs associated with infections of secondary endosymbionts in the cassava whitefly species *Bemisia tabaci*', *Journal of Pest Science* 91, 17–28.

Gore, M. A., Chia, J. M., Elshire, R. J., Sun, Q., Ersoz, E. S., Hurwitz, B. L., Peiffer, J. A., McMullen, M. D., Grills, G. S., Ross-Ibarra, J., Ware, D. H. and Buckler, E. S. (2009), 'A first-generation haplotype map of maize', *Science* 326, 1115–17.

Gowda, M., Das, B., Makumbi, D., Babu, R., Semagn, K., Mahuku, G., Olsen, M. S., Bright, J. M., Beyene, Y. and Prasanna, B. M. (2015), 'Genomewide association and genomic prediction of resistance to maize lethal necrosis disease in tropical maize germplasm', *Theoretical Applied Genetics* 128, 1957–68.

Guastella, D., Lulah, H., Tajebe, L. S., Cavalieri, V., Evans, G. A., Pedata, P. A., Rapisarda, C. and Legg, J. P. (2015), 'Survey on whiteflies and their parasitoids in cassava mosaic pandemic areas of Tanzania using morphological and molecular techniques', *Pest Management Science* 71, 383–94.

Hahn, S. K., Terry, E. R. and Leuschner, K. (1980), 'Breeding cassava for resistance to cassava mosaic disease', *Euphytica* 29, 673–83.

Harper, G., Osuji, J. O., Heslop-Harrison, J. S. P. and Hull, R. (1999), 'Integration of banana streak badnavirus into the *Musa* genome: Molecular and cytogenetic evidence', *Virology* 255, 207–13.

Harrison, B. D., Zhou, X., Otim-Nape, G. W., Liu, Y. and Robinson, D. J. (1997), 'Role of a novel type of double infection in the geminivirus-induced epidemic of severe cassava mosaic in Uganda', *Annals of Applied Biology* 131, 437–48.

Hema, M., Sreenivasulu, P., Patil, B. L., Kumar, P. L. and Reddy, D. V. R. (2014), 'Tropical food legumes: Virus diseases of economic importance and their control', *Advances in Virus Research* 90, 431–505.

Huguenot, C., Furneaux, M. T., Thottappilly, G., Rossel, H. W. and Hamilton, R. I. (1993), 'Evidence that Cowpea aphid-borne mosaic and Blackeye cowpea mosaic are two different potyviruses', *Journal of General Virology* 74, 335–40.

Isabirye, B. E. and Rwomushana, I. (2016), 'Current and future potential distribution of *Maize chlorotic mottle virus* and risk of maize lethal necrosis disease in Africa', *Journal of Crop Protection* 5, 215–28.

Jensen, S. G. (1985), 'Laboratory transmission of *Maize chlorotic mottle virus* by three species of corn rootworms', *Plant Disease* 69, 864–8.

Jensen, S. G., Wysong, D. S., Ball, E. M. and Higley, P. M. (1991), 'Seed transmission of *Maize chlorotic mottle virus*', *Plant Disease* 75, 497–8.

Jiang, X. Q., Meinke, L. J., Wright, R. J., Wilkinson, D. R. and Campbell, J. E. (1992), '*Maize chlorotic mottle virus* in Hawaiian-grown maize: Vector relations, host range and associated viruses', *Crop Protection* 11, 248–54.

Jones, G. H. and Mason, T. G. (1926), 'On two obscure diseases of cotton', *Annals of Botany* 40, 759–72.

Jones, M., Boyd, E. and Redinbaugh, M. (2011), 'Responses of maize (*Zea mays* L.) near isogenic lines carrying Wsm1, Wsm2, and Wsm3 to three viruses in the Potyviridae', *Theoretical Applied Genetics* 123, 729–40.

Khan, J. A., Lohuis, D., Goldbach, R. and Dijkstra, J. (1993), 'Sequence data to settle the taxonomic position of *Bean common mosaic virus* and *Blackeye cowpea mosaic virus* isolates', *Journal of General Virology* 74, 2243-9.

Kirkpatrick, T. W. (1931), 'Further studies on leaf-curl of cotton in Sudan', *Bulletin of Entomological Research* 22, 323-63.

Kiruwa, F., Feyissa, T. and Ndakidemi, P. (2016), 'Insights of maize lethal necrotic disease: A major constraint to maize production in East Africa', *African Journal of Microbiological Research*, 10: 271–9.

Kufferath, H. and Ghesquière, J. (1932), 'La mosaïque du manioc', *Compte-Rendu de la Société de Biologie Belge* 109, 1146–8.

Kumar, P. L., Akinbade, S. A., Dixon, A. G. O., Mahungu, N. M., Mutunda, M. P., Kiala, D., Londa, L. and Legg, J. P. (2009), 'First report of the occurrence of *East African cassava mosaic virus*-Uganda (EACMV-UG) in Angola', *Plant Pathology* 58, 402.

Kumar, P. L., Hanna, R., Alabi, O. J., Soko, M. M., Oben, T. T., Vangu, G. H. P. and Naidu, R. A. (2011), '*Banana bunchy top virus* in sub-Saharan Africa: Investigations on virus distribution and diversity', *Virus Research* 159, 171–82.

Kumar, P. L., Selvarajan, R., Iskra-Caruana, M.-L., Chabannes, M. and Hanna, R. (2015), 'Biology, etiology and control of virus diseases of banana and plantain', *Advances in Virus Research* 91, 229–69.

Legg, J. P. (1996), 'Host-associated strains within Ugandan populations of the whitefly *Bemisia tabaci* (Genn.), (Hom., Aleyrodidae)', *Journal of Applied Entomology* 120, 523–7.

Legg, J. P. (1999), 'Emergence, spread and strategies for controlling the pandemic of cassava mosaic virus disease in east and central Africa', *Crop Protection* 18, 627–37.

Legg, J. (2005), 'Whiteflies as vectors of plant viruses in cassava and sweetpotato in Africa: Introduction', in *Whiteflies and Whitefly-borne Viruses in the Tropics: Building a Knowledge Base for Global Action*, Anderson, P. K. and Morales, F. (Eds), Centro Internacional de Agricultura Tropical, Cali, Colombia, pp. 15–23.

Legg, J. P. and Ogwal, S. (1998), 'Changes in the incidence of African cassava mosaic geminivirus and the abundance of its whitefly vector along south-north transects in Uganda', *Journal of Applied Entomology* 122, 169–78.

Legg, J. P., Okao-Okuja, G., Mayala, R. and Muhinyuza, J.-B. (2001), 'Spread into Rwanda of the severe cassava mosaic virus disease pandemic and associated Uganda variant of *East African cassava mosaic virus* (EACMV-Ug)', *Plant Pathology* 50, 796.

Legg, J. P., French, R., Rogan, D., Okao-Okuja, G. and Brown, J. K. (2002), 'A distinct, invasive *Bemisia tabaci* (Gennadius) (Hemiptera: Sternorrhyncha: Aleyrodidae) genotype cluster is associated with the epidemic of severe cassava mosaic virus disease in Uganda', *Molecular Ecology* 11, 1219–29.

Legg, J. P., Ndjelassili, F. and Okao-Okuja, G. (2004), 'First report of cassava mosaic disease and cassava mosaic geminiviruses in Gabon', *New Disease Reports* 53, 232.

Legg, J. P., Owor, B., Sseruwagi, P. and Ndunguru, J. (2006), 'Cassava mosaic virus disease in East and Central Africa: Epidemiology and management of a regional pandemic', *Advances in Virus Research* 67, 355–418.

Legg, J. P., Jeremiah, S. C., Obiero, H. M., Maruthi, M. N., Ndyetabula, I., Okao-Okuja, G., Bouwmeester, H., Bigirimana, S., Tata-Hangy, W., Gashaka, G., Mkamilo, G., Alicai, T. and Lava Kumar, P. (2011), 'Comparing the regional epidemiology of the cassava mosaic and cassava brown streak pandemics in Africa', *Virus Research* 159, 161–70.

Legg, J. P., Sseruwagi, P., Boniface, S., Okao-Okuja, G., Shirima, R., Bigirimana, S., Gashaka, G., Herrmann, H.-W., Jeremiah, S. C., Obiero, H. M., Ndyetabula, I., Tata-Hangy, W., Masembe, C. and Brown, J. K. (2014a), 'Spatio-temporal patterns of genetic change amongst populations of cassava *Bemisia tabaci* whiteflies driving virus pandemics in East and Central Africa', *Virus Research* 186, 61–75.

Legg, J. P., Shirima, R., Tajebe, L. S., Guastella, D., Simon, B., Nsami, E., Chikoti, P. and Rapisarda, C. (2014b), 'Biology and management of *Bemisia* whitefly vectors of cassava virus pandemics in Africa', *Pest Management Science* 70, 1446–53.

Legg, J. P., Lava Kumar, P., Makeshkumar, T., Ferguson, M., Kanju, E., Ntawuruhunga, P., Tripathi, L. and Cuellar, W. (2015), 'Cassava virus diseases: Biology, epidemiology and management', *Advances in Virus Research* 91, 85–142.

Lukanda, M., Owati, A., Ogunsanya, P., Valimunzigha, K., Katsongo, K., Ndemere, H. and Kumar, P. L. (2014), 'First report of *Maize chlorotic mottle virus* infecting maize in the Democratic Republic of the Congo', *Plant Disease* 98, 1448.

Mahuku, G., Lockhart, B. E., Wanjala, B., Jones, M. W., Kimunye, J. N., Stewart, L. R., Cassone, B. J., Sevgan, S., Nyasani, J. O., Kusia, E., Kumar, P. L., Niblett, C. L., Kiggundu, A., Asea, G., Pappu, H. R., Wangai, A., Prasanna, B. M., and Redinbaugh, M. (2015a), 'Maize lethal necrosis (MLN), an emerging threat to maize-based food security in sub-Saharan Africa', *Phytopathology* 105, 956–65.

Mahuku, G., Wangai, A. W., Sadessa, K., Teklewold, A., Wegary, D., Adams, I., Smith. J., Braidwood, L., Feyissa, B., Regassa, B., Wanjala, B., Kimunye, J. N., Mugambi, C., Bottomley, E., Bryce, S., Ayalneh, D. and Prasanna, B. M. (2015b), 'First report of *Maize chlorotic mottle virus* and Maize lethal necrosis on maize in Ethiopia', *Plant Disease* 99, 1870.

Makone, S., Menge, D. and Basweti, E. (2014), 'Impact of maize lethal necrosis disease on maize yield: The case of Kisii, Kenya', *International Journal of Agricultural Extension* 2, 211–18.

Marenya, P. P., Erenstein, O., Prasanna, B., Makumbi, D., Bright, J. M. and Beyene, Y. (2018), 'Maize lethal necrosis disease: Evaluating agronomic and genetic control strategies for Ethiopia and Kenya', *Agricultural Systems* 162, 220–8.

Martin, D. P. and Shepherd, D. N. (2009), 'The epidemiology, economic impact and control of maize streak disease', *Food Security* 1, 305–15.

Maruthi, M. N., Hillocks, R. J., Mtunda, K., Raya, M. D., Muhanna, M., Kiozia, H., Rekha, A. R., Colvin, J. and Thresh, J. M. (2005), 'Transmission of *Cassava brown streak* virus by *Bemisia tabaci* (Gennadius)', *Journal of Phytopathology* 153, 307–12.

Maruthi, M. N., Jeremiah, S. C., Mohammed, I. U. and Legg, J. P. (2017), 'The role of the whitefly, *Bemisia tabaci* (Gennadius), and farmer practices in the spread of cassava brown streak ipomoviruses', *Journal of Phytopathology* 165, 707–17.

Masumba, E. A., Kapinga F., Mkamilo G., Salum G., Kulembeka H., Rounsley S., Bredeson J. V., Lyons J. B., Rokhsar D. S., Kanju E., Katari M. S., Myburg A. A., van der Merwe N. A. and Ferguson M. E. (2017), 'QTL associated with resistance to cassava brown streak and cassava mosaic diseases in a bi-parental cross of two Tanzanian farmer varieties, Namikonga and Albert', *Theoretical and Applied Genetics* 130, 2069-90.

Mbanzibwa, D. R., Tian, Y. P., Tugume, A. K., Mukasa, S. B., Tairo, F., Kyamanywa, S., Kullaya, A. and Valkonen, J. P. T. (2011), 'Simultaneous virus-specific detection of the two cassava brown streak-associated viruses by RT-PCR reveals wide distribution in East Africa, mixed infections and infections in *Manihot glaziovii*', *Journal of Virological Methods* 171, 394–400.

Mesfin, T. and Bosque-Pérez, N. A. (1998), 'Feeding behavior of *Cicadulina storeyi* China (Homoptera: Cicadellidae) on maize varieties susceptible or resistant to *Maize streak virus*', *African Entomol.* 6, 185–91.

Mesfin, T., Bosque-Pérez, N. A., Buddenhagen, I. W., Thottappilly, G. and Olojede, S. O. (1992), 'Studies of *Maize streak virus* isolates from grass and cereal hosts in Nigeria', *Plant Disease* 76, 789–95.

Mgbechi-Ezeri, J., Alabi, O., Naidu, R. and Kumar, P. (2008), 'First report of the occurrence of African cassava mosaic virus in soybean in Nigeria', *Plant Disease* 92, 1709–10.

Mignouna, H. D., Abang, M. M., Onasanya, A., Agindotan, B. and Asiedu, R. (2002), 'Identification and potential use of RAPD markers linked to *Yam mosaic virus* resistance in white yam (*Dioscorea rotundata* Poir.)', *Annals of Applied Biology* 140, 163-9.

Mulimbi, W., Phemba, X., Assumani, B., Kasereka, P., Muyisa, S., Ugentho, H., Reeder, R., Legg, J. P., Laurenson, L., Weekes, R. and Thom, F. E. F. (2012), 'First report of

Ugandan cassava brown streak virus on cassava in Democratic Republic of Congo', *New Disease Reports* 26, 11.

Nair, S. K., Babu, R., Magorokosho, C., Mahuku, G., Semagn, K., Beyene, Y., Das, B., Makumbi, D., Lava Kumar, P., Olsen, M. and Boddupalli, P. M. (2015), 'Fine mapping of *Msv1*, a major QTL for resistance to Maize Streak Virus leads to development of production markers for breeding pipelines', *Theoretical and Applied Genetics* 128, 1839–54.

Nault, L. R., Styer, W. P., Coffey, M. E., Gordon, D. T., Negi, L. S. and Niblett, C. L. (1978), 'Transmission of *Maize chlorotic mottle virus* by chrysomelid beetles', *Phytopathology* 68, 1071–4.

Ndunguru, J., Legg, J. P., Aveling, T. A. S., Thompson, G. and Fauquet, C. M. (2005), 'Molecular biodiversity of cassava begomoviruses in Tanzania: Evolution of cassava geminiviruses in Africa and evidence for East Africa being a center of diversity of cassava geminiviruses', *Virology Journal* 2, 21.

Nelson, S., Brewbaker, J. and Hu, J. (2011), 'Maize chlorotic mottle', *Plant Disease* 79, 1–6.

Neuenschwander, P., Hughes, J. d'A., Ogbe, F., Ngatse, J. M. and Legg, J. P. (2002), 'The occurrence of the Uganda Variant of *East African cassava mosaic virus* (EACMV-Ug) in western Democratic Republic of Congo and the Congo Republic defines the westernmost extent of the CMD pandemic in East/Central Africa', *Plant Pathology* 51, 384.

Ng, J. C. and Falk, B. W. (2006), 'Virus-vector interactions mediating nonpersistent and semipersistent transmission of plant viruses', *Annual Review of Phytopathology* 44, 183–212.

Niblett, C. and Claflin, L. (1978), 'Corn lethal necrosis, a new virus disease of corn in Kansas', *Plant Disease Reporter* 62, 15–19.

Nichols, R. F. W. (1947), 'Breeding cassava for virus resistance', *East African Agricultural Journal* 12, 184–94.

Nichols, R. F. J. (1950), 'The brown streak disease of cassava: Distribution, climatic effects and diagnostic symptoms', *East African Agricultural Journal* 15, 154–60.

Nkere, C. K., Oyekanmi, J., Silva, G., Bömer, M., Atiri, G. I., Onyeka, J., Maroya, N., Seal, S. E. and Kumar, P. L. (2018), 'Chromogenic detection of *Yam mosaic virus* by closed tube reverse transcription loop-mediated isothermal amplification (CT-RT-LAMP)', *Archives of Virology* 163, 1057–61.

Nombela, G. and Muñiz, M. (2010), 'Host plant resistance for the management of *Bemisia tabaci*: A multi-crop survey with emphasis on tomato', in *Bemisia: Bionomics and Management of a Global Pest*, Stansly, P. A. and Naranjo, S. E. (Eds), Springer, Dordreecht-Heidelberg-London-New York, pp. 257–83.

Odedara, O. and Kumar, P. (2016), 'Incidence and diversity of viruses in cowpeas and weds in the unmanaged farming systems of savanna zones in Nigeria', *Archives of Phytopathology and Plant Protection*, 50, 1–12.

Ogwok, E., Ilyas, M., Alicai, T., Rey, M. E. C. and Taylor, N. J. (2016), 'Comparative analysis of virus-derived small RNAs within cassava (*Manihot esculenta* Crantz) infected with cassava brown streak viruses', *Virus Research* 215, 1–11.

Okoth, V. A. O., Dabrowski, Z. T., Thottappilly, G. and van Emden, H. F. (1988), 'Comparative analysis of some parameters affecting *Maize streak virus* (MSV) transmission of various *Cicadulina* spp. populations', *Insect Science and its Application* 8, 445–61.

Otim, M., Legg, J., Kyamanywa, S., Polazsek, A. and Gerling, D. (2006), 'Population dynamics of *Bemisia tabaci* (Homoptera: Aleyrodidae) parasitoids on cassava mosaic disease-resistant and susceptible varieties', *Biocontrol Science and Technology* 16, 205–14.

Otim-Nape, G. W., Bua, A., Thresh, J. M., Baguma, Y., Ogwal, S., Semakula, G. N., Acola, G., Byabakama, B. and Martin, A. (1997), *Cassava Mosaic Virus Disease in Uganda: The Current Pandemic and Approaches to Control*, Natural Resources Institute, Chatham, UK. 65 pp.

Owor, B., Legg, J. P., Obonyo, R., Okao-Okuja, G., Kyamanywa, S. and Ogenga-Latigo, M. W. (2004), 'Field studies of cross protection with cassava in Uganda. *Journal of Phytopathology* 152, 243–9.

Pennisi, E. (2010), 'Armed and dangerous', *Science* 327, 804–5.

Pita, J. S., Fondong, V. N., Sangare, A., Otim-Nape, G. W., Ogwal, S. and Fauquet, C. M. (2001), 'Recombination, pseudorecombination and synergism of geminiviruses are determinant keys to the epidemic of severe cassava mosaic disease in Uganda', *Journal of General Virology* 82, 655–65.

Rabbi, I. Y., Hamblin, M. T., Lava Kumar, P. Gedil, M. A., Ikpan, A. S., Jannink, J. L. and Kulakow, P. A. (2014), 'High-resolution mapping of resistance to cassava mosaic geminiviruses in cassava using genotyping-by-sequencing and its implications for breeding', *Virus Research* 186, 87–96.

Reyes, H. E. and Castillo, L. J. (1988), 'Transmission of *Maize chlorotic mottle virus* (MCMV) by two species of *Diabrotica*, family *Chrysomelidae*', *Fitopatologia*, 23, 65–73.

Rosen, R., Kanakala, S., Kliot, A., Pakkianathan, B. C., Farich, B. A., Santana-Magal, N., Elimelech, M., Kontsedalov, S., Lebedev, G., Cilia, M. and Ghanim, M. (2015), 'Persistent, circulative transmission of begomoviruses by whitefly vectors', *Current Opinion in Virology* 15, 1–8.

Rybicki, E. P. (2015), 'A Top Ten list for economically important plant viruses', *Archives of Virology* 160, 17.

Samson, M., Dickson, M. and Evans, B. (2014), 'Impact of maize lethal necrosis disease on maize yield: The case of Kisii, Kenya', *International Journal of Agricultural Extension*, 2: 211–18.

Shepherd, D. N., Mangwende, T., Martin, D. P., Bezuidenhout, M., Kloppers, F. J, Carolissen, C. H., Monjane, A. L., Rybicki, E. P. and Thomson, J. A. (2007), 'Maize streak virus-resistant transgenic maize: A first for Africa', *Plant Biotechnology Journal* 5, 759–67.

Shoyinka, S. A., Thottappilly, G., Adebayo, G. G. and Anno-Nyako, F. O. (1997), 'Survey on cowpea virus incidence and distribution in Nigeria', *International Journal of Pest Management* 43, 127–32

Silva, M., Oyekanmi, J., Nkere, C. K., Bömer, M., Kumar, P. L. and Seal, S. E. (2018), 'Rapid detection of potyviruses from crude plant extracts', *Analytical Biochemistry* 546, 17–22.

Soto, P. E., Buddenhagen, I. W., Asnani, V. L. (1982), 'Development of streak virus-resistant maize populations through improved challenge and selection methods', *Annals of Applied Biology* 100, 539–46.

Sseruwagi, P., Legg, J. P., and Otim-Nape, G. W. (2005), 'Whiteflies as vectors of plant viruses in cassava and sweetpotato in Africa: Uganda', in *Whiteflies and Whitefly-borne Viruses in the Tropics: Building a Knowledge Base for Global Action*, Anderson, P. K. and Morales, F. (Eds), Centro Internacional de Agricultura Tropical, Cali, Colombia, pp. 46–53.

Sseruwagi, P., Maruthi, M. N., Colvin, J., Rey, M. E. C., Brown, J. K. and Legg, J. P. (2006), 'Colonisation of non-cassava plant species by cassava whiteflies (*Bemisia tabaci*)

(Gennadius) (Hemiptera: Aleyrodidae) in Uganda', *Entomologia Experimentalis et Applicata* 119, 145–53.

Storey, H. H. (1936a), 'Virus diseases of East African plants. V. Streak disease of maize', *East African Agricultural Journal* 1, 471–5.

Storey, H. H. (1936b), 'Virus diseases on East African plants. VI. A progress report on the studies of the diseases of cassava', *East African Agricultural Journal* 2, 34–9.

Storey, H. H. (1939), 'Transmission of plant viruses by insects', *Botanical Review* 5, 240–72.

Storey, H. H. and Howland, A. K. (1967), 'Inheritance of resistance in maize to the virus of streak disease in East Africa', *Annals of Applied Biology* 59, 429–36.

Storey, H. H. and Nichols, R. F. W. (1938), 'Studies on the mosaic of cassava', *Annals of Applied Biology* 25, 790–806.

Swanson, M. M. and Harrison, B. D. (1994), 'Properties, relationships and distribution of cassava mosaic geminiviruses', *Tropical Science* 34, 15–25.

Tay, W. T., Evans, G. A., Boykin, L. M. and De Barro, P. J. (2012), 'Will the real *Bemisia tabaci* please stand up?', *PLoS ONE* 7, e50550.

Tajebe, L. S., Guastella, D., Cavalieri, V., Kelly, S. E., Hunter, M. S., Lund, O. S., Legg, J. P. and Rapisarda, C. (2015), 'Diversity of symbiotic bacteria associated with *Bemisia tabaci* (Homoptera: Aleyrodidae) in cassava mosaic disease pandemic areas of Tanzania', *Annals of Applied Biology* 166, 297–310.

Terry, E. R. and Hahn, S. K. (1980), 'The effect of cassava mosaic disease on growth and yield of a local and an improved variety of cassava', *Tropical Pest Management* 26, 34–7.

Thomas, J. E., Iskra-Caruana, M. L., Kumar, P. L., Roux, N., Chabannes, M., Massart, S., Mathieu, Y., Chase, R. and Van den Houwe, I. (2015), 'Position paper on a strategy to distribute banana (Musa) germplasm with endogenous Banana streak virus genomes', https://agritrop.cirad.fr/579485/1/Position%20paper%20on%20BSV%20distribution%20strategies_versionWEB.pdf

Thottappilly, G. and Rossel, H. W. (1992), 'Virus diseases of cowpea in tropical Africa', *Tropical Pest Management* 38, 337–48.

Thottappilly, G., Bosque-Pérez, N. A. and Rossel, H. W. (1993), 'Viruses and virus diseases of maize in tropical Africa', *Plant Pathology* 42, 494–509.

Thresh, J. M., Otim-Nape, G. W. and Jennings, D. L. (1994), 'Exploiting resistance to *African cassava mosaic virus*', *Aspects of Applied Biology* 39, 51–60.

Time, I., Atiri, G. I., and Kumar, P. L. (2010), 'Viruses infecting soybean (*Glycine max* L. Merill) in Nigeria', *Phytopathology* 100, S126.

Turaki, A., Bömer, M., Silva, G., Kumar, P. L. and Seal, S. E. (2017), 'PCR-DGGE Analysis: Unravelling complex mixtures of badnavirus sequences present in yam germplasm', *Viruses* 9, 181.

Umber, M., Pichaut, J. P., Farinas, B., Laboureau, N., Janzac, B., Plaisir-Pineau. K., Pressat, G., Baurens, F. C., Chabannes, M., Duroy, P. O. and Guiougou, C. (2016), 'Marker-assisted breeding of *Musa balbisiana* genitors devoid of infectious endogenous *Banana streak virus* sequences', *Molecular Breeding* 36, 74.

Uyemoto, J. K. (1983), 'Biology and control of *Maize chlorotic mottle virus*', *Plant Disease* 67, 7–10.

Uyemoto, J. K., Bockelman, D. and Claflin, L. (1980), 'Severe outbreak of corn lethal necrosis disease in Kansas', *Plant Disease*, 64, 99–100.

Uyemoto, J. K., Claflin, L. and Wilson, R. (1981), Maize chlorotic mottle and maize dwarf mosaic viruses: Effect of single and double inoculations on symptomatology and yield', *Plant Disease* 65, 39–41.

Vanderschuren, H., Moreno, I., Anjanappa, R. B., Zainuddin, I. M. and Gruissem, W. (2012), 'Exploiting the combination of natural and genetically engineered resistance to cassava mosaic and cassava brown streak viruses impacting cassava production in Africa', *PLoS ONE* 7, e45277.

Vogel, W. O., Hennessey, R. D., Berhe, T. and Matungulu, K. M. (1993), 'Yield losses to maize streak disease and *Busseola fusca* (Lepidoptera: Noctuidae), and economic benefits of streak-resistant maize to small farmers in Zaïre', *International Journal of Pest Management* 39, 229–38.

Wagaba, H., Beyene, G., Aleu, J., Odipio, J., Okao-Okuja, G., Chauhan, R. D., Munga, T., Obiero, H., Halsey, M. E., Ilyas, M., Raymond, P., Bua, A., Taylor, N. J., Miano, D. and Alicai, T. (2017), 'Field level RNAi-mediated resistance to cassava brown streak disease across multiple cropping cycles and diverse East African agro-ecological locations', *Frontiers in Plant Science* 7, 2060.

Wangai, A. W., Redinbaugh, M. G., Kinyua, Z. M., Miano, D. W., Leley, P. K., Kasina, M., Mahuku, G., Scheets, K. and Jeffers, D. (2012a), 'First report of *Maize chlorotic mottle virus* and maize lethal necrosis in Kenya', *Plant Disease* 96, 1582.

Wangai, A., Sikinyi, E., Ochieng, J., Miyogo, S., Karanja, T., Odour, H., Kimani, E., Irungu, J., Kinyua, Z., Ngaruiya, P., Ligeyo, D. and Kipkemboi, S. (2012b), '*Joint assessment report: Report on status of maize lethal necrosis disease and general maize performance*', Ministry of Agriculture, Kenya. Retrieved August 2015, from http://www.fao.org/fileadmin/user_upload/drought/docs/Maize%20Let hal%20Necrotic%20Disease%20in%20Kenya_Joint%20Assessment %20Report%20(July%202012).pdf

Webb, M. D. (1987), 'Species recognition of *Cicadulina* leafhoppers (Hemiptera: Cicadellidae) vectors of pathogens of Gramineae', *Bulletin of Entomological Research* 77, 683–712.

Welz, H. G., Schechert, A., Pernet, A., Pixley, K. V. and Geiger, H. H. (1998), 'A gene for resistance to the *Maize streak virus* in the African CIMMYT maize inbred line CML202', *Molecular Breeding* 4, 147–54.

Were, H. K., Winter, S. and Maiss, E. (2004), 'Occurrence and distribution of cassava begomoviruses in Kenya', *Annals of Applied Biology* 145, 175–84.

Wolfe, M. D., Rabbi, I. Y., Egesi, C., Hamblin, M., Kawuki, R., Kulakow, P., Lozano, R., Carpio, D. P., Ramu, P. and Jannink, J. L. (2016), 'Genome-wide association and prediction reveals genetic architecture of cassava mosaic disease resistance and prospects for rapid genetic improvement', *Plant Genome* 9.

Wosula, E. N., Chen, W., Fei, Z. and Legg, J. P. (2017), 'Unravelling the genetic diversity among cassava *Bemisia tabaci* whiteflies using NextRAD sequencing', *Genome Biology and Evolution* 9, 2958–73.

Xia, Z., Zhao, Z., Chen, L., Li, M., Zhou, T., Deng, C., Zhou, Q. and Fan, Z. (2016), 'Synergistic infection of two viruses MCMV and SCMV increases the accumulations of both MCMV and MCMV-derived siRNAs in maize', *Scientifics Reports* 6, 20520.

Zhang, P., Vanderschuren, H., Fütterer, J. and Gruissem, W. (2005), 'Resistance to cassava mosaic disease in transgenic cassava expressing antisense RNAs targeting virus replication genes', *Plant Biotechnology Journal* 3, 385-97.

Zhou, X., Liu, Y., Calvert, L., Munoz, C., Otim-Nape, G. W., Robinson, D. J. and Harrison, B. D. (1997), 'Evidence that DNA-A of a geminivirus associated with severe cassava mosaic disease in Uganda has arisen by interspecific recombination', *Journal of General Virology* 78, 2101–11.

Part 2

Plant health in practice: managing threats to key African crops

Chapter 6

Identifying and managing plant health risks for key African crops: cassava

Muaka Toko *and* **Peter Neuenschwander**, *International Institute of Tropical Agriculture (IITA), Benin;* **J. Steve Yaninek**, *Purdue University, USA;* **Alejandro Ortega-Beltran**, *International Institute of Tropical Agriculture (IITA), Nigeria;* **André Fanou** *and* **Victor Zinsou**, *Université de Parakou, Benin;* **Kerstin D. Wydra**, *Erfurt University of Applied Sciences, Germany;* **Rachid Hanna** *and* **Appolin Fotso**, *International Institute of Tropical Agriculture (IITA), Cameroon; and* **Ouorou Douro-Kpindou**, *International Institute of Tropical Agriculture (IITA), Benin (corresponding author: m.toko@cgiar.org)*

1 Introduction

Cassava, a starchy root crop cultivated in tropical America for more than 5000 years, was introduced into Africa in the sixteenth century by Portuguese traders (Jones, 1959), and was found growing throughout the tropics by the end of the nineteenth century (CIAT, 1993). For over three centuries, cassava was regarded as a low value crop of too little prestige to be a subject of research or to draw attention of decision makers (Jones, 1959; Nestel, 1974; Cock, 1985). Cassava is a very hardy crop that can be grown on marginal land where other crops fail. It tolerates periods of drought, requires little input and allows tuberous roots to be left in the ground for many months until time for consumption or processing (FAO, 2013). These attributes coupled with the fact that the cassava leaves are consumed as green vegetables in many parts of Africa made cassava highly attractive to small-scale farmers and hence favoured its rapid adoption. It became the most important food security crop in areas prone to drought, famine, periods of civil disturbances, and when rural and urban populations dramatically increased in the twentieth century (Yaninek and Hanna, 2003). Today, cassava is the chief source of affordable carbohydrates for an estimated

http://dx.doi.org/10.19103/AS.2018.0043.07
Published by Burleigh Dodds Science Publishing Limited, 2019.

500 million Africans including the continent's poorest and provides income to ever more farmers interested in commercializing the numerous current and future cassava end products developed for human consumption, animal feed or industrial use (Grace, 1977; Donkor et al., 2017).

Cassava production in Africa has increased remarkably, almost doubling from 95.4 million tons in 2000 to 132.2 million tons in 2016. This resulted from an increase in harvested area from 11 to 17 million ha between 2000 and 2016, while yields increased only slightly from 8.2 tons/ha in 2000 to 9.2 tons/ha in 2016. There is great potential for yield to further increase from the current world average of 12.8 to 80 tons/ha, if production constraints such as the consequences of shortening the fallow periods and low quantity and quality of planting materials (Leihner, 1982; FAO, 2017) can be overcome by the introduction and adoption of an increased number of high-yielding varieties coupled with more efficient pest control.

Cassava is cultivated in continuously changing production systems as rising demands result in degradation of existing ecosystems, reduced traditional fallow periods and in cultivation extended into new and, often, marginal areas. Increased production demands together with finite agricultural resources pose a range of threats to the sustainability of cassava agro-ecosystems in Africa (Nweke et al., 2002). In these systems, the yield was generally low though this changed with the introduction of high-yielding varieties in the 1990s. Low yield is attributed to the aforementioned constraints including weeds, which are being treated in Chapter 14 of this book, but also to pests and diseases, the focus of this chapter.

Although cassava has had a long history of being relatively disease-free and tolerant to pests (Yonow et al., 2017), two exotic arthropod pests – the cassava mealybug *Phenacoccus manihoti* Mat.-Ferr. (Hemiptera: Pseudococcidae) and the cassava green mite, *Mononychellus tanajoa* Bondar (Acari: Tetranychidae) – invaded the continent in the 1970s (Neuenschwander and Herren, 1988) and changed that history. Their impact was superimposed on the damage already caused by the indigenous African root and tuber scale (ARTS) *Stictococcus vayssierei* Richard (Hemiptera: Stictococcidae) initially thought to be confined to the Congo Basin; the whitefly *Bemisia tabaci* (Gennadius) (Hemiptera: Aleyrodidae), known as vector of the African cassava mosaic virus disease treated in Chapter 5; the white scale *Aonidomytilus albus* (Cockerell) (Hemiptera: Diaspididae); the polyphagous grasshopper *Zonocerus variegatus* Linnaeus (Orthoptera: Pyrgomorphidae); several termites species (Isoptera: Termitidae), as well as several vertebrates and the exotic spiralling whitefly *Aleurodicus dispersus* Russell (Hemiptera: Aleyrodidae) only cited here but developed in Chapter 12 of this book. Individually and together they cause significant yield losses, particularly in the current unpredictable context of climate change (Bellotti and van Schoonhoven, 1978; James et al., 2000a; CABI, 2018). Besides

Published by Burleigh Dodds Science Publishing Limited, 2019.

these arthropod pests, cassava is also subject to diseases, among which the most important are the cassava mosaic virus disease (CMD) complex and the cassava brown streak virus disease (CBSD) both treated in Chapter 5 of this book, the cassava bacterial blight (CBB), cassava anthracnose disease (CAD), cassava root rot diseases (Msikita et al., 2000; Hillocks and Wydra, 2002; Zinsou et al., 2017), as well as cassava leaf spots caused by various pathogens. Many more cassava pests and diseases exist, particularly in South America (Bellotti, 2002). Thanks to quarantine measures they have up to now been excluded from Africa as described in Chapter 3.

In Africa, cassava plant protection effectively began in the 1930s with breeding for resistance against the African cassava mosaic virus (Legg and Thresh, 2000; Yaninek and Hanna, 2003). Thereafter, funding for cassava research and development gradually increased, particularly from the 1970s onward when the threats caused by the introduced pests across the continent required important funding. Examples of well-funded projects at IITA included the Africa-wide biological control of cassava mealybug and cassava green mite. This was followed by the Ecological Sustainable Cassava Plant Protection (ESCaPP) project, a joint effort between IITA, national plant protection staff, extension workers and farmers in Benin, Cameroon, Ghana and Nigeria, with a parallel component in Brazil under the Centro Internacional de Agricultura Tropical (CIAT) and Empresa Brasileira de Pesquisa Agropecuária (EMBRAPA) (Yaninek et al., 1994). ESCaPP was conceived to develop, test and adapt viable plant protection technologies through diagnostic surveys and farmers' participatory research to test appropriate technologies while women were professionally strengthened through postgraduate training.

2 The management of insect pests

2.1 Cassava mealybug

After its accidental introduction into Africa in the 1970s, cassava mealybug *Phenacoccus manihoti* spread from Congo Brazzaville and the nearby Democratic Republic of Congo (former Zaire) to new footholds in West and later Southern Africa, and from there across the entire continent's forest and savannah regions. In its wake, cassava production ceased and food security for up to 200 million Africans was endangered. Local insecticide applications proved useless and no varieties were found to be sufficiently resistant. Search for natural enemies in South America by an IITA-led coalition of institutions proved difficult. Finally, in 1981, cassava mealybug was found in Paraguay by a CIAT scientist and shortly thereafter the parasitoid *Anagyrus* (*Epidinocarsis, Apoanagyrus*) *lopezi* (de Santis) (Hymenoptera: Encyrtidae) was collected. Quarantine in England and mass production in Nigeria, together with biological and ecological studies, led to its release in Nigeria and beyond. Establishment

Published by Burleigh Dodds Science Publishing Limited, 2019.

and spread were quickly ascertained, and a large-scale training programme for scientists at national institutions and several local rearing facilities for *A. lopezi* were established (Herren and Neuenschwander, 1991). The programme did not stop at establishing natural enemies because it was clear from the beginning that a rigorous scientific impact assessment would be needed to demonstrate the effectiveness of this unusual and controversial CGIAR technology that did not rely on resistant varieties.

After mass production had been upgraded (Neuenschwander and Haug, 1992), a total of over 150 releases of adult *A. lopezi* were made in two dozen countries (Fig. 1), mostly from the ground, but in a few countries also by a novel release mechanism from the air. Releases usually comprised several hundred to a few thousand adults, set free under the best possible conditions. Practically in all release sites, *A. lopezi* was recovered after one year, which was interpreted as successful establishment. Releases were followed by surveys involving hundreds of randomly chosen fields each in a dozen countries across the continent (Fig. 1). In each field, cassava mealybug and *A. lopezi* populations and those of indigenous natural enemies were quantified, and climatic, soil and agricultural as well as plant parameters were characterized (e.g. Neuenschwander et al., 1990). Throughout and within a few years, cassava

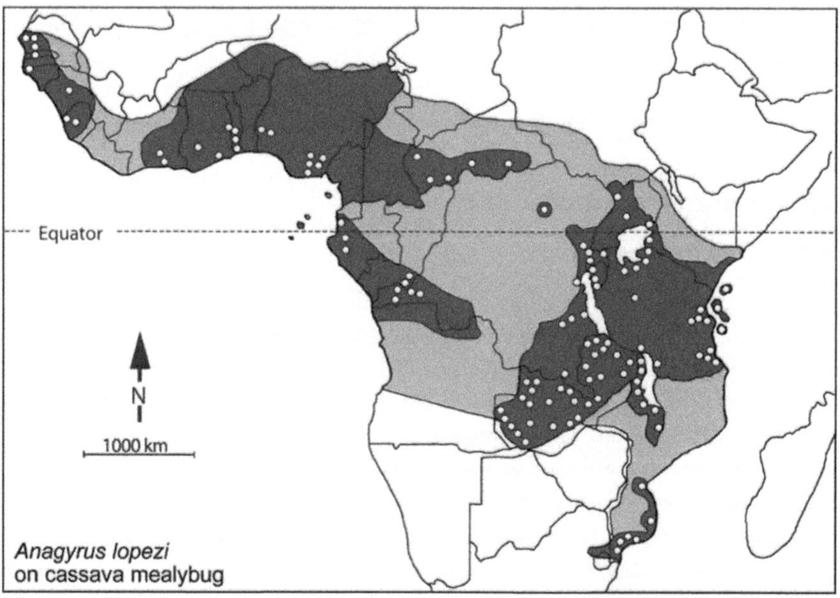

Figure 1 Spread of cassava mealybug across the cassava belt of Africa. Between 1981 and 1995, *Anagyrus lopezi* was released in about 150 locations (white dots, which sometimes cover more than one release) and spread (dark grey regions: known spread in the 1990s). Today, *A. lopezi* controls cassava mealybug across the entire cassava belt.

Published by Burleigh Dodds Science Publishing Limited, 2019.

mealybug densities and damage became low on 95% of all fields. A parallel nine-year field study of cassava mealybug, *A. lopezi,* and indigenous natural enemies demonstrated the long-term control by *A. lopezi* in Nigeria (Fig. 2; Hammond and Neuenschwander, 1990).

Several more exotic natural enemies, including some coccinellids, were introduced, but most did not establish and those that did remained rare (Neuenschwander, 2001). Of particular interest was *Anagyrus diversicornis* (Howard), which slightly preferred larger stages of cassava mealybug than *A. lopezi* (Pijls et al., 1990). Different sex allocation – too many males – was responsible for its competitive displacement by *A. lopezi* (Gutierrez et al., 1993). As a result, *A. diversicornis* could not establish in Africa.

The impact of *A. lopezi* on cassava mealybug was studied in the laboratory, in small fields and across regions (Neuenschwander, 1996): *A. lopezi* proved to be a special natural enemy indeed. Despite its narrow original range in South America, mainly in the de la Plata Valley, it established from the dry Sahel zone in Senegal, across the continent through the humid forest zone to the highlands of East Africa. Though it initially suffered from high pressure by hyper-parasitoids, it escaped them under the prevailing low host population levels. Adult wasps avoided high cassava mealybug populations with the corresponding dense, sticky mealybug clumps and aborted cassava leaves, but still managed to keep overall cassava mealybug populations low. Upon emergence, *A. lopezi* displayed a high level of dispersal and good density dependence in low to medium sized host populations. *A. lopezi*'s capacity to maintain cassava mealybugs at low levels, despite its low reproductive capacity

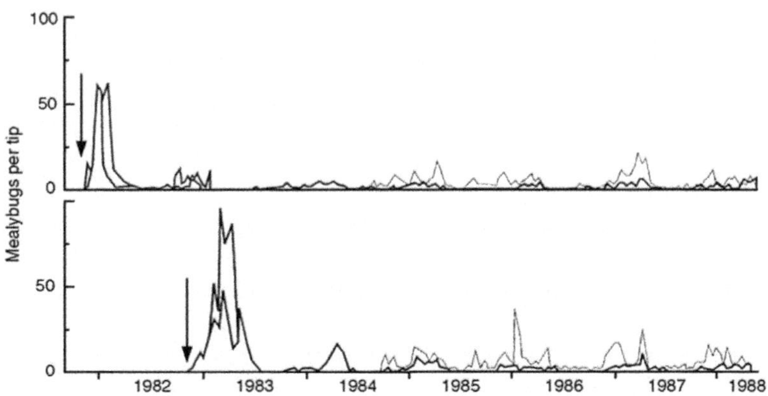

Figure 2 Cassava mealybug population density in two areas in Southwestern Nigeria following the release of *Anagyrus lopezi* (arrows), since 1984 on two cassava varieties with different susceptibilities to mealybug infestations (Lema and Herren, 1985; Hammond and Neuenschwander, 1990). Note the irregular eruptions of the mealybug depending on rainfall, which are brought under control immediately by *A. lopezi.*

(Neuenschwander et al., 1986), is attributed to its demonstrated host finding capacity, which is far greater than that of any of its competitors (Neuenschwander and Ajuonu, 1995). Although *A. lopezi* dispersed quickly after emergence, area-wide biological control still took 2–4 years depending on climate. A simulation model incorporating laboratory and experimental field results, including indigenous coccinellid natural enemies, was tested against the previously mentioned population dynamic study (Hammond and Neuenschwander, 1990). The model demonstrated that *A. lopezi* was responsible for lowering mealybug populations to below 10%, while indigenous lady beetles cut occasional peak mealybug populations by 20%, a result that no field study could have demonstrated (Gutierrez et al., 1988, 1999). Once mealybug populations were low, most coccinellids and other indigenous natural enemies disappeared from the fields.

Control differed, however, according to cassava variety and agronomic factors. From several country surveys, most notably from a 4-year study in Malawi (Neuenschwander et al., 1991) and across West Africa (Neuenschwander et al., 1990), it became clear that, despite the presence of *A. lopezi*, cassava mealybug populations were not reduced to satisfactory levels on about 5% of all fields that had naked, un-mulched soil preventing good development of cassava plants. Further releases of *A. lopezi* were not recommended by IITA, but sometimes made for political reasons. Good soil management with mulch to bring back good plant growth with low mealybug levels was proposed instead. The impact of this procedure was later demonstrated in a field experiment in Benin (Schulthess et al., 1997).

In Africa, up to now, no resurgences of cassava mealybug have been observed though this phenomenon was often claimed. In all cases of relatively high mealybug populations that we could study, the reason for crop loss always concerned bad soil management, insecticide treatments against grasshoppers with deleterious effects on natural enemies (P. Neuenschwander, unpublished data) or even herbicide treatments against weeds (J.S. Yaninek, unpublished data).

All these observations and experimental results on the impact on cassava yield were incorporated into an economic model, whereby results from different ecological regions in 27 countries and the timing of releases and observed impact were taken into consideration and contrasted to the estimated expenses. The resulting total gain from the cassava mealybug biological control programme in Africa was a stunning US$8–20 billion, and this despite significant initial costs, with a benefit: cost ratio of 200–500, depending on the scenario with different depreciation levels and overall durations (Zeddies et al., 2001).

The project was honoured with the King Baudouin Award together with CIAT in 1990 and Hans Herren, its initiator, received the World Food Prize

Published by Burleigh Dodds Science Publishing Limited, 2019.

in 1995. It was consistently supported over about 15 years by a vast donor community. Apart from the impact on farm productivity, the biological control programme against cassava mealybug also served as a training ground for a whole generation of African scientists, who later moved to higher administrative positions in their countries or in international organizations. From their positions, they influenced the politics of agricultural development, a benefit in the end as important as the achieved reduction of cassava crop loss.

In recent years, cassava mealybug invaded Southeast Asia threatening cassava, which is mainly an industrial cash crop in this region (Parsa et al., 2012). *A. lopezi* from the IITA-Benin culture was released and established in Thailand in collaboration with other institutions (Winotai et al., 2011). *A. lopezi* has since been distributed and is spreading across at least six Southeast Asian countries with similar impact as in Africa (Wyckhuys et al., 2018). Thus, by relying on the network established by the project to control exotic cassava pests and institutional linkages, the benefits from previous work in Africa were extended at relatively low cost to another continent.

2.2 Cassava green mite

The cassava green mite, *Mononychellus tanajoa*, was virtually unknown before it arrived in Africa nearly 50 years ago. After it was first reported in Uganda in 1972 (Z.M. Nyiira, unpublished results), 'presumably on illegally imported cassava cuttings' (Nyandarua Agricultural Research Station, unpublished results), little attention was given to this new pest on a low priority subsistence crop. Early literature tracked the incipient invasion (Lyon, 1973), described pest damage (Nyiira, 1973) and addressed species taxonomy (Flechtmann, 1978). As it spread and eventually spanned the African cassava belt in less than a decade, leaving in its wake damaged cassava with falling yields (Leuschner, 1980), the new invader got finally noticed (Yaninek, 1988).

The biological control efforts against *M. tanajoa* at IITA began in 1983 with the first shipments of natural enemies from Colombia and the recruitment of a mite specialist (Yaninek and Herren, 1988). At the time, not much was known about *M. tanajoa* or its natural enemies, so initial project efforts in Africa characterized the agroecological conditions that prompted pest outbreaks and high pest mite populations (Yaninek et al., 1989a,b, 1991), then applied the results using agrometeorological habitat matching to inform foreign exploration in South America (Yaninek and Bellotti, 1987). Phytoseiidae predatory mites were an obvious first choice given their close spatial and ecological relation with their spider mite prey (Yaninek and de Moraes, 1991).

Foreign exploration yielded 13 phytoseiid species from 44 geographically discrete populations from Colombia and Brazil that reached Africa in 126 shipments between 1983 and 1998. Eleven phytoseiid species were selected for

Published by Burleigh Dodds Science Publishing Limited, 2019.

experimental releases, and >11 million predators were shipped to 16 countries for release in 822 locations between 1984 and 2001 (Yaninek and Hanna, 2003). From this effort, three species became established, two spread well beyond their initial release sites and one became a dominant biological control agent now found throughout most of the cassava belt in Africa. Pathogenic fungi were also imported and evaluated as potential biological control agents. Virulent isolates of *Neozygites floridana* (Weiser and Muma), later renamed *N. tanajoae* Delalibera, Hajek and Humber (Entomophthorales: Neozygitaceae) (Delalibera et al., 2004), were first imported from Brazil in 1995 and continued with new isolates from Brazil, Benin, Ghana and Tanzania through 2007 (Agboton et al., 2009). Exotic pathogenic fungi released in several locations in Benin in 1999 demonstrated persistence and increased fungal epizootics in *M. tanajoa* host populations (Hountondji et al., 2002).

Neoseiulus idaeus Denmark and Muma (Acari: Phytoseiidae) was the first biological control agent widely distributed in Africa against *M. tanajoa*. Released in 13 countries, it became established in Benin and Kenya (Yaninek et al., 1991), persisted on cassava during the dry season, but disappeared from cassava during the wet season under low mite prey densities when it migrated to weedy plants in search of suitable alternate prey. Other mites like *Tetranychus urticae* Koch and *Oligonychus gossypii* (Oliggo) (Acari: Tetranychidae) were also accepted and allowed growth, development and oviposition. *N. idaeus* could not survive on whitefly nymphs alone, extrafloral plant exudate or pollen commonly found on cassava. In follow-up surveys, *N. idaeus* seemingly disappeared from cassava release fields for many years until Mutisya et al. (2017) recovered it from multiple locations in Kenya, including from sites well outside the original release fields.

Amblydromalus manihoti (Moraes)(=*Typhlodromalus limonicus sensu lato* Garman and McGregor) was released in 12 countries and recovered in Benin, Nigeria, Ghana and Rwanda (Yaninek et al., 1998). It spread slowly (ca. 3.3 km/year), but consistently on cassava grown in subhumid ecozones. This predator persisted on cassava during both the wet and dry season and was often found on other plant species surrounding cassava. *A. manihoti* also preferred tetranychid prey, particularly *M. tanajoa*, but persisted and developed on other diets including whitefly nymphs and pollen, although mite prey was required for oviposition (Gnanvossou et al., 2003). Evaluating the impact of *A. manihoti* in the field was initially difficult when predator densities were low and population distribution was limited. Subsequent studies revealed measurable impacts of *A. manihoti* on *M. tanajoa* when present together with other exotic and indigenous phytoseiids (Onzo et al., 2003).

Ten years into the campaign, *Typhlodromalus aripo* DeLeon (Acari: Phytoseiidae) became an unexpected breakthrough when it emerged as the leading biological control agent against *M. tanajoa* (Yaninek and Hanna, 2003). The first field release was made in October 1993 at Ikpinle, Benin. A few months

Published by Burleigh Dodds Science Publishing Limited, 2019.

after release, the predator was found in adjacent fields; by the end of that first season, *T. aripo* had spread 31 km from the original release field. This discovery prompted the project to rapidly mobilize national program partners to join a continent-wide release effort. Within three seasons of the first release, *T. aripo* had spread across the cassava belt from Cameroon to Ghana. *T. aripo* was eventually released in 16 countries, spread to five neighbouring countries and is now suspected in 24 of the 28 cassava-producing countries on the continent (Fig. 3).

M. tanajoa populations in the original release field crashed from 90 to <10 mites/leaf and persisted at low *M. tanajoa* densities for multiple seasons of cassava harvest and replanting (Fig. 4). Pest populations declined an average of 53% in countries across the cassava belt following releases of *T. aripo*. Exceptions were the highlands in Zambia (A. Chalwe, unpublished results) and Kenya (Mutisya et al., 2015) where long crop cycles and cool winters thwarted the establishment of *T. aripo*, as did cassava cultivars anywhere with relatively small, glabrous or otherwise missing cassava shoot tips where the phytoseiid

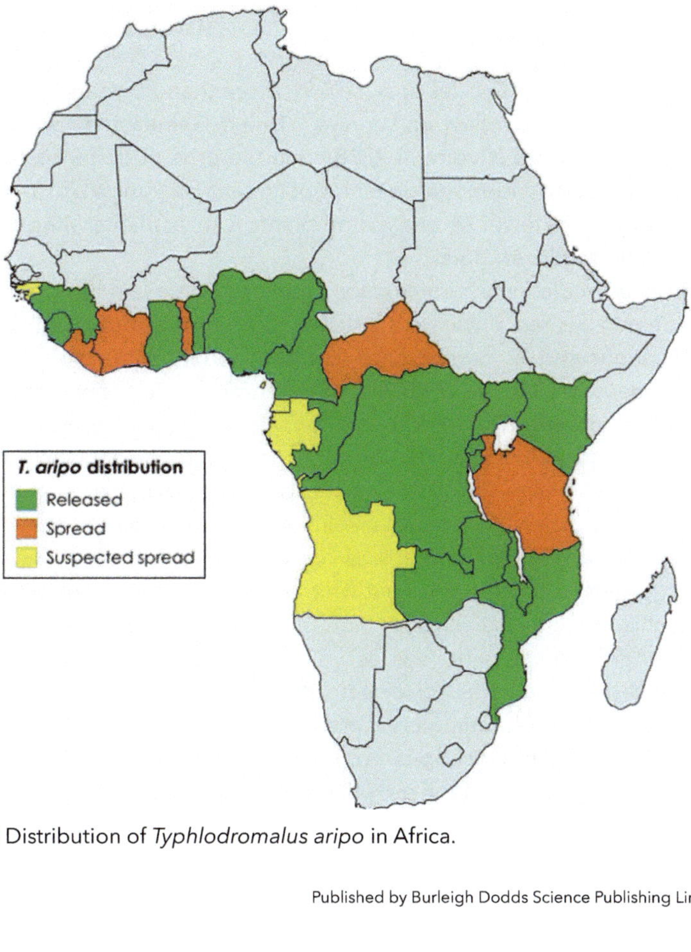

Figure 3 Distribution of *Typhlodromalus aripo* in Africa.

Published by Burleigh Dodds Science Publishing Limited, 2019.

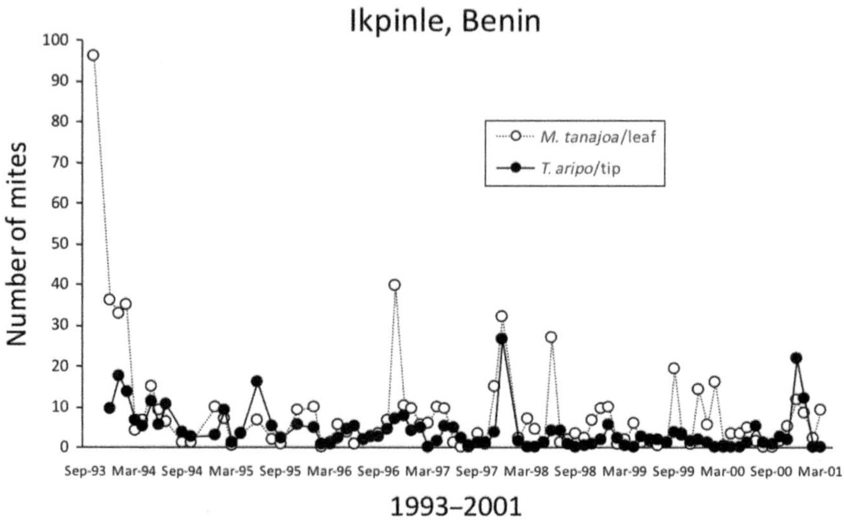

Figure 4 Multi-season population dynamics of *Mononychellus tanajoa* and *Typhlodromalus aripo* in Ikpinle, Benin.

shelters during the day (Zundel et al., 2009). More than 20 years after the first releases in each country, follow-up surveys in Benin, Guinea and Kenya in 2016, and in Nigeria and Côte d'Ivoire in 2018 confirmed the persistence of *T. aripo* associated with low to moderate densities of *M. tanajoa* [range of 18–43 mites/leaf; on average 43% fewer *M. tanajoa* on plants with vs plants without *T. aripo* (J.S. Yaninek, unpublished data)].

Differences in diets and prey exploitation largely explain the performance of the three established phytoseiid mites as predators and biological control agents. *N. idaeus* and *A. manihoti* are efficient predators, but poor biological agents. *N. idaeus* is the most voracious of the three predators with the highest intrinsic rate of increase when fed on *M. tanajoa*. However, because *N. idaeus* requires mite prey for growth, development and oviposition, seasonal fluctuations in *M. tanajoa* populations and prey overexploitation causes it to abandon cassava when pest mites are insufficient and search elsewhere for alternative prey. *A. manihoti* is nearly as voracious as *N. idaeus*, also with rapid population growth on *M. tanajoa,* but when prey populations are low or over-exploited, this predator survives on non-prey food, notably plant resources: maize, castor and *Leucaena* pollens, as well as cassava leaf exudates. It will stay on and near cassava until pressed to search for prey to reproduce. This probably explains why *A. manihoti* continues to be tracked on cassava in several countries, while *N. idaeus* has all but disappeared. *T. aripo*, by comparison, is an inefficient predator with modest population growth on *M. tanajoa*, but it can grow, develop and oviposit on the widest range of non-prey diets commonly found on cassava.

This allows the predator to persist on cassava during periods of highly variable *M. tanajoa* populations and intercept incipient pest mite populations.

The impact of *T. aripo* on cassava production was evaluated in multiple field trials soon after the initial establishments in West Africa. Plants with *T. aripo* had significantly higher average root yields (+36%) compared to plants with the predator selectively removed (Fig. 5); a comparable study some years later in Tanzania also found significantly higher root yields (+61%) from plants with *T. aripo* (Pallangyo et al., 2007). The economic value of *T. aripo* biological control on *M. tanajoa* was estimated at US$206/ha with a total of US$74 million for Benin, US$336/ha and US$383 million total for Ghana, and US$303/ha and US$1.7 billion total for Nigeria (Alene et al., 2005); estimates comparable to values found for the biological control of cassava mealybugs.

Research on pathogenic fungi began with a pre-release survey in Benin of pathogens associated with local cassava mite populations. Both *Hirsutella thompsonii* Fisher (Hypocreales: Ophiocordycipitaceae) and *Neozygites tanajoae* were found (Yaninek et al., 1996), but local fungal epizootics were relatively weak, prompting the importation of virulent *N. tanajoae* isolates as candidate biological control agents. Although laboratory assays failed to distinguish virulence differences between local and exotic isolates, field releases made in Benin in 1999 showed persistence and greater epizootics on cassava green mite populations that received the exotic *N. tanajoae* compared to those with local isolates (Hountondji et al., 2002).

The campaign against cassava green mite has been successful beyond expectations, but several issues still warrant attention. On-farm assessment and documentation of the efficacy and economics of the mite campaign are needed in most cassava-producing countries. This includes an evaluation of current cassava mite pests, particularly given today's agricultural market

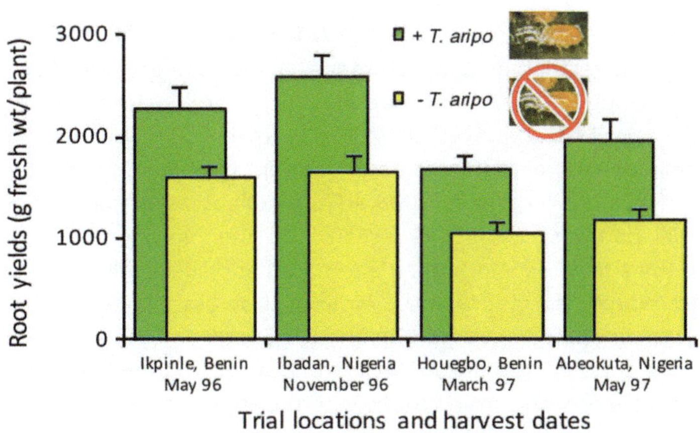

Figure 5 Average cassava root yields from plants with and without *Typhlodromalus aripo*.

pressures and evolving production practices, on-farm impact of released mite natural enemies on cassava production, and a comprehensive cost:benefit analysis of the campaign. Moreover, the ambiguous taxonomy about *M. tanajoa* in Africa continues to be muddled in the literature (Yaninek, 2007) and raises questions about the species and origin in South America, and now in Southeast Asia. Better diagnostic tools and a taxonomic revision of the mite species in the genus *Mononychellus* found on cassava would go a long way in sorting out the persistent confusion. Finally, work on fungal pathogens of *M. tanajoa* deserves a renewed effort with highly virulent isolates and practical molecular tools to diagnose and monitor field releases.

Interestingly, *T. aripo*, a natural enemy that was not a particularly efficient predator, turned out to be the most effective biological control agent. It was the best at becoming established, dispersing and persisting on cassava because of its ability to inhabit shoot tips for shelter, locate and survive on prey and non-prey food across a wide range of pest mite densities, and be present on cassava to thwart incipient *M. tanajoa* populations. This was the first time a mite pest was ever controlled continent-wide, and the campaign introduced the study of mites as a discipline and a pest management subject to scores of individuals and dozens of national programmes. Strategic collaborations filled critical gaps in project expertise, just as the partnerships with national programmes made the project relevant and meaningful to cassava-producing countries, and training left a legacy that continues to pay dividends. Finally, it was the long-term support from the donor community that created and sustained the momentum needed to make the project successful.

2.3 African root and tuber scale

The ARTS *Stictococcus vayssierei* is a subterranean insect indigenous to the humid forest zone of Central Africa where it has increasingly become a major pest of cassava since its description in 1971 (Richards, 1971). Initial research information was restricted to its presence in the Congo Basin (Cameroon, Republic of Congo, Democratic Republic of Congo) (Lutete et al., 1997; Mutsaers et al., 1981; Bani et al., 2003; Ngeve, 2003), but unfortunately, little information was available to assist farmers in managing this emerging pest (Ambe et al., 1999). Since 2003, IITA in collaboration with national and international partners conducted research on the biology and ecology of ARTS with the aim of developing an integrated and sustainable management strategy to reduce infestations and their impact on cassava productivity. Most research was conducted in Cameroon and DR Congo.

Extensive surveys established the presence of ARTS from Western Uganda to Equatorial Guinea and from Northwestern Cameroon to Southwestern DR Congo at elevations from 300 m to 1500 m (Tata-Hangy et al., 2006). ARTS had

Published by Burleigh Dodds Science Publishing Limited, 2019.

been considered the only subterranean stictococcid species until 2010, when a new species, *Stictococcus subterreus* Williams, Matile-Ferrero and Miller, was described from Southeastern Nigeria (Williams et al., 2010). The description of this new subterranean species, along with the breadth of ARTS distribution, its wide host range and the presence of ARTS 'hot spots' provided the basis to hypothesize about the existence of host/geographic-adapted biotypes or cryptic species with substantial implications for the development of sustainable management options - particularly host plant resistance. Morphological and molecular analyses on 213 ARTS samples collected from 14 host plants in 128 locations in seven countries of the Congo Basin showed a high genetic diversity of ARTS, possibly caused by its association with several species of endosymbionts known to affect speciation (O'Malley, 2015). Present evidence supports the conclusion of the existence in the ARTS complex of two cryptic species geographically separated: one species confined to the Western savannah highlands of Cameroon and the other a widespread species found throughout the humid forest of the Congo Basin, made up of at least four biotypes, with one of the biotypes associated with 'hot spots' of severe scale infestations in the areas bordering Cameroon, Equatorial Guinea and Gabon. This information supports specific control options, in particular the request to limit the movement of infested planting materials from hot spots to other areas.

To understand the factors that led to the relatively rapid emergence of ARTS from an unknown species to a major threat to cassava production in the Congo Basin, the initial research efforts focused on the effects of land use patterns. The length of fallows and the host plants found in these fallows were shown to be the primary contributors to ARTS outbreaks in cassava fields (Tata-Hangy et al., 2006). Fields in short fallows generally supported much larger ARTS densities than those in long fallows or secondary forest. Several indigenous host plants (16 in total) and particularly volunteer cassava were more common in short fallow vegetations and served as reservoir of the scale infesting newly planted cassava (Tindo et al., 2009). These patterns of ARTS abundance were not related to soil physical and chemical properties or to soil organic matter. In farmer participatory trials, the effects of removing leftover host plants and their residues prior to field establishment was established as an effective cultural control option. The second factor studied was the association of ARTS with the ant *Anoplolepis tenella* (Santschi) (Hymenoptera: Formicidae) that favours the proliferation of the pest (Dejean and Matile-Ferrero, 1996). The ant feeds on the honeydew secreted by the scale and in turn protects the scale from drowning in its own secretion. Exclusion experiments on ants showed that the scale cannot survive in nature without *A. tenella*, but this ant can still be found (though rarely) without the scale, indicating a very close, but facultative relationship between the ant and the scale (Hanna et al., 2015). Studies on *A. tenella* nest distribution showed that nests were in most cases located at the base of scale-infested crop plants, mainly cassava (94%). Ant workers disperse ARTS juvenile stages and are therefore the main cause of the spread of *S. vayssierei* (Fotso Kuate et al., 2015).

Host plant resistance, ant control, biological control, chemical control and cultural methods were evaluated as ARTS control methods. In various locations in Cameroon and DR Congo, hundreds of cassava clones were screened for resistance. While no clones showed resistance, several were identified as tolerant with other positive attributes (e.g. resistance to virus, fungal and bacterial diseases). Ant exclusion was an obvious approach to scale control given that the scale depends on *A. tenella* for its survival (Fotso Kuate et al., 2015). A 1% boric acid-sucrose solution significantly reduced ant density in treated fields, resulting in a reduction of scale infestation on cassava and a 16% yield increase (Hanna et al., 2015). No parasitoids were ever found associated with ARTS, but the possible role of soil predators was never quantified. Entomopathogenic nematodes and entomopathogenic fungi *Metarhizium anisopliae* Metschn. and *Beauveria bassiana* (Bals.-Criv.) Vuill. (both Hypocreales: Clavicipitaceae) were found associated with the scale during several surveys in Cameroon, but infection levels were typically very low (1–9%) (L. Wijnans et al., unpublished results; Ngo Kanga, 2012). They are unlikely to play a role in scale population dynamics, unless they are applied inundatively to soils, which cannot be economically justified for cassava in the zones affected by the scale. Chemical control is unpracticable for farmers' use, although reduction in scale density that significantly increased cassava root yield was achieved. For instance, scale exclusion alone increased yield of improved varieties and of farmers' varieties by 26% and 40%, respectively. By contrast, fertilizer alone in the presence of a heavy ARTS infestation benefitted only the local variety with a 91% increase in yield. Scale exclusion and soil fertilization further increased yield over the control: 54% for the improved variety against 225% for the local variety. Finally, cultural methods comprising vegetation management and soil fertility were tested. Farmers' participatory trials, using the best options to reduce ARTS were conducted in two villages in Southern Cameroon. They found that host plant removal and burning prior to planting reduced scale densities by up to 40% and increased cassava root yield by 39%. Although host plant residue removal has been shown to be an effective ARTS management option (Fig. 6), the practice has not been widely adopted, principally due to the necessary increase in labour.

2.4 Variegated grasshopper

Cassava is host to numerous species of grasshoppers in Africa and the Americas, but only the grasshoppers *Zonocerus variegatus* and its congenic *Zonocerus elegans* (Thumber) are major pests of cassava in Africa (Bellotti and van Schoonhoven, 1978). These two species may also disseminate *Xanthomonas axonopodis* pv. *manihotis* (Pammel) Dowson (Xanthomonadaceae), which causes cassava bacterial blight (Daniel et al., 1980; Zandjanjakou-Tachin

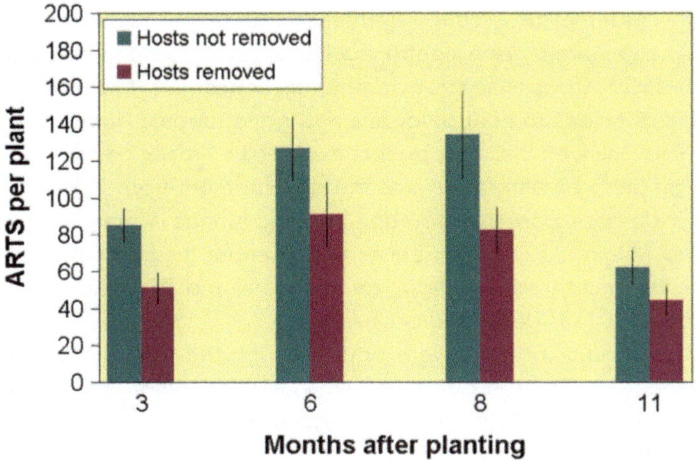

Figure 6 Effect of vegetation management on ARTS infestations over a period of 11 months.

et al., 2007). In West and Central Africa, *Z. variegatus* chews cassava leaves, petioles and green stems causing plants' defoliation and debarking of stems. Defoliation is of a particular concern in areas where cassava leaves are consumed as vegetables. Similarly, the debarking of cassava stems reduces the quantity and quality of planting materials used to plant new fields. Whether defoliation negatively affects tuber yield has been a subject of debate. Indeed, *Z. variegatus* generally attacks cassava at the end of the dry season, when the plants are generally already well developed and the growth of tubers is minimal. Moreover, after defoliation cassava usually recovers quickly. In such cases, there might be no yield loss, but a negative impact on the quality of tubers has been reported. In Southern Nigeria, for instance, over 50% of the cassava crop was estimated lost in years of high abundance of *Z. variegatus* (Mansaray et al., 2012). Economic damage is also caused when *Z. variegatus* attacks vegetables and other cash crops such as banana, citrus and coffee (D. Affoyon and J.M.Castel, unpublished results; Chiffaud and Mestre, 1990).

Spraying huge volumes of broad-spectrum chemical pesticides on vast tracts of land has been, for a long time, the most common control measure against locusts and grasshoppers used by farmers who can afford it and/or by governments. Exhumation of egg pods following the identification of oviposition sites, through community action by farmers was also recommended, given that 83% and 90% reductions in the populations were obtained in different studies in 1990 and 1994, respectively (Modder, 1995).

Because most, if not all, insecticides sprayed during the control campaign against *Z. variegatus* in the cassava ecosystem resulted in significant increase of cassava mealybug populations, by killing off the parasitoid *A. lopezi*

(Neuenschwander et al., 1986), alternative approaches were required. The LUBILOSA (Lutte biologique contre les Locustes et Sauteriaux) project was initiated in 1989 in response to environmental concerns over the heavy use of chemical insecticides to control locusts and grasshoppers during the 1986-9 plagues. This was a multi-donor project executed by many partners, including IITA. The project's breakthrough was the development from 1992 to 1997 of a biopesticide based on an entomopathogenic fungus, *Metarhizium acridum* (Driver and Milner) J.F. Bisch., Rehner and Humber, named Green Muscle®, which effectively controls grasshoppers in general and *Zonocerus* in particular (Lomer et al., 2001) as described in Chapter 13.

The best results in Benin were obtained with the strain I91 609 isolated from *Z. variegatus* (Douro-Kpindou et al., 1995). Indeed, in a field trial using cassava in 1994, this strain caused a significant decline in the population of *Z. variegatus*. In another comparative study to determine its efficacy on two grasshoppers namely *Z. variegatus* and the Senegalese grasshopper, *Oedaleus senegalensis* (Krauss) (Orthoptera: Acrididae), this strain killed *Z. variegatus* faster than *O. senegalensis* (Fig. 7). Once a grasshopper is killed, the development of hyphae causes the mummification of the corpse or cadaver.

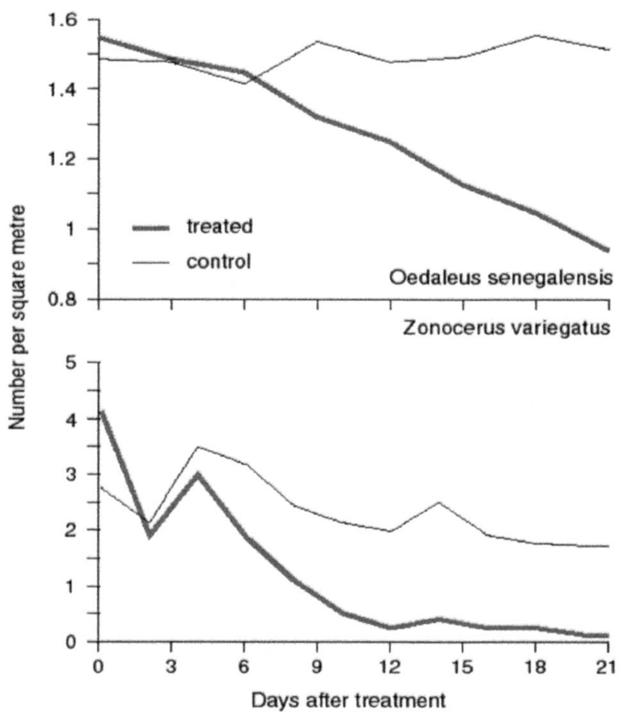

Figure 7 Comparative efficacy of *Metarhizium acridum* against *Oedaleus senegalensis* (above) and *Zonocerus variegatus* (below).

Figure 8 Manifestation of *Metarhizium acridum* on a grasshopper (a). Spores and mycelium of *Metarhizium acridum* growing on a cadaver of a mycosed *Zonocerus variegatus* (b).

Figure 9 Applying the biopesticide Green Muscle® by hand (a), on an adapted vehicle (b) and by airplane (c).

Thereafter *M. anisopliae* continue their development and emerge outside the corpse by releasing conidia (Fig. 8). These will in turn infect other individuals when the climatic conditions become favourable. *Metarhizium* can be sprayed using any standard ultra-low volume equipment (Fig. 9). It takes 14 days to achieve the maximum efficacy of the biopesticide against *Z. variegatus,* but longer to control *O. senegalensis*. Economic studies are now needed to support the use of Green Muscle® to control *Z. variegatus* on cassava.

3 Identifying and managing cassava diseases

The section covers four diseases – cassava bacterial blight (CBB), cassava anthracnose disease (CAD), cassava leaf spots, cassava root and stem rots and three exotic diseases found outside Africa, which have up to now been kept out by quarantine procedures.

3.1 *Cassava bacterial blight*

Cassava bacterial blight caused by *Xanthomonas axonopodis* pv. *manihotis* is one of the most serious diseases affecting cassava in Africa (Fanou et al., 2018).

It is present in all ecozones, though with variable incidences and severities (Wydra and Msikita, 1998; Wydra and Verdier, 2002). Our studies over more than two decades on the epidemiology and assessment of crop loss due to CBB showed yield losses of higher than 50% in a susceptible and 24% in a more resistant cassava variety (A. Fanou, unpublished results). Extent of losses depended on cassava varieties, virulence of the bacterial strains and environmental conditions. The CBB pathogen was disseminated by several means: It was detected in all parts of symptomatic cassava stem cuttings, but also from asymptomatic plants from infected fields, and even in stems of resistant varieties (Fessehaie et al., 1997; A. Fanou, unpublished results). Alternative hosts were not identified among weeds, but epiphytic survival occurred on several weed species frequently associated with cassava (Fanou et al., 2017). Also, cassava debris was identified as an important inoculum source, specifically after the dry season. Survival of CBB was longer in infected leaves on the soil surface than those buried in the soil. Even botanical seeds harboured the bacteria (Fanou and Wydra, 2015), and CBB symptoms developed already in emerging plants. Some insects frequently found in cassava fields, for example the grasshopper, *Zonocerus variegatus* and others, carried the pathogen on their legs, mandibles, in their intestines and faeces and were demonstrated to be vectors for *X. axonopodis* pv. *manihotis* (Zandjanakou-Tachin et al., 2007).

CBB can be managed properly by using cultural practices. These include soil preparation to avoid infested cassava debris on the soil before new planting, the use of disease-free cuttings collected from CBB-free plantations, weeding to avoid or suppress weeds that could lodge CBB, intercropping of cassava with maize (A. Fanou, unpublished data) or sorghum (Zinsou et al., 2004), early removal of symptomatic cassava leaves (Fanou and Wydra, 2014), botanical seed treatments (Fanou and Wydra, 2015), soil amendments (Zinsou et al., 2004), shift of planting date (Zinsou et al., 2004), choosing between early and late planting according to ecozones (Zinsou et al., 2004), control of insect vectors and the use of resistant varieties (A. Fanou, unpublished results; Zinsou et al., 2005; Wydra et al., 2007; Banito et al., 2008). The distribution of stomata and the formation of cassava leaf waxes were suggested to play a role in suppressing the disease (Zinsou et al., 2006), and the use of newly identified markers (QTL) for CBB resistance (Wydra et al., 2004) is recommended to breeders to select resistant genotypes.

3.2 Cassava anthracnose disease

Cassava anthracnose disease, caused by *Colletotrichum gloeosporioides* f. sp. *manihotis* Chevaug. (Glomerellaceae), is a severe disease that affects cassava across sub-Saharan Africa and has the potential to cause total crop failure (Fokunang et al., 2000a). CAD manifests itself with a variety of symptoms. It

Published by Burleigh Dodds Science Publishing Limited, 2019.

can appear as cankers in stems, branches and tubers, or as spots and die-back of leaf tips. Susceptibility to CAD depends on the cultivar and environmental conditions. Research efforts by IITA and its partners determined for the first time that the pathogen can be seed-borne and seed-transmitted (Fokunang et al., 1997). In addition, *Pseudotheraptus devastans* (Distant) (Hemiptera: Coreidae) plays a role in the transfer of the disease (Fokunang et al., 2000b). Inoculum thresholds necessary for infection and subsequent crop losses as well as environmental conditions promoting establishment of the disease were established (Fokunang et al., 2000c).

Chemical and cultural strategies to control CAD are often unavailable to farmers or, if implemented, are generally insufficient to reduce CAD incidences and severities. Therefore, breeding for resistance is the most promising strategy to control CAD. IITA developed rapid screening methodologies to evaluate resistance to CAD (Fokunang et al., 2002a,b). Later on, research shifted to complex, laborious screenings of 436 African landraces and 497 improved genotypes under field conditions during four years. That study revealed several landraces and improved genotypes that were consistently resistant to CAD in addition to several landraces completely immune to the disease (Owolade et al., 2005). Results from that study serve as the foundation to introgress CAD resistance genes into improved genotypes to diversify resistance to CAD. Subsequent studies determined that maternal and cytoplasmic inheritance is involved in CAD resistance and it was concluded that recurrent selection is the most appropriate method to accumulate genes for CAD resistance (Owolade et al., 2006). Molecular studies later identified DNA markers linked to CAD resistance genes and these are valuable tools for application in molecular breeding for resistance to CAD (Akinbo et al., 2007).

Future efforts should be directed towards understanding the genetic diversity of *C. gloeosporioides* f.sp. *manihotis* across Africa. This will allow testing susceptibilities of resistance genotypes to novel strains/races of the pathogen and designing appropriate CAD management strategies. Other efforts should focus on using the DNA markers identified by Akinbo et al. (2007) to detect CAD resistance in germplasm of interest.

3.3 Cassava leaf spot diseases

The most important leaf spot fungal diseases of cassava other than those caused by *C. gloeosporioides* f.sp. *manihotis* are brown leaf spot (BLS) caused by *Mycosphaerella henningsii* Allesch. and *Cercospora* leaf blight (CLB) caused by *Cercospora caribae* Cif. (both Mycosphaerellaceae) (Mahungu et al., 1994). CLB was detected by CIAT scientists during surveys of fungal and bacterial diseases conducted in 1978 in both Tanzania and Kenya (J. Persley, unpublished results). CLB was thereafter reported in both Benin and Ghana, although relatively sporadically and with minor importance (Wydra and Verdier, 2002). These authors also noticed that CLB was more severe in profusely branching varieties and was associated with higher rainfall, although its incidence decreased in plantations with mixed cultivars.

In the early 2000s, increased cassava cultivation occurred in Nigeria as a result of the presidential initiative for cassava expansion (Sanogo and Adetunji, 2008) and IITA's special project 'Preemptive management of cassava mosaic disease in Nigeria'. This led to extensive research on prevalence of cassava diseases in Nigeria. Both CLB and BLS were detected in all examined ecozones (Onyeka et al., 2004). These authors found that both diseases decreased in severity towards the savannah zone. The two diseases are suggested to pose minor threats to cassava production, but little is known about their interaction with other pathogens and pests, their spatio-temporal distribution and genetic diversity. Screening cultivars for resistance to both diseases and detecting markers linked to CLB and BLS resistance should be considered.

3.4 Cassava root rot diseases

Cassava root and stem rots are important cassava diseases in different ecozones of West Africa (Hillocks and Wydra, 2002; Wydra and Msikita, 1998; Msikita et al., 1997). Yield losses of up to 80% due to rots have been reported (R.L. Théberge et al., unpublished results). A study investigating the relationship between root rot disease severity and cassava yield losses also indicated up to 56% losses in susceptible cultivars (Onyeka et al., 2005a). Other studies identified *Lasiodiplodia* (*Botryodiplodia*) *theobromae* (Pat.) (Botryosphaeriaceae), *Sclerotium rolfsii* Sacc. (*incertae sedis*) and *Fusarium* spp. as important causal pathogens of the disease in Cameroon, Nigeria and Benin (L. Afouda and K. Wydra, unpublished reults). Surveys conducted in 1999 and 2000 in Togo and Benin showed *B. theobromae* as the most common cassava root rot pathogen in various moist savannah and forest zones of Benin and Togo. Eight more fungal species were identified as causal agents of root or stem rots (Banito et al., 2010; Zinsou et al., 2017). Based on the disease reaction after inoculation under controlled conditions, 88% of the highly virulent fungi were *B. theobromae* and occurred in 57% of all samples from the forest mosaic savannah in Benin (Onyeka et al, 2005b; Zinsou et al., 2017). Similar results were recorded in Togo (Banito et al., 2010).

Preliminary studies revealed that the antagonistic fungus *Trichoderma* sp. was capable of partially controlling root rots, besides its growth-enhancing effects (L. Afouda, unpublished results). During surveys in Benin, Msikita et al. (2005) frequently isolated *Nattrasia mangifera* Syd. (Botryosphaeriaceae), while *B. theobromae* and *Fusarium* spp. were the major root rot pathogens isolated in Nigeria (Onyeka et al., 2004). This led IITA to develop *in vitro* inoculation methods to rapidly detect resistance to *B. theobromae* among cassava genotypes (Onyeka et al., 2005c), while parallel studies from different geographical locations (Onyeka et al., 2004) revealed significant differences in aggressiveness between *B. theobromae* isolates (Onyeka et al., 2005b).

IITA then partnered with Kansas State University to examine cassava root rots in Nigeria and Cameroon emphasizing *Fusarium* species. Results revealed 13 groups of *Fusarium* species based on amplified fragment length polymorphism assays (Bandyopadhyay et al., 2006). Within those groups the two most prevalent were the *Fusarium oxysporum* Snyder and Hansen (Nectriaceae) and *Fusarium solani* (Mart.) Sacc. species complexes. In addition, a relationship between root rot pathogens and the weed *Chromolaena odorata* (L.) King and Robinson (Asteraceae: Eupatorieae), which is common in humid forest areas, was established. In Nigeria, nationwide evaluations in over 2500 locations determined differences in susceptibilities to root rot among 43 improved varieties, but found no resistant genotypes across environments (Okechukwu et al., 2009). Early harvest (9–11 months after planting) was therefore suggested as the best method to reduce yield losses associated with root rot diseases (Odedina et al., 2017).

Further research is needed to better understand the role and ability of the fungi in the colonization of the cassava root, and the interaction between species, isolates and genotypes of root rots and other pathogens and weeds, and to explore opportunities for biological control.

3.5 Cassava diseases outside Africa

Like the exotic cassava mealybug and cassava green mite, there exist some prominent cassava diseases outside Africa that should be prevented from entering the continent. These include frog skin disease, superelongation disease and the witches' broom disease. The latter has emerged as a major threat to cassava production in Southeast Asia where yield losses of up to 80% have been reported (Graziosi et al., 2016). Hence, being informed of such diseases and their causal agents, mode of spread as well as their impact on cassava production is important so that the existing tools (e.g. sensitization, phytosanitary measures, efficient diagnosis and surveillance) can be activated if accidental introduction occurs. The accidental introductions of the cassava mealybug and cassava green mites are still fresh in our mind: No farmer, extension worker or decision makers would like to experience new introductions into Africa.

4 Conclusion and future trends

About a dozen arthropod pests of cassava are capable of causing different levels of damage, contradicting the notion that cassava is a crop relatively free of pest attacks. In this chapter, we considered four arthropod pest species, namely the exotic cassava mealybug and cassava green mite, the indigenous ARTS and the indigenous and polyphagous *Zonocerus* grasshopper. We describe proven technologies developed to limit or contain them, which can be adopted and/or adapted to manage other major or emerging cassava pests.

The two exotic arthropod pests – cassava mealybug and cassava green mite – were managed through well-funded and carefully implemented biological control programmes because these pests could not be satisfactorily controlled by genetic improvement. Selecting for pest and disease resistance is a long process: a new variety would take seven years from seedlings to multiplication before being ready to be released, usually requiring the approval of a national release committee (IITA, 1990). Meanwhile the exotic invaders spread rapidly in every direction. The two well-documented programmes, by contrast, resulted in rapid establishment and spread of two biological control agents *Anagyrus lopezi* and *Typhlodromalus aripo* with measurable socio-economic impacts within a few years. Interestingly, the successes of these programmes are attributable to biological control agents that, in laboratory settings, would not be deemed efficient. The keys to success were the astonishing host finding capacity of *A. lopezi*, which was by far greater than that of any of its competitors, and the ability of *T. aripo* to inhabit shoot tips for shelter, to locate and survive on prey and non-prey food across a wide range of pest mite densities. Additionally, Brazilian isolates of the fungus *Neozygites tanajoae* were selected and screened against cassava green mites. Efficacious isolates could be an alternative option for cassava green mite control, important notably for consumers of cassava leaves in many parts of Africa where *T. aripo* failed to establish.

Cassava green mite control benefitted from experience gained from the cassava mealybug biological control programme and was later consolidated in the ESCaPP project. Despite generally successful control, local conditions, in which damage by cassava mealybug and green spider mite still exist, are well described. In Africa, no new exotic pests of cassava have since passed quarantine services. But should new introductions occur, the basis established by these two projects, namely the know-how and well-trained personnel on the continent, could easily be mobilized to face a new invasion.

The polyphagous grasshopper *Zonocerus variegatus* has been generally managed on various crops by indiscriminate sprays of chemical pesticides. Thanks to the LUBILOSA project, a biopesticide based on *Metarhizium anisopliae* was developed and found to be effective against grasshoppers generally including those found on cassava.

The indigenous ARTS has increasingly become a major pest of cassava in Central Africa since its initial description in 1971. Initially, little information was available to help farmers manage this emerging pest, and after more than 10 years of research, IITA with its national and international collaborators is still learning about the pest's biology, and concludes that the best immediate control action is the use of tolerant cassava varieties.

The diseases section of our chapter treats only the four most common cassava diseases, three of which are exotic - the viral diseases are addressed in Chapter 5. During the past two to three decades, IITA pathologists and their collaborators identified the causal agents and determined their epidemiology to understand the effects of cassava diseases on crop losses. Field and laboratory experiments conducted in diverse agro-ecozones in Africa yielded useful new knowledge which was exploited to develop management options essentially based on cultural practices for cassava bacterial blight and breeding for resistance for CBB and CAD. While research continues on root/stem rot diseases, three important causal pathogens and many other pathogen species were identified. Presently, early harvesting of cassava roots is the best method to reduce yield losses. The cassava leaf spot and exotic diseases treated here draw our attention to potential diseases that may threaten cassava production.

In the recent past, besides identifying and managing cassava pests and diseases as demonstrated above, there have been equally important issues that constrain yield and keep cassava from becoming more widely adopted as a cash crop. These include limited adoption of improved varieties, low use of herbicides and mechanization, limited use of fertilizers, irrigation, low storage life, low protein content of tubers, high labour use/cost, limited access to credit and high transportation costs. Additional challenges to sustainable cassava production in a changing world are likely to include managing water and soil nutrients, and climate change, for example temperature, relative humidity, CO_2 and so on (Burns et al., 2010). These production challenges need to be addressed cautiously by increasing the technical knowledge of cassava growers so that they enhance the integration of plant protection and plant production practices in order to grow a healthy crop (James et al., 2000b). Indeed, the expected booming demands required to nourish increasing populations will offer millions of cassava growers the opportunity to intensify production and earn higher incomes. However, how small-scale growers will choose to improve cassava productivity should be a matter of concern to policymakers in order to prevent them from making the same mistakes as during the Green Revolution in cereal production that jeopardized productivity at the expense of the agricultural resource base (FAO, 2013). With improved, sustainable management, there is little doubt that cassava yield will increase, and benefit cassava producers, particularly those who move beyond subsistence farming to become cash crop farmers.

5 Where to look for further information

This chapter has concentrated on the management (through four highlighted or mentioned interventions – host plant resistance, biological control, cultural or agronomic, and, indirectly, chemical methods) of selected cassava pests and diseases in the traditional cropping systems in which cassava had been cultivated since the work on cassava pest control started in the 1930s. However, the expected shift in cassava production from its current traditional to commercial and industrial crop to meet needs for food for increasing human population, animal feed and industrialization coupled with climate change will undoubtedly exacerbate attacks of existing cassava pests and diseases and even turn secondary into major pests/diseases, thereby requiring novel approaches in the way we will manage our crops' pests and diseases. For cassava, such novel approaches for research in view to reduce damage include, not exclusively, improved understanding of pests' bioecology, interactions between cassava-pests/diseases and climatic parameters notably temperature and humidity, modelling to predict pests/diseases outbreaks, cloning to produce resistant varieties, activation of cassava chemical defence, appropriate use of information technology to diffuse produced technologies and so on.

There are many hundreds of published items related to our chapter 'Identifying and managing plant health risks for key African crops: cassava'. However, the following review articles are good references for the cassava mealybug and cassava green mite while another article discusses cassava diseases and their control. We have also added two articles related to the topic for leisure reading.

1 Neuenschwander, P. 2001. Biological control of the cassava mealybug in Africa: a review. *Biological Control* 21: 214–229.
2 Yaninek, S and R. Hanna, 1993. Cassava green mite in Africa - a unique example of successful classical biological control of a mite pest on a continental scale. In: *Biological Control in IPM Systems in Africa*, P. Neuenschwander, C. Borgemeister and J. Langewald (eds) pp 61–75.
3 Lozano, J.C and E.R. Terry. 1976. Cassava diseases and their control. In: Proceedings of the fourth symposium of the International Society for Tropical Root Crops, held at CIAT, Cali, Colombia, 1–7 August 1976 (IDRC). J.H. Cock (ed) pp 156-160.
4 Makundi, R.H. 2006. Challenges in pest management in agriculture: African and global perspectives. In: *Management of some Selected Crops Pests in Tanzania*, R.H. Makundi (eds) 19 pp.
5 Bellotti, A, Vanessa Herrera Campo, B and Hyman, G. 2012. Cassava production and pest management: present and potential threats in a changing environment. *Tropical Plant Biology*. DOI 10. 1007/s12042-011-9091-4. 39 pp.

Centers of expertise include

- The International Institute of Tropical Agriculture, PMB 5320, Oyo Road State, Ibadan, Nigeria. www.iita.org/

Published by Burleigh Dodds Science Publishing Limited, 2019.

- The International Center for Tropical Agriculture, Km 17, Recta Cali-Palmira CP 763537, Apartado Aéro 6713, Cali, Colombia
- Food and Agriculture Organization of the United Nations (FAO). Viale delle Terme di Caracalla 00153, Rome, Italy, FAO-HQ@fao.org

6 References

Agboton, B. V., Delalibera, I., Hanna, R. and von Tiedemann, A. 2009. Molecular detection and differentiation of Brazilian and African isolates of the entomopathogen *Neozygites tanajoae* (Entomophthorales: Neozygitaceae) with PCR using specific primers. *Biocontrol Science and Technology* 19: 67–79.

Akinbo, O., Gedil, M., Ekpo, E., Oladele, J. and Dixon, A. G. 2007. Detection of RAPD markers-linked to resistance to cassava anthracnose disease. *African Journal of Biotechnology* 6:677.

Alene, A. D., Neuenschwander, P., Mangyong, V. M., Coulibaly, O. and Hanna, R. 2005. The impact of IITA-led biological of major pests in sub-Saharan African agriculture: A synthesis of milestones and empirical results. Impact Series, IITA, Ibadan, Nigeria, 26pp.

Ambe, J. T., Ntonifor, N. N., Awah, E. T. and Yaninek, J. S. 1999. The effect of planting dates on the incidence and population dynamics of the cassava root scale, *Stictococcus vayssierei* in Cameroon. *International Journal of Pest Management* 45: 125–30.

Bandyopadhyay, R., Mwangi, M., Aigbe, S. O. and Leslie, J. F. 2006. Fusarium species from the cassava root rot complex in West Africa. *Phytopathology* 96: 673–6.

Bani, G., Mapangu Divassa,. S., Nzemba, M. D. and Maguem, J.-M. 2003. Presence au Congo-Brazzaville de *Stictococcus vayssierei* Richard, ravageur du manioc (*Manihot esculenta* Crantz) (Hem., Stictococcidae). *Bulletin de la Sociéte entomologique de France* 108: 529–30.

Banito, A., Kpémoua, K. E. and Wydra, K. 2008. Expression of resistance and tolerance of cassava genotypes to bacterial blight determined by genotype x environment interactions. *Journal of Plant Diseases and Plant Protection* 115: 152–61.

Banito, A., Kpemoua, K. E., Bissang, B. and Wydra, K. 2010. Assessment of cassava root and stem rots in ecozones of Togo and evaluation of the pathogen virulence. *Pakistan Journal of Botany* 42: 2059–68.

Bellotti, A. C. 2002. Cassava arthropod pests. In: Hillocks, R. J., Thresh, J. M. and Bellotti, A. C. (Eds), *Cassava: Biology, Production and Utilization*. CABI-Publishing, Wallingford, UK, pp. 209–34.

Bellotti, A. C. and van Schoonhoven, A. 1978. Mite and insect pests of cassava. *Annual Review of Entomology* 23: 39–67.

Burns, A., Gleadow, R., Cliff, J., Zacarias, A. and Cavagnaro, T. 2010. Cassava: The drought, war and famine crop in a changing world. *Sustainability* 2: 3572–607.

CABI. 2018. *Stictococcus vayssierei* (Cassava brown root scale). Distribution map. CAB International, Wallingford, UK.

Centro Internacional de Agricultura Tropical (CIAT). 1993. Cassava: The latest facts about an ancient crop – A summary of information on 52 major cassava producing and consuming countries in Africa, Asia, Oceania, Latin America and the Caribbean. Aparado Aéreo 6713, Cali, Colombia. FAO Agrostat Database, 1992, World Bank, World Bank Tables 1992, CIAT, Cassava Program.

Chiffaud, J. and Mestre, J. 1990. Le Criquet puant *Zonocerus variegatus* (Linné, 1758). Essai de synthèse bibliographique. Ministère de la Coopération, Paris and CIRAD, Montpellier, 140pp.

Cock, J. 1985. *Cassava New Potential for a Neglected Crop*. Westview Press, Boulder, CO, 191pp.

Daniel, J. F., Boher, B. and Nkouka, N. 1980. Insect dissemination of *Xanthomonas campestris* pv. *manihotis* to cassava in the People's Republic of Congo. In: Terry, E. R., Oduro, K. A. and Caveness, F. (Eds), *Tropical Root Crops Research Strategies for the 1980s*, pp. 66–8.

Dejean, A. and Matile-Ferrero, D. 1996. How a ground-nesting ant species favors the proliferation of an endemic scale insect (Hymenoptera: Formicidae; Homoptera, Stictococcidae). *Sociobiology* 28: 183–95.

Delalibera, Jr., I., Hajek, A. E. and Humber, R. A. 2004. *Neozygites tanajoae sp. nov.*, a pathogen of the cassava green mite. *Mycologia* 96: 1002–9.

Donkor, E., Onakuse, S., Bogue, J. and de Los Rios Carmenado, I. 2017. The impact of the presidential cassava initiative on cassava productivity in Nigeria: Implication for sustainable food supply and food security. *Cogent Food and Agriculture* 3: 1368857.

Douro-Kpindou, O.-K., Godonou, I., Houssou, A., Lomer, C. and Shah, P. 1995. Control of *Zonocerus variegatus* by ultra-low volume application of an oil formulation of *Metarhizium flavoviride* conidia. *Biocontrol Science and Technology* 5: 131–9.

Fanou, A. A. and Wydra, K. 2014. Removal of symptomatic cassava leaves as cultural practice to control cassava bacterial blight. *International Journal of Phytopathology* 3: 117–24.

Fanou, A. A and Wydra, K. 2015. Physical and chemical treatment for the control of *Xanthomonas axonopodis* pv. *manihotis* in cassava seeds. *Journal of Experimental Biology and Agricultural Sciences* 3: 54–9.

Fanou, A. A., Zinsou, A. V. and Wydra, K. 2017. Survival of *Xanthomonas axonopodis* pv. *manihotis* in weed species and in cassava debris: Implication in the epidemiology of cassava bacterial blight. *International Journal of Advanced Research* 5: 2098–112. doi:10.21474/IJAR01/4057.

Fanou, A. A., Zinsou, V. A. and Wydra, K. 2018. Cassava bacterial blight: A devastating disease of cassava. In: *Cassava*. IntechOpen. doi:10.5772/intechopen.71527. Available from: https://www.intechopen.com/books/cassava/cassava-bacterial-blight-a-devastating-disease-of-cassava

Fessehaie, A., Wydra, K. and Rudolph, K. 1997. Development of a new semi-selective medium for isolating *Xanthomonas campestris* pv. *manihotis* from plant material and soil. *Phytopathology* 89: 591–7.

Flechtmann, C. H. W. 1978. The cassava mite complex: Taxonomy and identification. In: Brekelbaum, T., Bellotti, A. C. and Lozano, J. C. (Eds), *Cassava Protection Workshop* (1977, Cali, Colombia). Proceedings. Centro Internacional de Agricultura Tropical (CIAT), Cali, CO (Serie CE-14), pp. 143–53.

Fokunang, C., Ikotun, T., Dixon, A. and Akem, C. 1997. First report of *Colletotrichum gloeosporioides* f. sp. *manihotis*, cause of cassava anthracnose disease, being seed-borne and seed-transmitted in cassava. *Plant Disease* 81: 695.

Fokunang, C., Ikotun, T., Dixon, A. and Akem, C. 2000a. Field reaction of cassava genotypes to anthracnose, bacterial blight, cassava mosaic disease and their effects on yield. *African Crop Science Journal* 8: 179–86.

Fokunang, C., Akem, C., Ikotun, T., Dixon, A. and Tembe, E. 2000b. Role of the insect vector, *Pseudotheraptus devastans*, in cassava anthracnose disease development. *European Journal of Plant Pathology* 106: 319–27.

Fokunang, C., Ikotun, T., Akem, C., Dixon, A., Tembe, E. and Koona, P. 2000c. Investigation of inoculum threshold and latent infection in *Colletotrichum gloeosporioides* f.sp. *manihotis* in cassava cultivars. *Pakistan Journal of Biological Science* 3: 713–16.

Fokunang, C., Dixon, A., Ikotun, T., Akem, C. and Tembe, E. 2002a. Rapid screening method of cassava cultivars for resistance to *Colletotrichum gloeosporioides* f.sp. *manihotis*. *Journal of Phytopathology* 150: 6–12.

Fokunang, C., Dixon, A., Ikotun, T., Asiedu, R., Tembe, E. and Akem, C. 2002b. *In vitro*, greenhouse and field assessments of cassava lines for resistance to anthracnose disease caused by *Colletotrichum gloeosporioides* f.sp. *manihotis*. *Mycopathologia* 154: 191–8.

Food and Agricultural Organization of the United Nations (FAO). 2013. Save and grow cassava – A guide to sustainable production intensification. FAO, Rome, 142pp.

Food and Agricultural Organization of the United Nations (FAO). 2017. FAOSTAT – Statistics Division, Rome. www.fao.org/faostat/en/#data

Fotso Kuate, A., Hanna, R., Nagel, P. and Tindo, M. 2015. Transport and dispersal of Stictococcus vayssierei (Homoptera, Stictococcidae) by *Anoplolepis tenella* (Hymenoptera, Formicidae). *Journal of Insect Behavior* 28: 426–35.

Gnanvossou, D., Yaninek, J. S., Hanna, R. and Dicke, M. 2003. Effects of prey mite species on life history of the phytoseiid predators *Typhlodromalus manihoti* and *Typhlodromalus aripo*. *Experimental and Applied Acarolology* 30: 265–78.

Grace, M. R. 1977. Cassava processing. FAO Plant Production and Protection Series No. 3 0259-2525. The Chief Editor, FAO, Rome.

Graziosi, I., Minato, N., Alvarez, E., Ngo, D. T., Hoat, T. X., Aye, T. M., Pardo, J. M., Wongtiem, P. and Wyckhuys, K. A. G. 2016. Emerging pests and diseases of South-east Asian cassava: A comprehensive evaluation of geographic priorities, management options and research needs. *Pest Management Science* 72: 1067–274.

Gutierrez, A. P., Neuenschwander, P., Schulthess, F., Herren, H. R., Baumgaertner, J. U., Wermelinger, B., Löhr, B. and Ellis, C. K. 1988. Analysis of biological control of cassava pests in Africa: Cassava mealybug *Phenacoccus manihoti. Journal of Applied Ecology* 25, 921–40.

Gutierrez, A. P., Neuenschwander, P. and van Alphen, J. J. M. 1993. Factor affecting biological control of cassava mealybug by exotic parasitoids: A ratio-dependent supply-demand driven model. *Journal of Applied Ecology* 30: 706–21.

Gutierrez, A. P., Yaninek, J. S., Neuenschwander, P. and Ellis, C. K. 1999. A physiologically-based tritrophic metapopulation model of the African cassava food web. *Ecological Modelling* 123: 225–42.

Hammond, W. N. O. and Neuenschwander, P. 1990. Sustained biological control of the cassava mealybug *Phenacoccus manihoti* (Hom.: Pseudococcidae) by *Epidinocarsis lopezi* (Hym.: Encyrtidae). *Entomophaga* 35: 515–26.

Hanna, R., Fotso Kuate, A., Nanga Nanga, S., Tindo, M. and Nagel, P. 2015. Boric acid for the suppression of the ant *Anoplolepis tenella* in cassava fields. *Crop Protection* 74: 131–7.

Herren, H. R. and Neuenschwander, P. 1991. Biological control of cassava pests in Africa. *Annual Review of Entomology* 36: 257–83.

Hillocks, R. J. and Wydra, K. 2002. Bacterial, fungal and nematode diseases. In: Hillocks, R. J., Thresh, J. M. and Bellotti, A. C. (Eds), *Cassava: Biology, Production and Utilization*. CABI-Publishing, Wallingford, UK, pp. 261–80.

Hountondji, F. C. C., Lomer, C. J., Hanna, R., Cherry, A. J. and Dara, S. K. 2002. Field evaluation of Brazilian isolates of *Neozygites floridana* (Entomophthorales: Neozygitaceae) for the microbial control of cassava green mite in Benin, West Africa. *Biocontrol Science and Technology* 12: 361–70.

International Institute of Tropical Agriculture (IITA). 1990. Cassava in tropical Africa - A reference manual. Edited and designed by Chayce Publication Services, United Kingdom, pp. 176.

James, B., Yaninek, J., Neuenschwander, P., Cudjoe, A., Modder, W., Echendu, N. and Toko, M. 2000a. Pest control in cassava farms - IPM field guide for extension agents. International Institute of Tropical Agriculture (IITA), ISBN 978-131-134-184. Modern Design Ltd, Lagos, Nigeria, 20pp.

James, B., Yaninek, J., Tumantey, A., Maroya, N., Dixon, A., Salawu, R. and Kwarteng, J. 2000b. Starting a cassava farm- IPM field guide for extension agents. International Institute of Tropical Agriculture (IITA), ISBN 978-131-134-184. Modern Design Ltd, Lagos, 12pp.

Jones, W. O. 1959. *Manioc in Africa*. Stanford University Press, Stanford, CT, 315pp.

Legg, J. P and Thresh, J. M. 2000. Cassava mosaic virus disease in East Africa: A dynamic disease in a changing environment. *Virus Research* 71: 135–49.

Leihner, D. E. 1982. Current practices in the production of cassava planting material. In: Cock, J. H. (Ed.), *Global Workshop on Root and Tuber Crops Propagation*. Proceedings of a Regional Workshop held in Cali, Colombia, 13–16 September, pp. 41–6.

Lema, K. M. and Herren, H. R. 1985. Release and establishment in Nigeria of *Epidinocarsis lopezi* a parasitoid of the cassava mealybug, *Phenacoccus manihoti*. *Entomologia Experimentalis et Applicata* 38: 171–5.

Leuschner, K. 1980. Biology, ecology and control of the green spider mite and cassava mealybug. In: Ezumah, H. C. (Ed.), *Cassava Production and Extension in Central Africa*. Proceedings of a workshop held in Zaire, 19–22 May 1980. IITA, pp. 123–8.

Lomer, C. J., Bateman, R. P., Johnson, D. L., Langewald, J. and Thomas, M. B. 2001. Biological control of locusts and grasshoppers. *Annual Review of Entomology* 46: 667–702.

Lutete, D., Tata-Hangy, K. and Kau, T. 1997. Présence au Zaïre de *Stictococcus vayssierei* (Homoptera: Stictococcidae), un ravageur du manioc (*Manihot esculenta*). *Journal of African Zoology* 111: 71–3.

Lyon, W. F. 1973. A plant-feeding mite *Mononychellus tanajoa* (Bondar)(Acarina: Tetranychidae) new to the African continent threatens cassava (*Manihot esculenta* Crantz) in Uganda, East Africa. *Pest Articles & News Summaries* 19: 36–7.

Mahungu, N., Dixon, A. G. and Kumbira, J. 1994. Breeding cassava for multiple pest resistance in Africa. *African Crop Science Journal* 2: 539–52.

Mansaray, A., Sundufu, A. J., Samura, A. E., Massaquoi, F. B., Quee, D. D., Fomba, S. N. and Moseray, M. T. 2012. Cassava genotype evaluation for grasshopper *Zonocerus variegatus* (L.) (Orthoptera: Pyrgomorphidae) susceptibility in Southern Sierra Leone. *International Journal of Agriculture and Forestry* 2: 294–9.

Modder, W. W. E. 1995. Control of the variegated grasshopper on cassava. *African Crop Science Journal* 2: 391–406.

Msikita, W., Yaninek, J., Ahounou, M., Baimey, H. and Fagbemissi, R. 1997. First report of *Nattrassia mangiferae* root and stem rot of cassava in West Africa. *Plant Disease* 81: 1332.

Msikita, W, James, B., Nnodu, E, Legg, J., Wydra, K. and Ogbe, F. 2000. Disease control in cassava farms - IPM field guide for extension agents. International Institute of Tropical Agriculture (IITA), ISBN 978-131-134-184. Modern Design Ltd, Lagos, 26pp.

Msikita, W., Bissang, B., James, B., Baimey, H., Wilkinson, H., Ahounou, M. and Fagbemissin, R. 2005. Prevalence and severity of *Nattrassia mangiferae* root and stem rot pathogen of cassava in Benin. *Plant Disease* 89: 12–16.

Mutisya, D. L., El-Banhawy, E. M., Khamala, C. P. M. and Kariuki, C. W. 2015. Management of cassava green mite *Mononychellus progresivus* (Acari: Tetranychidae) in different agro-ecological zones of Kenya. *Systematic and Applied Acarology* 20: 39–50.

Mutisya, D. L., El-Banhawy, E. S. M., dos Santos, V. V., Kariuki, C. W., Khamala, C. P. M. and Tixier, M. 2017. Predatory phytoseiid mites associated with cassava in Kenya, identification key and molecular diagnosis (Acari: Phytoseiidae). *Acarologia* 57: 541–54.

Mutsaers, H., Mbouémboué, J. W. and Boyomo, M. P. 1981. Traditional food crops growing in the Yaoundé area (Cameroon). Part II. Crop production, yields and fertility aspects. *Agro-ecosystems* 6: 289–303.

Nestel, B. L. 1974. *Current Trends in Cassava Research*. International Development Research Centre (IDRC), Ottawa, 32pp.

Neuenschwander, P. 1996. Evaluating the efficacy of biological control of three exotic Homopteran pests in tropical Africa. *Entomophaga* 41: 405–24.

Neuenschwander, P. 2001. Biological control of the cassava mealybug: A review. *Biological Control* 21: 214–29.

Neuenschwander, P. and Haug, T. 1992. New technology for rearing *Epidinocarsis lopezi*, a biological control agent against the cassava mealybug. In: Anderson, T. A. and Leppla, N. (Eds), *Advances and Applications in Insect Rearing*. Westview Press, Boulder, CO, pp. 353–77.

Neuenschwander, P. and Ajuonu, O. 1995. Measuring host finding capacity and arrestment of natural enemies of the cassava mealybug *Phenacoccus manihoti*, in the field. *Entomologia Experimentalis et Applicata* 77: 47–55.

Neuenschwander, P. and Herren, H. R. 1988. Biological control of the cassava mealybug, *Phenacoccus manihoti*, by the exotic parasitoid *Epidinocarsis lopezi* in Africa. *Philosophical Transactions of the Royal Society of London Series B (Biological Sciences)* 318: 319–33.

Neuenschwander, P., Schulthess, F. and Madojemu, E. 1986. Experimental evaluation of the efficiency of *Epidinocarsis lopezi*, a parasitoid introduced into Africa against the cassava mealybug *Phenacoccus manihoti*. *Entomologia Experimentalis et Applicata* 42: 133–8.

Neuenschwander, P., Hammond, W. N. O., Ajuonu, A., Gado, A., Echendu, N., Bokonon-Ganta, A. H., Allomasso, R. and Okon, I. 1990. Biological control of the cassava mealybug, *Phenacoccus manihoti* (Hom., Pseudococcidae) by *Epidinocarsis lopezi* (Hym., Encyrtidae) in West Africa, as influenced by climate and soil. *Agriculture, Ecosystems and Environment* 32: 39–55.

Neuenschwander, P., Borowka, R., Phiri, G., Hammans, H., Nyirenda, S., Kapeya, E. H. and Gadabu, A. 1991. Biological control of the cassava mealybug *Phenacoccus manihoti*

(Hom., Pseudococcidae) by *Epidinocarsis lopezi* (Hym., Encyrtidae) in Malawi. *Biocontrol Science and Technology* 1: 297–310.

Ngeve, J. M. 2003. The cassava root mealybug (*Stictococcus vayssierei*) (Homoptera: Stictococcidae): A threat to cassava production and utilization in Cameroon. *International Journal of Pest Management* 49: 327–33.

Ngo Kanga, F., Waeyenberge, L., Hauser, S. and Moens, M. 2012. Distribution of entomopathogenic nematodes in Southern Cameroon. *Journal of Invertebrate Pathology* 109: 41–51.

Nweke, F. I., Spencer, D. S. C. and Lynam, J. K. 2002. *The Cassava Transformation: Africa's Best Kept Secret*. Michigan State University Press, Lansing, MI, 273pp.

Nyiira, Z. 1973. Bioecological studies on the cassava mite *Mononychellus tanajoa* (Bondar) (Acarina: Tetranychidae). In: Terry, E. R. (Ed.), *Proceedings of the 3rd Symposium of the International Society for Tropical Root Crops*. IITA, Ibadan, Nigeria, pp. 415–18.

Odedina, J. N., Adigbo, S. O., Kulakow, P., Iluebbey, P., Fabunmi, T. O., Afolabi, C. G. and Osipitan, A. 2017. Effect of cropping system and age of plant at harvest on tuber rot and performance of elite cassava varieties in derived savannah. *Acta Agriculturae Slovenica* 109: 165–73.

Okechukwu, R., Dixon, A., Akoroda, M., Mwangi, M. and Bandyopadhyay, R. 2009. Root rot resistance in new cassava varieties introduced to farmers in Nigeria. *Experimental Agriculture* 45: 15–24.

O'Malley, M. A. 2015. Endosymbiosis and its implications for evolutionary theory. *Proceedings National Academy of Science USA* 112: 10270–7.

Onyeka, T., Dixon, A., Bandyopadhyay, R., Okechukwu, R. and Bamkefa, B. 2004. Distribution and current status of bacterial blight and fungal diseases of cassava in Nigeria. International Institute of Tropical Agriculture (IITA), Ibadan, Nigeria, 37pp.

Onyeka, T., Dixon, A. and Ekpo, E. 2005a. Field evaluation of root rot disease and relationship between disease severity and yield in cassava. *Experimental Agriculture* 41: 357–63.

Onyeka, T., Ekpo, E. and Dixon, A. 2005b. Virulence and host-pathogen interaction of *Botryodiplodia theobromae* isolates of Cassava root rot disease. *Journal of Phytopathology* 153: 726–9.

Onyeka, T., Dixon, A. and Ekpo, E. 2005c. Identification of levels of resistance to cassava root rot disease (*Botryodiplodia theobromae*) in African landraces and improved germplasm using in vitro inoculation method. *Euphytica* 145: 281-8.

Onzo, A., Hanna,. R. and Sabelis, S. W. 2003. Interactions in an acarine predator guild: Impact on *Typhlodromalus aripo* abundance and biological control of cassava green mite in Benin, West Africa. *Experimental and Applied Acarology* 31: 225–41.

Owolade, O., Dixon, A., Adeoti, A. and Osunlaja, S. 2005. Sources of resistance to cassava anthracnose disease. *African Journal of Biotechnology* 4: 570–2.

Owolade, O., Dixon, A. G. and Adeoti, A. 2006. Diallel analysis of cassava genotypes to anthracnose disease. *World Journal of Agricultural Science* 2: 98–104.

Pallangyo, B., Hanna,. R., Toko,. M., Gnanvossou,. D., Mgoo,. V., Otema,. M., Onzo,. A., Hountondji,. F., Nsami, E. and Mfugale, O. 2007. Biological control of cassava green mite in Tanzania. In: Mahungu, N. M and Manyong, V. M.(Eds), *Proceedings of 9th Triennial Symposium of the International Society for Tropical Root Crops – Africa Branch (ISTRC – AB)*, Mombasa, Kenya, 1-5 November 2004, pp. 597–604.

Parsa, S., Kondo, T. and Winotai, A. 2012. The cassava mealybug (*Phenacoccus manihoti*) in Asia: First records, potential distribution, and an identification key. *PLoS ONE* 7: e47675.

Pijls, J. W. A. M., Kofker, K. D., van Staalduinen, M. J. and van Alphen, J. J. M. 1990. Interspecific host discrimination and competition by *Epidinocarsis lopezi* and *E. diversicornis*, parasitoids of the cassava mealybug, *Phenacoccus manihoti*. *Medelingen van de Faculteit voor Landbouwwetenschappen van de Rijksuniversiteit Gent* 55: 404–16.

Richards, C. 1971. Contribution à l'étude morphologique et biologique des Stictoccocinae (Hom. Coccoidea). *Annales de la Société Entomologique de France (n.s.)* 7: 571–609.

Sanogo, D. and Adetunji, O. 2008. Presidential initiatives on cassava in Africa: Case studies of Ghana and Nigeria. IITA Books and Book Chapters, NEPAD Pan African Cassava Initiative, IITA/NEPAD, Malawi, 73pp.

Schulthess, F., Neuenschwander, P. and Gounou, S. 1997. Multitrophic interactions in cassava, *Manihot esculenta*, cropping systems in the subhumid tropics of West Africa. *Agriculture, Ecosystems and Environment* 66: 211–22.

Tata-Hangy, K., Legg, J., Hanna, R., Toko, M., Lema, K. M., Dixon, A. and Mahungu, N. M. 2006. Incidence and distribution of cassava diseases and pests in the Democratic Republic of Congo. In: Mahungu, N. M. and Manyong, V. M. (Eds), *Proceedings of 9th ISTRC-AB symposium: Advances in Root and Tuber Crops Technologies for Sustainable Food Security, Improved Nutrition, Wealth and Environmental Conservation in Africa*, Mombasa, Kenya, pp. 614–22.

Tindo, M., Hanna, R., Georgen, G., Zapfack, L., Tata-Hangy, K. and Attey, A. 2009. Host plants of *Stictococcus vayssierei* Richard (Stictococcidae) in non-crop vegetation in the Congo Basin and implications for developing scale management options. *International Journal of Pest Management* 55: 339–45.

Williams, D. J., Matille-Ferrero, D. and Miller, D. R. 2010. A study of some species of the genus *Stictococcus* Cockerell (Hemiptera: Sternorrhyncha: Coccoidea: Stictococcidae), and a discussion on *Stictococcus vayssierei* Richard, a species injurious to cassava in equatorial Africa with a description of a new species from Nigeria. *Zootaxa* 2527: 1–27.

Winotai, A., Goergen, G., Tamò, M. and Neuenschwander, P. 2011. Cassava mealybug has reached Asia. *Biocontrol News and Information* 31: 10N–11N.

Wyckhuys, K. A. G., Wongtiem, P., Rauf, A., Thancharoen, A., Heimpel, G. E. Le, N. T. T. Fanani, M. Z., Gurr, G. M., Lundgren, J. G., Burra, D. D., Palao, L. K., Hyman, G., Graciozi, I., Le, V. X., Cock, M. J. W., Tscharntke, T., Wratten, S. D., Nguyen, L. V., You, M., Lu, Y., Ketelaar, J. W., Goergen, G. and Neuenschwander, P. 2018. Continental-scale suppression of an invasive pest by a host-specific parasitoid underlines both environmental and economic benefits of arthropod biological control. *PeerJ6:e5796*. doi: 10.7717/peerj.5796, 24pp.

Wydra, K. and Msikita, W. 1998. Overview of the present situation of cassava diseases in West Africa. In: Akoroda, M. O. and Ekanayake, I. J. (Eds), *Root Crops for Poverty Alleviation*. Proceedings of the 6th Triennial Symposium International Society for Tropical Root Crops Africa Branch (ISTRC-AB), Malawi, 22–28 October 1995. ISTRC, IITA and Government of Malawi, pp. 198–206.

Wydra, K. and Verdier, V. 2002. Occurrence of cassava diseases in relation to environmental, agronomic and plant characteristics. *Agriculture, Ecosystems and Environment* 93: 211–26.

Wydra, K., Zinsou, V., Jorge, V. and Verdier, V. 2004. Identification of pathotypes of *Xanthomonas axonopodis* pv. *manihotis* in Africa and detection of quantitative trait loci and markers for resistance to bacterial blight of cassava. *Phytopathology* 94: 1084–93.

Wydra, K., Banito, A. and Kpémoua, K. E. 2007. Characterization of resistance of cassava genotypes to bacterial blight by evaluation of symptom types in different ecozones. *Euphytica* 155: 337–48.

Yaninek, J. S. 1988. Continental dispersal of the cassava green mite, an exotic pest in Africa, and implications for biological control. *Experimental and Applied Acarology* 4: 211–24.

Yaninek, J. S. 2007. Biological control of the cassava green mite in Africa: Overcoming challenges to implementation. In: Vincent, C., Goettel, M. S. and Lazarovits, G. (Eds), *Biological Control: A Global Perspective*. CAB International Publisher, Wallingford, UK, pp. 28–37.

Yaninek, J. S. and Bellotti, A. C. 1987. Exploration for natural enemies of cassava green mites based on agrometeorological criteria. In: Rijks, D. A. and Mathys, G. (Eds), *Proceedings of the Seminar on Agrometeorology and Crop Protection in the Lowland Humid and Sub-Humid Tropics*. Cotonou, Benin, 7–11 July 1986. World Meteorological Organization, Geneva, Switzerland, pp. 69–75.

Yaninek, J. S. and de Moraes, G. J. 1991. A synopsis of classical biological control of mite pests in agriculture. In: Dusbabek, F. and Bukva, V. (Eds), *Modern Acarology*. Academia, Prague and SPB Academic Publishing BV, The Hague, the Netherlands, 1: 133–49.

Yaninek, J. S. and Hanna, R. 2003. Cassava green mite in Africa- a unique example of successful classical biological control of a mite on a continental scale. In: Neuenschwander, P., Borgemeister, C. and Langewald, J. (Eds), *Biological Control in IPM Systems in Africa*. CABI Publishing, Wallingford, UK, pp. 61–75.

Yaninek, J. S. and Herren, H. R. 1988. Introduction and spread of the cassava green mite *Mononychellus tanajoa* (Bondar) (Acari: Tetranychidae), an exotic pest in Africa, and the search for appropriate control methods: A review. *Bulletin of Entomological Research* 78: 1–13.

Yaninek, J. S., Herren, H. R. and Gutierrez, A. P. 1989a. Dynamics of *Mononychellus tanajoa* (Acari: Tetranychidae) in Africa: Seasonal factors affecting phenology and abundance. *Environmental Entomology* 18: 625–32.

Yaninek, J. S., Gutierrez, A. P. and Herren, H. R. 1989b. Dynamics of *Mononychellus tanajoa* (Acari: Tetranychidae) in Africa: Experimental evidence of the effects of temperature and host plant on population growth rates. *Environmental Entomology* 18: 633–40.

Yaninek, J. S., Mégevand, B., de Moraes, G. J., Bakker, F., Braun, A. and Herren, H. R. 1991. Establishment of the Neotropical predator *Amblyseius idaeus* (Acari: Phytoseiidae) in Benin, West Africa. *Biocontrol Science and Technology* 1: 323–30.

Yaninek, J. S., Braima, B. D. and Bieler, P. 1994. Ecologically Sustainable Cassava Plant Protection (ESCaPP): A model for environmentally sound pest management in Africa. *African Crop Science Journal* 2: 553–62.

Yaninek, J. S., Saizonou, S., Onzo, A., Zannou, I. and Gnanvossou, D. 1996. Seasonal and habitat variability in the fungal pathogens, *Neozygites cf floridana* and *Hirsutella thompsonii*, associated with cassava mites in Benin, West Africa. *Biocontrol Science and Technology* 6: 23–33.

Yaninek, J. S., Megevand, B., Ojo, B., Cudjoe, A. R., Abole, E., Onzo, A. and Zannou, I. 1998. Establishment and spread of *Typhlodromalus manihoti* (Acari: Phytoseiidae), an introduced phytoseiid predator of *Mononychellus tanajoa* (Acari: Tetranychidae) in Africa. *Environmental Entomology* 27: 1496–505.

Published by Burleigh Dodds Science Publishing Limited, 2019.

Yonow, T., Kriticos, D. J. and Ota, N. 2017. The potential distribution of cassava mealybug (*Phenacoccus manihoti*), a threat to food security for the poor. *PLoS ONE* 12(3): e0173265.

Zandjanakou-Tachin, M., Fanou, A. A., LeGall, P. and Wydra, K. 2007. Detection, survival and transmission of *Xanthomonas axonopodis* pv. *manihotis* and *X. axonopodis* pv. *vignicola*, causal agents of cassava and cowpea bacterial blight, respectively, in/by insect vectors. *Journal of Phytopathology* 155: 159–69.

Zeddies, J., Schaab, R. P., Neuenschwander, P. and Herren, H. R. 2001. Economics of biological control of cassava mealybug in Africa. *Agricultural Economics* 24: 209–19.

Zinsou, V., Wydra, K., Ahohuendo, B. and Hau, B. 2004. Effect of soil amendments, intercropping and planting time in combination on the severity of cassava bacterial blight and yield in two ecozones of West Africa. *Plant Pathology* 53: 585–95.

Zinsou, V., Wydra, K., Ahohuendo, B. and Hau, B. 2005. Genotype× environment interactions in symptom development and yield of cassava genotypes with artificial and natural cassava bacterial blight infections. *European Journal of Plant Pathology* 111: 217–33.

Zinsou, V., Wydra, K., Ahohuendo, B. and Schreiber, L. 2006. Leaf waxes of cassava (*Manihot esculenta* Crantz) in relation to ecozone and resistance to *Xanthomonas* blight. *Euphytica* 149: 189–98.

Zinsou, V. A., Afouda, L. A. C., Ahohuendo, B. and Wydra, K. 2017. Importance of fungal root rot pathogens of cassava in Benin. *Pakistan Journal of Botany* 111: 2023–8.

Zundel, C., Nagel, P., Hanna, R., Korner, F. and Scheidegger, U. 2009. Environment and host-plant genotype effects on the seasonal dynamics of a predatory mite on cassava in sub-humid tropical Africa. *Agricultural and Forest Entomology* 11: 321–31.

Chapter 7

Identifying and managing plant health risks for key African crops: maize

Ranajit Bandyopadhyay, *International Institute of Tropical Agriculture (IITA), Nigeria;* **Kitty F. Cardwell**, *Oklahoma State University, USA;* **Alejandro Ortega-Beltran**, *International Institute of Tropical Agriculture (IITA), Nigeria;* **Fritz Schulthess**, *Rheinstrasse 160, Switzerland;* **William Meikle**, *USDA-ARS, USA;* **Mamoudou Sétamou**, *Texas A&M University, USA; and* **Peter J. Cotty**, *USDA-ARS, USA (corresponding author: r.bandyopadhyay@cgiar.org)*

1 Introduction

Wheat, rice and maize are considered as the top three agricultural commodities that the world depends on for food, feed and industrial raw material (Awika, 2011). It is projected that the world population would increase by slightly more than 1 billion people from the 2017 level, reaching 8.6 billion in 2030, and will increase further to 9.8 billion in 2050, a 13% increase from 2017. The populations of 26 African countries are projected to reach at least double their current size between 2017 and 2050 (United Nations, Department of Economic and Social Affairs, Population Division, 2017). Meeting the food and nutritional demand of this rapidly growing population remains a major global concern. Maize, a crop domesticated in the Central Balsas Valley of Mexico around 9000 YBP (Matsuoka et al., 2004), is an important staple food and a major source of daily intake of calories in many countries in sub-Saharan Africa

http://dx.doi.org/10.19103/AS.2018.0043.08
Published by Burleigh Dodds Science Publishing Limited, 2019.

(SSA). Maize was introduced into Africa in different events, at various times in the sixteenth century (Miracle, 1965). Among all major cereals contributing to calories in African diets, the share of maize is 45% in East and Southern Africa and 21% in West and Central Africa (Shiferaw et al., 2011).

The use of maize in SSA has diversified from principally being a food security crop to a cash crop for meeting the growing demand of raw material for animal feed (especially poultry), food processing and nutraceutical industries. In SSA, 70.6 million tons of maize are harvested from 36.6 million ha, with Nigeria (6.54 million ha), Tanzania (4.04 million ha) and Kenya (2.33 million ha) being the three largest producers of the crop in 2016 (FAOSTAT, 2018). Currently the productivity level in Africa is 1.9 tons/ha, and it is expected that maize production will fall short of the accelerating demand unless measures are taken to improve productivity (Shiferaw et al., 2011). However, productivity increase should go hand in hand with curtailing biotic and abiotic constraints and reducing post-harvest losses. About 30% or more of post-harvest losses of maize are recorded annually at the farm level in SSA (APHLIS, 2014; Hodges et al., 2011). Post-harvest losses of maize include direct losses resulting from physical and quantitative loss or indirect losses due to decline in quality and nutritive values (Pantenius, 1988; Kitinoja et al., 2011) that occur from harvest until consumption. Land, energy, water, fertilizer and labour invested in producing the lost grain is also wasted. Mycotoxin formation in grains that begins at the pre-harvest stages further contributes to post-harvest losses and food safety issues. Reducing post-harvest losses of maize in SSA could be a rapid, cost-effective and sustainable solution for increasing food availability and improving human and animal health.

There are several types of biotic factors limiting the full potential of maize. These include insect pests, parasitic weeds, nematodes, fungi, bacteria, viruses and cell-wall-deficient bacteria that attack one or more plant parts. Climate change and increase in maize cultivation both in the existing traditional domains and extension into the non-traditional dry Sahelian agro-ecozone (AEZ) have increased vulnerability of maize to insects and pathogens. Invasive pests continue to devastate maize cultivation in Africa. A case in point is the justifiable panic created by the aggressive damage and alarming rate of spread of fall armyworm *Spodoptera frugiperda* (J.E. Smith) (Lepidoptera: Noctuidae). The International Institute of Tropical Agriculture (IITA) has placed considerable emphasis on research to understand and manage several biotic constraints that reduce the quantity and quality of maize. Students and visiting scientists from several universities and national programs in Africa have enriched plant health research at IITA. The following sections in this chapter reflect the rich scientific productivity of the maize plant health research program of the past 50 years. The name, author and taxonomic positions (where appropriate) of insects and pathogens dealt in this chapter are mentioned in Table 1. This chapter omits

Published by Burleigh Dodds Science Publishing Limited, 2019.

Table 1 Names of insect pests and fungal/bacterial pathogens of maize on which IITA plant health specialists have worked during the last 50 years

Pathogens	Insect pests
Bipolaris maydis ((Miskado and Miyake) Shoeaken)	*Busseola fusca* Fuller (Lepidoptera: Noctuidae)
Physoderma maydis (Myiabi)	*Eldana saccharina* Walker (Lepidoptera: Pyralidae)
Puccinia sorghi (Schw.)	*Sesamia calamistis* Hampson (Lepidoptera: Noctuidae)
Rhizoctonia solani (Kuhn)	*Chilo partellus* Swinhoe (Lepidoptera: Crambidae)
Stenocarpella macrospora ((Earle) Sutton)	*Telenomus isis* Polaszek (Hymenoptera: Scelionidae)
Cercospora zeae-maydis (Tehon and Daniels)	*Cotesia sesamiae* Cameron (Hymenoptera: Braconidae)
Peronosclerospora sorghi ((Westfall and Uppal) Shaw)	*Mussidia nigrivenella* Ragonot (Lepidoptera: Pyralidae)
Fusarium verticillioides ((Sacc.) Nirenberg)	*Cryptophlebia leucotreta* Meyrick (Lepidoptera: Tortricidae)
Aspergillus flavus (Link) and other *Aspergillus* spp.	*Sitophilus zeamais* Motschulsky (Coleoptera; Curculionidae)
Penicillium spp.	*Cathartus quadricollis* Guerin-Meneville (Coleoptera; Silvanidae)
Wolbachia	*Carpophilus dimidiatus* Fabricius (Coleoptera: Nitidulidae)
	Tribolium castaneum Herbst (Coleoptera: Tenebrionidae)
	Prostephanus truncatus Horn (Coleoptera: Bostrichidae)

treatment of fall armyworms (Goergen et al.: Chapter 4), viruses (Legg et al.: Chapter 5), nematodes (Coyne et al.: Chapter 15) and agricultural and other parasitic weeds (Chikoye et al.: Chapter 14) which are discussed in other chapters of this book.

2 Exploratory years in maize research

When IITA was established in 1967, the International Maize and Wheat Improvement Center (CIMMYT) had the global mandate for maize germplasm development. In 1980, IITA established a regional mandate for maize research for West and Central Africa. Most early work focussed on development of robust, open-pollinated maize varieties, with resistance to the prevailing major diseases in the humid forest and moist savannas of West and Central Africa. In 1990, IITA surpassed CIMMYT as the source for maize varieties in the region (Manyong et al., 2000). The regional focus not only improved the status of maize germplasm across the region, but in general strengthened the institutional capabilities of the National Agriculture Research programmes through training and coordination. The Maize Research Program did not have a plant protection focus per se, but as a matter of course, considered stemborers, the parasitic plant *Striga hermonthica* (Del.) Benth., root, foliar and ear insects and pathogens as part of the overall yield-improvement breeding objective. When maize health was threatened by new or emerging pests and diseases, the Institute was able to acquire the needed expertise to respond quickly and decisively. An example of this includes the successful development and introgression of resistance to the Maize Streak Mosaic virus into regionally adapted maize varieties. For this, IITA was awarded the King Baudouin award in 1986. Plant health specialists worked on maize with/for the plant breeders, as was the CGIAR model at that time.

With the reorganization of IITA, and the development of the Plant Health Division in 1990, plant protection specialists were grouped, which resulted in new and different synergies. Where needed, maize breeders continued to have access to, and partnerships with, entomologists and plant pathologists. However, the research focus for these plant health scientists was also allowed to shift towards process-level understanding of systems of crop plants and the hierarchical relationships with soil and environmental biota. The strength of this approach led to fundamental breakthroughs in understanding interactions between plants, biota, environment and cropping systems to increase yield and improve the quality of the harvested maize. Problems that had been intractable through plant breeding, such as resistance to stemborers, post-harvest pests or to *Aspergillus* infection of the ear, were examined with new perspectives from which different sets of solutions arose.

3 Diagnostic surveys and in-country plant health analysis

Knowledge of biological factors limiting maize production is crucial for setting priorities according to the pest organism to be tackled and for development of plant health strategies. During 1993–7, work began to retrain the plant protection specialists of the national agricultural research systems to evaluate and prioritize constraints to maize productivity across all AEZs in Ghana, Cameroon and Uganda. The country teams were comprised of an entomologist and a plant pathologist from IITA, and agronomists, entomologists and plant pathologists from the respective national system. A system approach was designed to evaluate crop health by survey and evaluation of maize fields (Cardwell et al., 1997b; Ngoko et al., 2002). We describe below the research methods and outcomes for Cameroon, which was a good example of the approach.

Diagnostic surveys of roadside maize fields covering several AEZs were undertaken in 1993, 1995, 1996 and 1997 in Cameroon to identify the major biotic constraints to maize production and to assess yield losses on a countrywide basis. Field conditions (weediness, soil type, soil moisture), insects,

Published by Burleigh Dodds Science Publishing Limited, 2019.

diseases, plant growth and yield variables were assessed in 164 fields. Yield was assessed destructively by harvesting cobs, or if non-destructive sampling was required, cob weight was estimated using a multiple regression equation that included growth stage, % grain fill, cob width and length. Principal component analyses and stepwise regressions showed which biotic and abiotic factors had a significant negative or positive impact on yield (Cardwell et al., 1997b).

Increased soil fertility (expressed as plant vigour and absence of foliar symptoms of N and K deficiencies) favoured the fungi *Bipolaris maydis* and *Physoderma maydis* and the stemborer *Busseola fusca*. Positive relationships existed between soil organic matter, *B. fusca*, *Eldana saccharina* and *B. maydis attacks*. As organic matter decreased, percentage leaf area lost to pathogens, particularly *Puccinia sorghi*, increased. Maize streak virus, stem diseases, *B. maydis, Rhizoctonia* sheath blight and *P. sorghi* consistently had significant negative impacts on cob weight. Average reduction in cob weight from foliar and stem diseases in the different AEZs and seasons ranged between 10 and 12 g/plant. In further surveys that concentrated on maize diseases in 1995–7, maximum reductions in potential yield were estimated at 68% due to *B. maydis* and 46% due to *Stenocarpella macrospora* in the humid forest zone in 1995. In the Western Highlands, an epidemic of *Cercospora zeae-maydis* caused yield reductions of up to 79% in 1996 (Ngoko et al., 2002).

Among all the stemborer species found on maize in Cameroon, *B. fusca* was most prevalent (95%) followed by *E. saccharina* and *Sesamia calamistis*. In the first cropping season, similar borer infestation levels (43%) occurred in lowland and highland, but borer incidence increased during the second cropping season. In both low- and mid-altitude fields, around 55% of the plants were damaged, resulting in a calculated cob weight loss of 9 g/plant, with additional plant loss from dead hearts. Similar observations were reported in surveys in 1994–7 (Ndemah et al., 2001a). The predominance of *B. fusca* in all zones is in stark contrast to its distribution in West Africa where it is the predominant species in the savannah zone on sorghum. *Sesamia calamistis*, *Sesamia botanephaga* and *E. saccharina* are the key pests of maize in the humid lowland zones (Schulthess et al., 1997; Buadu et al., 2003). In East and Southern Africa, *B. fusca* is the predominant species of the cooler regions and of importance 1000 m above sea level while in the hot lowland, the invasive *Chilo partellus* is the economically most important species (Smithers 1960; Van Rensburg, 1988; Overholt et al., 1994; Phiri, 1995). Phylogenetic studies by Sezonlin et al. (2006) verified the existence of several *B. fusca* populations in Africa that occupy different ecological niches.

There was a strong and inverse relationship between wild grasses around a field and stemborers in the field. In Cameroon, *B. fusca* was exceedingly scarce on wild grasses and the insect was detected in a few, uncommon wild grass species that were sampled (Ndemah et al., 2007). Thus, wild habitats surrounding a field might simply act as a barrier rather than trap plants. Stemborer incidence was lower in maize fields that were intercropped with companion crops that were not hosts of the pest. This will have some implications for the efficiency of habitat management technologies which was verified by further research (Ndemah et al., 2003) (see section Stemborers: Researching control options for intractable problems).

In the lowland humid tropics in Cameroon, parasitism of eggs by *Telenomus* spp. (Hym.: Scelionidae) had a significant impact on *B. fusca* densities. Parasitism was on average 20% during the first versus 67% during the second season explaining the lower pest densities during the second growing season. This was also reflected in maize yields, which significantly increased with egg parasitism. A similar phenomenon of increased egg parasitism in the second growing season with attendant lowered population of a different borer species (*S. calamistis*) was also observed in Benin (Schulthess et al., 2001). One of the parasitoid species, *Telenomus isis* Polaszek, is endemic in West Africa including Cameroon (Chabi-Olaye et al., 2005a). It has never been reported from East and Southern Africa and was thus later introduced and released in Kenya (Bruce et al., 2009).

Similarly, a comparison with other regions of Africa showed that other species and populations of parasitoids are lacking in West Africa (Schulthess et al., 1997). In Cameroon, the countrywide surveys yielded the braconid *Cotesia sesamiae* Cameron (Hymenoptera: Braconidae) from other noctuid species on wild grasses but never from *B. fusca* on maize (Ndemah et al., 2001b, 2007). By contrast, *C. sesamiae* is the most common larval parasitoid found on *B. fusca* and *S. calamistis* in East and Southern Africa. Thus, there appears to be ecological and genetic differences in *C. sesamiae* populations between Cameroon and eastern Africa. Several studies showed that this was mostly due to differences in symbiotic polyDNA viruses, which enable the parasitoid to overcome the immune system of its host, and in the bacterial genus *Wolbachia*, which can induce reproductive incompatibilities (Ngi-Song et al., 1995; Mochiah et al., 2002; Gitau et al., 2010; Branca et al., 2011).

The diagnostic surveys thus helped to prioritize the plant disease to be tackled and opened new avenues for solving intractable pests such as *B. fusca*.

4 Downy mildew eradication programme, southern Nigeria 1995

Peronosclerospora sorghi, the causal agent of downy mildew of cultivated and wild sorghum [(*Sorghum bicolor* and *Sorghum halepense* (L.) Pers. (Poaceae)], as well as maize, is a problem on sorghum, but one of the most destructive diseases of maize, able to cause total losses in unprotected fields. The pathogen is obligate (must have living host tissue to grow) and cannot be cultured. *P. sorghi* produces oospores and conidia. Oospores are thick-walled, long-lived structures that survive in soil and serve as primary inoculum when there has been a fallow period between crops. On sorghum, the leaves become shredded and full of oospores, but on maize, no oospores are formed in the leaves. Conidia, copiously produced on both hosts, are thin-walled, ephemeral spores, which emerge from the stomata of living leaf tissue between 3 am and 4 am under high humidity (Cardwell et al., 1997a; Bock et al., 1998b). Conidia are responsible for epidemic build-up of the disease within a season. Adenle and Cardwell (2000) showed that in maize the pathogen could be seed-borne and that 'crazy-tops', one of the most destructive manifestations of the disease (Fig. 1), were a potential source of soil oospores if the deformed structures are not removed from the field. However, the primary inoculum in maize was considered to be air-borne conidia from infected plants (Olanya and Fajemisin, 1993; Cardwell, 1995).

Two pathotypes of *P. sorghi* have been reported, one capable of infecting sorghum and maize and the other infecting maize only. The sorghum and maize infecting strain has been reported in the tropics and subtropics worldwide (Williams, 1984) and throughout Africa and in Northern Nigeria (Olanya and Fajemisin, 1993; Bock et al., 1998a; Adenle and Cardwell, 2000; Bigirwa et al., 2000). Around 1975, a maize-specific downy mildew, identified as *P. sorghi*, appeared in two fields of Ondo State in Nigeria (Olanya and Fajemisin, 1993). Unlike most *P. sorghi* that causes disease in sorghum, Johnson grass and maize (Bigirwa et al., 2000) the pathogen in Ondo State was only infectious

Figure 1 Two maize plants with 'crazy tops' symptom of downy mildew.

Published by Burleigh Dodds Science Publishing Limited, 2019.

on maize. Previous reports of a maize-specific strain of downy mildew in Africa were anecdotal at best, although there were confirmed reports of such a strain in Thailand, Indonesia and Central African Republic (Bock and Jeger, 1996; Bock et al., 1998a). Subsequent identification related this phenotype to a maize-specific strain, tentatively named *P. zeae*. The strain in southern Nigeria appeared to be *P. zeae*. Considering the highly obligate nature of this southern Nigeria strain, and the short-lived nature of its conidia [mature conidia germinate or disintegrate within an hour of dislodging from the conidiophore (Cardwell et al., 1997a)], it was unclear exactly how the pathogen had arrived in maize fields in south central Nigeria (Fig. 2). Perhaps it was brought in seed from a source that had either come directly from Thailand or indirectly from Thailand via another source in the Central African Republic. Thus, it was highly likely that the southern Nigerian strain was introduced and is not an endemic pathogen, making an eradication strategy possible if the epidemiology could be constrained.

Beginning as a point source of introduction in Ondo State, the pathogen found the moist conditions and continual maize cropping to be conducive to establishment and spread in southern Nigeria. Throughout the 1980s, the disease remained localized in the inland valleys of Ondo and Osun States, but by 1990, the disease began to spread exponentially, up to 100 km/year (Cardwell, 1995). By 1995, maize downy mildew was found in seven states in south and central Nigeria (Cardwell, 1995), where incidence of systemic infection exceeded 90% at several sites. Systemically infected plant produced no ears, thus a field with 90% infected plants would correspond to a 90% yield loss.

Figure 2 A map of Nigeria showing status of downy mildew occurrence in the late 1990s.

Published by Burleigh Dodds Science Publishing Limited, 2019.

To break an epidemiological cycle that was dependent primarily on plant-to-plant spread, one tactic could be to interrupt the disease cycle by mandating a cessation in planting. Considering that in southern Nigeria maize was grown continually throughout the year by smallholder farmers, this solution was unlikely to be practicable. Alternative tactics were to render the crop resistant through plant breeding (Kling et al., 1994; Kim et al., 1995) or chemical treatment (Bock et al., 2000). IITA's maize breeding programme had been introgressing downy mildew resistance (DMR) (using lines from CIMMYT, Thailand) into local Tropical *Zea* (rustic, open pollinated maize streak virus resistant) varieties since 1980 (Kim et al., 1995). A new field screening technique, inoculation and planting of pre-germinated susceptible spreader row seed, eliminated infection variability, cut 20 days from each screening cycle and improved resistance by up to 35% per selection cycle (Cardwell et al., 1997a). With this new screening method, resistance in existing DMR varieties could be brought to 90–95% in 2–3 years (Kling et al., 1994). As the epidemic accelerated, IITA made rapid progress in maize breeding (Kling et al., 1994; Kim et al., 1995; Cardwell et al., 1997a). However, it was imperative to coordinate scale up of seed production and distribution with the Nigerian National Seed Service (NSS), private seed producers supported by State Agriculture Development Projects (ADP), NGOs and the World Bank. Scaling up of production of resistant varieties was estimated by IITA and the NSS to require 4.5 metric tons of breeder seed and 240 tons of foundation seed to meet an estimated need of 14 400 tons of DMR improved seed for farmers by 1998 (Usman, 1995; Cardwell, 1995). In 1995, existing resistant seed stock was 0.75 ton breeder seed, 60 tons foundation seed and 2.4 tons of grower seed (Usman, 1995; Cardwell, 1995), which highlights the scope of the challenge.

Ciba-Geigy (Syngenta, Switzerland) had meanwhile formulated and packaged a seed-dressing fungicide, Apron plus (metalaxyl, carboxin and furathiocarb), for small-scale producers in Africa. One 10-g sachet of Apron plus mixed with 1 kg of seed just before planting rendered complete protection to *P. sorghi*. As early as 1991, the Nigerian government began to purchase Apron plus for distribution. However, the concern was that the epidemic was outpacing the production and distribution of resistant seed and procurement and distribution of the seed treatment. The strategy would be to provide enough seed treatment to stop the epidemic advance while scaling up DMR seed production and distribution.

The impending risk of widespread food insecurity due to the rapid spread of the disease caused IITA and the Federal Department of Agriculture (FDA) of Nigeria to convene a workshop in 1994 to coordinate tactics for the eradication of maize downy mildew. Food and Agriculture Organization (FAO) of the United Nations and the World Bank were approached for support to mount an eradication campaign (Cardwell, 1995). IITA, together with the Nigerian

Extension Service, the ADPs in each state, were to coordinate the outreach and information campaign - so that farmers would recognize the disease, remove infected plants and crazy-tops from the field, seek resistant seed and/or use seed treatment. The FDA and IITA were to work with the World Bank to procure enough Apron plus to blanket the disease fronts. Trucks with seed treaters would be deployed to each region to travel from village to village where farmers would bring their seed to be treated. Finally, the NSS and the ADPs would train and provide inputs for farmers to scale up the production of resistant seed. The programme was rolled out immediately in 1995. One contretemps was that the ship transporting Apron plus from Europe was struck by a storm that swept the containers with 1 million 10-g sachets of Apron plus off the ship into the ocean, where the containers broke open and the contents washed ashore on the beaches of France and Holland. The public outcry made the World Bank withdraw support for the chemical treatment aspect of the programme. They proffered instead to allow funding to be used to distribute seed as it was being produced. The strategy was operationalized, and despite the setback, it was ultimately successful. Spore trap catches in the IITA farm showed the gradual diminution of downy mildew spores until the counts became zero. By 2000, the downy mildew epidemic was no more. By 2015, the disease could no longer be found in the affected States.

5 Biological control of stemborers

Diagnostic surveys in Cameroon identified several technologies to control stemborers (Cardwell et al., 1997b; Ndemah et al., 2002). These have been researched in Benin and Cameroon.

To understand the role of planting patterns with hosts and non-hosts, the effects of intercropping of maize with cassava and several leguminous crops, which are non-hosts of *B. fusca*, on infestation levels, parasitism of *B. fusca* and yield were investigated in the humid forest zone in Cameroon during 2002 and 2003. Depending on the crop association and planting pattern, intercrops reduced egg numbers by 41–55% and larval densities by 44–62% compared to monocropped maize (Chabi-Olaye et al., 2005b). Larval densities were 21–48% higher in the monocrops than in intercrops, and they tended to be higher in alternating rows than alternating hills arrangements. Hence, maize yield losses were 2–3 times higher in monocrops than in intercrops. Similarly, in a two-season study at IITA in the forest–savannah transition zone in the Republic of Benin, intercropping with cassava reduced egg and immature numbers of *S. calamistis* by 67% and 83%, respectively (Schulthess et al., 2004). Yield losses were up to four times higher in mono- compared to intercropped maize. Intercropping maize with non-host companion crops had no effect on the stemborer *E. saccharina* and the cob borer pyralid *Mussidia nigrivenella*

Published by Burleigh Dodds Science Publishing Limited, 2019.

Ragonot, nor on the tortricid *Cryptophlebia leucotreta* that attack maize at or after the tasseling stage as well as for pests on late maturing, tall maize cultivars prevalent in the Western Highlands in Cameroon (Schulthess et al., 2004). This suggests that companion crops produce a `masking' effect on young plants (i.e. visual camouflage or masking of host odour) (Finch and Collier 2012), which affects host finding of the moth, and that this effect disappears as the maize plant grows taller. During the short rainy season, up to two-fold higher levels of parasitism of eggs of the prevalent key pest by *Telenomus* spp. were recorded compared to the dry season. The higher parasitism was, however, attributed to density dependence effects rather than the effect of the presence of non-host plants.

Intercropping maize with non-host plants substantially reduced stemborer infestations of maize in both AEZs. This was mainly a result of reduced oviposition rather than migration-related mortality of *B. fusca*. All intercropping systems offered the additional advantage that land productivity was higher than with monocrops as shown by land equivalent ratios of 1.1–1.8. The maize–cassava intercrop was the most efficient in terms of land use and productivity. Thus, this intercropping system is especially recommended to land-constrained poor farmers, who cannot afford external inputs such as fertilizer or insecticides.

In multi-seasonal trials in Cameroon and Benin, conducted between 1996 and 2002, influences of several grass species planted as border rows had no conclusive effect on pest and yields. In Cameroon in 1996–7, *Pennisetum purpureum* Moench had no effect on the numbers of *B. fusca* eggs deposited on maize but it significantly lowered larva numbers and in some cases increased yields (Ndemah et al., 2002). In trials in 2002, soil water content was significantly higher under the grass borders than in maize plots and nitrogen uptake by maize tended to be highest in plots surrounded by *P. purpureum* (Ndemah et al., 2006). During a prolonged drought in 1997, Ndemah et al. (2002) observed less wilting and more vigorous plants in crops surrounded by *P. purpureum*. Thus, nitrogen uptake and higher soil water content in fields with *P. purpureum* rather than differences in pest densities explained the occasional yield differences observed in those trials. In 1998 in Benin, two grass species (*Pennisetum polystachion* L., *Panicum maximum* Jacq.) reduced larval densities of *S. calamistis* and *M. nigrivenella*, and enhanced parasitism of eggs of *S. calamistis* by *Telenomus* spp. and of larvae by the braconid *C. sesamiae*, which had been introduced from Kenya (Gounou et al., 2009; Hailemichael et al., 2009). However, this population suppression had no beneficial effect on yields partly because rodent damage to maize was aggravated by the grassy border rows.

The inconclusive results on the effect of grassy border rows corroborate results by Matama-Kauma et al. (2000) in Uganda, where *C. partellus* was the predominant pest. Since for oligophagous borers such as *B. fusca* few grass species are natural hosts (Le Rü et al., 2006; Ndemah et al., 2007), most grasses appear not to be trap plants as proposed by Khan et al. (2000), but they rather act as a physical obstruction (barrier) to the pest (Finch and Collier, 2012). It can be concluded that planting grassy border rows is not a reliable and economically meaningful control option. Grasses, however, appear to enhance the activity of parasitoids of polyphagous species such as *S. calamistis* and help perennating parasitoids during the off-season, when oligophagous species such as *B. fusca* living on the crop are diapausing (Chabi-Olaye et al., 2006a). Thus, the uncontrolled burning to clear land from bush fallow should have a considerable negative impact especially on natural enemies. Hence, rather than planting border rows natural habitats should be left intact.

In order to assess the relationship between borer population density and nutritional status of the plant, multi-seasonal trials were conducted in the rainforest of Cameroon to evaluate the effect of different combinations of N-K fertilizer on yields and *B. fusca* infestations of maize. In addition, continuous maize production systems were compared with crop sequence systems, in which maize followed a grain legume (cowpea or soybean), a cover crop (mucuna or pigeon pea) or a bush fallow. In the fertilizer trials, whereas K had no effect, stemborer numbers increased with increasing N level at the early growth stage (Chabi-Olaye et al., 2008); similar results were obtained in Benin for *S. calamistis* by Sétamou et al. (1995). In seeming contradiction, grain yield losses in Cameroon were 11–18 times higher with 0 kg N/ha compared to 120 kg N/ha. In both the insecticide-free and insecticide treatments, the highest net benefits were obtained with 120 kg/ha N – 0 kg K applications. Similar to the fertilizer trial, maize in the legume–maize rotation systems had 1.5–2 times higher borer densities than in the continuous maize production systems. Whereas average yields increased in the crop sequences, grain yield losses were 2–3 times higher in the continuous maize production system than in the crop sequences of grain legumes with maize. Compared to maize after legumes, yield losses were also 5–11 times higher in the continuous maize production system (Chabi-Olaye et al., 2005c). It was concluded that an increased nutritional status of the plants led to an increase in borer attacks at the early stage of plant growth, but it also improved plant vigour, resulting finally in a net benefit for the plant and grain yields. Increasing soil N also increases larval parasitism probably by enhancing the nutritional quality of the plant and its insect host (Jiang and Schulthess, 2005). Improving soil fertility can thus be an effective means of compensating for damage by *B. fusca*. For land-constrained farmers, who cannot afford external inputs such as fertilizers or insecticides, a mixed cropping system with cassava preceded by a grain legume is recommended.

While the use of fertilizer and novel cropping systems reduced borer damage and increased yields, it was not clear whether these improved technologies were beneficial to the farmers. Field trials were set up in the humid forest zone of Cameroon to compare (1) a traditional maize–cassava–groundnut system (farmers' practice) with the best bets of the above studies, namely (2) a maize-cassava intercrop + 120 kg N/ha treatment, (3) a rotation system in which maize-cassava intercrop followed a mucuna fallow, (4) a maize monocrop grown after mucuna fallow, (5) a maize monocrop grown with 120 kg N/ha, (6) a maize monocrop without fertilizer (Chabi-Olaye et al., 2006b). All the practices were compared with or without insecticide application. The differences in net benefits between insecticide-treated and untreated plots were minor. The results further showed that borer densities can be considerably reduced, yields improved and total net benefits increased without drastic changes of the system already used.

Because of the importance and differences in the ecology of *C. sesamiae* on *B. fusca* between Cameroon and eastern Africa, a biological control project to introduce *C. sesamiae* populations from Kenya against *B. fusca* into Cameroon was proposed by Schulthess et al. (1997). Thus, between 2006 and 2008, seven populations of *C. sesamiae* obtained from *B. fusca* in Kenya were introduced into the humid forest zone and the Western Highlands of Cameroon.

In order to avoid negative effects on the locally occurring parasitoid population only specific populations whose polyDNA Viruses (PDVs) and *Wolbachia* bacteria were compatible with those occurring in *C. sesamiae* in Cameroon were chosen (Dupas et al., 2008; Branca et al., 2009; Gitau et al., 2010; Branca et al., 2011). Post-release surveys were carried out on both maize and wild grasses. In all, 393 individuals - one from each *C. sesamiae* cocoon mass - were genotyped for 11 microsatellite loci. Cluster analysis of the microsatellite data showed that the parasitoids recovered from maize almost all belonged to the same genetic cluster as those introduced from Kenya. The parasitoids recovered from wild grasses post-release clustered with the endemic population found on wild host plants pre-release, indicating absence of non-target population effects caused by the introduction of the Kenyan populations (R. Ndemah, pers. comm.). The results show that the Kenyan populations are permanently established in Cameroon. Biological control using *C. sesamiae* populations is both environmentally and economically sound and, if working, is self-perpetuating and sustainable with little effort from the farmer. As already shown by Kipkoech et al. (2006) and Midingoyi et al. (2016), benefits of biological control of stemborers in Africa by a *Cotesia* sp. can be considerable in terms of food security and reduction in the use of agrochemicals.

6 Interactions between *Fusarium verticillioides* and maize pests

Fusarium verticillioides is a common pathogen of maize capable of infecting almost all plant parts, sometimes spending its life cycle in an endophytic phase in the plant. It causes seed decay, root rot, stalk rot, ear rot and produces fumonisin, a potent mycotoxin associated with oesophageal cancer, neural tube defect and child stunting (Blacutt et al., 2018). Various studies in temperate zones report that insects are vectors of *F. verticillioides* and fungal incidence is therefore higher on attacked plants (Christensen and Schneider, 1950; Chiang and Wilcoxon, 1961; Bottalico and Logrieco, 1988; Munkvold and Desjardins, 1997). Similarly, in a survey carried out by IITA in southern Benin, *F. verticillioides* was the most common endophytic fungus inhabiting maize stalks (Baba-Moussa, 1998). The fungus was cultured from stems of up to 80% of plants attacked by *S. calamistis* and *E. saccharina*.

A field experiment to compare ear- and grain-quality characteristics, interactions with *Aspergillus flavus* and *F. verticillioides* infections and insect ear infestation was thus conducted over two growing seasons at IITA, Ibadan, Nigeria (Cardwell et al., 2000). The purpose of the study was to evaluate IITA maize breeding genotypes with differing cycles of selection for yield and insect resistance, and for reactions and interactions to the two common toxigenic fungi, *A. flavus* and *F. verticillioides*. Maize genotypes did not influence the number of insects or fungal incidence. By contrast, *F. verticillioides* in the ear increased while *A. flavus* suppressed ear borer and beetle infestations (Fig. 3). The cucurlionid *Sitophilus zeamais*, the silvanid *Cathartus quadricollis*, the nitidulid *Carpophilus* sp., the tenebroid *Tribolium castaneum* and six other species of beetles were always highest in the *F. verticillioides* infected cobs (Fig. 3). Depending on the maize genotype, *F. verticillioides* was also significantly associated with higher infestations of the stemborers *S. calamistis*, *E. saccharina* and the ear borer *M. nigrivenella*. As shown by Sétamou et al. (1998) in Benin, damage by *M. nigrivenella* predisposes maize to pre- and post-harvest infestation of storage beetles and to infection by *A. flavus* and subsequent aflatoxin contamination. Stemborer damage is thus partly the cause for higher incidence of beetles and *A. flavus* in the ear.

In subsequent field experiments at the IITA-Benin station, stemborers and beetle numbers were evaluated on plants inoculated with a conidial suspension of *F. verticillioides*, on plants grown from surface sterilized/fungicide treated seed (for *Fusarium* suppression), as well as untreated plants (Schulthess et al., 2002). Similar to Cardwell et al. (2000), *F. verticillioides* infections significantly increased attack by *E. saccharina*, *S. calamistis*, *C. leucotreta*, the pyralids *Chilo* spp. and *M. nigrivenella*, and beetles in general. It was hypothesized that some lepidopterous and coleopteran pests are attracted by and survive longer (or have lower mortality) on plants infected with *F. verticillioides*.

These hypotheses were tested in preference and life table studies in the laboratory (Ako et al., 2003). In a choice experiment, *E. saccharina* and *M. nigrivenella* laid, respectively, 3–4 times and 4–5 times more eggs on inoculated than protected maize stems. Both larval survival and fecundity of *E. saccharina*

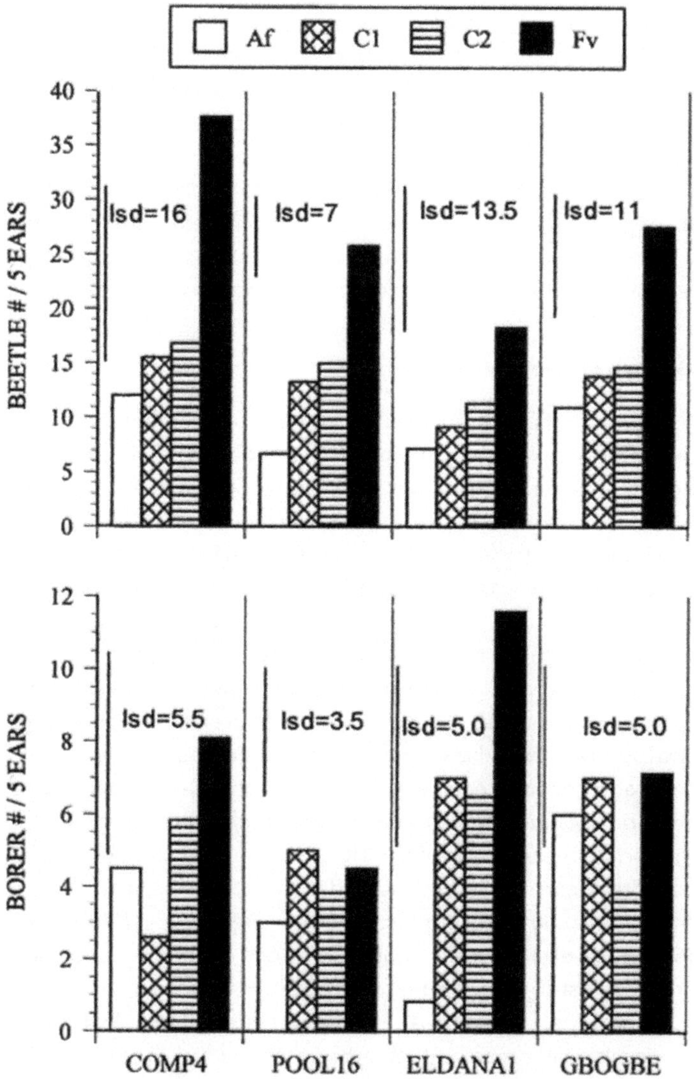

Figure 3 Effect of inoculations with *Aspergillus flavus* (Af) and *Fusarium verticillioides* (Fv), and controls with no fungal inoculations (C1 and C2, respectively) on the number of beetles and ear borers per five ears in four maize genotypes (Cardwell et al., 2000). In general, maize ears inoculated with *F. verticillioides* are more prone to beetle and borer attack compared to uninoculated controls.

were around 40% higher on inoculated than protected plants. In olfactory experiments, the nitidulid *Carpophilus dimidiatus* was significantly attracted to infected grain, while its development time was shorter, and fecundity was higher on infected grain, allowing for a faster population build-up. It was concluded that the higher pest densities found in the stem and ear of field-grown maize was due to a higher attraction to and higher immature survival and adult fecundity on *F. verticillioides* infected plants. Whether or not the higher attraction could be in part due to emission of volatile compounds by *F. verticillioides* (Munkvold, 2003) is unclear.

Interactions with the two fungi had not heretofore been considered in stem borer and/or beetle-resistance breeding programmes. Hence, different levels of inherent susceptibility to and/or temporal variations in infection by the fungi must be accounted and controlled for in breeding for insect resistance.

7 Post-harvest losses

The harvested maize undergoes several post-harvest operations such as transport, de-husking, threshing, cleaning, drying and storage that, if improperly done, can all be potential sources of post-harvest losses (Hodges and Maritime, 2012).

Post-harvest losses of maize are affected by both abiotic (temperature, humidity, rain) and biotic factors (Abedin et al., 2012). As both factors tend to vary with AEZ, and since the activity of organisms affecting maize is weather dependent, the type and magnitude of post-harvest losses may be region dependent. Insects, fungi and rodents are the major biotic factors affecting maize during storage, but insect infestation (Holst et al., 2000) and fungal contamination (Suleiman et al., 2013), either alone or often together, are the major contributors to post-harvest losses in most cases in SSA.

The storage insect pests while feeding on maize grains create small holes and produce fine dust as excrements, rendering the grains unsuitable for human and animal consumption. There are several species of post-harvest beetle pests that can cause sufficient damage to warrant control, including *Rhyzopertha dominica*, *Tribolium castaneum*, *S. zeamais* and *P. truncatus* (Kossou et al., 1992; Holst et al., 2000). While most such beetles are cosmopolitan in occurrence, the distribution of the fourth pest, *P. truncatus*, was confined until the early 1980s largely to Meso-America and the southern US. However, it was known to occasionally cause high post-harvest losses in its native range (Barnes et al., 1959; Ramírez Genel, 1960) and caused severe post-harvest losses in Nicaragua in the 1970s (Giles, 1977). *P. truncatus* was particularly problematic because of its ability to feed on both cereals and pulses (Shires, 1977).

Maize contamination by fungi is another important cause of post-harvest losses that poses a significant threat to human and animal health by producing poisons known as mycotoxins. Many mycotoxin-producing fungi including species of *Aspergillus*, *Fusarium* and *Penicillium* have been reported in stored maize in SSA (Suleiman et al., 2013). However, maize contamination by these fungi and their mycotoxins begins pre-harvest in the fields (Sétamou et al., 1997). As shown in Benin by Sétamou et al. (1998), there are complex interactions between field and storage insects, and mycotoxin-producing fungi in maize. For instance, pre-harvest infestations by the ear borer *M. nigrivenella* predisposes maize to pre-harvest *Aspergillus* infection and aflatoxin contamination, as well as post-harvest pest infestation (Sétamou et al., 1998). Hell et al. (2000) also found a positive correlation between insect damage and aflatoxin levels in stored maize. Thus, effective pre- and post-harvest insect pest management is critical for mitigation of fungal and mycotoxin contamination and ensuring the quality of maize.

Lack of adequate storage facilities forces farmers to sell their crops shortly after harvest leading to a glut in the market with attendant lowered prices and significant revenue losses. Maize is produced across all major AEZs in SSA spanning the wet and humid forest zone to the drier Sudan Savannah (Sétamou et al., 1997; Hell et al., 2000; Schneider et al., 2004). In all these AEZs maize is mainly cultivated by small- and medium-scale farmers who lack infrastructure for adequate post-harvest operations, along with a lack of information about best management practices. Most of these farmers utilize traditional granaries for maize storage (Pantenius, 1988). Hell et al. (2000) reported that over 93% of farmers in Benin preferred using their traditional granaries over the improved storage 'Crib' promoted by FAO. As these granaries are made of local materials, storage systems vary by AEZ. In SSA, maize is generally stored in bulk in giant baskets made of woven raffia palms, tree branches, bamboo, straw or granaries made of mud, clay or bricks (Fig. 4). Maize is mainly stored with husks or de-husked in these traditional granaries, or as shelled kernels in bags. Hell et al. (2000) reported that over 70% of maize is stored on the cob with or without husks in all AEZs in Benin, with generally less than 30% of the crop being stored in bags as grains. The duration of maize storage varies with the size of the farm and the AEZ, but most farmers in Benin stored their maize for 3–10 months (Hell et al., 2000).

Figure 4 Traditional grain storage structures in Burkina Faso.

The development of adapted solutions to reduce post-harvest losses requires relatively modest investment that can result in higher returns compared to only field-based efforts. The use of hermetic bags (Baoua et al., 2014; Williams et al., 2017) and metal silos (Tefera et al., 2011) for shelled maize is increasingly becoming popular. Willingness-to-pay studies have demonstrated metal silos to be cost-effective in Kenya (Gitonga et al., 2013). Major scaling up efforts, such as the YieldWise initiative of the Rockefeller Foundation (www.rockefellerfoundation.org/YieldWise) and several other programmes funded by various donors, are underway in SSA with major successes (Baributsa et al., 2014), but challenges remain (Nouhoheflin et al., 2017).

8 Larger grain borer and other beetles

The situation for post-harvest storage in SSA changed drastically in the early 1980s, with the accidental introduction of *P. truncatus* first to East Africa (Hodges et al., 1983; Hill et al., 2002). Heavy post-harvest losses were reported within a short period, prompting researchers to develop viable control options quickly and those options were mainly based on chemical control (Golob et al., 1985; Giles et al., 1995). Shortly after its introduction to East Africa, *P. truncatus* was found in West Africa. Grain stores in West Africa were particularly susceptible to attack by *P. truncatus*, because of the higher ambient relative humidity during most of the year, particularly in coastal regions, leading to higher grain moisture content and higher beetle reproduction and survivorship. At this point several research institutes, including IITA, launched programmes to identify ways of reducing pest damage. Research on stored product beetle pests in general, and on *P. truncatus* in particular, was focussed along several lines.

Prostephanus truncatus is a bostrichid, called powder-post beetle for the dust produced by its boring into wood (Hill et al., 2002). The same behaviour when applied to the rich substrate of maize stores often resulted in a complete loss of produce. Y-olfactometer and electroantennogram studies compared the reaction of the beetle to volatiles from different sources, including maize kernels and the insect's own pheromone. If the beetles responded strongly to volatiles from food sources, it may be possible to trap or confuse dispersing beetles (Scholz et al., 1997b). However, the strength of the pheromone response was far higher than to any other volatile. It was clear that, once a male had found a food source and begun emitting pheromone, large numbers of beetles could be expected to arrive (Scholz et al., 1997b,c).

Constructing a population simulation model of *P. truncatus* was seen as an important step in developing a 'basket of options' for beetle control. A model would provide a framework that links beetle development rates, survivorship, reproduction and dispersal to different ambient conditions and control strategies and thus help forecast population growth (Meikle et al., 1998b). *P. truncatus* was found to reproduce much faster and tolerate much higher densities of conspecifics before emigrating compared to pests such as *S. zeamais* (Meikle et al., 1998b; Scholz et al., 1997a). Models of other grain store insects and of damage and economic loss were also constructed (Meikle et al., 1999; Holst et al., 2000) and used to develop a rapid scouting programme for *P. truncatus* (Meikle et al., 2000, 2002a).

Prostephanus truncatus is considered a forest insect that comparatively recently became a stored product pest (Hill et al., 2002). Pheromone traps caught *P. truncatus* in large numbers in forests in Meso-America (Rees, 1987; Rees et al., 1990; Tigar et al., 1994; Ramírez-Martínez et al., 1994), in East Africa (Hill et al., 2003; Nang'ayo et al., 1993) and in West Africa (Borgemeister et al., 1997b, 1998c; Nansen et al., 1999, 2002; Nansen and Meikle, 2002), far from any grain stores. *P. truncatus* was recovered from the branches of forest trees girdled by cerambycid beetles in Meso-America (Rees et al., 1990) as well as in West Africa (Nansen and Meikle, 2002). Girdling of branches by cerambycid beetles provides a habitat for development of their larvae, which are then protected from the tree's sap-based defence mechanisms while providing nutritious substrate because the girdling prevents the tree from withdrawing amino acids and sugars from the dying branch (Forcella, 1982; Borgemeister et al., 1998b).

Controlling *P. truncatus* infestations in the early years of its invasion in SSA mainly involved pesticides, but in an effort to hold down pest control costs, reduce pesticide use and abuse, and account for local methods of grain storage (e.g. storage of maize on the cob with the husk intact), other methods of pest management were considered. These methods included, among others, changing the timing of harvest (Borgemeister et al., 1998a), the use of resistant maize varieties (Meikle et al., 1998a) and scouting for pests (Meikle et al., 2002a), as well as combining methods such as sorting ears prior to storage and using resistant varieties (Adda et al., 2002). Although those methods probably had a much lower impact on *P. truncatus* control than other factors, they could, with modification, be applied against other emerging stored product pests.

As with other IPM approaches, exploratory work for natural enemies was conducted in the native range of *P. truncatus* (Hill et al., 2002). Pathogens, including protozoans, microsporidians and fungi, were identified from *P. truncatus* specimens (Meikle et al., 2002b) and/or tested in field experiments (Meikle et al., 2001). Several species of pteromalid wasps were found associated with *P. truncatus*, but as those species already had cosmopolitan distributions, they were not considered for mass release in Africa. In a collaborative effort by the Natural Resources Institute (NRI) and governmental agencies in Mexico and Costa Rica, the histerid predator *Teretrius nigrescens* (Lewis) (Coleoptera:

Histeridae) was found to have the greatest potential for success as a possible control agent. Through a collaborative project between governmental agencies and the Gesellschaft für Techniches Zusammenarbeit (GTZ), field releases of *T. nigrescens* began in Togo in 1991 (Richter et al., 1997) and subsequently in many other countries in SSA including Benin, Ghana, Guinea, Kenya, Mozambique, Tanzania, Zambia and so on. The releases were undertaken in collaboration with national agricultural partners of the host countries (Giles et al., 1996; Borgemeister et al., 1997a; Hill et al., 2003). The source population of the releases in Benin came from Costa Rica while those in Kenya came from Mexico. These two populations were subsequently found to belong to two distinct clades (Omondi et al., 2011a). Studies on its host-finding behaviour (Borgemeister et al., 1998b; Scholz et al., 1998) showed that *T. nigrescens* is strongly attracted to *P. truncatus* aggregation pheromone and thrived on *P. truncatus* larvae. But *T. nigrescens* does have a broad prey range (Ayertey et al., 1999; Oussou et al., 1999).

Indications by the late 1990s offered reasons for optimism (Borgemeister et al., 1997a; Schneider et al., 2004) and for caution, as the somewhat slow population growth rate of *T. nigrescens* compared to its prey suggested that it was unlikely to provide sufficient control of *P. truncatus* in infested grain stores (Holst and Meikle, 2003; Meikle et al., 2002a; Muatinte and Cugala, 2015). This inference could have been due to the influence of environmental factors on population growth (Omondi et al., 2011b). In West Africa the status of *P. truncatus* as a major pest has declined a great deal. In the absence of studies on the causes of *P. truncatus* mortality, the question remains whether it is effectively controlled by *T. nigrescens*, or whether unknown predators, parasitoids or pathogens, either native or inadvertently introduced, also play important roles. Nevertheless, while the interactions between beetle populations in forests and grain stores have not been clearly defined, *T. nigrescens* appeared to have succeeded as a significant predator of *P. truncatus* on a regional basis, across both environments. In East Africa and some parts of West Africa and Mozambique, the situation is more complex. In these areas, *T. nigrescens* appears to have become extinct in some of the higher and cooler areas where it was introduced while the range of *P. truncatus* has increased (Omondi et al., 2011a). This poor establishment may be due to an inadequate match of original *T. nigrescens* biotype from Meso-America to the local ecology (Omondi et al., 2014). The more successful release in West Africa was attributed to the introduction of genetically more diverse *T. nigrescens* in higher numbers and more sites thereby possibly increasing chances of establishment compared to Kenya (Omondi et al., 2014). Since ecotypes in *P. truncatus* exists (Guntrip et al., 1996), more introductions of genetically diverse *T. nigrescens* population may help to select more efficient predators for suppressing the larger grain borer.

9 Mycotoxins

Contamination of food by toxin-producing fungi and the mycotoxins they produce are of common occurrence in Africa. Chronic and lifelong exposure of a large proportion of African populations to foodborne mycotoxins, beginning *in utero*, is a reality and a serious problem. There is an increasing mass of literature that documents the wide-ranging health impact of mycotoxins in humans and animals that includes cancer, impairment of growth and development of children, immune system suppression, reduced nutrient uptake and utilization, birth defects and death. Due to their severe effects on health, several countries have adopted legislation barring trade in agricultural products exceeding permissible levels of mycotoxins. However, most countries in Africa lack capacity to enforce food safety regulations, and consequently face frequent rejection of exportable products leading to economic loss (Marasas et al., 2008).

Since the early 1990s, a multidisciplinary research team in partnership with advanced research institutes in the United States and Europe as well as many national programmes in Africa began work to understand the problem and develop viable options to reduce mycotoxin contamination in staple crops and its impact on health and economy.

While assessing the principle problems in maize in West and Central Africa in early 1990s, exceptionally high levels of toxigenic fungi and mycotoxin were noticed in maize in all agroecological zones (Udoh et al., 2000; Hell et al., 2000; Cardwell, 2001; Ngoko et al., 2001, 2008; Sétamou et al., 1997) signifying the importance of mycotoxins. An assessment of relative severity of aflatoxin and fumonisin in maize, sorghum and pearl millet in Nigeria showed that maize is most prone to both toxins followed by sorghum and then pearl millet (Bandyopadhyay et al., 2007; Vismer et al., 2015). Due to high dependence on maize as a staple diet in West Africa, a large proportion of children and adults in various countries of the region have high levels of aflatoxin or its biomarker in the blood (Gong et al., 2003; Watson et al., 2015), urine (Ezekiel et al., 2014) and breast milk (Adejumo et al., 2013). Advancements in analytical procedures have enabled simultaneous measurement of more than 120 fungal and bacterial metabolites - both regulated and unregulated. Multi-mycotoxin assays revealed frequent co-occurrence of several regulated mycotoxins in maize (Warth et al., 2012) while 62% of the commercial poultry feed sampled from markets had unsafe levels of aflatoxins for productive poultry performance (Ezekiel et al., 2012). More mycotoxin prevalence data over space and time at country-level are required for informing management and policy decisions to contain mycotoxin in food and feed systems.

The frequent occurrence of aflatoxins at high levels in maize was recognized as a matter of significant concern since maize is the primary weaning food across the West African region (Egal et al., 2005). IITA hosted an international workshop on 'Mycotoxins in Food in Africa' in Cotonou to discuss the scope of the mycotoxin problem in Africa (Cardwell, 1996). The question placed before international toxicology and medical experts was, aside from the known relationship to hepatocellular carcinomas (Cardwell and Henry, 2003; Wild and Gong, 2010), could chronic exposure to aflatoxin from early in life be impacting infant health, perhaps even be part of the intractable pre-five mortality in SSA? The output of the workshop was a position paper that became the basis for a project to evaluate the impact of aflatoxin exposure on infants in West Africa. The findings of the subsequent longitudinal and cohort studies in Benin and Togo were startlingly clear (Fig. 5). Almost 100% of the children in Benin and Togo had detectable blood aflatoxin levels due to dietary exposure to aflatoxin-contaminated foods at weaning, which strikingly led to a significant dose-related growth faltering (Gong et al., 2002, 2004; Cardwell and Henry, 2003). This study, piloted by IITA, was the seminal evidence of the severe impact of dietary exposure to aflatoxin in weaning foods on child health that was subsequently corroborated around the world (Khlangwiset et al., 2011; Turner et al., 2007).

Aspergillus species is a highly diverse group of fungi and consists of several morphotypes, chemotypes and genotypes. Certain *Aspergillus* species produce only B aflatoxins while others produce both B and G types. For example, *A. flavus*, the most common causal agent of contamination, produces only B aflatoxins (Bandyopadhyay et al., 2016). The species is subdivided into L and S morphotypes, which differ in several traits including aflatoxin-producing

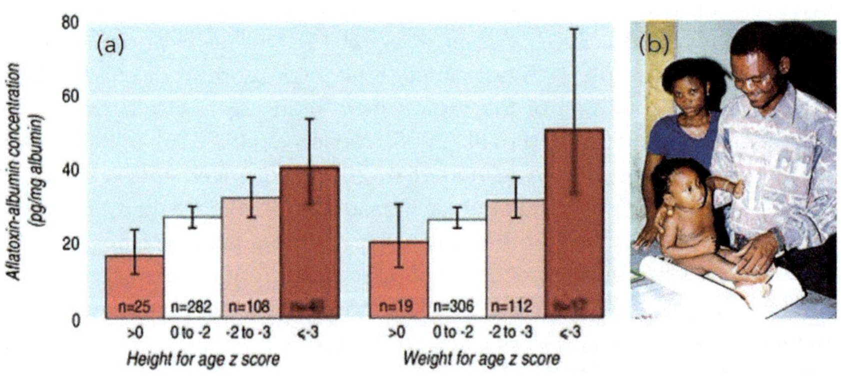

Figure 5 (a) Geometric mean concentrations of aflatoxin-albumin adduct categorized into four groups for height for age and weight for age z scores based on the WHO classification of malnutrition (z score ≤2) and severe malnutrition (≤3). Height for age and weight for age z scores were significantly negatively associated with aflatoxin-albumin concentration (Gong et al., 2002). This is the first study to report dose-dependent response of aflatoxin on child stunting. (b) Weight of a child being measured by a medical professional.

Published by Burleigh Dodds Science Publishing Limited, 2019.

potentials (Cotty, 1989). Acknowledging the importance of assessing *Aspergillus* diversity, IITA and US Department of Agriculture – Agricultural Research Service (USDA-ARS) started collaboration in 1996 to examine fungal communities associated with maize and soil in Benin. Results revealed the presence of a fungal group resembling the *A. flavus* S morphotype but capable of producing high levels of both B and G aflatoxins (Cotty and Cardwell, 1999; Cardwell and Cotty, 2002). This group, referred to as the unnamed taxon S_{BG}, was subsequently found to be prevalent particularly in the hot, dry areas of Burkina Faso, Nigeria, Senegal and Ghana, but not in Central and East Africa (Atehnkeng et al., 2008b; Donner et al., 2009; Diedhiou et al., 2011; Probst et al., 2014; Agbetiameh et al., 2018). Further studies showed that the Kenyan isolates responsible for death of more than 125 people in Kenya's Eastern Province resembled *A. flavus* S morphotype and had extremely high aflatoxin-producing potential (Probst et al., 2007), but were genetically distinct from other lineages, including the unnamed taxon S_{BG} (Probst et al., 2011, 2012). This initial examination in various countries helped to design appropriate aflatoxin management strategies and revealed thousands of atoxigenic *A. flavus* isolates with potential as biocontrol agents.

In SSA, the *A. flavus* L morphotype consistently dominates on maize. However, other *Aspergillus* genotypes prevail in certain maize agroecosystems (e.g. the Kenyan lineage mentioned above) (Probst et al., 2007, 2012) or even at low frequencies play major roles in contamination events (e.g. unnamed taxon S_{BG}) (Cotty and Cardwell, 1999; Atehnkeng et al., 2008b; Diedhiou et al., 2011). On the other hand, studies examining both maize and groundnut in Zambia revealed that *A. parasiticus* is more frequently associated with groundnut (Kachapulula et al., 2017). That was not the case in Ghana, where *A. flavus* dominated both maize and groundnut (Agbetiameh et al., 2018). Other species are infrequently reported and have minor and often insignificant roles in maize contamination events.

IITA and partners conducted substantial work in further characterizing *Aspergillus* diversity at the genetic level. *Aspergillus* species (and morphotypes) are further subdivided in numerous vegetative compatibility groups (VCGs) that vary in physiological and genetic traits (Bayman and Cotty, 1991; Grubisha and Cotty, 2010). VCGs are clonal lineages and their members are more closely related among them than to members of other VCGs (Leslie, 1993). VCGs are composed of members with i) high aflatoxin-producing potential, ii) either low or high aflatoxin potential or iii) no aflatoxin-producing ability, that is 'atoxigenic'. Understanding community composition at VCG level and VCG efficacy to limit aflatoxin content in the field provided powerful information to select the most superior strains to constitute the registered biocontrol products Aflasafe in Nigeria (Atehnkeng et al., 2008a; Atehnkeng et al., 2014), Aflasafe KE01 in Kenya (Probst et al., 2011), and Aflasafe SN01 in Senegal and Aflasafe BF01 in Burkina Faso (Adhikari et al., 2016).

Atoxigenicity in some *A. flavus* strains occurs due to the non-functional nature of one or more genes in the aflatoxin biosynthetic gene cluster (Yu et al., 2004). The genetic defects/lesions causing atoxigenicity in African *Aspergillus* communities were examined in the population from Nigeria by individually monitoring defects in certain aflatoxin biosynthesis genes (Donner et al., 2010). Subsequently, the work shifted towards characterizing complete aflatoxin gene clusters, which revealed shared ancestry of aflatoxigenicity in strains from different geographic origin and provided evidence that there is no danger of atoxigenic strains to gain aflatoxin-producing ability (Adhikari et al., 2016).

IITA has a long-term (1994–present), intensive collaboration with the USDA-ARS for the development of aflatoxin biocontrol strategy in Africa centred on the use of atoxigenic *A. flavus* strains to competitively displace aflatoxin producers. The idea of biological control of aflatoxin in Africa was introduced during the 'Workshop on Mycotoxins in Food in Africa' in Cotonou, Benin (Cardwell, 1996). At that time, most research in Benin focussed on effects of aflatoxins on child health while there was less emphasis on biocontrol was less. Then biocontrol research efforts shifted from Benin to Nigeria where collaboration among IITA, USDA-ARS, University of Ibadan and University of Bonn commenced in 2003.

The biocontrol technology, initially developed by USDA-ARS, involves competitive displacement of toxigenic *Aspergillus* population in the environment with a carefully selected and widely distributed native atoxigenic strain of *A. flavus*. The atoxigenic strain is applied to the soil before the normal increase of the *Aspergillus* population in the field. Early application creates a founding population of the atoxigenic strain that gets the first opportunity to occupy substrates (e.g. organic matter, decaying leaves, dead insects etc.) normally colonized by *Aspergillus* in the field. This founding population subsequently increases in the field instead of the toxigenic strains thereby changing the community composition in favour of the atoxigenic strain. The atoxigenic strains infect the crops in place of the toxigenic strains resulting in reduction in aflatoxin formation in the crop (Mehl et al., 2012).

During 2004–6, more than 4200 *Aspergillus* isolates associated with maize and soils from major AEZs of Nigeria were examined and >1000 atoxigenic *A. flavus* isolates were identified (Atehnkeng et al., 2008b; Donner et al., 2009). Subsequent studies determined VCG diversity and environmental distribution (Atehnkeng et al., 2016) and reasons for atoxigenicity in *Aspergillus* populations from Nigeria (Donner et al., 2010). In parallel, eleven atoxigenic genotypes were tested for their ability to decrease aflatoxin content when challenged with highly toxigenic strains in competition experiments in laboratory conditions (Atehnkeng et al., 2008a). Subsequently, efficacy of mixture of atoxigenic strains was tested in maize fields of four AEZs of Nigeria and aflatoxin content

Published by Burleigh Dodds Science Publishing Limited, 2019.

in treated crops ranged from 67% to 95% less in comparison to non-treated crops. An application of 10 kg of the strain mixture/ha (Fig. 6) protected the crop from the field to post-harvest stages (Atehnkeng et al., 2014). Information from VCG composition, environmental distribution, competitiveness, ecological fitness and efficacy were used to select four atoxigenic genotypes for constituting Aflasafe, a biocontrol product for aflatoxin mitigation of maize and groundnut in Nigeria. The approach followed in Nigeria was replicated in other nations where Aflasafe products are registered (i.e. Kenya, Senegal, The Gambia and Burkina Faso) or will soon be registered (i.e. Ghana, Tanzania, Zambia, Mozambique and Malawi) (Bandyopadhyay et al., 2016). A major effort is currently underway to scale up Aflasafe products to reach millions of farmers in Africa (Schreurs et al.: Chapter 13).

Until recently in the United States, aflatoxin biocontrol products contained a single atoxigenic *A. flavus* strain as active ingredient (Cotty et al., 2007; Dorner, 2009). The African biocontrol programme from the beginning pioneered the use of mixtures of atoxigenic genotypes in product formulations. Using atoxigenic strain mixtures has a synergistic effect that allows for long-term maintenance of *Aspergillus* communities with low aflatoxin-producing potentials (Mehl et al., 2012). Use of multiple strains of diverse geographic origin that belong to widely distributed genetic groups with local adaptation to cropping systems in target areas makes Aflasafe products effective in diverse ecosystems (Atehnkeng et al., 2016).

Research is underway to develop new products for use in Rwanda, Burundi, Uganda, Zimbabwe, Mali and Cameroon while product development is

Figure 6 (a) Field workers applying Aflasafe (a biological control product for aflatoxin mitigation) 2–3 weeks before flowering of maize in the field. (b) Formulated Aflasafe product (heat-killed carrier sorghum grains coated with a mixture of atoxigenic strains of *Aspergillus flavus* suspended in water, a polymer sticker and a blue food colourant). (c) Atoxigenic strains sporulating on a sorghum carrier a few days after Aflasafe was applied in the field. These atoxigenic spores spread in the field and displace the toxigenic strains through 'founder effect'.

required in several other nations. In such new nations where strain development is beginning or to begin later, the objective should be to search for atoxigenic VCGs that are co-distributed in several nations regionally to reduce product development time and increase commercialization opportunities of the same product in several countries. There is a need to generate empirical data on the impact of biological control on economic and health benefits of farmers beyond the ex ante desk study conducted earlier (Wu and Khlangwiset, 2010).

IITA and partners have demonstrated that there are areas consistently affected by aflatoxin contamination and others in which environmental and crop conditions favouring aflatoxin formation occur infrequently (Cardwell and Cotty, 2002; Probst et al., 2007, 2014; Atehnkeng et al., 2008b; Donner et al., 2009; Diedhiou et al., 2011; Kachapulula et al., 2017; Agbetiameh et al., 2018). However, the risk of crops becoming contaminated with unsafe aflatoxin levels is latent and we recommend implementing all available aflatoxin management strategies in both aflatoxin hot spots and areas frequently affected by contamination events (Bandyopadhyay et al., 2016).

A successful aflatoxin management system must include a package of complementary technologies, both traditional and innovative, to ultimately reduce aflatoxins to a safe level. Traditional strategies include good agricultural practices (planting date, spacing, irrigation scheme, fertilization, insect control, appropriate harvesting, among others), grain drying, improved storage structure, optimal post-harvest processing, sorting, monitoring and crop destruction, and dietary interventions (Bandyopadhyay et al., 2016; Seetha et al., 2017; Udomkun et al., 2017). However, relying only on traditional strategies does not guarantee production of aflatoxin-safe crops. Unsafe aflatoxin levels may still occur even with slight environmental changes.

Innovative management strategies include the use of atoxigenic *A. flavus* strains as biocontrol agents (Mehl et al., 2012; Bandyopadhyay et al., 2016). Over 100 000 ha cropped to maize and groundnut have been treated in Nigeria, Senegal, The Gambia and Kenya and over 95% of treated maize contained safe aflatoxin levels (<20 ppb). Tailored aflatoxin management programmes are in the process of being implemented at scale in other African nations (Bandyopadhyay et al., 2016). Although germplasm with reduced susceptibilities to aflatoxin accumulation have been identified by IITA and USDA-ARS (Brown et al., 2016), no commercially adoptable maize with aflatoxin resistance is available (Brown et al., 2013).

Use of atoxigenic strains to displace toxigenic fungi should be the central component of a comprehensive management strategy, which must include all available practices in the target nation/area. The strategy must include sensitization, training, institutional (e.g. inclusion of aflatoxins in national health and agricultural agendas, development of aflatoxin-conscious markets) and policy (e.g. disposal systems, testing facilities) actions. Only a holistic aflatoxin

Published by Burleigh Dodds Science Publishing Limited, 2019.

management system approach will result in reduced food insecurity, decreased health disorders, increased productivity, and increased income and trade (Udomkun et al., 2017).

The first holistic aflatoxin management system in Africa is being practised by farmers in Nigeria with good outcome. Over 20 000 farmers grouped by agricultural enterprises (known as implementers) have adopted technologies to produce maize with low aflatoxin levels, with the biocontrol product Aflasafe as a central component through the AgResults Aflasafe Pilot project (www.agresults.org) (Bandyopadhyay et al., 2016). Farmers also receive training on productivity-enhancing practices, and have access to inputs and aflatoxin testing support. Implementers also participate in innovation platforms to link with quality conscious buyers of maize. The project itself has increased both aflatoxin awareness and adoption of aflatoxin management strategies through training, monitoring and feedback sessions. A tangible result of the project is the production of more than 100 000 tons of maize of high quality allowing farmers to obtain significant premium over maize market prices. In addition, farmers retained a portion of their harvest for home consumption, which results in provision of safe food for their families. Increased productivity, economic incentives by industries, policy and institutional support, and health benefits to farmer families have resulted after implementation of a holistic, successful aflatoxin management strategy (Bandyopadhyay et al., 2016). Similar strategies should be pursued in other SSA nations.

10 Conclusions and future trends

Plant health problems arising from insect pests and fungal pathogens continue to be major issues for maize cultivation in SSA. IITA has a rich history of strategic, applied, adaptive and problem-solving research on maize insect pests and diseases that have plagued maize. Some disease problems could be easily resolved with a resistance breeding approach alone in which breeders and plant protection specialists collaborated. A few problems such as downy mildew required a combined breeding and other management approaches. However, some other biotic constraints (e.g. post-harvest insects and stemborer) were not readily amenable to a breeding solution and required habitat management and biological control approaches rooted to science. Some problems, such as downy mildew, were reduced from a threat to being 'of no consequence'. Major advances in our understanding of toxigenic fungi and aflatoxins have led to biological control solutions being adapted for several African nations. Biological control of aflatoxins is currently being scaled out and forms the pillar of aflatoxin management systems in various African nations. Future research efforts should focus on monitoring *Aspergillus* genotypes both spatially and temporally in areas with or without aflatoxin management programmes. Such efforts will aid in designing a second generation of aflatoxin management options at a wider scale across regions.

All pest problems were overshadowed in 2016, when the fall armyworm *S. frugiperda* invaded Africa out of South America, spread across Africa and devastated maize culture. Following its identification (Goergen et al., 2016), IITA became an important partner in a worldwide coalition of organizations that is now attempting to tackle this invasive species, first by use of appropriate insecticides and cultural methods, but eventually concentrating on biological control (Prasanna et al., 2018). Just as other seemingly intractable problems were overcome by problem-solving research of IITA plant health specialists and their partners, we are confident that very soon the fall armyworm will be managed to economically harmless levels.

11 References

Abedin, M., Rahman, M., Mia, M. and Rahman, K. (2012), In-store losses of rice and ways of reducing such losses at farmers' level: An assessment in selected regions of Bangladesh, *J. Bangladesh Agric. Univ.*, 10, 133–44.

Adda, C., Borgemeister, C., Biliwa, A., Meikle, W. G., Markham, R. H. and Poehling, H.-M. (2002), Integrated pest management in post-harvest maize: A case study from the Republic of Togo (West Africa), *Agric. Ecosyst. Environ.*, 93, 305–21.

Adejumo, O., Atanda, O., Raiola, A., Somorin, Y., Bandyopadhyay, R. and Ritieni, A. (2013), Correlation between aflatoxin M1 content of breast milk, dietary exposure to aflatoxin B1 and socioeconomic status of lactating mothers in Ogun State, Nigeria, *Food Chem. Toxicol.*, 56, 171–7.

Adenle, V. O. and Cardwell, K. F. (2000), Seed transmission of maize downy mildew (*Peronosclerospora sorghi*) in Nigeria, *Plant Pathol.*, 49, 628–34.

Adhikari, B. N., Bandyopadhyay, R. and Cotty, P. J. (2016), Degeneration of aflatoxin gene clusters in *Aspergillus flavus* from Africa and North America, *AMB Express*, 6, 62. doi:10.1186/s13568-016-0228-6.

Agbetiameh, D., Ortega-Beltran, A., Atehnkeng, J., Awuah, R. T., Cotty, P. J. and Bandyopadhyay, R. (2018), Prevalence of aflatoxin contamination in maize and groundnuts in Ghana: Population structure, distribution, and toxigenicity of the etiologic agents, *Plant Dis.* doi:10.1094/PDIS-05-17-0749-RE.

Ako, M., Schulthess, F., Gumedzoe, M. Y. D. and Cardwell, K. F. (2003), The effect of *Fusarium verticillioides* Sacc. (Nirenberg) on oviposition behaviour and bionomics of lepidopteran and coleopteran pests attacking the stem and cobs of maize in West Africa, *Entomol. Exp. Appl.*, 106, 201–10.

APHLIS (2014), Africa Postharvest Losses Information System. Losses tables; Estimated Postharvest losses (%) 2003-2014, Online available at: http://www.aphlis.net/?form=home (accessed on 22 February 2018).

Atehnkeng, J., Ojiambo, P. S., Ikotun, T., Sikora, R. A., Cotty, P. J. and Bandyopadhyay, R. (2008a), Evaluation of atoxigenic isolates of *Aspergillus flavus* as potential biocontrol agents for aflatoxin in maize, *Food Addit. Contam.*, 25, 1266–73.

Atehnkeng, J., Ojiambo, P. S., Donner, M., Ikotun, T., Sikora, R. A., Cotty, P. J. and Bandyopadhyay, R. (2008b), Distribution and toxigenicity of *Aspergillus* species

Published by Burleigh Dodds Science Publishing Limited, 2019.

isolated from maize kernels from three agro-ecological zones in Nigeria, *Int. J. Food Microbiol.*, 122, 74–84.

Atehnkeng, J., Ojiambo, P. S., Cotty, P. J. and Bandyopadhyay, R. (2014), Field efficacy of a mixture of atoxigenic *Aspergillus flavus* Link: Fr vegetative compatibility groups in preventing aflatoxin contamination in maize (*Zea mays* L.), *Biol. Control*, 72, 62–70.

Atehnkeng, J., Donner, M., Ojiambo, P. S., Ikotun, B., Augusto, J., Cotty, P. J. and Bandyopadhyay, R. (2016), Environmental distribution and genetic diversity of vegetative compatibility groups determine biocontrol strategies to mitigate aflatoxin contamination of maize by *Aspergillus flavus*, *Microb. Biotechnol.*, 9, 75–88.

Awika, J. M. (2011), Major cereal grains production and use around the world, *in* J. Awika, S. Bean and V. Piironen (Eds), *Advances in Cereal Science: Implications to Food Processing and Health Promotion* (pp. 1–13), Washington DC: American Chemical Society.

Ayertey, J. N., Meikle, W. G., Borgemeister, C., Camara, M. and Markham, R. H. (1999), Studies on predation of *Prostephanus truncatus* (Horn) (Col., Bostrichidae) and *Sitophilus zeamais* Mots. (Col., Curculionidae) at different densities on maize by *Teretriosoma nigrescens* Lewis (Col., Histeridae), *J. Appl. Entomol.*, 123, 265–71.

Baba-Moussa, A.-A. M. T. (1998), La microflore associée aux dégâts des Lépidoptères foreurs de tiges et mineurs d'épis de maïs (*Zea mays*) dans la région sud du Bénin avec référence spécial à *Fusarium moniliforme* Sheld, Mémoire d'Ingénieur Agronome, Université du Bénin, Lomé, Togo.

Bandyopadhyay, R., Kumar, M. and Leslie, J. F. (2007), Relative severity of aflatoxin contamination of cereal crops in West Africa, *Food Add. Contam.*, 24, 1109–14.

Bandyopadhyay, R., Ortega-Beltran, A., Akande, A., Mutegi, C., Atehnkeng, J., Kaptoge, L., Senghor, A. L., Adhikari, B. N. and Cotty, P. J. (2016), Biological control of aflatoxins in Africa: Current status and potential challenges in the face of climate change, *World Mycotoxin J.*, 9, 771–89.

Baoua, I. B., Amadou, L., Ousmane, B., Baributsa, D. and Murdock, L. L. (2014), PICS bags for post-harvest storage of maize grain in West Africa, *J. Stored Prod. Res.*, 58, 20–8.

Baributsa, D., Abdoulaye, T., Lowenberg-DeBoer, J., Dabiré, C., Moussa, B., Coulibaly, O. and Baoua, I. (2014), Market building for post-harvest technology through large-scale extension efforts, *J. Stored Prod. Res.*, *58*, 59–66.

Barnes, D., De las Casas, E. and Ramírez Genel, M. (1959), Fumigación de los granos almacenados [Fumigation of stored grains], *Agricultura Técnica en México*, 8, 7–10.

Bayman, P. and Cotty, P. J. (1991), Vegetative compatibility and genetic diversity in the *Aspergillus flavus* population of a single field, *Can. J. Bot.*, 69, 1707–11.

Bigirwa, G., Adipala, E., Esele, J. P. and Cardwell, K. F. (2000), Reaction of maize, sorghum and Johnson grass to *Peronosclerospora sorghi*, *Int. J. Pest Manage.*, 46, 1–6.

Blacutt, A. A., Gold, S. E., Voss, K. A., Gao, M. and Glenn, A. E. (2018), *Fusarium verticillioides*: Advancements in understanding the toxicity, virulence, and niche adaptations of a model mycotoxigenic pathogen of maize, *Phytopathology*, 108, 312–26.

Bock, C. H. and Jeger, M. J. (1996), Downy mildew of sorghum, *Int. Sorghum Millets News*, 37, 33–51.

Bock, C. H., Jeger, M. J., Mughogho, L. K., Cardwell, K. F., Adenle, V., Mtisi, E., Akpa, A. D., Kaula, G., Mukasmbina, D. and Blair-Myers, C. (1998a), Occurrence and distribution of *Peronosclerospora sorghi* [Weston and Uppal (Shaw)] in selected countries of West and Southern Africa, *Crop Prot.*, 17(5), 427–39.

Bock, C. H., Jeger, M. J., Mughogho, L. K., Mtisi, E. and Cardwell, K. F. (1998b), Production of conidia by *Peronosclerospora sorghi* on sorghum crops in Zimbabwe, *Plant Pathol.*, 47, 243–51.

Bock, C. H., Jeger, M. J., Cardwell, K. F., Mughogho, L. K. and Sherington, J. (2000), Control of sorghum downy mildew of maize and sorghum in Africa, *Trop. Sci.*, 40, 47–57.

Borgemeister, C., Djossou, F., Schneider, H., Djomamou, B., Degbey, P., Azoma, K. and Markham, R. H. (1997a), Establishment, spread and impact of *Teretriosoma nigrescens* (Coleoptera: Histeridae), an exotic predator of the larger grain borer in southwestern Benin, *Environ. Entomol.*, 26, 1405–15.

Borgemeister C., Meikle, W. G., Scholz, D., Adda, C., Degbey, P. and Markham, R. H. (1997b), Seasonal and weather factors influencing the annual flight cycle of *Prostephanus truncatus* (Col.: Bostrichidae) and its predator *Teretriosoma nigrescens* (Col.: Histeridae) in Benin, *Bull. Entomol. Res.*, 87, 239–46.

Borgemeister, C., Adda, C., Sétamou, M., Hell, K., Djomamou, B., Markham, R. H. and Cardwell, K. F. (1998a), Timing of harvest in maize: Effects on post harvest losses due to insects and fungi in central Benin, with particular reference to *Prostephanus truncatus* (Horn) (Coleoptera: Bostrichidae), *Agric. Ecosyst. Environ.*, 69, 233–42.

Borgemeister, C., Goergen, G., Tchabi, A., Awande, S., Markham, R. H. and Scholz, D. (1998b), Exploitation of a woody host plant and Cerambycid-associated volatiles as host-finding cues by the larger grain borer (Coleoptera: Bostrichidae), *Ann. Ent. Soc. Am.*, 91, 741–7.

Borgemeister, C., Tchabi, A. and Scholz, D. (1998c), Trees or stores? The origin of migrating *Prostephanus truncatus* collected in different ecological habitats in Southern Benin, *Entomol. Exp. Appl.*, 87, 285–94.

Bottalico, A. and Logrieco, A. (1988), Osservazioni sulla fusariosi del mais in Basilicata: Influenza di alcuni fattori colturali, *Inf. Fitopatol.*, 2, 55–8.

Branca, A., Vavre, F., Silvain, J.-F. and Dupas, S. (2009), Maintenance of adaptive differentiation by *Wolbachia* induced bidirectional cytoplasmic incompatibility: The importance of sib-mating and genetic systems, *BMC Evol. Biol.*, 9, 185–98.

Branca, A., Le Ru, B., Vavre, F., Silvain, J.-F. and Dupas, S. (2011), Intraspecific specialization of the generalist parasitoid *Cotesia sesamiae* revealed by polyDNAvirus polymorphism and associated with different *Wolbachia* infections, *Mol. Ecol.*, 20, 959–71.

Brown, R. L., Menkir, A., Chen, Z.-Y., Bhatnagar, D., Yu, J., Yao, H. and Cleveland, T. E. (2013), Breeding aflatoxin-resistant maize lines using recent advances in technologies – a review, *Food Addit. Contam. Part A*, 30, 1382–91.

Brown, R. L., Williams, W. P., Windham, G. L., Menkir, A. and Chen, Z.-Y. (2016), Evaluation of African-bred maize germplasm lines for resistance to aflatoxin accumulation, *Agron. J.*, 6, 24.

Bruce, Y. A., Schulthess, F., Mueke, J. and Gohole, L. (2009), Egg parasitoids of noctuid stemborers on maize in Kenya, *Int. J. Biodivers. Sci. Manag.*, 5, 174–80.

Buadu, E. J., Gounou, S., Cardwell, K. F., Mochiah, B., Botchey, M., Darkwa, E. and Schulthess, F. (2003), Distribution and relative importance of insect pests and diseases in Southern Ghana, *Afr. Plant Prot.*, 8, 3–11.

Cardwell, K. F. (1995), The nature of the downy mildew disease of maize caused by *Peronosclerospora sorghi*: A technical overview, *in* K. F. Cardwell (Ed.), *Proceedings of the Maize Downy Mildew Disease Eradication Campaign Conference and*

Published by Burleigh Dodds Science Publishing Limited, 2019.

Workshop, 22–25 February 1994, IITA/FAO/Nigerian Federal Department of Agriculture (pp. 15–25), Ibadan, Nigeria: International Institute of Tropical Agriculture.

Cardwell, K. F. (Ed.) (1996), *Proceedings of the Workshop on Mycotoxin in Foods in Africa*, 6–10 November 1995 (p. 77), Cotonou, Benin: International Institute of Tropical Agriculture.

Cardwell, K. F. (2001), Mycotoxin contamination of foods in Africa: Anti-nutritional factors, *Food Nutr. Bull.*, 21, 488–92.

Cardwell, K. F. and Cotty, P. J. (2002), Distribution of *Aspergillus* section Flavi among field soils from the four agroecological zones of the republic of Bénin, West Africa, *Plant Dis.*, 79, 1039–45.

Cardwell, K. F. and Henry, S. (2003), Risk of exposure to and mitigation of effect of aflatoxin on human health: A West African example, *J. Toxicol.: Toxin Rev.*, 23, 217–47.

Cardwell, K. F., Kling, J. G. and Bock, C. (1997a), Methods for screening maize against downy mildew *Peronosclerospora sorghi*, *Plant Breed.*, 116, 221–6.

Cardwell, K. F., Schulthess, F., Ndemah, R. and Ngoko, Z. (1997b), A systems approach to assess crop health and maize yield losses due to pests and diseases in Cameroon, *Agric. Ecosyst. Env.*, 65, 33–47.

Cardwell, K. F., Kling, J. G., Maziya-Dixon, B. and Bosque-Perez, N. A. (2000), Interactions between *Fusarium verticillioides, Aspergillus flavus*, and insect infestation in four maize genotypes in lowland Africa, *Phytopathology*, 90, 276–84.

Chabi-Olaye, A., Nolte, C., Schulthess, F. and Borgemeister, C. (2005a), Abundance, dispersion and parasitism of the noctuid stem borer *Busseola fusca* (Fuller) in mono- and intercropped maize in the humid forest zone of southern Cameroon, *Bull. Entomol. Res.*, 95, 169–77.

Chabi-Olaye, A., Nolte, C., Schulthess, F. and Borgemeister, C. (2005b), Relationships of intercropped maize, stem borer damage to maize yield and land-use efficiency in the humid forest of Cameroon, *Bull. Entomol. Res.*, 95, 417–27.

Chabi-Olaye, A., Nolte, C., Schulthess, F. and Borgemeister, C. (2005c), Effects of grain legumes and cover crops on maize yield and plant damage by *Busseola fusca* (Fuller) (Lepidoptera: Noctuidae) in the humid forest of southern Cameroon, *Agric. Ecosyst. Environ.*, 108, 17–28.

Chabi-Olaye, A., Nolte, C., Schulthess F. and Borgemeister, C. (2006a), The role of inland valleys in the invasion of upland maize fields by stem borers and their natural enemies in the humid forest zone of Cameroon, *Env. Entomol.*, 35, 282–92.

Chabi-Olaye, A., Nolte, C., Schulthess, F. and Borgemeister, C. (2006b), Relationships of soil fertility and stem borers damage to yield in maize-based cropping system in Cameroon, *Ann. Soc. Entomol. Fr.*, 42, 471–9.

Chabi-Olaye, A., Schulthess, F. and Borgemeister, C. (2008), Effects of nitrogen and potassium combinations on yields and infestations of maize by *Busseola fusca* (Fuller) (Lepidoptera: Noctuidae) in the humid forest of Cameroon, *J. Econ. Entomol.*, 101, 90–8.

Chiang, H. C. and Wilcoxon, R. D. (1961), Interaction of the European corn borer and stalk rot in corn, *J. Econ. Entomol.*, 54, 850–2.

Christensen, J. J. and Schneider, C. L. (1950), European corn borer (*Pyrausta nubilalis* Hbn.) in relation to shank, stalk and ear rot of corn, *Phytopathology*, 40, 284–91.

Cotty, P. J. (1989), Virulence and cultural characteristics of two *Aspergillus flavus* strains pathogenic on cotton, *Phytopathology*, 79, 808–14.

Cotty, P. J. and Cardwell, K. F. (1999), Divergence of West African and North American communities of *Aspergillus* section Flavi, *Appl. Environ. Microbiol.*, 65, 2264–6.

Cotty, P. J., Antilla, L. and Wakelyn, P. J. (2007), Competitive exclusion of aflatoxin producers: Farmer driven research and development, *in* C. Vincent, N. Goettel and G. Lazarovits (Eds), *Biological Control: A Global Perspective* (pp. 242–53), Oxfordshire, UK: CAB International.

Diedhiou, P. M., Bandyopadhyay, R., Atehnkeng, J. and Ojiambo, P. S. (2011), *Aspergillus* colonization and aflatoxin contamination of maize and sesame kernels in two agro-ecological zones in Senegal, *J. Phytopathol.*, 159, 268–75.

Donner, M., Atehnkeng, J., Sikora, R. A., Bandyopadhyay, R. and Cotty, P. J. (2009), Distribution of *Aspergillus* section Flavi in soils of maize fields in three agroecological zones of Nigeria, *Soil Biol. Biochem.*, 41, 37–44.

Donner, M., Atehnkeng, J., Sikora, R. A., Bandyopadhyay, R. and Cotty, P. J. (2010), Molecular characterization of atoxigenic strains for biological control of aflatoxins in Nigeria, *Food Addit. Contam.*, 27, 576–90.

Dorner, J. W. (2009), Biological control of aflatoxin contamination in corn using a nontoxigenic strain of *Aspergillus flavus*, *J. Food Prot.*, 72, 801–4.

Dupas, S., Gitau, C. W., Branca, A., Le Ru, B. P. and Silvain, J. (2008), Evolution of a polydnavirus gene in relation to parasitoid- host species immune resistance, *J. Hered.*, 99, 491–9.

Egal, S., Hounsa, A., Gong, Y. Y., Turner, P. C., Wild, C. P., Hall, A. J., Hell, K. and Cardwell, K. F. (2005), Dietary exposure to aflatoxin from maize and groundnut in young children from Benin and Togo, West Africa, *Int. J. Food Microbiol.*, 104, 215–24.

Ezekiel, C. N., Bandyopadhyay, R., Sulyok, M., Warth, B. and Krska, R. (2012), Fungal and bacterial metabolites in commercial poultry feed from Nigeria, *Food Add. Contam. Part A*, 29, 1288–99.

Ezekiel, C. N., Ogara, I. M., Abia, W. A., Ezekiel, V. C., Atehnkeng, J., Sulyok, M., Turner, P. C., Tayo, G. O., Krska, R. and Bandyopadhyay, R. (2014), Mycotoxin exposure in rural residents in northern Nigeria: A pilot study using multi-urinary biomarkers, *Environ. Int.*, 66, 138–45.

FAOSTAT (2018), Food and Agriculture Organization of the United Nations (FAO), FAOSTAT Database, http://www.fao.org/faostat/en/#data/QC (accessed on 6 March 2018).

Finch, S. and Collier, R. H. (2012), The influence of host and non-host companion plants on the behaviour of pest insects in field crops, *Entomol. Exp. Appl.*, 142, 87–96.

Forcella, F. (1982), Why twig-girdling beetles girdle twigs, *Naturwissenschaften*, 69, 398–400.

Giles, P. H. (1977), Bean storage problems in Nicaragua, *Trop. Stored Prod. Info.*, 34, 63–7.

Giles, P., Hill, G., Nang'ayo, F., Farrell, G., Stabrawa, A. and Wekesa, P. (1995), Entomological and socio-economic investigations for the development of integrated pest management strategies against *Prostephanus truncatus*, Natural Resources Institute, Kent, UK, p. 273.

Giles, P. H., Hill, M. G., Nang'ayo, F. L. O., Farrell, G. and Kibata, G. N. (1996), Release and establishment of the predator *Teretriosoma nigrescens* Lewis for the biological control of *Prostephanus truncatus* (Horn) in Kenya, *Afr. Crop. Sci. J.*, 4, 325–37.

Gitau, C. W., Schulthess, F. and Stephane, D. (2010), An association between host acceptance and virulence status of different populations of *Cotesia sesamiae*, a braconid larval parasitoid of lepidopteran cereal stem borers in Kenya, *Biol. Control*, 54, 100–6.

Gitonga, Z. M., De Groote, H., Kassie, M. and Tefera, T. (2013), Impact of metal silos on households' maize storage, storage losses and food security: An application of a propensity score matching, *Food Pol.* 43, 44–55.

Goergen, G., Kumar, P. L., Sankung, S. B., Togola, A. and Tamò, M. (2016), First report of outbreaks of the fall armyworm *Spodoptera frugiperda* (J E Smith) (Lepidoptera, Noctuidae), a new alien invasive pest in West and Central Africa, *PLoS One* 11(10), e0165632. doi:10.1371/journal.pone.0165632.

Golob, P., Changjaroen, P., Ahmed, A. and Cox, J. (1985), Susceptibility of *Prostephanus truncatus* (Horn) (Coleoptera: Bostrichidae) to insecticides, *J. Stored Prod. Res.*, 21, 141–50.

Gong, Y. Y., Cardwell, K. F, Hounsa, A., Egal, S., Turner, P. C., Hall, A. J. and Wild, C. P. (2002), Dietary aflatoxin exposure and impaired growth in young children from Benin and Togo: Cross sectional study, *Br. Med. J.*, 325, 20–1.

Gong, Y. Y., Egal, S., Hounsa, A., Turner, P. C., Hall, A. J., Cardwell, K. F. and Wild, C. P. (2003), Determinants of aflatoxin exposure in young children from Benin and Togo, West Africa: The critical role of weaning, *Int. J. Epidemiol.*, 32, 556–62.

Gong, Y., Hounsa, A., Egal, S., Turner, P. C., Sutcliffe, A. E., Hall, A. J., Cardwell, K. F. and Wild, C. P. (2004), Postweaning exposure to aflatoxin results in impaired child growth: A longitudinal study in Benin, West Africa, *Environ. Health. Perspect.*, 112, 1334–8.

Gounou, S., Jiang, N. and Schulthess, F. (2009), Long-term seasonal fluctuations of lepidopteran cereal stem borers and their natural enemies on maize and wild host plants in southern Benin, *Insect Sci.*, 16, 329–41.

Grubisha, L. C. and Cotty, P. J. (2010), Genetic isolation among sympatric vegetative compatibility groups of the aflatoxin-producing fungus *Aspergillus flavus*, *Mol. Ecol.*, 19, 269–80.

Guntrip, J., Silby, R. M. and Smith, R. H. (1996), A phenotypic and genetic comparison of egg to adult life history traits between and within two strains of the Larger Grain Borer *Prostephanus truncatus* (Horn) (Coloeptera: Bostrichidae), *J. Stored Prod. Res.*, 32, 213–23.

Hailemichael, Y., Schulthess, F., Smith Jr., J. W. and Overholt, W. A. (2009), Physiological suitability of six West African gramineous borers (Lepidoptera: Noctuidae, Pyralidae) for development of *Cotesia* species complex (Hymenoptera: Braconidae), *Int. J. Trop. Insect Sci.*, 29, 130–40.

Hell, K., Cardwell, K. F., Sétamou M. and Poehling, H.-M. (2000), The influence of storage on aflatoxin contamination in stored grains in four agroecological zones in Benin, West Africa, *J. Stored Prod. Res.*, 36, 365–82.

Hill, M. G., Borgemeister C. and Nansen, C. (2002), Ecological studies on the larger grain borer, *Prostephanus truncatus* (Horn) (Coleoptera: Bostrichidae) and their implications for integrated pest management, *Integr. Pest Manag. Rev.*, 7, 201–21.

Hill, M. G., Nang'ayo, F. L. and Wright, D. J. (2003), Biological control of the larger grain borer *Prostephanus truncatus* (Horn) (Coleoptera: Bostrichidae) in Kenya using a predatory beetle *Teretrius nigrescens* (Coleoptera: Histeridae), *Bull. Ent. Res.*, 93, 299–306.

Hodges, R. and Maritime, C. (2012), Post-harvest weight losses of cereal grains in sub-Saharan Africa, http://www.erails.net/FARA/aphlis/aphlis/weightlosses-review/.

Hodges, R. J., Dunstan, W. R., Magazini, I. A. and Golob, P. (1983), An outbreak of *Prostephanus truncatus* (Horn) (Coleoptera: Bostrichidae) in East Africa, *Prot. Ecol.*, 5, 183–94.

Hodges, R. J., Buzby, J. C. and Bennett, B. (2011), Postharvest losses and waste in developed and less developed countries: Opportunities to improve resource use, *J. Agric. Sci.*, 149, 37–45.

Holst, N. and Meikle, W. G. (2003), *Teretrius nigrescens* against larger grain borer *Prostephanus truncatus* in African maize stores: Biological control at work?, *J. Appl. Ecol.*, 40(2), 307–19.

Holst, N., Meikle, W. G. and Markham, R. H. (2000), Grain injury models for *Prostephanus truncatus* (Coleoptera: Bostrichidae) and *Sitophilus zeamais* (Coleoptera: Curculionidae) in rural maize stores in West Africa, *J. Econ. Entomol.*, 93, 1338–46.

Jiang, N. and Schulthess, F. (2005), The effect of nitrogen fertilizer to maize on the bionomics of *Chilo partellus* (Swinhoe) (Lepidoptera: Crambidae) and performance of its larval parasitoid *Cotesia flavipes* Cameron (Hymenoptera: Braconidae), *Bull. Entomol. Res.*, 95, 495–504.

Kachapulula, P. W., Akello, J., Bandyopadhyay, R. and Cotty, P. J. (2017), Aflatoxin contamination of groundnut and maize in Zambia: Observed and potential concentrations, *J. Appl. Microbiol.*, 122, 1471–82.

Khan, Z. R., Pickett, J. A., Van den Berg, J., Wadhams, L. J. and Woodcock, C. M. (2000), Exploiting chemical ecology and species diversity: Stem borer and striga control for maize and sorghum in Africa, *Pest Manag. Sci.*, 56, 957–62.

Khlangwiset, P., Shephard, G. S. and Wu, F. (2011), Aflatoxins and growth impairment: A review, *Crit. Rev. Toxicol.*, 41, 740–55.

Kim, S. K., Kling, J. G., Iken, J. E., Cardwell, K. F., Adenola, A. O. and Adenle, V. (1995), Breeding maize for downy mildew resistance, *in* K. F. Cardwell (Ed.), *Proceedings of the Maize Downy Mildew Disease Eradication Campaign Conference and Workshop*, 22–25 February 1994, IITA/FAO/Nigerian Federal Department of Agriculture. Ibadan, Nigeria: International Institute of Tropical Agriculture.

Kipkoech, A. K., Schulthess, F., Yabann, W. K., Maritim, H. K. and Mithoefer, D. (2006), Biological control of cereal stemborers in Kenya: A cost benefit approach, *Ann. Soc. Entomol. Fr.*, 42, 519 528.

Kitinoja, L., Saran, S., Roy, S. K. and Kader, A. A. (2011), Postharvest technology for developing countries: Challenges and opportunities in research, outreach and advocacy, *J. Sci. Food Agric.*, 91, 597–603.

Kling, J. G., Cardwell, K. F. and Kim, S. K. (1994), Advances in screening methods and breeding for Downy Mildew (*Peronosclerospora sorghi*) resistance of maize, *4th Eastern and Southern Africa Regional Maize Conference: Maize Research for Stress Environments*, Harare, Zimbabwe, 28 March – 1 April 1994 (pp. 164–68).

Kossou, D. K., Bosque-Pérez, N. A. and Mareck, J. H. (1992), Effects of shelling maize cobs on the oviposition and development of *Sitophilus zeamais* Motschulsky, *J. Stored Prod. Res.*, 28, 187–92.

Le Rü, B. P., Ong'amo, G. O., Moyal, P., Ngala, L., Musyoka, B., Abdullah, Z., Cugala, D., Defabachew, B., Haile, T. A., Kauma Matama, T., Lada, V. Y., Negassi, B., Pallangyo, B., Ravololonandrianina, J., Sidumo, A., Omwega, C. O., Schulthess, F., Calatayud, P. A. and Silvain, J. F. (2006), Geographic distribution and host plant ranges of East African noctuid stem borers, *Ann. Soc. Entomol. Fr.,* 41, 353–61.

Leslie, J. F. (1993), Fungal vegetative compatibility, *Annu. Rev. Phytopathol.*, 31, 127–50.

Manyong, V. M., Kling, J. G., Makinde, K. O., Ajala, S. O. and Menkir, A. (2000), Impact of IITA-improved germplasm on maize production in West and Central Africa, IITA, Ibadan, Nigeria, p. 16.

Marasas, W. F., Gelderblom, W. C., Shephard, G. S. and Vismer, H. F. (2008), Mycotoxins: A global problem, *in* J. F. Leslie, R. Bandyopadhyay and A. Visconti (Eds), *MYCOTOXINS: Detection Methods, Management, Public Health and Agricultural Trade* (pp. 29–36), Wallingford, UK: CABI Publishing.

Matsuoka, Y., Vigouroux, Y., Goodman, M. M., Sanchez, G. J. and Doebley, J. (2004), A single domestication for maize shown by multilocus microsatellite genotyping, *Proc. Natl. Acad. Sci. USA*, 99, 6080–4

Matama-Kauma, T., Schulthess, F., Mueke, J. M., Omwega, C. O. and Ogwang, J. A. (2000), Impact of wild grasses planted as border rows on stemborer infestations in maize in Uganda, *Ann. Soc. Entomol. Fr.*, 42, 455–60.

Mehl, H. L., Jaime, R., Callicott, K. A., Probst, C., Garber, N. P., Ortega-Beltran, A., Grubisha, L. C. and Cotty, P. J. (2012), *Aspergillus flavus* diversity on crops and in the environment can be exploited to reduce aflatoxin exposure and improve health, *Ann. N. Y. Acad. Sci.*, 1273, 7–17.

Meikle, W. G., Adda C., Azoma, K., Borgemeister, C., Degbey, P., Djomamou, B. and Markham, R. H. (1998a), Varietal effects on the density of *Prostephanus truncatus* (Horn) (Col.: Bostrichidae) and *Sitophilus zeamais* Motsch. (Col.: Curculionidae) in grain stores in Benin Republic, *J. Stored Prod. Res.*, 34(1), 45–58.

Meikle, W. G., Holst, N., Scholz, D. and Markham, R. H. (1998b), A simulation model of *Prostephanus truncatus* (Horn) (Coleoptera: Bostrichidae) in rural grain stores in Benin, *Environ. Entomol.*, 27, 59–69.

Meikle, W. G., Holst, N. and Markham, R. H. (1999), A simulation model of *Sitophilus zeamais* Motschulsky (Coleoptera: Curculionidae) in rural grain stores in Benin, *Environ. Entomol.*, 28, 836–44.

Meikle, W. G., Holst, N., Degbey, P. and Oussou, R. (2000), An evaluation of sequential sampling plans for the larger grain borer (Coleoptera: Bostrichidae) and the maize weevil (Coleoptera: Curculionidae) and of visual grain assessment in West Africa, *J. Econ. Entomol.*, 93(6), 1822–31.

Meikle, W. G., Cherry, A., Holst, N., Hounna, B. and Markham, R. H. (2001), The effects of an entomopathogenic fungus, *Beauveria bassiana* (Balsamo) Vuillemin (Hyphomycetes), on *Prostephanus truncatus* (Horn) (Col.: Bostrichidae), *Sitophilus zeamais* (Col.: Curculionidae) and grain losses in stored maize in the Benin Republic, *J. Invert. Pathol.*, 77(3), 198–205.

Meikle, W. G., Markham, R. H., Nansen, C., Holst, N., Degbey, P., Azoma, K. and Korie, S. (2002a), Pest management in traditional maize stores in West Africa: A farmer's perspective, *J. Econ. Entomol.*, 95(5), 1079–88.

Meikle W. G., Rees D. and Markham, R. H. (2002b), Biological control of the larger grain borer, *Prostephanus truncatus* (Horn) (Coleoptera: Bostrichidae), *Integr. Pest Manag. Rev.*, 7, 123–38.

Midingoyi, S.-K. G., Affognon, H. D., Macharia, I., Ong'amo, G., Abonyo, E., Ogola, G., De Groote, H. and Le Ru, B. (2016), Assessing the long- term welfare effects of the biological control of cereal stemborer pests in East and Southern Africa: Evidence from Kenya, Mozambique and Zambia, *Agric. Ecosyst. Environ.*, 230, 10–23.

Miracle, M. P. (1965), The introduction and spread of maize in Africa, *J. Afr. Hist.*, 6, 39–55.

Mochiah, M. B., Ngi-Song, A. J., Overholt, W. A. and Stouthamer, R. (2002), Variation in encapsulation sensitivity of *Cotesia sesamiae* (Cameron) (Hymenoptera: Braconidae) biotypes to *Busseola fusca* Fuller (Lepidoptera: Noctuidae), *Entomol. Exp. Appl.*, 105, 111–8.

Muatinte, B. L. and Cugala, D. R. (2015), Monitoring the establishment and dispersal of *Teretrius nigrescens* Lewis (Coleoptera: Histeridae), a predator of *Prostephanus truncatus* Horn (Coleoptera: Bostrichidae) in Manica Province, Mozambique, *Afr. Entomol., 23*, 250–4.

Munkvold, G. P. (2003), Epidemiology of Fusarium diseases and their mycotoxins in maize ears, *Eur J. Plant Pathol.*, 109, 705–13.

Munkvold, G. P. and Desjardins, A. E. (1997), Influence of inoculation method on systemic *Fusarium moniliforme* infection of maize plants grown from infected seeds, *Plant Dis.*, 81, 211–16.

Nang'ayo, F. L. O., Hill, M. G., Chandi, E. A., Chiro, C. T., Nzeve, D. N. and Obiero, J. (1993), The natural environment as a reservoir for the larger grain borer *Prostephanus truncatus* (Horn) (Coleoptera: Bostrichidae) in Kenya, *Afr. Crop Sci. J.*, 1, 39–47.

Nansen C. and Meikle, W. G. (2002), The biology of the larger grain borer, *Prostephanus truncatus* (Horn) (Coleoptera: Bostrichidae), *Integr. Pest Manag. Rev.*, 7, 91–104.

Nansen, C., Korie, S. and Meikle, W. G. (1999), CANOCO and SADIE – software packages for spatial analysis of pheromone trap catches of maize storage pest and its biological control agent, *J. Tropic. For. Res.*, 15(2), 146–58.

Nansen, C., Meikle, W. G. and Korie, S. (2002), Spatial analysis of *Prostephanus truncatus* (Horn) (Coleoptera: Bostrichidae) flight activity near maize stores and in different forest types in southern Benin, West Africa, *Entomol. Soc. Am.*, 95(1), 66–74.

Ndemah, R., Schulthess, F., Korie, S., Borgemeister, C. and Cardwell, K. F. (2001a), Distribution, relative importance and effect of lepidopterous borers on maize yields in the forest zone and mid-altitude of Cameroon, *J. Econ. Entomol.*, 94, 1434–44.

Ndemah, R., Schulthess, F., Poehling, H.-M. and Borgemeister, C. (2001b), Natural enemies of lepidopterous borers on maize and elephant grass in the forest zone of Cameroon with special reference to *Busseola fusca* (Fuller) (Lepidoptera: Noctuidae), *Bull. Ent. Res.*, 91, 205–12.

Ndemah, R., Gounou, S. and Schulthess, F. (2002), The role of wild grasses in the management of lepidopterous cereal stemborers on maize in western Africa, *Bull. Ent. Res.*, 92, 507–19.

Ndemah, R., Schulthess, F., Korie, S., Borgemeister, C., Poehling, M. and Cardwell, K. F. (2003), Factors affecting infestations of the stalk borer *Busseola fusca* (Fuller) on maize in the forest zone of Cameroon with special reference to scelionid egg parasitoids, *Environ. Entomol.*, 32, 61–70.

Ndemah, R., Schulthess, F. and Nolte, C. (2006), The effect of grassy field margins and fertilizer on soil water, plant nutrient levels, stemborer attacks and yield of maize in the humid forest of Cameroon, *Ann. Soc. Entomol. Fr.*, 42, 461–70.

Ndemah, R., Schulthess, F., Le Rü, B. and Bame, I. (2007), Lepidopteran cereal stemborers and associated natural enemies on maize and wild grass hosts in Cameroon, *J. Appl. Entomol.*, 131, 658–68.

Ndemah, R., Le Ru, B., Capdevielle, C., Schulthess, F., Kaiser, L., Hanna, R., Depoilly, A., Obonyo, J., Calatayud, P., Abang, A. and Dupas, S., Environmentally safe agriculture: Managing genetic interactions between wild and cultivated compartment in introductions of *Cotesia sesamiae* for the biological control cereal stemborer in Cameroon, *Agric. Ecosyst. Environ*. (*submitted*).

Ngi-Song, A. J., Overholt, W. A. and Ayertey, J. N. (1995), Suitability of African gramineous stemborers for development of *Cotesia flavipes* and *C. sesamiae* (Hymenoptera: Braconidae), *Environ. Entomol*, 24, 978–84.

Ngoko, Z., Marasas, W. F. O., Rheeder, J. P., Shepherd, G. S., Wingfield, M. J. and Cardwell, K. F. (2001), Fungal infection and mycotoxin contamination of maize in the Humid Forest and the Western Highlands of Cameroon, *Phytoparasitica* 29, 352–60.

Ngoko, Z., Cardwell, K. F., Marasas, W. F. O., Wingfield, M., Ndemah, R. and Schulthess, F. (2002), Biological and physical constraints on maize production in the Humid Forest and Western Highlands of Cameroon, *Eur. J. Plant Pathol.*, 108, 893–902.

Ngoko, Z., Daoudou, I. H., Kamga, P. T., Mendi, S., Mwangi, M., Bandyopadhyay, R. and Marasas, W. F. O. (2008), Fungi and mycotoxins associated with food commodities in Cameroon, *J. Appl. Biosci.*, 6, 164–8.

Nouhoheflin, T., Coulibaly, J. Y., D'Alessandro, S., Aitchédji, C. C., Damisa, M., Baributsa, D. and Lowenberg-DeBoer, J. (2017), Management lessons learned in supply chain development: The experience of PICS bags in West and Central Africa, *Int. Food Agribusiness Manag. Rev.*, 20, 427–38.

Olanya, O. M. and Fajemisin, J. M. (1993), Host range and survival studies of *Peronosclerospora sorghi* at two locations in Nigeria, *Int. J. Pest Manag.*, 39, 13–18.

Omondi, B. A., van den Berg, J., Masiga, D. and Schulthess, F. (2011a), Phylogeographic structure of *Teretrius nigrescens (Coleoptera:Histeridae) predator of the invasive postharvest pest Prostephanus truncatus (Coleoptera: Bostrichidae), Bull. Entomol. Res.*, 101, 521–32.

Omondi, B. A., Jiang, N., van den Berg, J. and Schulthess, F. (2011b), The flight activity of *Prostephanus truncatus* (Horn) (Coleoptera: Bostrichidae) and *Teretrius nigrescens* Lewis (Coleoptera: Histeridae) in Kenya, *J. Stored Prod. Res.*, 47, 13–9.

Omondi, B. A., van den Berg, J., Masiga, D. and Schulthess, F. (2014), Molecular markers reveal narrow genetic base and culturing-associated genetic drift in *Teretrius nigrescens* Lewis populations released for the biological control of the larger grain borer in Africa, *Bull. Entomol. Res.*, 104(2), 143–54.

Oussou, R., Meikle, W. G. and Markham, R. H. (1999), Factors affecting the development and survivorship of larvae of *Teretriosoma nigrescens* Lewis, *Insect Sci. Appl.*, 18(1), 53–8.

Overholt, W. A., Ogedah, K. and Lammers, P. M. (1994), Distribution and sampling of Chilo partellus (Lepidoptera: Pyralidae) in maize and sorghum on the Kenyan coast, *Bull. Entomol. Res.*, 84, 367–78.

Pantenius, C. (1988), Storage losses in traditional maize granaries in Togo, *Int. J. Trop. Insect Sci.*, 9, 725–35.

Phiri, G. S. N. (1995), Interaction of the spotted stem borer *Chilo partellus* (Swinhoe) with some alternative hosts and its larval parasitoid *Cotesia sesamiae* (Cameron) in Malawi, PhD dissertation, Reading University, Reading, UK.

Prasanna, B. M., Huesing, J. E., Eddy, R. and Peschke, V. M. (Eds) (2018), *Fall Armyworm in Africa: A Guide for Integrated Pest Management*, 1st ed. Mexico, CDMX: CIMMYT.

Probst, C., Njapau, H. and Cotty, P. J. (2007), Outbreak of an acute aflatoxicosis in Kenya in 2004: Identification of the causal agent, *Appl. Environ. Microbiol.*, 73, 2762–4.

Probst, C., Bandyopadhyay, R., Price, L. E. and Cotty, P. J. (2011), Identification of atoxigenic *Aspergillus flavus* isolates to reduce aflatoxin contamination of maize in Kenya, *Plant Dis.*, 95, 212–8.

Probst, C., Callicott, K. A. and Cotty, P. J. (2012), Deadly strains of Kenyan *Aspergillus* are distinct from other aflatoxin producers, *Eur. J. Plant Pathol.*, 132, 419–29.

Probst, C., Bandyopadhyay, R. and Cotty, P. J. (2014), Diversity of aflatoxin-producing fungi and their impact on food safety in sub-Saharan Africa, *Int. J. Food Microbiol.*, 174, 113–22.

Published by Burleigh Dodds Science Publishing Limited, 2019.

Ramírez Genel, M. (1960), Protectores de grano. Toxicidad comparativa de cuatro materiales. [Grain Protectants. Comparative toxicity of four compounds], *Agricultura Técnica en México*, 10, 36–8.

Ramírez-Martínez, M., de Alba-Avila, A. and Ramírez-Zurbía, R. (1994), Discovery of the larger grain borer in a tropical deciduous forest in Mexico, *J. Appl. Entomol.*, 118, 354–60.

Rees, D. P. (1987), Laboratory studies on predation by *Teretriosoma nigrescens* Lewis (Col.: Histeridae) on *Prostephanus truncatus* (Horn) (Col.: Bostrichidae) infesting maize cobs in the presence of other maize pests, *J. Stored Prod. Res.*, 23(4), 191–5.

Rees, D. P., Rodríguez Rivera, R. and Herrera Rodríguez, F. J. (1990), Observations of the ecology of *Teretriosoma nigrescens* Lewis (Col.: Histeridae) and its prey *Prostephanus truncatus* (Horn) (Col.: Bostrichidae) in the Yucatan peninsula, Mexico, *Trop. Sci.*, 30(2), 153–65.

Richter, J., Biliwa, A., Helbig, J. and Henning-Helbig, S. (1997), Impact of *Teretriosoma nigrescens* Lewis (Coleoptera: Histeridae) on *Prostephanus truncatus* (Horn) (Coleoptera: Bostrichidae) and losses in traditional maize stores in southern Togo, *J. Stored Prod. Res.*, 33, 137–42.

Schneider, H., Borgemeister, C., Sétamou, M., Affognon, H., Bell, A., Zweigert, M., Poehling, H.-M. and Schulthess, F. (2004), Biological control of the larger grain borer *Prostephanus truncatus* (Horn) (Coleoptera: Bostrichidae) by its predator *Teretrius nigrescens* (Lewis) (Coleoptera: Histeridae) in Togo and Benin, *Biol. Control*, 30(2), 241–55.

Scholz, D., Borgemeister, C., Markham, R. H. and Poehling, H.-M. (1997a), Flight initiation in *Prostephanus truncatus*: Influence of population density and aggregation pheromone, *Entomol. Exp. Appl.*, 85, 237–45.

Scholz, D., Borgemeister, C., Meikle, W. G., Markham, R. H. and Poehling, H.-M. (1997b), Infestation of maize by *Prostephanus truncatus* initiated by male-produced pheromone, *Entomol. Exp. Appl.*, 83, 53–61.

Scholz, D., Tchabi, A., Borgemeister, C., Markham, R. H., Poehling, H.-M. and Lawson, A. (1997c), Host finding behaviour of *Prostephanus truncatus*: Primary attraction or random attack? *J. Appl. Entomol.*, 121, 261–9.

Scholz, D., Borgemeister, C. and Poehling, H.-M. (1998), EAG and behavioural responses of the larger grain borer, *Prostephanus truncatus* and its predator, *Teretriosoma nigrescens*, to the borer-produced aggregation pheromone, *Physiol. Entomol.*, 23, 265–73.

Schulthess, F., Bosque-Pérez, N. A., Chabi-Olaye, A., Gounou, S., Ndemah, R. and Goergen, G. (1997), Exchanging natural enemies species of lepidopterous cereal stemborers between African regions, *Insect Sci. Appl.* 17, 97–108.

Schulthess, F., Chabi-Olaye, A. and Goergen, G. (2001), Seasonal fluctuations of noctuid stemborer egg parasitism in southern Benin with special reference to *Sesamia calamistis* Hampson (Lepidoptera: Noctuidae) and *Telenomus* spp. (Hymenoptera: Scelionidae) on maize, *Biocontrol Sci. Technol.*, 11, 765–77.

Schulthess, F., Cardwell, K. F. and Gounou, S. (2002), The effect of endophytic *Fusarium verticillioides* on infestation of two maize varieties by lepidopterous stemborers and coleopteran grain feeders, *Phytopathology*, 92, 120–8.

Schulthess, F., Chabi-Olaye, A. and Gounou, S. (2004), Multi-trophic level interactions in a cassava-maize relay cropping system in the humid tropics of West Africa, *Bull. Entomol. Res.*, 94, 261–72.

Seetha, A., Munthali, W., Msere, H. W., Swai, E., Muzanila, Y., Sichone, E., Tsusaka, T. W., Rathore, A. and Okori, P. (2017), Occurrence of aflatoxins and its management in diverse cropping systems of central Tanzania, *Mycotoxin Res.*, 33, 323–31.

Sétamou, M., Schulthess, F., Bosque-Pérez, N. A. and Thomas-Odjo, A. (1995), The effect of stem and cob borers on maize subjected to different nitrogen treatments with special reference to *Sesamia calamistis* Hampson (Lepidoptera: Noctuidae), *Entomol. Exp. Appl.*, 77, 205–10.

Sétamou, M., Cardwell, K. F., Schulthess, F. and Hell, K. (1997), *Aspergillus flavus* infection and aflatoxin contamination of preharvest maize in Benin, *Plant Dis.*, 81, 1323–7.

Sétamou, M., Cardwell, K. F., Schulthess, F. and Hell, K. (1998), Effect of insect damage to maize ears, with special reference to *Mussidia nigrivenella* (Lepidoptera; Pyralidae), on *Aspergillus flavus* (Deuteromycetes; Moniliales) infection and aflatoxin production in maize before harvest in the Republic of Benin, *J. Econ. Entomol.*, 91, 433–8.

Sezonlin, M., Dupas, S., Le Ru, B., Le Gall, P., Moyal, P., Calatayud, P.-A., Faure, N. and Silvain, J.-F. (2006), Phylogeography and population genetics of the maize stalk borer *Busseola fusca* (Lepidoptera, Noctuidae) in sub-Saharan Africa, *Mol. Ecol.*, 15, 407–20.

Shiferaw, B., Prasanna, B. M., Hellin, J. and Bänziger, M. (2011), Crops that feed the world 6. Past successes and future challenges to the role played by maize in global food security, *Food Sec.*, 3, 307–27.

Shires, S. W. (1977), Ability of *Prostephanus truncatus* (Horn) (Coleoptera: Bostrichidae) to damage and breed on several stored food commodities, *J. Stored Prod. Res.*, 13, 205–8.

Smithers, C. N. (1960), Some recent observations on *Busseola fusca* (Fuller) (Lep.: Noctuidae) in southern Rhodesia, *Bull. Entomol. Res.*, 50, 809–19.

Suleiman, R., Rosentrater, K. and Bern, C. (2013), Effects of deterioration parameters on storage of maize: A Review, *J. Nat. Sci. Res.* 3, 147–65.

Tefera, T., Kanampiu, F., De Groote, H., Hellin, J., Mugo, S., Kimenju, S., Beyene, Y., Prasanna, M. B., Shiferaw, B. and Banziger, M. (2011), The metal silo: An effective grain storage technology for reducing post-harvest insect and pathogen losses in maize while improving smallholder farmers' food security in developing countries, *Crop Prot.*, 30, 240–5.

Tigar, B. J., Osborne, P. E., Key, G. E., Flores-S, S. M. E. and Vazquez-Arista, M. (1994), Distribution and abundance of *Prostephanus truncatus* (Coleoptera: Bostrichidae) and its predator *Teretriosoma nigrescens* (Coleoptera: Histeridae), *Bull. Ent. Res.*, 84, 555–65.

Turner, P. C., Collinson. A. C., Cheung, Y. B., Gong, Y., Hall, A. J., Prentice, A. M. and Wild, C. P. (2007), Aflatoxin exposure in utero causes growth faltering in Gambian infants, *Int. J. Epidemiol.*, 36, 1119–25.

Udoh, J. M., Ikotun, T. and Cardwell, K. F. (2000), Storage structures and aflatoxin content of maize in five agro-ecological zones of Nigeria, *J. Stored Prod. Res.*, 36, 187–201.

Udomkun, P., Wiredu, A. N., Nagle, M., Bandyopadhyay, R., Müller, J. and Vanlauwe, B. (2017), Mycotoxins in Sub-Saharan Africa: Present situation, socio-economic impact, awareness, and outlook, *Food Control*, 72, 110–22.

United Nations, Department of Economic and Social Affairs, Population Division (2017), World Population Prospects: The 2017 Revision, Key Findings and Advance Tables, ESA/P/WP/248.

Usman, I. (1995), The role of the National Seed Service in strategies of maize Downy Mildew control in Nigeria, *in* K. F. Cardwell (Ed.), *Proceedings of the Maize Downy Mildew Disease Eradication Campaign Conference and Workshop*, 22–25 February 1994, IITA/FAO/Nigerian Federal Department of Agriculture (pp. 34–40). Ibadan, Nigeria: International Institute of Tropical Agriculture.

Van Rensburg, J. B. J. (1988), Aspects of the injuriousness of the maize stalk borer, *Busseola fusca* (Fuller) (Lepidoptera: Noctuidae), *Bull. Entomol. Res.*, 78, 101–10.

Vismer, H. F., Shephard, G. S., Rheeder, J. P., Westhuizen, L., Bandyopadhyay, R. and Leslie, J. F. (2015), Relative severity of fumonisin contamination of cereal crops in West Africa, *Food Add. Contam. Part A*, 32, 1952–8.

Warth, B., Parich, A., Atehnkeng, J., Bandyopadhyay, R., Schuhmacher, R., Sulyok, M. and Krska, R. (2012), Quantitation of mycotoxins in food and feed from Burkina Faso and Mozambique using a modern LC-MS/MS multitoxin method, *J. Agric. Food Chem.*, 60, 9352–63.

Watson, S., Diedhiou, P. M., Atehnkeng, J., Bandyopadhyay, R., Srey, C., Routledge, M. N. and Gong, Y. Y. (2015), Seasonal and geographical differences in aflatoxin exposures in Senegal, *World Mycotoxin J.*, 8, 525–31.

Wild, C. P. and Gong, Y. Y. (2010), Mycotoxins and human disease: A largely ignored global health issue, *Carcinogenesis*, 31, 71–82.

Williams, R. J. (1984), Downy mildews of tropical cereals, *Adv. Plant Pathol.*, 2, 1–103.

Williams, S. B., Murdock, L. L. and Baributsa, D. (2017), Storage of maize in Purdue improved crop storage (PICS) bags, *PLoS One*, 12(1), p.e0168624.

Wu, F. and Khlangwiset, P. (2010), Health economic impacts and cost-effectiveness of aflatoxin-reduction strategies in Africa: Case studies in biocontrol and post-harvest interventions, *Food Add. Contam., Part A*, 27, 496–509.

Yu, J., Chang, P.-K., Ehrlich, K., Cary, J., Bhatnagar, D., Cleveland, T., Payne, G., Linz, J., Woloshuk, C. and Bennett, J. (2004), Clustered pathway genes in aflatoxin biosynthesis, *Appl. Environ. Microbiol.*, 70, 1253–62.

Published by Burleigh Dodds Science Publishing Limited, 2019.

Chapter 8

Identifying and managing plant health risks for key African crops: yam, taro and cocoyam

Babatima Djana Mignouna, *IITA, Benin;* **P. Lava Kumar**, *IITA, Nigeria;* **Danny Coyne**, *IITA, Kenya; and* **Ranajit Bandyopadhyay**, **Alejandro Ortega-Beltran**, **Ranjana Bhattacharjee** *and* **David De Koeyer**, *IITA, Nigeria (corresponding author: d.mignouna@cgiar.org)*

1 Introduction

The importance of crops like yam, taro and cocoyam in Africa, in terms of volume and value of production cannot be overemphasized (Ubalua, 2016). These crops have received very little attention from agricultural researchers and government policymakers. Moreover, various constraints, mainly disease and pests, affect their production in the region. The impacts of these challenges are exacerbated by their vegetative mode of propagation which supports transmission of diseases from one generation to the next. These crops are often recycled from farmer's fields, compounding the disease and pest build-up and leading to significant yield losses.

The production area for yam and taro (cocoyam) in Africa stagnated during the 1960s and 1970s, but increased substantially after the 1970s (Fig. 1). While the amount of land allocated to yam crops is growing, the area allocated to taro cultivation has remained virtually unchanged. Therefore, increases in yield are often difficult to achieve in the region due to a complex combination of factors. Specific limitations to yield increases include pest and disease build-up, lack of field-water management strategies especially in the drier areas, absence of irrigation infrastructure, low nutrient application and absence of fallows to restore soil fertility levels, lack of availability of suitable high-yielding varieties, and the lack of farmer expertise in appropriate agronomic practices.

http://dx.doi.org/10.19103/AS.2018.0043.09
Published by Burleigh Dodds Science Publishing Limited, 2019.

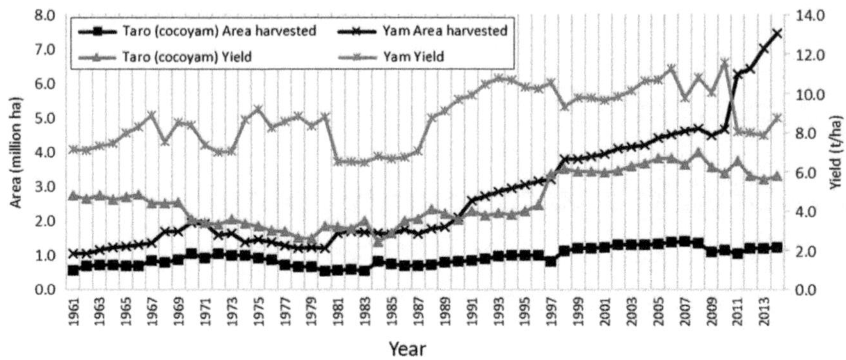

Figure 1 Trends in yam and taro (cocoyam) area and yields in Africa, 1961–2014. Source: FAO (2017).

Yam is a multi-species crop produced for its starchy edible tubers, and is from the genus *Dioscorea* L. (Dioscoreaceae) (APG II, 2003). There are up to 600 species and 95% of these grow in Africa. According to Coursey (1967), the word 'yam' could have been derived from African tribes such as Mande 'niam' to mean 'taste' adapted into Portuguese as 'ynhame', Spanish as 'name', French as 'igname' and English as 'yam' (Burkill, 1938). Among the cultivated yam species the white yam (*Dioscorea rotundata* Poir.) and the yellow yam (*D. cayenensis* Lam.) are native to West Africa (Bradshaw, 2010) and are the most important cultivated yams. Other major yam species cultivated are the water yam 'Khamba aalu' (*D. alata* L.), the bitter yam or trifoliate (*D. dumetorum* (Kunth) Pax) and the aerial yam 'Pita aalu' (*D. bulbifera* L.).

As stated by Yamaguchi (1983), artefacts dated about 50000 BC from West Africa indicate wild yams were used for food. The growing of yams appears to have developed in West Africa about 3000 BC. The crop plays a very important part in the food security and livelihoods of at least 100 million people in Africa. It is cultivated mostly in the Derived and Southern Guinea Savanna (Mignouna et al., 2015). Within the region, yam accounts for about 97% of global production (FAO, 2017) from about 7 million ha, mainly in five countries: Benin, Côte d'Ivoire, Ghana, Nigeria and Togo. Its prominence as an arable crop in the region is intricately linked to its substantial contribution to dietary protein, value as a ceremonial and traditional crop, and wide consumption as an important delicacy (Mignouna et al., 2015). The crop has the highest production value among all African staple crops (FAO, 2017). Despite the socioeconomic importance of yam, it faces much pressure from a range of pests, including fungal and viral diseases, as well as nematodes. Moreover, the crop has been neglected for a considerable length of time (Mignouna et al., 2015). Traditionally, seed yams or setts are used for propagation through double harvesting or 'milking', with cut setts or small whole tubers sorted from a ware crop (Aighewi et al., 2015). Nowadays, modern propagation techniques such as minisett technique, microsetts and microtubers, tissue and organ culture techniques, vine cuttings, aeroponics system, temporary immersion bioreactor system and botanic seeds, are being adopted to avoid transferring disease (Maroya et al., 2014a). Despite the high labour requirements and production costs, demand for yams is still high, making the yam business quite profitable (Mignouna et al., 2015). In Africa, the high value of yam gives its farmers and consumers a prestige and grandeur that are equalled only by taro.

The common names of taro and cocoyam add further to the confusion because the two crops look similar and produce similar tubers. Both crops are known as taro in some areas whilst both are called cocoyam in other areas (Onyilagha et al., 1987). We, therefore, treat them together as in the FAO database or separate them when applicable.

Taro [*Colocasia esculenta* (L.) Schott (Araceae)], which is used as a vegetable for its corms, leaves and petioles, seems to be one of the oldest crops on earth and is believed to have originated many centuries ago in southern or Southeast Asia. The crop was domesticated more than 10 000 years ago in Asia, Southeast Asia and Melanesia (Bradshaw, 2010). Its leaves and corms are a rich source of vitamins A and C, carbohydrate and protein. Other species of taro include giant taro, swamp taro and arrowleaf elephant's ear. It has been postulated that taro was well established in Senegambia by AD 1500, long before Portuguese navigators reached West Africa (Blench, 2009). Taro has gained a high importance mainly in countries like Nigeria, Cameroon, Ghana, Madagascar, Central African Republic, Egypt, Burundi, Guinea, Gabon and Côte d'Ivoire. Taro has acquired considerable importance in social rites of passage and traditional ceremonies. It is also grown and consumed on a subsistence basis, and as a cash crop, thereby playing a role in poverty alleviation. Taro has not been well studied in comparison with other crops (Perrier et al., 2011; Roullier et al., 2013; Sanou et al., 2015). It serves as an important food, especially in the dry season or before the yam and cassava harvest in West Africa. In addition to its food use, taro is used in treating constipation, tuberculosis and haemorrhoids. Its starch serves in cosmetic formulations like face powder, and in dusting preparations which use aerosol dispensing systems, as well as in the manufacture of biodegradable plastics. The crop's petioles are sometimes used to make dots or semicircle designs on cloth (Ahmed and Khan, 2013). Its industry provides meaningful employment to many people, mainly in rural areas of West Africa. Taro is exported mostly for Europe especially for African migrants and provides additional avenues for poverty alleviation. Among all root crops, taro and cocoyam produce the lowest average yields (FAO, 2017).

Cocoyam [*Xanthosoma sagittifolium* (L.) Schott (Araceae)] is an important food crop, being ranked third after cassava and yam among the root and tuber crops that are cultivated and consumed in many African countries. As discussed earlier, cocoyam is a common name for more than one tropical root and vegetable crop belonging to the Arum family (Aroids) (Bown, 2000) and many refer to Taro as old cocoyam and Malanga as new cocoyam. According to Bown (2000), cocoyam was introduced to Africa between the sixteenth and seventeenth centuries. Cocoyams that are cultivated as food crops originated from the region of Southeast Asia (Rao et al., 2010) or Central and South America (https://plantvillage.org/topics/cocoyam/infos). They are generally composed of a large spherical corms and are mainly grown in Nigeria, Ghana, Cameroon and Côte d'Ivoire. Cocoyam is vegetatively propagated from headsetts or suckers which establish quickly and give the highest rate of survival. Larger corms and bigger yields tend to be produced by larger headsetts and suckers, but these should only be taken from healthy plants in order to protect yields and prevent the spread of diseases (https://plantvillage.org/topics/cocoyam/infos). Cocoyams are superior to both cassava and yams in terms of protein, mineral and vitamin contents. In 2016, the continent accounted for about 73% of global cocoyam production (FAO, 2017), with the bulk of this production occurring in the Central and West African sub-region. One important aspect of the crop is that it is essentially being produced by resource-poor smallholder farmers, mostly female farmers with very limited agricultural input. In addition, diseases and pests are depleting the diversity in the already narrow genetic base of cocoyam because of the high susceptibility of most farmer's cultivars, which results in their inability to survive in the field. Also, cocoyam has a different ecological distribution than yams, being found in bas-fonds (inland valleys) and humid forests.

2 Major threats to yam, taro and cocoyam and management options

2.1 Nematodes

Regarding nematode pests of these crops, there is much more information available for yams, than there is for taro and cocoyam. There is very little information pertaining to these latter two crops, save for a few sporadic reports from field surveys and occasional efficacy studies. As for most crops, root-knot nematodes, *Meloidogyne* spp. (Hoplolaimidae), are reported in all crops but without significant damage to taro and cocoyam in general. Four species have been recorded affecting taro: *M. incognita* (Kofoid & White) Chitwood, *M. javanica* (Treub) Chitwood, *M. arenaria* (Neal) Chitwood and *M. hapla* Chitwood, although often the species have not been indicated (Coyne and Affokpon, 2018). The same four species of root-knot have been recovered from cocoyam, but in Uganda, *M. hapla* occurred in greater mean densities than other species (Coyne et al., 2003).

In general, cocoyam appears to be a poor host of *Meloidogyne,* although high initial soil densities can reduce corm weight (Caveness et al., 1981). The spiral nematode, *Helicotylenchus multicinctus* (Cobb) Golden (Hoplolaimidae), was regularly recovered from cocoyam in Uganda (Coyne et al., 2003) as well as in Ghana (Addoh, 1971), but without any damage associated with infection.

For yams, the situation is somewhat different and nematodes are a clearly defined problem for the crop. Two types of nematodes constitute a major constraint to yam production in Africa. The two types, root-knot nematodes and lesion nematodes, can occur in combination, but tend to be present as one or the other. Between them though, they result in considerable damage and losses to yam production. The yam nematode, *Scutellonema bradys* (Steiner & LeHew) Andrassy (Hoplolaimidae), is not a true lesion nematode but its sub-dermal feeding activity on tubers results in lesion damage. It is indigenous to Africa and is omnipresent with yam cultivation in West Africa (Kolombia et al., 2017a). Outside of West and Central Africa, it is less common, if not absent, and was detected for the first time in Tanzania and Kenya only recently (Coyne et al., 2016). Low densities may be undetectable, causing slight lesioning below the tuber cortex. As densities become higher and the feeding activity increases, the lesions expand in size, depth and colour. Brown lesions indicate the presence of nematodes when the yam skin is broken with a knife or thumbnail. The colour will darken as the nematodes proliferate, eventually turning black once the tissue has been killed. Over time, such as during lengthy storage periods, infected tubers can become completely destroyed. Other signs of infection include a flaky appearance of the skin and a cracked tuber surface. The yam nematode is mainly tuber borne, but can also be field borne, especially if yams are cropped on the same land following an infected crop. *Pratylenchus coffeae* (Zimmermann) Filipjev and Schuurmans Stekhoven (Pratylenchidae) and *P. sudanensis* Loof & Yassin (Mudiope et al., 2004), are true lesion nematodes, which cause very similar symptoms and damage to *S. bradys*. *Pratylenchus coffeae*, however, has yet to be found causing such damage to yams in Africa, although it is present. In Uganda, *P. sudanensis* has been recovered from yams with dry rot symptoms (Coyne et al., 2003; Mudiope et al., 2008). *Pratylenchus brachyurus* (Godfrey) Filipjev & Schuurmans Stekhoven (O'Bannon and Taylor, 1968), is also regularly found associated with yam rhizosphere, but has yet to be directly associated with any damage to yams (Kolombia et al., 2017b). Theoretically, the yam nematode should be relatively simple to manage, through the use of clean healthy planting material. A challenge however, is that low levels of tuber infection, which is difficult to detect, can lead to large losses during storage. Similarly, low levels of infection on planting material may go undetected, resulting in the perpetuation of the problem when such seed yams are planted without treatment.

Root-knot nematodes are the other prominent nematode pests. *Meloidogyne incognita* and *M. javanica* seem to be the key species, but *M. arenaria* and recently *M. enterolobii* Yang & Eisenback (=*M. mayaguensis*

Rammah and Hirschmann, 1988) have also been identified (Kolombia et al., 2017b). These nematodes are also tuber (seed) borne, however there is considerable infection from the field and infested soil. Affected tubers become unmarketable due to their physical appearance. However, the flesh of affected tubers also becomes corky and tough and so more of the tuber needs to be peeled when preparing for cooking. Infected tubers will also provide inoculum to infect the following crop if planted untreated. Although root-knot nematodes have always been associated with yams, over the last 10–20 years or so, the levels of tuber damage appear to have escalated in West Africa (Jatala and Bridge, 1990; Coyne and Affokpon, 2018).

2.2 Insects

A number of arthropod pests also attack these tuber crops causing unsightly physical damage, reducing the edible proportion and affecting marketability. Damage by various arthropod pests also impacts on the quality and viability of seed material, through reduced germination and vigour. The incidence and level of damage vary considerably, but can be locally severe at times. On yams, four species of mealybugs and scale insects colonize the tuber surface in the field but continue to cause damage during storage. Termites (Isoptera: Termitidae) especially *Ancistrotermes* spp. and *Microtermes* spp. tunnel within tubers and completely consume tubers under heavy infestation (Wood et al., 1980). Yam beetles [*Heteroligus meles* Bilb (Coleoptera: Dynastidae)] cause sizeable holes on the tuber surface while in the field (Knoth, 1993). Geographically sporadic in occurrence, where they occur they can cause havoc leading to the abandonment of yam crops. They appear to be emerging as a serious threat to yams in forest areas in Ghana and Nigeria. In West Africa, yam moths, *Euzopherodes vapidella vapidella* Mann (Lepidoptera: Pyralidae) and *Dasyses rugosella* Stainton (Lepidoptera: Tineidae) have also been observed causing damage to stored yams (Ashamo and Odeyemi, 2004).

2.3 Fungal diseases

Several fungal diseases reduce productivity and tuber quality of the different *Dioscorea* species. However, yam anthracnose, caused by *Colletotrichum gloeosporioides* (Penz.) Penz. & Sacc., is the most destructive fungal disease, and it can cause yield losses of up to 90% when conditions are favourable (Winch et al., 1984). Therefore, IITA has placed major emphasis on research to control this disease in comparison to other fungal diseases. Management strategies include deployment of resistant or moderately resistant cultivars as the central component (Hahn et al., 1987, 1989). IITA has identified both resistant germplasm to the disease, as well as molecular markers associated with the resistance that

Published by Burleigh Dodds Science Publishing Limited, 2019.

can be used in breeding programs (Mignouna et al., 2001, 2002; Narina et al., 2011). In addition, suitable sites were determined within Nigeria to evaluate yam susceptibilities to the disease (Egesi et al., 2009) and have developed rapid screening methods to identify resistant germplasm (Nwadili et al., 2017). Pathogenic and genetic diversity, epidemiology, and secondary metabolite profiles of *C. gloeosporioides* populations were assessed from different zones in Nigeria (Abang et al., 2002, 2003, 2006, 2009). This research was able to contribute both to a better understanding of the disease and the design of improved management strategies. Also, major tuber rot pathogens affecting yams were determined in three agroecological zones of Nigeria, and different techniques evaluated to control these pathogens. Techniques included the use of biocontrol agents (*Trichoderma* Pers., *Bacillus* Cohn., and *Pseudomonas* Migula. strains) and botanicals such as neem (azadirachtin) and pyrethrum (Dania et al., 2014, 2015, 2016). Since both taro (*Colocasia esculenta*) and cocoyam (*Xanthosoma* spp.) are not mandate crops of IITA, relatively little research efforts on identification and management of their fungal diseases have been conducted. Those efforts include describing the most common causal agents of disease in cocoyam, making the first reports of taro leaf blight (TLB), caused by the fungus-like organism *Phytophthora colocasiae* Racib., in Nigeria (Bandyopadhyay et al., 2011) and Ghana (Omane et al., 2012), and designing strategies to manage outbreaks of TLB in Nigeria along with collaborators. IITA and partners have conducted TLB research activities in Cameroon that include elucidating the best culture conditions to produce *P. colocasiae* inoculum for pathogenicity assays (Mbong et al., 2015) and determining suitable chemical regimes to manage TLB in three agroecological zones (Manju et al., 2017). However, there are still plenty of unexplored fungal pathosystems of yam, taro, and cocoyam.

2.4 Viruses

While there is very limited information pertaining to taro and cocoyam regarding viruses of these crops in Africa, much work has been done on *Dioscorea* yam, and the information provided in this part is mainly from the seed health point of view which is very relevant to yams. 'Mosaic disease' caused by viruses is most prevalent in the West African yam belt that stretches from the west of Cameroon to Côte d'Ivoire supported by the Bill & Melinda Gates Foundation. Yam mosaic virus (YMV) of the genus Potyvirus is the most prevalent, and is known to reduce yields by up to 60% in most of the yam varieties. Surveys conducted in the 2000s confirmed virus incidence and common exceeding 80% in white yam in the yam belt. Aphid-transmitted YMV is the most prevalent, detected in over 80% of the white yam (*D. rotundata*) grown in West Africa (Eni et al., 2008). Other viruses infecting yams in West Africa included Yam mild mosaic virus (YMMV) of the genus Potyvirus, and a complex of yam badnaviruses, but these were not

Published by Burleigh Dodds Science Publishing Limited, 2019.

known to be economically significant. Due to high prevalence and incidence, YMV is the most important virus threat for decreasing yields of yams, and also a major phytosanitary bottleneck for clean seed production and the international distribution of yams. Vegetative propagation, sequential reuse of farmer-saved seed yams, farmer-to-farmer seed exchange and informal seed trade are the major contributing factors for the wide dissemination of virus - infected seed yams, and consequent spread of yam viruses in West Africa. Virus perpetuation remains unabated due to a lack of virus-free yams. High resistance to YMV in *D. rotundata* is yet to be identified. Consequently, greater emphasis has been placed on production and dissemination of virus-free seed yams. There has been a concerted effort to develop procedures for the production of virus-free yam stocks (Maroya et al., 2014b; Balogun et al., 2017), improved diagnostic tools for seed health testing (Nkere et al., 2018; Silva et al., 2018), procedures for high-ratio multiplication of virus-free seed yams, and quality management protocols for sustaining the seed yam quality. Virus-free seed yams developed are being systematically propelled through public and private sector seed producers to replace infected stock, and are expected to reduce virus disease incidence and enhance crop yields. In 2015, an integrated approach to reduce virus infection in the farmer's fields through a 'positive selection' approach was introduced. This method, piloted in Nigeria and Ghana, was found to be effective in reducing virus disease incidence as well as severity. Sustainable control of YMV is best achieved through the use of resistant varieties. In 2015, IITA improved capacity for yam breeding in West Africa and simultaneously breeding for resistance to major biotic threats to yams, including yam breeding for YMV resistance. As part of this program, a wide range of genetic resources are being screened for YMV resistance to identify high levels of resistance to YMV, and develop genomic tools to fast track breeding for YMV resistance.

3 Status of pest and disease control measures

3.1 Clean seed systems

As with all vegetatively propagated crops, the provision of healthy planting material through sustainable healthy seed systems is essential for yam, cocoyam and taro production. A key restriction to improved productivity of these crops is the availability of and access to healthy planting material. Pests and diseases are a major source of loss and damage to these crops. The provision, therefore, of seed systems that can sustainably provide healthy material to farmers is critical. For yam in particular, which is heavily affected by seed-borne nematodes and viruses, the lack of availability of quality planting material is viewed as the greatest restriction to improved production (Aighewi et al., 2015). To overcome virus infections, it is necessary to generate planting material from virus-free stocks, such as through tissue culture, and then use this material to

Published by Burleigh Dodds Science Publishing Limited, 2019.

propagate virus-free material. This approach has been adopted at IITA, using novel techniques such as vine cuttings, aeroponics, temporary immersion bioreactor systems, and semi-autotrophic hydroponics (Maroya et al. 2014a; Aighewi et al., 2015). Seed material derived from tissue culture is also suitable for preventing transmission of nematode-infected material. However, healthy planting material can be generated by disinfecting seed tubers using hot water (Speijer, 1996; Coyne et al., 2010) and chemical treatment (Claudius-Cole et al., 2017). Hot water therapy, at 50–53°C for 20 min is more environmentally acceptable than using pesticides, but can result in initial losses and is more cumbersome and technically challenging to implement. Seed setts cut from treated tubers significantly increased the yields of seed tubers. Such healthy tubers also stored better, had higher germination, and provided higher yields of ware yams, when used as planting material compared to non-treated tubers (Claudius-Cole et al., 2017). Seed tubers produced in this manner can then be used to maintain production of healthy seed stocks, preferably with periodic retreatment to overcome any field infections.

3.2 Host resistance against biotic stresses

The mechanism of plant resistance/tolerance to various pests and pathogens involves an array of morphological, genetic, biochemical and molecular processes. In the classic gene-for-gene model (also known as receptor–ligand model) of host–pathogen interactions, R gene products recognize pathogen elicitors encoded by avirulence (Avr) genes (Fig. 2). This is a host-specific defence and can only be activated when both R gene and corresponding Avr gene are present (Staskawicz et al., 1995). The absence of either component results in disease, which is typically associated with damage and a reduction in yield of the host plant.

Similar to other crops, the major challenge in yam, taro and cocoyam improvement is to breed for resistance against several pests and diseases. Problems include anthracnose, viruses and nematodes in yams, TLB in taro, and cocoyam root rot disease (CRRD). A challenging aspect of resistance breeding is the introgression of resistance genes into locally preferred varieties. The effectiveness of harnessing host–pathogen resistance for disease control, needs a thorough understanding of the variations within the host plant populations and the pathogen, in addition to standard operating procedures for screening the host for disease or pest reactions (Parlevliet, 1993). The isolates of the pathogen causing yam anthracnose disease (YAD), *Colletotrichum gloeosporioides* Penz. (Abang et al., 2002; Aduramigba-Modupe et al., 2010) represent high genetic and pathogenic variations, suggesting strong evolutionary potential of the pathogen to develop new virulent strains that may break existing resistance (McDonald and Linde, 2002; Nwadili et al., 2017). Despite the variance and heterogeneity of *C. gloeosporioides*, breeders

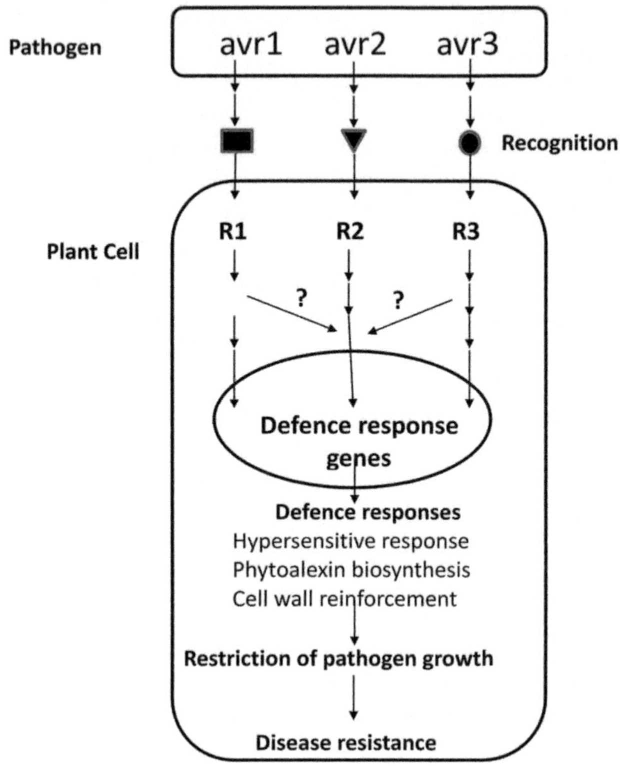

Figure 2 The R-Avr combinations leading to disease resistance.

at IITA, Centre de Coopération Internationale en Recherche Agronomique pour le Développement (French Agricultural Research Centre for International Development) (CIRAD) and national programmes, developed anthracnose-resistant varieties and identified QTL(s) (quantitative trait loci) using RAPD and SSR markers (Mignouna et al., 2002; Petro et al., 2011; Bhattacharjee et al., 2018). Several mapping populations have been developed at IITA and CIRAD to verify the segregations and validate the QTL(s) using next-generation sequencing techniques, for use in marker-assisted breeding for anthracnose resistance in yams. For nematodes, there is no confirmed evidence for complete resistance against *S. bradys*, however there are possibilities for developing resistance against *Meloidogyne* spp. (Bridge et al., 2005). The use of resistant varieties represents the most effective ways to control the disease or the vector carrying a virus. Several improved varieties were developed at IITA which were considered resistant, and a QTL was identified (Mignouna et al., 2001). However, during field screening, the presence of virus disease symptoms were recorded in these varieties. For taro, DNA markers were used to reveal a very high diversity in the *Phytophthora colocasiae* isolates which cause TLB, suggesting a rapid evolutionary capacity of the pathogen and that resistance was isolate-specific (Lebot et al., 2003). The viruses are not considered as detrimental as TLB, however studies have shown that there are as many as five viruses that infect taro. A few hybrids have been generated through breeding efforts in the Solomon Islands and Papua Guinea, these were heavily affected but did later show some recovery (Ivancic et al., 1993).

It is well known that pests and diseases have major effects on agricultural production. Good management and agronomic practices such as positive selection of planting materials, instant removal of on-farm diseased materials, crop rotation and intercropping, adequate management of plant population, and irrigation practices were encouraged to keep plants healthy and productive. The development of strategies based on a plant's own defence mechanisms for pest and disease control is also critical for sustaining agricultural production and improving our environment and health. There is a need for research efforts in these crops to dissect the genetic basis of pest and disease resistance, and of host-pathogen interactions. This could allow the manipulation of disease resistance genes in practical breeding programmes, and the deployment of high-yielding genetically resistant crop cultivars. There is a huge potential for of using genomics and transgenics approaches to accelerate the rate of improvement without affecting production and productivity. As signalling cascades and metabolic pathways are elucidated in model systems and crop plants, key regulatory genes can be targeted for silencing, overexpression or knock-out to provide broad-spectrum resistance. It will, therefore, be necessary to employ different approaches including transcriptomics, proteomics, metabolomics and mutational analysis, in combination with genomics-assisted breeding, to improve understanding of plant-pathogen or plant-pest interactions.

4 Conclusion and future trends

Despite the importance of yam, taro and cocoyam in Africa, these crops have received very little attention from researchers and various development agencies. Comparatively, more attention has been given to yams than to taro and cocoyam regarding plant health. The crops are highly susceptible to risk since their vegetative mode of propagation compounds transmission of diseases and pests, resulting in significant yield losses in the region.

Therefore, major research actions were undertaken by IITA and partners to better understand and address the several health risks to these crops. We compiled the outcomes of IITA's 50 year involvement in designing and applying disease and pest management strategies for yams, cocoyams and taro. These include but are not limited to the state of knowledge regarding pests and disease as major threats to the studied crops. From the initial efforts made via breeding, to deployment of tolerant/resistant cultivars as the central research components, they include good agronomic practices for soil quality improvement, water usage enhancement, crop' management and environmental improvement.

Research options have been developed at IITA to generate planting material from virus-free stocks via tissue culture, novel techniques like vine cuttings, aeroponics, temporary immersion bioreactors and semi-autotrophic hydroponics. However, there are plenty of unexplored fungal pathosystems of yams, taro, and cocoyams and emphasis is needed on conducting more disease surveys to clarify the current status of the most common diseases affecting yams, taro and cocoyams. Other aspects of research focus include understanding the spatiotemporal distribution and genetic diversity of the most common causal agents of diseases in the sub-region, as well as screening cultivars for resistance to protect the crops, and identifying markers linked to other fungal diseases in yams, taro and cocoyams.

In short, future research efforts will focus on (1) more integrated approaches to controlling the risk to crops; (2) surveys to clarify the current status of some common diseases and pests affecting those crops yet not well known; and (3) supporting breeding with genetic engineering and genome editing.

5 References

Abang, M. M., Winter, S., Green, K. R., Hoffmann, P., Mignouna, H. D. and Wolf, G. A. (2002). Molecular identification of *Colletotrichum gloeosporioides* causing yam anthracnose in Nigeria. *Plant Pathology*, 51, 63–71.

Abang, M. M., Winter, S., Mignouna, H. D., Green, K. R. and Asiedu, R. (2003). Molecular taxonomic, epidemiological and population genetic approaches to understanding yam anthracnose disease. *African Journal of Biotechnology*, 2, 486–96.

Abang, M. M., Asiedu, R., Hoffmann, P., Wolf, G. A., Mignouna, H. D. and Winter, S. (2006). Pathogenic and genetic variability among *Colletotrichum gloeosporioides* isolates from different yam hosts in the agroecological zones in Nigeria. *Journal of Phytopathology*, 154, 51-61.

Abang, M. M., Abraham, W.-R., Asiedu, R., Hoffmann, P., Wolf, G. and Winter, S. (2009). Secondary metabolite profile and phytotoxic activity of genetically distinct forms of *Colletotrichum gloeosporioides* from yam (*Dioscorea* spp.). *Mycological Research*, 113, 130–40.

Addoh, P. G. (1971). The distribution and economic importance of plant parasitic nematodes in Ghana. *Ghana Journal of Agricultural Science*, 4, 21–32.

Aduramigba-Modupe, A. O., Odebode, A. C. and Asiedu, R. (2010). Distribution of yam anthracnose disease in Nigeria. *African Journal of Root and Tuber Crops*, 8(1), 51-5.

Ahmed, A. and Khan. F. (2013). Extraction of starch from Taro (*Colocasia esculenta*) and evaluating it and further using Taro starch as disintegrating agent in tablet formulation with over all evaluation. *Inventi Rapid: Novel Excipients*, 2, 1-5.

Aighewi, B. A., Asiedu, R., Maroya, N. and Balogun, M. (2015). Improved propagation methods to raise the productivity of yam (*Dioscorea rotundata* Poir.). *Food Security*, 7, 823–34.

Akem, C. N. (1999). Yam die-back and its principal cause in the yam belt of Nigeria. *Pakistan Journal of Biological Sciences*, 2, 1106–09.

Angiosperm Phylogeny Group – APG II (2003). An update of the Angiosperm Phylogeny Group classification for orders and families of flowering plants: APG II. *Botanical Journal of the Linnean Society*, 161(2), 399–436.

Published by Burleigh Dodds Science Publishing Limited, 2019.

Ashamo, M. O. and Odeyemi, O. O. (2004). Effect of temperature on the development of the yam moth, *Dasyses rugosella* Stainton (Lepidoptera: Tineidae). *Journal of Stored Products Research*, 40, 95–102.

Balogun, M. O., Maroya, N., Augusto, J., Ajayi, A., Kumar, P. L., Aighewi, B. and Asiedu, R. (2017). Relative efficiency of positive selection and tissue culture for generating pathogen-free planting materials of yam (*Dioscorea* spp.). *Czech Journal of Genetics and Plant Breeding*, 53(1), 9–16. doi:10.17221/117/2016-CJGPB.

Bandyopadhyay, R., Sharma, K., Onyeka, T. J., Aregbesola, A. and Kumar, P. L. (2011). First report of Taro (*Colocasia esculenta*) leaf blight caused by *Phytophthora colocasiae* in Nigeria. *Plant Disease*, 95, 618.

Bhattacharjee, R., Nwadili, C., Saski, C., Paterne, A., Scheffler, B. E., Augusto, J., Lopez-Montes, A., Onyeka, J. T., Lava Kumar, P. and Bandyopadhyay, R. (2018). An EST-SSR based genetic linkage map and identification of QTLs for anthracnose disease resistance in water yam. *BioRxiv: The Preprint Server for Biology*, posted 9 May 2018. https://www.biorxiv.org/content/early/2018/05/09/318378

Blench, R. (2009). Bananas and plantains in Africa: Re-interpreting the linguistic evidence. *Ethnobotany Research & Applications*, 7, 363–80.

Bown, D. (2000). *Aroids: Plants of the Arum Family*. Timber Press, Portland, 392pp.

Bradshaw, J. E. (Ed.). (2010). *Root and Tuber Crops. Handbook of Plant Breeding 7*. Springer Science & Business Media, 298pp.

Bridge, J., Coyne, D. L. and Kwoseh, C. K. (2005). Nematode parasites of tropical root and tuber crops (excluding potatoes). In: Luc, M., Sikora, R. A. and Bridge, J. (Eds), *Plant Parasitic Nematodes in Subtropical and Tropical Agriculture*, 2nd ed. CAB International, Wallingford, UK, pp. 221–58.

Burkill, I. H. (1938). The contact of the Portuguese with African food-plants which gave words such as 'yam' to European languages. *Proceedings of the Linnean Society of London*, 150, 84–95.

Caveness, F. E., Wilson, J. E. and Terry, R. (1981). Root-knot nematodes on tannia (*Xanthosoma sagittifolium*). *Nematologia Mediterranea*, 9, 201–3.

Claudius-Cole, A. O., Kenyon, L. and Coyne, D. L. (2017). Effect of pre-plant treatments of yam (*Dioscorea rotundata*) setts on the production of healthy seed yam, seed yam storage and consecutive ware tuber production. *Journal of Agriculture and Rural Development in the Tropics and Subtropics*, 118, 297–306.

Coursey, D. G. (1967). *Yams*. Longmans, London, 230pp.

Coyne, D. L. and Affokpon, A. (2018). Nematode parasites of tropical root and tuber crops. In: Sikora, R. A., Coyne, D. L., Hallman, J. and Timper, P. (Eds), *Plant Parasitic Nematodes in Subtropical and Tropical Agriculture*, 3rd ed. CAB International, Wallingford, UK pp. 252–89.

Coyne, D. L., Talwana, H. A. L. and Maslen, N. R. (2003). Plant-parasitic nematodes associated with root and tuber crops in Uganda. *African Plant Protection*, 9, 87–98.

Coyne, D. L., Claudius-Cole, A. O., Kenyon, L. and Baimey, H. (2010). Differential effect of hot water treatment on whole tubers versus cut setts of yam (*Dioscorea* spp.). *Pest Management Science*, 66, 385–9.

Coyne, D. L., Kolombia, Y. A., Kariuki, G., Luambano, N. and Bert, W. (2016). First report of dry rot disease of yam caused by *Scutellonema bradys* in East Africa. *Plant Disease*. doi:10.1094/PDIS-12-15-1524-PDN.

Dania, V. O., Fadina, O. O., Ayodele, M. and Kumar, P. L. (2014). Efficacy of *Oryza sativa* husk and *Quercus phillyraeoides* extracts for the *in vitro* and *in vivo* control of fungal rot disease of white yam (*Dioscorea rotundata* Poir). *SpringerPlus*, 3, 711.

Dania, V. O., Fadina, O. O., Ayodele, M. and Kumar, P. L. (2015). Allelopathic potential of some biocontrol agents for the control of fungal rot of yellow yam (*Dioscorea cayenensis* Lam). *African Journal of Biotechnology*, 14, 474–81.

Dania, V. O., Fadina, O. O., Ayodele, M. and Kumar, P. L. (2016). Evaluation of isolates of Trichoderma, Pseudomonas and Bacillus species as treatment for the control of post-harvest fungal rot disease of yam (*Dioscorea* spp.). *Archives of Phytopathology and Plant Protection*, 49, 456–70.

Egesi, C. N., Onyeka, T. J. and Asiedu, R. (2009). Environmental stability of resistance to anthracnose and virus diseases of water yam (*Dioscorea alata*). *African Journal of Agricultural Research*, 4, 113–18.

Eni, A. O., Hughes, J. d'A. and Rey, M. E. C. (2008). Survey of the incidence and distribution of five viruses infecting yam in the major yam producing zones in Benin. *Annals of Applied Biology*, 153, 223–32.

FAO (2017). Food and Agriculture Organization of the United Nations. On-line and Multilingual Database. http://faostat.fao.org/ (accessed 12 February 2018).

Hahn, S. K., Osiru, D. S. O., Akoroda, M. O. and Otoo, J. A. (1987). Yam production and its future prospects. *Outlook on Agriculture*, 16, 105–10.

Hahn, S. K, Isoba, J. C. G. and Ikotun, T. (1989). Resistance breeding in root and tuber crops at the International Institute of Tropical Agriculture (IITA), Ibadan, Nigeria. *Crop Protection*, 8, 147–68.

Ivancic, A., Liloqula, R., Levela, H., Saelea, J. and Wagatora, D. (1993). Genetic resistance to alomae-bobone virus complex, the lethal disease of taro (*Colocasia esculenta* L. Schott). In: Ferentinos, L. (Ed.), *Proceedings of the Sustainable Taro Culture for the Pacific Conference*. College of Tropical Agriculture and Human Resources, University of Hawaii, pp. 48–54.

Jatala, P. and Bridge, J. (1990). Nematode parasites of root and tuber crops. In: Luc, M., Sikora, R. A. and Bridge, J. (Eds), *Plant-parasitic Nematodes in Subtropical and Tropical Agriculture*. CAB International, Wallingford, UK, pp. 137–80.

Knoth, J. (1993). *Traditional Storage of Yams and Cassava and Its Improvement*. GTZ, Hamburg, Germany.

Kolombia, Y. A., Karssen, G., Viaene, N., Kumar, P. L., Joos, L., Coyne, D. L. and Bert, W. (2017a). Morphological and molecular characterisation of *Scutellonema* species from yam (*Dioscorea* spp.) and a key to the species of the genus. *Nematology*, 19, 751–87.

Kolombia, A. Y., Karssen, G., Viaene, N., Kumar, P. L., De Sutter, N., Joos, L., Coyne, D. L. and Bert, W. (2017b). Diversity of root-knot nematodes associated with tubers of yam (*Dioscorea* spp.) established using isozyme analysis and mitochondrial DNA-based identification. *Journal of Nematology*, 49, 177–88.

Lebot, V., Herail, C., Pardales, J., Gunua, T., Prana, M., Thongjiem, M. and Viet, N. (2003). Isozyme and RAPD variation among *Phytophthora colocasiae* isolates from Southeast Asia and the Pacific. *Plant Pathology*, 52, 303–13.

Manju, E. B., Fokunang, C. N., Mbong, G. A., Tima, T. S., Suh, C., Tembe-Fokunang, E. A. and Hanna, R. (2017). Impact of fungicide application on taro leaf blight disease in three regions of Cameroon. *Journal of Experimental Agriculture International*, 17, 1–23.

Maroya, N. G., Balogun, M., Asiedu, R., Aighewi, B., Kumar, P. L. and Augusto, J. (2014a). Yam propagation using 'aeroponics' technology. *Annual Research & Review in Biology*, 4 (24), 3849–903.

Maroya, N. G., Asiedu, R., Kumar, P. L., Mignouna, D., Lopez-Montes, A., Kleih, U., Phillips, D., Ikeorgu, J. and Otoo, E. (2014b). Yam improvement for income and food security in West Africa: Effectiveness of a multi-disciplinary and multi-institutional team-work. *Journal of Root Crops*, 40(1), 85–92.

Mbong, G. A, Fokunang, C. N., Manju, E. B., Njukeng, A. P., Tembe-Fokunang, E. A. and Hanna, R. (2015). Mycelia growth and sporulation of *Phytophthora colocasiae* isolates under selected conditions. *American Journal of Experimental Agriculture*, 8, 193–201.

McDonald, B. A. and Linde, C. (2002). Pathogen population genetics, evolutionary potential and research advances in disease management. *Tropical Agriculture*, 53–7.

Mignouna, H. D., Abang, M. M., Green, K. R. and Asiedu, R. (2001). Inheritance of resistance in water yam (*Dioscorea alata*) to anthracnose (*Colletotrichum gloeosporioides*). *Theoretical and Applied Genetics*, 103, 52–5.

Mignouna, H. D., Abang, M. M., Onasanya, A. and Asiedu, R. (2002). Identification and application of RAPD markers for anthracnose resistance in water yam (*Dioscorea alata*). *Annals of Applied Biology*, 141, 61–6.

Mignouna, D. B., Abdoulaye, T., Akinola, A., Alene, A. and Nweke, F. (2015). Factors influencing the use of selected inputs in yam production in Nigeria and Ghana. *Journal of Agriculture and Rural Development in the Tropics and Subtropics*, 116(2), 131–42.

Mudiope, J., Coyne, D. L., Adipala, E. and Sikora, R. A. (2004). Monoxenic culture of *Pratylenchus sudanensis* on carrot discs, with evidence of differences in reproductive rates between geographical isolates. *Nematology*, 6, 617–19.

Mudiope, J., Speijer, P. R., Coyne, D. L., Maslen, R. N. and Adipala, E. (2008). Association of *Meloidogyne* spp. and *Pratylenchus sudanensis* with post-harvest damage on yam (*Dioscorea* spp.) in Central and Eastern Uganda. *African Journal of Crop Science*, 15, 93–9.

Mudiope, J., Coyne, D. L., Adipala, E. and Talwana, H. A. L. (2012). Damage to yam (*Dioscorea* spp.) by root-knot nematode (*Meloidogyne* spp.) under field and storage conditions in Uganda. *Nematropica*, 42, 137–45.

Narina, S. S., Buyyarapu, R., Kottapalli, K. R., Sartie, A. M., Ali, M. I., Asiedu, R., Mignouna, H. D., Sayre, B. L. and Scheffler, B. E. (2011). Generation and analysis of expressed sequence tags (ESTs) for marker development in yam (*Dioscorea alata* L.). *BMC Genomics*, 12, 100.

Nkere, C. K., Oyekanmi, J., Silva, G., Bömer, M., Atiri, G. I., Onyeka, J., Maroya, N., Seal, S. E. and Kumar, P. L. (2018). Chromogenic detection of Yam mosaic virus by closed tube reverse transcription loop-mediated isothermal amplification (CT-RT-LAMP). *Archives of Virology*, 163, 1057–61.

Nwadili, C. O., Augusto, J., Bhattacharjee, R., Atehnkeng, J., Lopez-Montes, A., Onyeka, T. J., Kumar, P. L., Asiedu, R. and Bandyopadhyay, R. (2017). Comparative reliability of screening parameters for anthracnose resistance in water yam (*Dioscorea alata*). *Plant Disease*, 101, 209–16.

O'Bannon, J. H. and Taylor, A. L. (1968). Migratory endoparasitic nematodes reared on carrot discs. *Phytopathology*, 58, 385.

Omane, E, Oduro, K. A., Cornelius, E. W., Opoku, I. Y., Akrofi, A. Y., Sharma, K., Kumar, P. L. and Bandyopadhyay, R. (2012). First report of leaf blight of taro (*Colocasia esculenta*) caused by *Phytophthora colocasiae* in Ghana. *Plant Disease*, 96, 292–92.

Onyilagha, J. C., Omenyi, A. S., Illoh, H. C. and Lowe, J. (1987). *Colocasia esculenta* (L.) Schott, *Colocasia antiquorum* Schott, how many species? A preliminary investigation. *Euphytica*, 36, 687–92.

Parlevliet, J. E. (1993). What is durable resistance? A general outline. In: Jacobs, T. and Parlevliet, J. E. (Eds), *Durability for Disease Resistance*. Kluwer Academic Publishers, Dordrecht, the Netherlands, pp. 23–39.

Perrier, X., De Langhe, E., Donohue, M., Lentfer, C., Vrydaghs, L. and Bakry, F. (2011). Multidisciplinary perspectives on banana (*Musa* spp.) domestication. *Proceedings of the National Academy of Sciences of the United States of America*, 108(28), 11311–8.

Petro, D., Onyeka, T. J., Etienne, S. and Rubens, S. (2011). An intraspecific genetic map of water yam (*Dioscorea alata* L.) based on AFLP markers and QTL analysis for anthracnose resistance. *Euphytica*, 179, 405–16.

Rammah, A. and Hirschmann, H. (1988). *Meloidogyne mayaguensis* n. sp. (Meloidogynidae), a root-knot nematode from Puerto Rico. *Journal of Nematology*, 20, 58 69.

Rao, V. R., Matthews, P. J., Eyzaguirre, P. B. and Hunter, D. (Eds) (2010). *The Global Diversity of Taro: Ethnobotany and Conservation*. Bioversity International, Rome.

Roullier, C., Benoit, L., McKey, D. B. and Lebot, V. (2013). Historical collections reveal patterns of diffusion of sweet potato in Oceania obscured by modern plant movements and recombination. *Proceedings of the National Academy of Sciences of the United States of America*, 110(6), 2205–10.

Sanou, J., Bayala, J., Teklehaimanot, Z. and Bazie, P. (2012). Effect of shading by baobab (*Adansonia digitata*) and néré (*Parkia biglobosa*) on yields of millet (*Pennisetum glaucum*) and taro (*Colocasia esculenta*) in parkland systems in Burkina Faso, West Africa. *Agroforestry Systems*, 85(3), 431–41.

Sanou, H., Angulo-Escalante, M. A., Martinez-Herrera, J., Kone, S., Nikiema, A. and Kalinganire, A. (2015). Loss of genetic diversity of *Jatropha curcas* L. through domestication: Implications for its genetic improvement. *Crop Science*, 55(2), 749–59.

Silva, M., Oyekanmi, J., Nkere, C. K., Bömer, M., Kumar, P. L. and Seal, S. E. (2018). Rapid detection of potyviruses from crude plant extracts. *Analytical Biochemistry*, 546, 17–22.

Speijer, P. R. (1996). Hot water treatment: A possible strategy for farmer groups and NGOs. *African Journal of Root and Tuber Crops*, 1, 38–42.

Staskawicz, B. J., Ausubel, F. M., Baker, B. J., Ellis, J. B. and Jones, J. D. (1995). Molecular genetics of plant disease resistance. *Science*, 268, 661–7.

Ubalua, O. A. (2016). Cocoyam (taro and tannia): Staples with untapped enormous potential – A review. *Plant Knowledge Journal*, 5(1), 27–35.

Winch, J. E., F. J. Newhook, G. V. H. Jackson and J. S. Cole. (1984). Studies of *Colletotrichum gloeosporioides* disease on yam, *Dioscorea alata*, in Solomon Islands. *Plant Pathology*, 33, 467–77.

Wood, T. G., Smith, R. W., Johnson, R. A. and Komolafe, P. O. (1980). Termite damage and crop loss studies in Nigeria – pre-harvest losses to yams due to termites and other soil pests. *Tropical Pest Management*, 26, 355–70.

Yamaguchi, M. (1983). Yam. In: *World Vegetables*. Springer, Dordrecht, the Netherlands, pp. 139–47.

Chapter 9

Identifying and managing plant health risks for key African crops: banana and plantain

Stefan Hauser, *IITA, Nigeria;* **Clifford Gold**, *Nevada City, USA;* **Cornelia Pasberg-Gauhl** *and* **Friedhelm Gauhl**, *FG-Inter-Agro-Consult KG, Austria;* **Juliet Akello**, *IITA, Zambia;* **Kim Jacobsen**, *Royal Museum for Central Africa, Belgium;* **Lindsey Norgrove**, *Bern University of Applied Sciences, Switzerland;* **Danny Coyne**, *IITA, Kenya;* **P. Lava Kumar**, *IITA, Nigeria;* **George Mahuku**, **Manoj Kaushal**, *IITA, Tanzania;* **Valentine Nakato**, *IITA, Uganda; and* **Leena Tripathi** *and* **Jaindra Tripathi**, *IITA, Kenya (corresponding author: s.hauser@cgiar.org)*

1 Introduction

Bananas and plantains (*Musa* spp.) are major food crops throughout the humid tropical and subtropical regions of the world. Dessert bananas are, globally, the fourth most important fruit crop. In 2014, sub-Saharan Africa (SSA) produced 38 million tons of banana and plantain (FAO, 2017), being about one-third of the global production. Only 4% of the West African and 1% of the East and Southern African production were exported (Stellenbosch University, 2017), highlighting their importance as staple food crops locally. Banana and plantain provide 25% of the food energy requirements to >100 million Africans.

Plantains and bananas have been grown as crops in SSA for millennia. Phytolith evidence from Cameroon shows that *Musa* were already cultivated

http://dx.doi.org/10.19103/AS.2018.0043.10
Published by Burleigh Dodds Science Publishing Limited, 2019.

approximately 2500 years ago (Mbida et al., 2000). In East Africa, introductions were as early as 2000 years ago (Kitavi et al., 2016). Economically important bananas and plantains are predominantly triploid genotypes resulting from evolutionary crosses within and between *Musa acuminata* Colla (A genome) and *Musa balbisiana* Colla (B genome). Cultivars grown today are classified by their genomic composition into three major groups: AAA – predominantly dessert and East African Highland cooking and beer bananas; AAB – plantains; and ABB – cooking bananas. Plantains are more common in West and Central Africa; cooking or highland bananas are more common in East Africa. This chapter will not discuss export dessert bananas as they are not vital to food security and income generation in SSA smallholder systems.

Growth habit, fruit properties and genetics of cooking banana and plantain are different. Although different *Musa* species suffer from a wide range of common pests and diseases, their geographical separation leads to some pests and diseases being specific to the *Musa* type or the environment. East African Highland bananas are distinguished by being grown above 1000 metres above seal level and thus have a pest and disease complex specific to that elevation range.

In this chapter we are presenting work carried out at the International Institute of Tropical Agriculture (IITA) to deal with banana bunchy top disease (BBTD) and banana streak virus (BSV), which were no threat to banana and plantain production prior to the 1990s; the banana *Xanthomonas* wilt (BXW), a disease originating from *Ensete* in Ethiopia and known for a long time; the black leaf streak disease (BLSD), which appeared in Zambia in 1973 (Ploetz, 2000) and has since spread across all of Africa; *Fusarium* wilt (FOC), specifically Tropical Race 4, which was detected in Africa more recently; the nematode *Radopholus similis* Cobb and some species currently gaining importance; and the banana weevil *Cosmopolites sordidus* (Germar) (Coleoptera: Curculionidae) the most important insect pest on *Musa* in Africa.

At IITA, plant health research on *Musa* started in 1991 in Nigeria and Uganda. Parallel to directly developing pest and disease management options, known control methods were integrated with agronomy research, to disseminate farmer-friendly options to treat planting material (Hauser et al., 2010) and systems research on limiting damage from BLSD (Norgrove and Hauser, 2013).

This chapter will not treat all pests and diseases of *Musa* but focus on the most important ones, reporting on the research process leading to decision-making on most appropriate methods to prevent and combat pests and diseases.

2 Viruses

Of about 20 viruses infecting *Musa* worldwide, BBTV (genus *Babuvirus*) responsible for bunchy top and banana streak viruses (BSVs, genus

Published by Burleigh Dodds Science Publishing Limited, 2019.

Badnavirus) responsible for 'leaf streak', banana bract mosaic virus (BBrMV, genus *Potyvirus*) responsible for 'bract mosaic' and cucumber mosaic virus (CMV, genus *Cucumovirus*) responsible for 'leaf chlorosis' result in economically significant damage to banana and plantain (Kumar et al., 2015). BBTV, BSVs and CMV are known to occur in all or several *Musa* producing countries in Africa; BBrMV is confined to a few countries in Asia. IITA research focused on BSV and BBTV. In this chapter we are only presenting *Musa*-specific aspects of these viruses, whereas their biology and epidemiology are described in detail in Chapter 5.

2.1 Banana streak virus

In May 1991, plants expressing severe chlorotic virus streak symptoms on the leaves were observed for the first time at Onne, SE Nigeria. A wider range of severe virus symptoms was identified on leaves, petioles, pseudostems and bunches, some plants died due to virus attack. Banana streak badnavirus (BSV) was identified in symptomatic samples. BSV was threatening IITA's *Musa* breeding programme and little was known about the nature of the disease. In October 1994, the recommended stop of all movement of *Musa* hybrid germplasm and landraces out of Onne Station was implemented and it was advised to destroy all off-station hybrid germplasm. At Onne Station all symptomatic plants had been destroyed by the end of 1994. As later demonstrated, BSV cannot be treated, because BSV genomic sequences are integrated into genomic DNA of *Musa* and *Ensete* (LaFleur et al., 1996).

BSV disease incidence in landraces on farmers' fields varies strongly. In some sites the disease appears to be serious, while elsewhere it is confined to a few plants of certain cultivars. BSV incidence was highest when infected source plants had been vegetatively propagated. High virus incidence seemed to be particularly related to newly developed hybrid germplasm, multiplied through in vitro propagation. Hybrids were apparently more susceptible to virus infection than landraces. However, different hybrids seemed to exhibit different levels of virus susceptibility. Unfortunately, visual assessment of symptom expression did not necessarily identify all plants and genotypes susceptible to virus infection and multiplication.

Virus incidence was not directly significantly correlated with rainfall, solar radiation, minimum or maximum temperatures. During the warmer dry season, virus symptoms were rare or absent. Symptom incidence built up during the cooler, rainy season when symptoms became easily detectable. Thus, apparent incidence was generally much higher in the rainy and cooler season (Fig. 1). However, it may be that maximum temperatures below a certain threshold or the changes in temperature triggered/facilitated BSV symptom development.

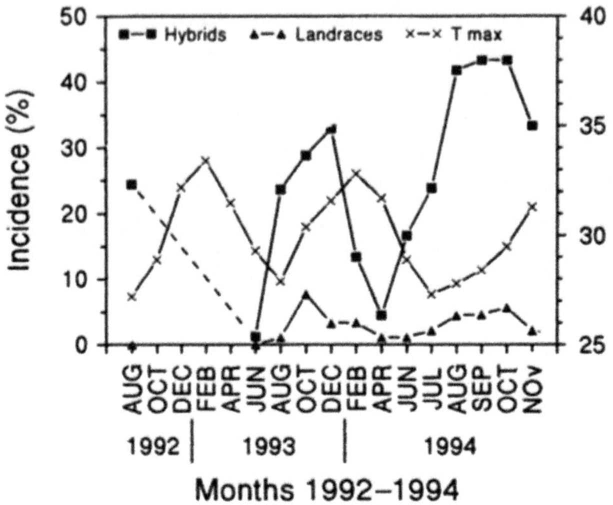

Figure 1 Average virus symptom incidence of eight hybrids and ten landraces and monthly mean maximum temperature at Onne Station, southeast Nigeria, 1992–4. Source: Pasberg-Gauhl and Gauhl (1996).

When bunch initiation coincided with this temperature-dependent onset of increased virus replication and symptom expression, the resulting bunches were small, did not emerge normally and might have had distorted fingers. By contrast, plants that initiated flowering when temperature-dependent symptom expression was repressed, produced bunches of normal size and appearance, even though some leaves might have shown pronounced disease symptoms. Bunch size and fruit quality tended to be significantly less affected by BSV in intensively managed plantings than under low-input management regimes, where plants also suffered from nutrient or moisture stress.

2.2 Banana bunchy top virus

The BBTV is responsible for the most devastating disease, first recorded during an epidemic on Cavendish banana (AAA) in 1889 in Fiji (Magee, 1927). The virus is transmitted by the aphid *Pentalonia nigronervosa* (Coquerel) (Hemiptera: Aphididae), the common banana aphid, which is the only known vector of the virus (Fig. 2). Because banana cultivars fully resistant to BBTV are not available, BBTV management is based on integration of several approaches such as eradication of infected plants by regular rogueing, planting virus-free plants and exclusion (quarantine) measures to prevent disease spread (Kumar et al., 2015). However, the implementation of measures to control a viral disease in a perennial crop can be marred by financial and operational bottlenecks, including farmers' resistance to eradicate infected mats. Meanwhile, extensive efforts

Published by Burleigh Dodds Science Publishing Limited, 2019.

Figure 2 Left to right: Bunchy top disease affected *Musa* mat and banana aphids on the pseudostem of *Musa*. Photos: L. Kumar.

are ongoing to implement an integrated pest management (IPM) approach through an 'ALLIANCE' approach (see Chapter 5), including advocacy and awareness campaigns such as 'Stop Bunchy Top' in the threatened countries. This approach is being tested in nine BBTD-affected pilot sites in Benin, Burundi, Cameroon, Congo Brazzaville, Democratic Republic of Congo, Gabon, Malawi and Nigeria (Kumar et al., 2016, unpublished results). Implementation steps are: (1) baseline studies mapping the disease spread, (2) gender and social analysis of households and communities with banana cropping systems for effective mobilization of communities in BBTD management, (3) eradication of infected sources, manually or by herbicides, vector control with insecticides, (4) a banana-free 2–3 months period prior to new planting and (5) replanting with virus-free farmer-preferred cultivars. Protocols have been developed to monitor presence of banana and aphids during the banana-free period and colonization by aphids and re-infection by BBTV of new fields (Kumar et al., 2011). The effectiveness of this approach, its profitability and sustainability are under evaluation. In at least three sites, replanted fields have grown for up to 3 years without re-infection, while in other sites re-infection occurred within 6 months. Preliminary observations suggest the strong influence of sociocultural aspects, incentives (free planting material) and sustained commitment of researchers and farmer groups as critical to the success of this programme.

3 Banana *Xanthomonas* wilt

The most important bacterial disease in East and Central Africa (ECA) is banana *Xanthomonas* wilt (BXW) caused by the bacterium *Xanthomonas campestris* pv. *musacearum* (Xcm) (Nakato et al., 2018; Tripathi et al., 2009), which can cause up to 100% yield loss (Kagezi et al., 2006; Ssekiwoko et al., 2006; Tripathi et al., 2009). BXW was reported in Ethiopia in 1968 on *Ensete* (Yirgou and Bradbury,

1968) and later on banana in 1974 (Yirgou and Bradbury, 1974). The disease has reached Uganda (Tushemereirwe et al., 2003), Democratic Republic of Congo (Ndungo et al., 2004), Rwanda, Tanzania (S.R.B. Mgenzi et al., 2006, unpublished results), Kenya (Mbaka et al., 2009) and Burundi (Carter et al., 2010). The earliest symptoms of BXW appear within 14–30 days post infection, depending on the entry point of the pathogen and age of the plants (Ssekiwoko et al., 2006; Welde-Michael et al., 2008; Addis et al., 2010; Ocimati et al., 2013a).

BXW control starts with destroying infected plants to avoid spreading within and to neighbouring farms. Additional measures are to limit access of workers and animals, remove diseased plants and regularly disinfect farm tools. De-budding bunches to avoid insects contaminating plants when visiting flowers is a standard practice in ECA (Mwangi and Nakato, 2007). Weed control, through changes of the micro-climate and organic matter amendments, through higher biological activity, can suppress the pathogen. Only clean planting material should be used to re-establish banana.

In Ethiopia, four bacterial antagonists were found that reduced disease incidence by 56–75%. *Herbaspirillum* spp., *Enterobacter* spp. and *Burkholderia* spp. isolates from Uganda attained similar levels of BXW suppression (E. Were, 2016, unpublished results). Mburu et al. (2016) applied 200 mg Si per plant per week and enhanced resistance to BXW, which may augment existing control measures.

Due to the absence of known resistant germplasm, conventional breeding has not been successful. So far only genetic engineering has produced germplasm with resistance to Xcm by transferring the plant ferrodoxin-like protein (*Pflp*) and hypersensitive resistance-assisting proteins (*Hrap*) from *Capsicum annuum* (Tripathi et al., 2010; Namukwaya et al., 2012). Transgenic banana lines carrying the *Pflp* or *Hrap* genes showed enhanced resistance against Xcm in glasshouse and confined field trials (Namukwaya et al., 2012; Tripathi et al., 2010, 2014), without altering the plant physiology and agronomic traits such as yields.

4 Fungal diseases

The greatest threat particularly to lowland *Musa* production in West and Central Africa has probably been black leaf streak disease (BLSD) also known as black sigatoka. *Fusarium* wilt (FOC), Tropical Race 4, recently arrived in Africa, could become the greatest future threat to banana if left unchecked.

4.1 *Leaf diseases*

Leaf diseases are spread by wind-dispersed spores, making vain any efforts to contain them through plant quarantine measures.

Published by Burleigh Dodds Science Publishing Limited, 2019.

Despite its economic importance, little was known about the epidemiology of the causal agent of BLSD, *Pseudocercospora fijiensis* (Morelet) previously called *Mycosphaerella fijiensis*, nor on host plant/pathogen interactions. BLSD causes symptoms similar to those of yellow sigatoka (YS) caused by *M. musicola* (Leach) (Fig. 3). Worldwide, both diseases are the economically most important leaf diseases on banana and plantain. In large banana plantations, control is achieved by up to 60 aerial fungicide applications per year (Abadie et al., 2009) and by leaf pruning.

BLSD is more aggressive and virulent and affects a wider range of *Musa* genotypes than YS. It develops more rapidly and causes more severe defoliation on bananas and plantain. BLSD destroys the leaves, leading to yield reduction and premature fruit ripening.

Both diseases originated between Papua New Guinea and the Solomon Islands in the South Pacific, which is also considered to be the area of origin of the genus *Musa*. BLSD depends upon the aerial dispersal of conidia and ascospores for local and regional spread and on humans for intercontinental spread (Drenth and Guest, 2016). BLSD has spread rapidly to all major banana and plantain growing areas and is still spreading. In most lowland ecologies, BLSD has quickly replaced YS, but YS is still important in highland areas such as in East Africa. BLSD is a more pronounced problem in the humid forest zone than in the lower rainfall derived savannah areas (Zandjanakou-Tachin et al., 2013).

The first authenticated report of BLSD in Africa is from Gabon in 1978 (Frossard, 1980), most likely introduced with planting material from Asia. Within 10 years, BLSD had spread to Cameroon in 1980, Congo, Côte d'Ivoire, Democratic Republic of Congo, Nigeria, Ghana and Togo. In East Africa the earliest report of BLSD is from the island of Pemba (Tanzania) in 1987, being most probably another independent introduction to Africa. From Pemba, BLSD spread to all of Eastern Africa within the following 15 years.

Figure 3 Typical symptoms of BLSD on a plantain leaf of the False Horn cv. Agbagba (left) and close up of necrotic tissue on the False Horn cv. Ebang (right). Photos: S. Hauser.

In 1991, the presence of *P. fijiensis* was confirmed at Onne Station, SE Nigeria, while YS was not detected. Apparently BLSD had replaced YS in almost all lowland areas. However, the continued presence of *M. musicola* in lowland *Musa* production systems was later proved (Pasberg-Gauhl and Gauhl, 1996). The population dynamics, population drift(s), genetic make-up, genetic variability and mutation frequency of *P. fijiensis* were virtually unknown. Therefore BLSD single spore isolates from different sites in southern Nigeria were obtained. It was not possible to distinguish strains or races of *P. fijiensis* by phenotype or morphology.

Disease development is closely correlated with rainfall. During the rainy season, symptom development is faster than during the dry season (Fig. 4). Seasonality of disease development and fungal reproduction were reflected in ascospore counts, peaking during the rainy season (Fig. 5). Conidiospore counts were low all year round, indicating that conidiospores play a secondary role in BLSD spread (Gauhl and Pasberg-Gauhl, 1994; Jones, 2018).

Disease management, through fungicides, reduces yield losses considerably. However, fungicides are expensive and require equipment and proper field management. *Musa* in Africa are grown mainly by resource-poor farmers, unable to afford fungicides. Hence, the development of tolerant or resistant cultivars as a component of an IPM programme becomes an important

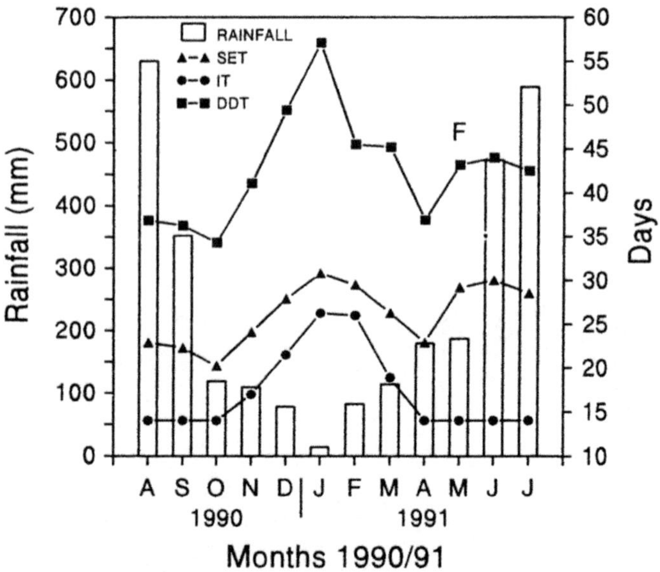

Figure 4 Seasonal variation of incubation time (IT), symptom evolution time (SET) and disease development time (DDT = IT + SET) of black sigatoka on 110 plantain landraces before and after flowering (F) at Onne, Nigeria, 1990/91. Source: Pasberg-Gauhl and Gauhl (1996).

Published by Burleigh Dodds Science Publishing Limited, 2019.

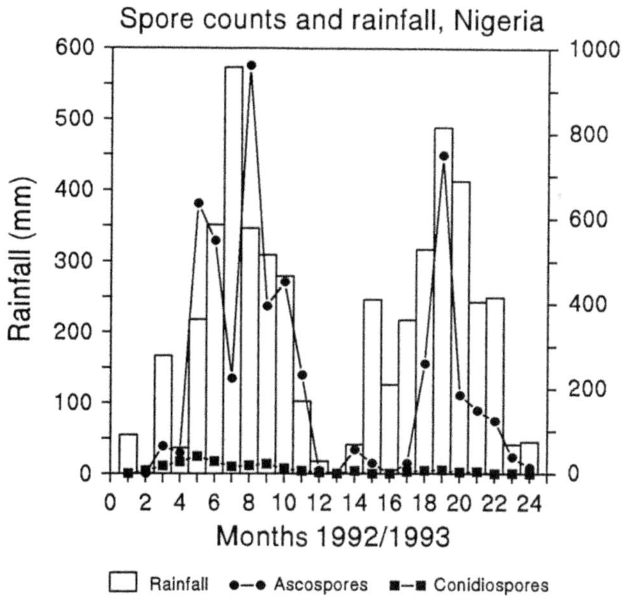

Figure 5 Monthly release of ascospores of the *Pseudocercospora fijiensis/M. musicola* type, conidiospores of *P. fijiensis* in a False Horn plantain plantation and rainfall at Onne Station, southeast Nigeria, 1992/93. Source: Pasberg-Gauhl and Gauhl (1996).

component for BLSD management (Norgrove et al., 2012). In 1987, IITA started breeding plantain and highland banana for durable host plant resistance to BLSD. Today a range of banana and plantain cultivars are available with high levels of tolerance to BLSD.

BLSD and YS cause less damage under shade; however, the mechanism linking severity with light intensity is not understood (Donzelli and Churchill, 2007). In a mesocosm experiment comparing 33% and 67% with 100% light, Norgrove et al. (2012) showed that leaf area attacked by BLSD on plants grown at 33% and 67% light was significantly less than when unshaded. A large-scale agroforestry experiment with *Terminalia ivorensis* timber trees at high and low stand densities showed greater leaf retention and reduced severity of BLSD in the high timber tree density (Norgrove and Hauser, 2013). However, leaf production rates were reduced under shade, increasing planting to harvest time, a parameter negatively related to bunch mass (Norgrove and Hauser, 2002), resulting in no yield gain from reduced BLSD (Norgrove and Hauser, 2013).

4.2 Fusarium wilt

Also known as Panama disease, *Fusarium* wilt is caused by different races of the soil-borne fungus *Fusarium oxysporum* f. sp. *cubense* (FOC) (Jones, 2018).

Published by Burleigh Dodds Science Publishing Limited, 2019.

Wilt-like symptoms had been observed earlier across West Africa. At Onne Station, FOC symptoms were never observed and the presence of the pathogen was never confirmed. In Nigeria, plants with *Fusarium* wilt symptoms were not detected, except once in 1994, on a single banana plant and confirmed as *Fusarium oxysporum* f. sp. *cubense*. In Africa it was first recorded in South Africa and damaged 'Gros Michel' (dessert banana, AAA-group) severely, whereas 'Bluggoe' (cooking banana, ABB-group) appeared highly resistant. Although FOC does not seem to be a major constraint to *Musa* production in West Africa, the recent accidental introduction to Mozambique of the aggressive Tropical Race 4 shows the necessity to observe the disease for possible outbreaks, particularly when new germplasm is introduced. Most recent research in Australia produced FOC Tropical Race 4-resistant bananas through genetic engineering (Dale et al., 2017). IITA has the capacity to transform banana and plantain, and thus is well positioned to introduce resistance genes into African *Musa* germplasm. The future challenge is thus raising awareness to allow rapid measures being taken when FOC symptoms are observed (see Chapter 3).

5 Nematodes

Musa are affected by several species of plant parasitic nematodes, which often occur simultaneously (McSorley and Parrado, 1986; Adiko, 1988; Bridge et al., 1995; Gowen et al., 2005, Osei et al., 2013). Consequently, efforts to control one species may inadvertently enhance population densities of other nematode species (Desaeger and Rao, 2000; Luc et al., 2005). Although a range of nematode species have been associated with *Musa*, just a handful are viewed as threats to production. Due to their microscopic size and rather cryptic life, most farmers are unaware of nematodes (Hauser and Amougou, 2010) and attribute damage to more visible causes.

Controlled yield loss studies from Nigeria have demonstrated the interaction between pests and nutrient management. Inoculating plants with nematodes reduced yield by more than 50% (Norgrove and Hauser, 2014), while the addition of mulch was able to overcompensate for the yield loss due to nematodes (Fig. 6).

5.1 Radopholus similis

The burrowing nematode, *Radopholus similis* (Thorne) (Nematoda: Pratylenchidae), is viewed as the most important nematode pest of *Musa* globally, largely due to its occurrence and damage caused on commercial dessert banana plantations (Gowen et al., 2005). In SSA, *R. similis* was documented as the most important nematode pest on *Musa* across West Africa and East Africa (Fogain, 2001a), but lately its prevalence in West Africa has

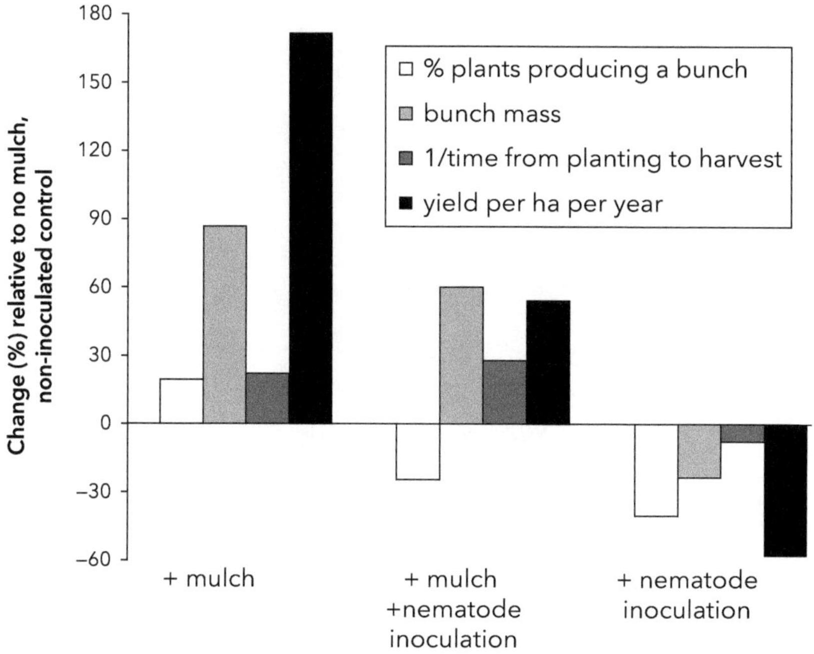

Figure 6 Relative change (%) in the proportion (%) of plants producing a bunch, the bunch mass per plant, 1/time from planting to harvest and the fresh bunch yield (per ha and per year) in False Horn plantain cv. Agbagba inoculated with nematodes and/or mulched relative to a non-mulched, non-inoculated control in SE Nigeria after hot water treatment. Sources: Norgrove and Hauser (2014), after O. Rotimi (2003, unpublished results) and Coyne et al. (2005).

declined where *Pratylenchus coffeae* (Goodey) (Nematoda: Pratylenchidae) appears to be replacing it (Coyne, 2009). On East African Highland bananas, *R. similis* remains the dominant and most serious nematode threat (Price, 2006). *R. similis* is sensitive to low temperatures, thus less prevalent at higher altitudes (>1400 masl) (Kashaija et al., 1994). *Radopholus similis* infection causes decay of the root cortex indicated by a dark purple discolouration of the otherwise white tissue (Fig. 7).

Plantain is an important staple in southern Cameroon, where farmers still have access to secondary forest and long fallows, which are likely free of common pests and diseases of banana and plantain. However, Fogain (2001b) reported most fields below 700 masl in the Littoral, Central, South and South-West provinces being infested by *R. similis* with >50% of plantations showing severe root necrosis. A later survey found that farmers' attitude towards sourcing planting material was very lax and did not consider plant health concerns. As nematodes are microscopic, farmers attribute symptoms of nematode damage

Figure 7 Longitudinally cut plantain roots, with various stages of *R. similis* infection.

to uprooting and wind damage. Farmers ordinarily take suckers from any plant, including uprooted ones (Hauser and Amougou, 2010) to establish new fields, inadvertently spreading the pest.

Farmers not having access to pest and disease-free planting material need affordable and simple protection methods. In plantain, paring suckers by peeling a thin layer off the corm and removing discoloured tissue is a simple method to reduce nematode infection and to control for weevil galleries. Some farmers were found to apply this recommendation, but the efficiency of this method alone was low (Hauser, 2007). Trials on conventional suckers and on macro-propagated (corm fragment method) plants showed that nematicide (carbofuran) application is ineffective in controlling nematodes and avoiding yield losses, while thermal control methods showed significant yield increases at low cost (Hauser, 2007).

Submerging suckers in 52°C hot water for 20 min was tested successfully in Cameroon (Hauser, 2000; Hauser et al., 2008; Mekoa and Hauser, 2010) and other countries (Coyne and Hauser, 2010). However, hot water treatment requires a high-volume tank to maintain the temperature for a long time, as well as temperature control and regulated heating (gas burner) to maintain a constant temperature (Fig. 8). While suited to large-scale plantings, the system proved to be expensive, time consuming and cumbersome, thus not appropriate for preparing small numbers of suckers as is usually done by smallholders.

A simplified, rapid, scale-independent method, not requiring technical equipment was developed using boiling water. Testing variable lengths

Figure 8 Left - a hot water tank with steps to facilitate moving the cage with suckers inside the tank. A central duct inside the tank is split into six pipes to insert a gas burner to maintain the temperature. Right - a used oil drum and a wicker basket on a stick to submerge suckers into boiling water. Photos: S. Hauser.

of submergence (10–60 s) confirmed that the boiling water method was suitable to clean plantain suckers of nematodes and weevils without killing the sucker. No nematode survived on suckers treated in boiling water for 30 s, while survival of the suckers was still excellent (Hauser and Messiga, 2010).

At the time the boiling water method was being developed and tested, tissue culture (TC) banana and plantain plants became an alternative option, yet facilities to mass produce were lacking. Nevertheless, rather than promoting thermal methods of nematode, weevil and disease control, government programmes preferred to invest in TC technology and facilities to produce plantlets from corm fragments (CFP). The CFP facilities were supported and subsidized with cheap nematicide supplies (carbofuran) and farmers were given the plantlets free of charge, but had to buy nematicide at subsidized rates to apply in their fields. Interested, yet inexperienced farmers asked IITA Cameroon to assist in establishing plantain fields from CFP material. The propagation had started in the middle of the year, thus plantlets were ready for transplanting just by the end of the rainy season. Field establishment was acceptable and differences between hot water treated suckers (88%) and CFP material (79%) were small. However, dry season survival and yields were much lower for the CFP material with a high proportion of CFP plants not reaching flowering and bunch production. This example shows that without a high level of coordination, new technologies are prone to fail. Simple methods such as boiling water treatment are not exposed to such problems as they are decentralized and flexible in time and space.

5.2 Pratylenchus goodeyi

Pratylenchus goodeyi (Sher & Allen) (Nematoda: Pratylenchidae) is indigenous to Africa, thermophobic, exhibiting an Afromontane distribution (Price and Bridge, 1995; Ssango et al., 2004; Gowen et al., 2005; Price, 2006) and commonly known as a pathogen of *Musa* in smallholder fields in the highlands of Africa (Speijer et al., 1993; Bridge et al., 1995; Speijer and De Waele, 1997; Talwana et al., 2003). More recently however, this species was found in hotter, tropical conditions in East Africa (Seshu Reddy et al., 2007), raising questions on its ecological plasticity, or its true identity (Coyne et al., 2018). *Pratylenchus goodeyi* has a relatively narrow host range, particularly in comparison to other *Pratylenchus* spp. Of 123 screened plant species, just ~10% were classified as good hosts: four food crops (banana, plantain, onion and watermelon), one medicinal crop, one fodder crop and six weeds (Price, 1994a; Mbwana et al., 1995; Prasad et al., 1999; Namaganda et al., 2000).

Pratylenchus goodeyi is a migratory endoparasite similar to *R. similis,* yet with a lower reproductive capacity (R. Fogain, 1995, unpublished results). At higher altitudes, *Musa* growth cycles tend to be longer and thus the period for nematode population build-up and related damage is longer. So, despite the lower reproductive rate, *P. goodeyi* can cause considerable damage in the highlands of southwestern Cameroon (Jacobsen et al., 2004). Bridge (1988) identified *P. goodeyi* as the most important nematode on banana in the Kagera region of Tanzania. Talwana et al. (2003) found higher damage caused by simultaneous infection with *R. similis* and *P. goodeyi* than with *P. goodeyi* alone. Speijer and De Waele (1997) found little impact of *P. goodeyi* on East African Highland cooking bananas when concomitantly infected with *Helicotylenchus multicinctus* (Cobb) (Nematoda: Hoplolaimidae), *Meloidogyne* sp. and *R. similis*. On Pisang Awak (*Musa* group ABB), however, *P. goodeyi* proved the best single predictor of damage, with *R. similis* and *H. multicinctus* each contributing less, indicating a genotype-mediated response to nematode infection (Speijer and De Waele, 1997).

5.3 Pratylenchus coffeae

The *Pratylenchus coffeae* lesion nematode has gradually increased in prevalence across SSA and indeed, has now replaced *R. similis* as the most damaging nematode species on *Musa* in many areas of West Africa (Coyne, 2009). The reason behind this rather dramatic shift is unknown, but has major implications regarding the deployment of host resistance against nematode pests because resistance conferred against one nematode species does not necessarily work against other species (Dubois and Coyne, 2011). In East Africa the shift is less dramatic: in Tanzania and Kenya, *P. coffeae* is being more frequently recovered but not in Uganda (Seshu Reddy et al., 2007; Coyne, 2009).

Published by Burleigh Dodds Science Publishing Limited, 2019.

In addition, it is possible that more than one species of *Pratylenchus* is involved, as they are difficult to distinguish and often require molecular techniques. For instance, after reporting the effects of an aggressive population of *P. coffeae* on plantain in Ghana (Brentu et al., 2004), it was later discovered to be a separate, undescribed species, which was morphologically very similar *to P. coffeae* and named *Pratylenchus speijeri* n. sp (Nematoda: Pratylenchidae) after the late IITA nematologist Paul Speijer (De Luca et al., 2012).

5.4 Other nematode species

Meloidogyne spp. are frequently found infecting *Musa* roots, causing stunted and swollen roots (Fig. 9), which become necrotic. Above-ground symptoms include stunted growth, thinner pseudostems and smaller fruit bunches (Sudha and Prabhoo, 1983; Lin and Tsay, 1985). Brentu et al. (2004) and Coyne et al. (2013) observed yield reductions ranging from 26% to 57%, although other studies report no yield loss or effect on growth parameters due to *Meloidogyne* infection (Adiko, 1989). *Meloidogyne* spp. have a wide host range (CABI, 2002).

Helicotylenchus multicinctus is regularly associated with *Musa* (McSorley and Parrado, 1986; Bridge et al., 1995). As it most often occurs in combination with other nematodes, especially *R. similis*, its status has been traditionally overlooked, although it has more recently been recognized as a key pest in its own right (e.g. Ssango et al., 2004; Coyne et al., 2013). It appears to be omnipresent with *Musa* (Coyne and Kidane, 2018) and occurs on plantain in the West African lowlands regularly in combination with *Meloidogyne* spp.

Figure 9 *Meloidogyne* spp. infecting *Musa* roots. Above: typical galling of banana roots, below: necrotic tissue surrounding the female nematode. Photos: D. Coyne.

resulting in quite severe losses (Caveness and Badra, 1980). Root damage due to *H. multicinctus* infection has been reported elsewhere (McSorley and Parrado, 1986; Chau et al., 1997; Moens et al., 2006) and yield reductions range from 19% to 34% (Speijer and Fogain, 1999; Brentu et al., 2004).

Hoplolaimus pararobustus (Sher) (Nematoda: Hoplolaimidae) is often recovered from *Musa,* but has seldom been associated with any serious damage. In the Cameroon Highlands, it was the third most frequent nematode species, after *Meloidogyne* spp. and *P. goodeyi,* in more than 75% of the samples studied (K.S. Jacobsen, 2010, unpublished results). *Musa* is the main host of *H. pararobustus* (CABI, 2002) and while Price (1994b) was unable to show that it was pathogenic to banana, Coyne et al. (2013) showed over 50% plantain yield loss in the mother crop cycle following inoculation with 2000 nematodes per plant in Nigeria.

6 Banana weevil

IITA initiated work on banana weevil (*Cosmopolites sordidus*) in 1991. Primary partners included national programmes (Uganda, Ghana, Kenya, Rwanda); ICIPE; CABI; and universities. Diagnostic surveys and yield loss trials confirmed the banana weevil to be a serious pest on East African Highland bananas and plantains. Most farmers had limited understanding of the pest and how to manage it.

6.1 Biology, pest distribution and pest status

Adult banana weevils are free living but they rarely fly. Hence, dissemination into new fields is mostly through infested planting material. Eggs (about one per week) are inserted in the base of the plant. The larvae bore into the corm, reducing nutrient uptake and weakening the stability of the plant (Fig. 10).

Highest rates of egg hatching and larval development occur at 25–30°C (Traoré et al., 1993, 1996). Extended periods with low night-time temperatures (12°C for eggs and 10°C for larvae (L. Traoré, 1995, unpublished results)) at higher elevations may reduce population development. In East Africa, banana weevil damage was greatest below 1300 masl and negligible above 1600 masl. However, climate change and particularly increasing temperatures may affect pest distribution and incidence.

Damage causes reduced bunch weights, plant and mat loss and shortened stand life. Damage and yield loss increased with time, for example from 5% at crop establishment to 44% in the fourth cycle (Rukazambuga et al., 2002). Gold et al. (2004b) found negligible yield loss in the first three crop cycles and 60% in the sixth crop cycle, suggesting that single cycle trials may give inaccurate

Published by Burleigh Dodds Science Publishing Limited, 2019.

Figure 10 Plantain corm with weevil larvae galleries in the centre (left) and highland banana corm with extensive tunnelling and secondary rot in the periphery of the weevil galleries. Photos: S. Hauser (left) and C. Gold (right).

yield loss measurements. No single control strategy provides complete control of the banana weevil. Therefore, a broad IPM approach is proposed.

6.2 Clean planting material

Infested planting material is the principal transfer pathway for banana weevils into newly planted fields. Use of clean planting material reduces initial weevil numbers and delays population build-up. Methods include selection of weevil-free suckers, paring, hot or boiling water treatment and/or heat sterilization of suckers and TC plantlets.

In Uganda, paring of the outer corm surface of suckers removed >90% of weevil eggs but had little effect on weevil larvae (Gold et al., 1998). Farmers were therefore encouraged to reject pared suckers with larval tunnels. Immersion of suckers in hot water baths has been recommended for weevil and nematode control. Suckers immersed in hot water (54°C for 20 min) remained nematode-free for years, yet the treatment failed to kill a high percentage of weevil larvae, which were able to inflict substantial damage already in the first ratoon crop.

6.3 Cropping Systems

Application of coffee husks at the base of banana mats reduced weevil damage, provided nutrients and resulted in higher yields (Okech et al., 2004). However, cover crops with reported insecticidal properties (*Canavalia ensiformis* (L.(DC)), *Mucuna pruriens* (L.(DC)), *Tephrosia vogelii* (Hook.f.) [all Fabaceae]) had no effect on weevil adult numbers or damage (McIntyre et al., 2002).

6.4 Mulching

Mulching increased banana yields by 18–27% (Rukazambuga et al., 1998; Gold et al., 2006). Grass mulches may, however, increase weevil damage by creating a favourable cool and moist environment for adult weevils (Rukazambuga et al., 2002). Contrary to that risk, the positive benefits of mulching offset any increase in weevil damage in most cases.

6.5 Residue management

Crop residues can provide adult shelters and oviposition sites for weevils (Gold et al., 2004a; Masanza et al., 2009). Gold et al. (2004a) found 25–32% of adult weevils associated with cut or fallen residues, while 10–12% were in standing stumps in old banana plots. Residues of resistant cultivars may support large numbers of larvae, contributing to population build-up. Oviposition, development rates and larval survival were greatest on residues <30 days old (Masanza et al., 2009), suggesting that sanitation practices should be implemented soon after harvest. In an on-farm study, increasing sanitation levels significantly reduced adult populations, lowered corm damage and increased yields (Masanza et al., 2004).

6.6 Trapping

In Ugandan farms with high banana weevil levels, weevil populations declined by 61% with researcher-managed trapping (one pseudostem trap/mat/month), by 53% in farmer-managed trapping (trap intensity at farmers' discretion) and by 38% in controls (no trapping) during the course of the study (Gold et al., 2004b), displaying a weak relationship between the number of adults removed and the change in population density. Although farmers were convinced that trapping was beneficial, subsequent adoption was low due to resource requirements.

6.7 Pheromone-based

Chemtica International (Costa Rica) produces commercial pitfall traps (Cosmolure+) using sordidin, a banana weevil male aggregation pheromone. The recommended density is 4/ha with lures changed and traps moved every month. The weevils drown after entering the traps. By contrast, pseudostem traps presented above need to be placed at the base of as many mats as possible, last only 3–7 days, require frequent visits and manual destruction of weevils.

In Uganda, pheromone-baited traps captured 18 times more weevils/day than pseudostem traps (W. Tinzaara, 2005, unpublished results). Nevertheless, four or eight pheromone traps/ha applied for 21 months failed to reduce weevil populations and damage on-farm. In a field experiment, lures attracted 14% of marked weevils released within 0.5 m of the trap and 7% captured when released 4 m away, suggesting a limited efficacy of pheromones in attracting weevils to traps. These results suggest that pseudostem trapping will remain the more cost-effective strategy for most farmers.

6.8 Chemical control

Chemical control can be highly effective at reducing banana weevil populations. For example, field application of carbofuran close to the base of the plant effectively controls adults and weevil larvae inside the corm (Gold et al., 1999). Although highly effective, pesticide applications are costly, have detrimental environmental effects and have resulted in widespread resistance in banana weevil (Collins et al., 1991; Gold et al., 1999, 2001).

6.9 Biological control

Banana weevil was introduced into Africa from its area of origin in southeast Asia. The insect's biology suggests biological control would be best directed at the destructive immature stages within the plant. Early classical biological control attempts using generalist predators from Asia were unsuccessful (Gold et al., 2001). Recent work considered egg parasitoids, myrmicine ants, entomopathogens (*Beauveria bassiana* and *Metarhizium anisopliae*) and endophytes.

Myrmicine ants in the genera *Pheidole* and *Tetramorium* reportedly exert partial control of banana weevil in Cuba (Perfecto and Castineiras, 1998). Surveys in Uganda identified 11 species of ants feeding on exposed banana weevil larvae in field plots, 17 species foraging in pseudostems and 34 species foraging in corms of live plants and residues (Abera et al., 2006). Two species, *Odontomachus troglodytes* Santschi and *Pheidole* sp., removed 33–68% of eggs from naturally infested corms. However, in a 30-month field study *O. troglodytes* and *Pheidole* sp. failed to reduce larval numbers and damage in ant-enhanced plots compared to plots where ants were removed with the insecticide Amdro (Abera et al., 2008).

Microbial control approaches of the banana weevil include entomopathogenic fungi (e.g. *Beauveria bassiana*) and endophytes (e.g. non-pathogenic *Fusarium* spp. and introduced strains of *B. bassiana*). Field applications of entomopathogenic fungi target adult weevils, while endophytes target immature stages.

Although a number of strains of the biopesticide *B. bassiana* have shown promise in the laboratory and in preliminary field studies, efficient

Table 1 Effect of endophyte treatment on the proportion of banana weevil adults recovered, relative to the number released

Weevils	Control	*B. bassiana*
Adults recovered (%)	77.6	73.6
Dead adults (%)	25.8	62.0[a]
Mycosed adults (%)	0.0	57.9[a]

Source: Akello et al. (2008a).
[a]Significantly different from control.

and economically viable mass production and delivery systems still need to be developed. In Uganda, for example, application of 500 g of *B. bassiana*-colonized maize bran on topsoil around banana mats led to 48% weevil mortality 1 month after application (C.M. Nankinga, 1999, unpublished results). Weevil mortality declined to 20% after 5 months and 1–5% after 8 months, suggesting the need for repeated applications.

Hence, efforts were directed to finding alternative application methods, for example introducing the biopesticide as an endophyte (Akello et al., 2007, 2009). Greenhouse experiments showed that *B. bassiana* pre-inoculation of TC banana plantlets caused high mortalities to weevil immatures and adults (Akello et al., 2008a,b) (Table 1) leading to reduced weevil damage (Fig. 11). Though promising, the performance of microbial control agents against the banana weevil as biopesticide or endophyte under different agro-ecological conditions still requires further research.

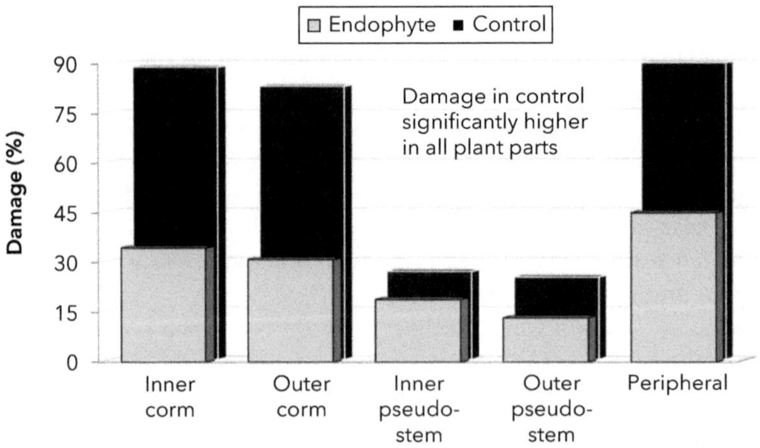

Figure 11 Weevil damage reduction in 2-month-old *Beauveria bassiana*-inoculated tissue cultured banana (cv. Kibuzi, *Musa* spp., genome group AAA-EA) plants, 45 days after infestation with female banana weevils (Akello et al., 2008b).

7 Conclusion

On 16 January 2003, the UK-based *Guardian* newspaper reported that 'Yes - in 10 years we may have no bananas' (Meek, 2003). These comments by the then Director-General of INIBAP (the International Network for the Improvement of Banana and Plantain) painted a bleak picture for the future of bananas worldwide, all caused by diseases that spread rapidly around the world and admittedly had caused dramatic damage largely to the dessert banana industry. Although the devastation caused by Panama disease (*Fusarium oxysporum* f. sp. *cubense*) was catastrophic, the article did not point out that the focus on a single cultivar, Gros Michel, thus neglecting varietal diversity, contributed to the destruction. Even the variety replacing Gros Michel, Cavendish, cannot be grown without fungicides and is today threatened by a new strain of the fungus called Tropical Race 4. This strain has arrived at Mozambique and poses major challenges to the banana industry. More stringent quarantine rules and a higher level of farmers' awareness have so far limited its spread.

Musa are highly susceptible to a wide range of pests and diseases and this is aggravated by the relatively narrow genetic basis. Breeding for tolerance and resistance has made progress and genetic modification has opened new opportunities to introduce tolerance or resistance genes into a crop that has not been sexually propagating for thousands of years. The work carried out so far by IITA and partners suggests that integrated pest (and disease) management, considering different agro-ecological conditions and requirements, rather than (un)controlled applications of pesticides is the way forward. Vast plantations of genetically uniform bananas are an invitation to pests and diseases to cause havoc. Smallholder banana and plantain crops may also be at risk, but owing to the heterogeneous mosaic of small fields, fallows and forests common in Africa, it is less likely that a single pest or disease can spread fast and wipe out the crop. The constraint of a narrow genetic base, and the ensuing relative high uniformity, can only be tackled through the introduction of improved varieties or GM materials. Most banana consumers may not like the idea, yet the biology of the banana offers few options.

The spread of many pests and diseases of *Musa* can be controlled by the rigorous application of the already existing laws on the movement of plants and seeds between countries. However, on the African continent, where one country (DR Congo) stretches almost entirely from west to east, quarantine regulations alone are unlikely to slow down the spread of pests and diseases. As discussed in detail in Chapter 3, awareness and effective control measures need to be in place to prevent epidemics. IITA, in collaboration with other CG centres, is continuing to pursue the development of appropriate diagnostic tools and smallholder-friendly control recommendations.

Thus today, YES, there are still bananas on local markets across Africa and in the supermarkets across the world. Not only did they not disappear but they still look and taste the same.

8 Where to look for further information

A very good textbook is Robinson and Galán Saúco (2010), *Bananas and Plantains*. Jones, D. R. (Ed.) (2018). *Handbook of Diseases of Banana, Abaca and Enset*. CABI Publishing, Wallingford, UK, 656pp. Those with an interest in dessert bananas *'Achieving Sustainable Cultivation of Bananas Volume 1 Cultivation Techniques'* by Gert Kema (Ed.), 2018, Burleigh Dodds Science Publishing, 378pp.

Other comprehensive books are:

Acta Horticulturae 540: I International Symposium on Banana: I International Conference on Banana and Plantain for Africa.

Acta Horticulturae 879: V International Symposium on Banana: International Conference on Banana and Plantain in Africa: Harnessing International Partnerships to Increase Research Impact.

Acta Horticulturae 986: VII International Symposium on Banana: ISHS – ProMusa Symposium on Bananas and Plantains: Towards Sustainable Global Production and Improved Use.

Acta Horticulturae 1196: X International Symposium on Banana: ISHS – ProMusa Symposium on Agroecological Approaches to Promote Innovative Banana Production Systems.

More *Acta Horticulturae* issues containing banana and plantain articles can be found at https://www.actahort.org/books/.

CG Centers with expertise on *Musa* are:

The International Institute of Tropical Agriculture (IITA); www.iita.org
Bioversity International; www.bioversityinternational.org

The CGIAR's Research Program on Roots Tubers and Bananas (RTB); www.rtb.org which includes a number of current research projects designed to support banana and plantain farmers in Africa through improving the varieties, the yields through improved agronomy, better marketing through value chain analyses and processing techniques and innovative dissemination approaches.

9 References

Abadie, C., Chilin, C. Y., Huat, J., Salmon, F., Pignolet, L., Carlier, L., Lescot, T., Côte, F. and Jenny, C. (2009). New approaches to select cultivars of banana with durable resistance to Mycosphaerella leaf spot diseases. *Acta Horticulturae*, 828, 171–8.

Abera, A., Gold, C. S., van Driesche R. and Ragama, P. E. (2006). Composition, distribution and relative abundance of ants in banana farming systems in Uganda. *Biological Control*, 40, 168–78.

Abera, A. M. K., Gold, C. S. and van Driesche, R. (2008). Experimental evaluation of the impacts of two ant species on banana weevil in Uganda. *Biological Control*, 46, 147–57.

Addis, T., Turyagyenda, L. F., Alemu, T., Karamura, E. and Blomme, G. (2010). Garden tool transmission of Xanthomonas campestris pv. *Musacearum* on banana (*Musa* spp.) and enset in Ethiopia. *Acta Horticulturae*, 879, 367–72.

Adiko, A. (1988). Plant-parasitic nematodes associated with plantain, *Musa* paradisiaca (AAB), in the Ivory Coast. *Revue de Nématologie*, 11, 109–13.

Adiko, A. (1989). Effect of *Meloidogyne incognita* on plantain, *Musa* AAB. *International Nematology Network Newsletter*, 6, 27–30.

Akello, J., Dubois, T., Gold, C., Coyne, D., Nakavuma, J. and Paparu, P. (2007). *Beauveria bassiana* (Balsamo) Vuillemin as an endophyte in tissue culture banana (*Musa* spp.). *Journal of Invertebrate Pathology*, 96, 34–42.

Akello, J., Dubois, T., Coyne, D. and Kyamanywa, S. (2008a). Effect of endophytic Beauveria bassiana on populations of the banana weevil, Cosmopolites sordidus, and their damage in tissue-cultured banana plants. *Entomologia Experimentalis et Applicata*, 129, 157–65.

Akello, J., Dubois, T., Coyne, D. and Kyamanywa, S. (2008b). Endophytic *Beauveria bassiana* in banana (*Musa* spp.) reduces banana weevil (*Cosmopolites sordidus*) fitness and damage. *Crop Protection*, 11, 1437–41.

Akello, J., Dubois, T., Coyne, D. and Kyamanywa, S. (2009). The effects of *Beauveria bassiana* dose and exposure duration on colonization and growth of tissue cultured banana (*Musa* sp.) plants. *Biological Control*, 49, 6–10.

Brentu, C. F., Speijer, P. R., Green, K. R., Hemeng, B. M. S., De Waele, D. and Coyne, D. (2004). Micro- plot evaluation of the yield reduction potential of *Pratylenchus coffeae, Helicotylenchus multicinctus* and *Meloidogyne javanica* on plantain cv. Apantu-pa (*Musa* spp., AAB-group) in Ghana. *Nematology*, 6, 455–62.

Bridge, J. (1988). Plant nematode pests of banana in East Africa with particular reference to Tanzania. In: INIBAP (Ed.), *Nematodes and the Borer Weevil in Bananas: Present Status of Research and Outlook*. Proceedings of a workshop held in Bujumbura, Burundi, 7–11 December 1987, pp. 35–9.

Bridge, J., Price, N. S. and Kofi, P. W. (1995). Plant parasitic nematodes of plantain and other crops in Cameroon, West Africa. *Fundamental and Applied Nematology*, 18, 251–60.

CABI (2002). *Crop Protection Compendium – Global Module*, 4th ed. CABI Publishing, Wallingford, UK.

Carter, B. A., Reeder, R., Mgenzi, S. R., Kinyua, Z. M., Mbaka, J. N., Doyle, K., Nakato, V., Mwangi, M., Beed, F., Aritua, V., Lewis Ivey, M. L., Miller, S. A. and Smith, J. J. (2010). Identification of *Xanthomonas vasicola* (formerly *X. campestris* pv. *Musacearum),* causative organism of banana Xanthomonas wilt, in Tanzania, Kenya and Burundi. *Plant Pathology*, 59, 403.

Caveness, F. E. and Badra, T. (1980). Control of *Helicotylenchus multicinctus* and *Meloidogyne javanica* in established plantain and nematode survival as influenced by rainfall. *Nematropica*, 10, 10–14.

Chau, N. N., Thanh, N. V., De Waele, D. and Geraert E. (1997). Plant-parasitic nematodes associated with banana in Vietnam. *International Journal of Nematology*, 7, 122–6.

Collins, P. J., Treverrow, N. L. and Lambkin, T. M. (1991). Organophosphorus insecticide resistance and its management in the banana weevil borer, *Cosmopolites sordidus*, Germar (Coleoptera: Curculionidae), in Australia. *Crop Protection*, 10, 215–21.

Coyne, D. (2009). Pre-empting plant-parasitic nematode losses on banana in Africa: Which species do we target? *Acta Horticulturae*, 828, 227–36.

Coyne, D. and Hauser, S. (2010). Rapid treatment of banana suckers for nematode management. *South African Journal of Plant and Soil*, 27, 259.

Coyne, D. and Kidane, S. (2018). Nematode pathogens. In: Jones, D. (Ed.), *Diseases of Banana, Abacá and Enset,* 2nd ed. CABI Publishing, Wallingford, UK pp. 429–461.

Coyne, D. L., Rotimi, O., Speijer, P., De Schutter, B., Dubois, T., Auwerkerken, A., Tenkouano, A. and De Waele, D. (2005). Effects of nematode infection and mulching on the yield of plantain (*Musa* spp., AAB-group) ratoon crops and plantation longevity in southeastern Nigeria. *Nematology*, 7, 531–41.

Coyne, D., Omowumi, A., Rotifa, I. and Afolami, S. O. (2013). Pathogenicity and damage potential of five plant-parasitic nematode species on plantain (*Musa* spp., AAB genome) cv. Agbagba. *Nematology*, 15, 589–99.

Coyne, D., Cortada, L., Dalzell, J., Claudius-Cole, A., Haukeland, S., Luambano, N. and Talwana, H. (2018). Plant-Parasitic Nematodes and Food Security in Sub-Saharan Africa. *Annual Review of Phytopathology*, 56, 381–403.

Dale, J., James, A., Paul, J.-Y., Khanna, H., Smith, M., Peraza-Echeverria, S., Garcia-Bastidas, F., Kema, G., Waterhouse, P., Mengersen, K. and Harding, R. (2017). Transgenic Cavendish bananas with resistance to Fusarium wilt tropical race 4. *Nature Communications*, 8, 1496. doi:10.1038/s41467-017-01670-6. www.nature.com/naturecommunications

De Luca, F., Troccoli, A., Duncan, L. W., Subbotin, S. A., Waeyenberge, L., Coyne, D. L., Brentu, F. C. and Inserra, R. N. (2012). *Pratylenchus speijeri* n. sp., a new root-lesion nematode pest of plantain in West Africa. *Nematology*, 14, 987–1004.

Desaeger, J. and Rao, M. R. (2000). Parasitic nematode populations in natural fallows and improved cover crops and their effects on subsequent crops in Kenya. *Field Crops Research*, 65, 41–56.

Donzelli, B. G. G. and Churchill, A. C. L. (2007). A quantitative assay using mycelial fragments to assess virulence of *Mycosphaerella fijiensis*. *Phytopathology*, 97, 916–29

Drenth, A. and Guest, D. I. (2016). Fungal and oomycete diseases of tropical tree fruit crops. *Annual Review of Phytopathology*, 54, 373–95.

Dubois, T. and Coyne, D. (2011). Integrated pest management of banana. In: Pillay, M. and Tenkouano, A. (Eds), *Banana Breeding: Constraints and Progress*. CRC Press, Boca Raton, FL, pp. 121–44

FAO (2017). Production data on banana and plantain in Africa by region. http://www.fao.org/faostat/en/#data/QC (accessed on 27 September 2017).

Fogain, R. (2001a). Effect of *Radopholus similis* on plant growth and yield of plantains (*Musa*, AAB). *Nematology*, 2, 129–33.
Fogain, R. (2001b). Nematodes and weevil of bananas and plantains in Cameroon: Occurrence, importance and host susceptibility. *International Journal of Pest Management*, 47, 201–5.
Frossard, P. (1980). Apparition d'une nouvelle et grave maladie folaire des bananiers et plantains au Gabon: la maladie des raies noires, *Mycosphaerella fijiensis* Morelet. *Fruits*, 35, 519–27.
Gauhl, F. and Pasberg-Gauhl, C. (1994). Epidemiology of black sigatoka disease on plantain in Nigeria. *Phytopathology*, 84, 1080.
Gold, C. S., Night, G., Abera, A. and Speijer, P. R. (1998). Hot-water treatment for control of banana weevil, *Cosmopolites sordidus* Germar (Coleoptera: Curculionidae) in Uganda. *African Entomology*, 6, 215–21.
Gold, C. S., Bagabe, M. J. and Ssendege, R. (1999). Banana weevil, *Cosmopolites sordidus* (Germar) (Coleoptera: Curculionidae) tests for suspected resistance to carbofuran and dieldrin in Masaka District, Uganda. *African Entomology*, 7, 189–96.
Gold, C. S., Pena, J. E. and Karamura, E. B. (2001). Biology and integrated pest management for the banana weevil, *Cosmopolites sordidus* (Germar) (Coleopteran: Curculionidae). *Integrated Pest Management Reviews*, 6, 79–155.
Gold, C. S., Night, G., Ragama, P. E., Kagezi, G. H., Tinzaara, W. and Abera, A. M. K. (2004a). Field distribution of banana weevil (*Cosmopolites sordidus* (Germar)) adults in cooking banana stands in Uganda. *Insect Science and its Application*, 24, 242–8.
Gold, C. S., Kagezi, G. H., Night, G. and Ragama, P. E. (2004b). The effects of banana weevil, *Cosmopolites sordidus* (Germar), damage on highland banana growth, yield and stand duration in Uganda. *Annals of Applied Biology*, 145, 263–9.
Gold, C. S., Ragama, P. E., Coe, R. and Rukazambuga, N. D. T. M. (2005). Selection of assessment methods for evaluating banana weevil *Cosmopolites sordidus* (Germar) (Coleoptera: Curculionidae) damage on highland cooking banana (*Musa* spp., genome group AAA-EA). *Bulletin of Entomological Research*, 95, 115–23.
Gold, C. S., Okech, S. H., McIntyre, B. D., Kagezi, G., Ragama, P. E. and Night, G. (2006). Effects of mulch on banana weevil *Cosmopolites sordidus* (Germar) populations and damage in Uganda. *Crop Protection*, 25, 1153–60.
Gowen, S., Quénéhervé, P. and Fogain, R. (2005). Nematode parasites of bananas and plantains. In: Luc, M., Sikora, R. A. and Bridge, J. (Eds), *Plant-parasitic Nematodes in Subtropical and Tropical Agriculture*, 2nd ed. CABI Publishing, Wallingford, UK, pp. 611–43.
Hauser, S. (2000). Effects of fertilizer and hot-water treatment upon establishment survival and yield of plantain (*Musa* spp. AAB French). *Field Crops Research*, 66, 213–23.
Hauser, S. (2007). Plantain (*Musa* spp. AAB) bunch yield and root health response to combinations of mechanical, thermal and chemical nematode control measures on suckers. *African Plant Protection*, 13, 1–15.
Hauser, S. and Amougou, D. (2010). Plantain (*Musa* spp.) cropping systems of southern Cameroon. *Acta Horticulturae*, 879, 495–508.
Hauser, S. and Messiga, F. N. K. (2010). Nematode control on plantain suckers (*Musa* spp. AAB Genome) through submergence in boiling water: Emergence rates, early growth, bunch yield and root health. *Acta Horticulturae*, 879, 323–31.
Hauser, S., Mekoa, C. and Jacobsen, K. (2008). Bunch yield response of two cultivars of plantain (*Musa* spp., AAB, subgroups French and False horn) to hot-water treatment

and fertilizer application planted after forest and bush/grass fallow. *Archives of Agronomy and Soil Science*, 54, 541–56.

Hauser, S., Amougou, D., Bengono, B., Ngo Kanga, F. and Pekeleke, M. (2010). On-farm demonstration, testing and dissemination of boiling water treatment for plantain (*Musa* spp.) sucker sanitation in southern Cameroon. *Acta Horticulturae*, 879, 509–15.

Jacobsen, K., Fogain, R., Mouassom, H. and De Waele, D. (2004). Musa-based cropping systems of the Cameroon highlands: A case study of the West and North West provinces of Cameroon, with emphasis on nematodes. *Fruits,* 59(5), 311–18. doi:10.1051/fruits:2004029.

Jones, D. R. (Ed.) (2018). *Handbook of Diseases of Banana, Abaca and Enset*. CABI Publishing, Wallingford, UK, 656pp.

Kagezi, G. H., Kangire, A., Tushmereirwe, W., Bagamba, F., Kikulwe, E. Gold, C. S., Ragama, P. and Kubiriba, J. (2006). Banana bacterial wilt incidence in Uganda. *African Crop Science Journal*, 14, 83–92.

Kashaija, I. N., Speijer, P. R, Gold, C. S. and Gowen, S. (1994). Occurrence, distribution and abundance of plant parasitic nematodes of bananas in Uganda. *African Crop Science Journal*, 2, 99–104.

Kitavi, M., Downing, T., Lorenzen, J., Karamura, D., Onyango, M., Nyine, M., Ferguson, M. and Spillane, C. (2016). The triploid East African Highland Banana (EAHB) genepool is genetically uniform arising from a single ancestral clone that underwent population expansion by vegetative propagation. *Theoretical and Applied Genetics*, 129, 547–61.

Kumar, P. L., Hanna, R., Alabi, O. J., Soko, M. M., Oben, T. T., Vangu, G. H. P., and Naidu, R. A. (2011). *Banana bunchy top virus* in sub-Saharan Africa: Investigations on virus distribution and diversity. *Virus Research*, 159, 171–82.

Kumar, P. L., Selvarajan, R., Iskra-Caruana, M.-L., Chabannes, M. and Hanna, R. (2015). Biology, etiology and control of virus diseases of banana and plantain. *Advances in Virus Research*, 91, 229–69.

LaFleur, D. A., Lockhart, B. E. L. and Olszewski, N. E. (1996). Portions of the banana streak badnavirus genome are integrated into the genome of its host *Musa*. *Phytopathology*, 86, 100–1.

Lin, Y. Y. and Tsay, T. T. (1985). Studies on banana root-rot disease in the central area of Taiwan. *Journal of the Chinese society for Horticultural Science*, 31, 44–6.

Luc, M., Sikora, R. A., and Bridge, J. (2005). *Plant-parasitic Nematodes in Subtropical and Tropical Agriculture*, 2nd ed. CABI Publishing, Wallingford, UK, 871pp.

Magee, C. J. (1927). Investigation on the bunchy top disease of bananas. *Bulletin of the Council for Scientific and Industrial Research in Australia*, 30, 64.

Masanza, M., Gold, C. S., van Huis, A., Ragama, P. E. and Okech, S. H. O. (2004). Effect of sanitation on banana weevil *Cosmopolites sordidus* (Germar) (Coleoptera: Curculionidae) population and crop damage in farmers' fields in Uganda. *Crop Protection*, 24, 275–83.

Masanza, M., Gold, C. S., van Huis, A. and Ragama, P. E. (2009). Influence of plant and residue age on attraction, acceptance and larval survival of the banana weevil *Cosmopolites sordidus* (Coleoptera: Curculionidae). *International Journal of Tropical Insect Sciences*, 29, 171–9.

Mbaka, J. N., Nakato, V. G., Auma, J. and Odero, B. (2009). Status of banana *Xanthomonas* wilt in Western Kenya and factors enhancing its spread. *African Crop Science Conference Proceedings*, 9, 673–6.

Mbida, C. M., Van Neer, W., Doutrelepont, H. and Vrydaghs, L. (2000). Evidence for banana cultivation and animal husbandry during the first millennium BC in the forest of southern Cameroon. *Journal of Archaeological Science*, 27, 151–62.

Mburu, K., Oduor, R., Mgutu, A. and Tripathi, L. (2016). Silicon application enhances resistance to *Xanthomonas* wilt disease in banana. *Plant Pathology*, 65, 807–15.

Mbwana, A. S. S., Waudo, S. W. and Seshu Reddy, K. V. (1995). Host range of the lesion nematode *Pratylenchus goodeyi* commonly found in highland bananas of East Africa. *International Journal of Pest Management*, 41, 46–9.

McIntyre, B. D., Gold, C. S., Kashaija, I. N., Ssali, H., Night, G. and Bwamiki, D. P. (2002). Effects of legume intercrops on soil-borne pests, biomass, nutrients and soil water in banana. *Biology and Fertility of Soils*, 34, 342–8.

McSorley, R. and Parrado, J. L. (1986). *Helicotylenchus multicinctus* on bananas: An international problem. *Nematropica*, 16, 73–91.

Meek, J. (2003). Yes – in 10 years we may have no bananas. *The Guardian*, 16 January 2003.

Mekoa, C. and Hauser, S. (2010). Survival and yield of the plantain 'Ebang' (*Musa* spp., AAB Genome, 'False Horn') produced from corm fragment initiated plants and suckers after hot water treatment in southern Cameroon. *Acta Horticulturae*, 879, 527–35

Moens, T., Araya, M., Swennen, R. and De Waele, D. (2006). Reproduction and pathogenicity of *Helicotylenchus multicinctus, Meloidogyne incognita* and *Pratylenchus coffeae*, and their interaction with *Radopholus similis* on *Musa*. *Nematology*, 8, 45–58.

Mwangi, M. and Nakato, V. (2007). Key factors responsible for the banana Xanthomonas wilt pandemic on banana in East and Central Africa. *Acta Horticulturae*, 828, 395–404.

Nakato, V., Mahuku, G. and Coutinho, T. (2018). *Xanthomonas campestris* pv. *Musacearum*: A major constraint to banana, plantain and enset production in central and east Africa over the past decade. *Molecular Plant Pathology*, 19, 525–36. doi:10.1111/mpp.12578.

Namaganda, J. M., Karamura, E., Namanya, P., Gowen, S. and Bridge, J. (2000). Studies on the host range of the banana lesion nematode, *Pratylenchus goodeyi*. In: Craenen, K., Ortiz, R., Karamura, E. and Vuylsteke, D. (Eds), *Proceedings of the First International Conference on Banana and Plantain for Africa*. *Acta Horticulturae* 540, International Society for Horticultural Science, Leuven, Belgium, pp. 419–25.

Namukwaya, B., Tripathi, L., Tripathi, J. N., Arinaitwe, G., Mukasa, S. B. and Tushemereirwe, W. K. (2012). Transgenic banana expressing *Pflp* gene confers enhanced resistance to Xanthomonas wilt disease. *Transgenic Research*, 12, 855–65.

Ndungo, V., Bakelana, K., Eden-Green, S. and Blomme, G. (2004). An outbreak of banana Xanthomonas wilt (*Xanthomonas campestris* pv. *Musacearum*) in the Democratic Republic of Congo. *InfoMusa*, 13, 43–4.

Norgrove, L. and Hauser, S. (2002). Yield of plantain grown under different tree densities and 'slash and mulch' versus 'slash and burn' management in an agrisilvicultural system in southern Cameroon. *Field Crops Research*, 78, 185–95.

Norgrove, L. and Hauser, S. (2013). Black leaf streak disease and plantain fruit characteristics as affected by tree density and biomass management in a tropical agroforestry system. *Agroforestry Systems*, 87, 349–54.

Norgrove, L. and Hauser, S. (2014). Improving plantain (*Musa* spp. AAB) yields on smallholder farms in West and Central Africa. *Food Security*, 6, 501–14.

Norgrove, L., Tueche, J. R., Jacobsen, K., Nkakwa-Attey, A. and Holmes, K. (2012). Tackling black leaf streak disease and soil fertility constraints to enable the expansion of plantain production to grassland in the humid tropics. *International Journal of Pest Management*, 58, 175–81.

Ocimati, W., Ssekiwoko, F., Karamura, E., Tinzaara, W., Eden-Green, S. and Blomme, G. (2013a). Systemicity of *Xanthomonas campestris* pv. *Musacearum* and time to disease expression after inflorescence infection in East African highland and 'Pisang Awak' bananas in Uganda. *Plant Pathology*, 62, 777–85.

Okech, S. H. O., Gold, C. S., Abele, S., Nankinga, C. M., Wetala, P. M., van Asten, P., Nambuye, A. and Ragama, P. (2004). Agronomic, pests and economic factors influencing sustainability of banana-coffee systems of Western Uganda and potentials for improvement. *Ugandan Journal of Agricultural Sciences*, 9, 432–44.

Osei, K., Mintah, P., Dzomeku, B. M., Braimah, H., Adomako, J., Mochiah, M. B., Asiedu, E., Darkey, S. and Danso, Y. (2013). Nematode pests of plantain: A case study of Ashanti and Brong Ahafo regions of Ghana. *Journal of Soil Science and Environmental Management*, 4, 6–10.

Pasberg-Gauhl, C. and Gauhl, F. (1996). *Musa* research in the plant health management division at IITA: Activities at the High Rainfall Station, Onne in Nigeria. In: Ortiz, R. and Akoroda, M. O. (Eds), *Plantain and Banana: Production and Research in West and Central Africa*. Proceedings of a regional workshop held at High Rainfall Station, Onne, Nigeria, 24–28 September 1995. IITA, Ibadan, Nigeria, pp. 7–14.

Perfecto, I. and Castineiras, A. (1998). Development of the predaceous ants and their conservation in agroecosystems. In: Barbosa, P. (Ed.), *Conservation Biological Control*. Academic Press, San Diego, CA, pp. 269–89.

Ploetz, R. (2000). Black sigatoka. *Pesticide Outlook*, 19–23 February 2000.

Prasad, J. S., Seshu Reddy, K. V. and Sikora, R. A. (1999). Life cycle of *Pratylenchus goodeyi* in banana roots. *Nematologia Mediterranea*, 27, 111–14.

Price, N. S. (1994a). Alternate cropping in the management of *Radopholus similis* and *Cosmopolites sordidus*, two important pests of banana and plantain. *International Journal of Pest Management*, 40, 237–44.

Price, N. S. (1994b). Field trial evaluation of *Musa* varieties and of other crops as hosts of *Pratylenchus goodeyi* in Cameroon. *Afro-Asian Journal of Nematology*, 4, 11–16.

Price, N. S. (2006). The banana burrowing nematode, *Radopholus similis* (Cobb) Thorne, in Lake Victoria region of East Africa: Its introduction, spread and impact. *Nematology*, 8, 801–17.

Price, N. S. and Bridge, J. (1995). *Pratylenchus goodeyi* (Nematoda: Pratylenchidae): A plant-parasitic nematode of the montane highlands of Africa. *Journal of African Zoology*, 109, 435–42.

Robinson, J.C. and Galán Saúco, V. (2010). *Bananas and Plantains*. CABI, Wallingford, UK, 311pp.

Rukazambuga, N. D. T. M., Gold, C. S. and Gowen, S. R. (1998). Yield loss in East African highland banana (*Musa* spp., AAA-EA group) caused by the banana weevil, *Cosmopolites sordidus* Germar. *Crop Protection*, 17, 581–9.

Rukazambuga, N. T. D. M., Gold, C. S. and Gowen, S. (2002). The influence of crop management on banana weevil, *Cosmopolites sordidus* (Coleoptera: Curculionidae) populations and damage to highland cooking banana (*Musa* spp. genome group AAA-EA) in Uganda. *Bulletin of Entomological Research*, 92, 413–21.

Seshu Reddy, K. V., Prasad, J. S., Speijer, P. R., Sikora, R. A. and Coyne, D. L. (2007). Distribution of plant-parasitic nematodes on *Musa* in Kenya. *InfoMusa*, 16, 18–23.

Speijer, P. R. and De Waele, D. (1997). Screening of *Musa* germplasm for resistance and tolerance to nematodes. INIBAP Technical guidelines N°1. INIBAP, Montpellier, France.

Speijer, P. R. and Fogain, R. (1999). *Musa* and Ensete nematode pest status in selected African countries. In: Frison, E. A., Gold, C. S., Karamura, E. B. and Sikora, R. A. (Eds), *Mobilizing IPM for Sustainable Banana Production in Africa*. Proceedings of a Workshop on Banana IPM held in Nelspruit, South Africa, 23–8 November 1998. INIBAP, Montpellier, France, pp. 99–108.

Speijer, P. R, Budenberg, W. J. and Sikora, R. A. (1993). Relationships between nematodes, weevils, banana and plantain cultivars and damage. *Annals of Applied Biology*, 123, 517–25.

Ssango, F., Speijer, P. R., Coyne, D. and De Waele, D. (2004). Path analysis: A novel approach to determine the contribution of nematode damage to East African Highland banana (*Musa* spp., AAA) yield loss under two crop management practices in Uganda. *Field Crops Research*, 90, 177–87.

Ssekiwoko, F., Taligola, H. K. and Tushmereirwe, W. K. (2006). *Xanthomonas campestris* pv. *Musa*cearum host range in Uganda. *African Crop Science Journal*, 14, 111–20.

Sudha, S. and Prabhoo, N. R. (1983). *Meloidogyne* (Nematoda: Meloidogynidae) induced root galls of the banana plant *Musa paradisiaca* – a study of histopathology. *Proceedings of the Indian Academy of Animal Sciences*, 92, 467–73.

Stellenbosch University (2017). http://www.sun.ac.za/english/faculty/agri/plant-pathology/ac4tr4/foc-tr4-in-africa/banana-production-in-africa (accessed on 11 November 2017).

Talwana, H. A. L., Speijer, P. R., Gold, C. S., Swennen, R. L. and De Waele, D. (2003). A comparison of the effects of the nematodes *Radopholus similis* and *Pratylenchus goodeyi* on growth, root health and yield of an East African highland cooking banana (*Musa* AAA-group). *International Journal of Pest Management*, 49, 199–204.

Traoré, L., Gold, C. S., Pilon, J. G. and Boivin, G. (1993). Effects of temperature on embryonic development of banana weevil *Cosmopolites sordidus* Germar (Coleoptera: Curculionidae). *African Crop Science Journal*, 2, 111–16.

Traoré, L., Gold, C. S., Boivin, G. and Pilon, J. G. (1996). Développement postembryonaire du charançon du bananier. *Fruits*, 51, 105–13.

Tripathi, L., Mwangi, M., Abele, S., Aritua, V., Tushemereirwe, W. K. and Bandyopadhyay, R. (2009). Xanthomonas wilt: A threat to banana production in East and Central Africa. Plant Disease, 93, 440–51. doi:10.1094/PDIS-93-5-0440.

Tripathi, L., Mwaka, H., Tripathi, J. N., and Tushemereirwe, W. K. (2010). Expression of sweet pepper Hrap gene in banana enhances resistance to Xanthomonas campestris pv. *Musa*cearum. *Molecular Plant Pathology*, 11, 721–31.

Tripathi, L., Tripathi, J. N., Kiggundu, A., Kori, S., Shotkoski, F. and Tushemereirwe, W. K. (2014). Field trial of *Xanthomonas* wilt disease-resistant bananas in East Africa. *Nature Biotechnology*, 32, 868–70. doi:10.1038/nbt.3007.

Tushemereirwe, W., Kangire, A., Smith, J., Ssekiwoko, F., Nakyanzi, M., Kataama, D., Musiitwa, C. and Karyaija, R. (2003). An outbreak of bacterial wilt on banana in Uganda. *InfoMusa*, 12, 6–8.

Welde-Michael, G., Bobosha, K., Addis, T., Blomme, G., Mekonnen, S. and Mengesha, T. (2008). Mechanical transmission and survival of bacterial wilt on enset. *African Crop Science Journal*, 16, 97–102.

Yirgou, D. and Bradbury, J. F. (1968). Bacterial wilt of enset (*Ensete ventricosum*) incited by *Xanthomonas Musacearum* sp. n. *Phytopathology*, 58, 111–12.

Yirgou, D. and Bradbury, J. F. (1974). A note on wilt of banana caused by the enset wilt organism *Xanthomonas Musacearum*. *East African Agriculture and Forestry Journal*, 40, 111–14.

Zandjanakou-Tachin, M., Ojiambo, P. S., Vroh-Bi, I., Tenkouano, A., Gumedzoe, Y. M. and Bandyopadhyay, R. (2013). Pathogenic variation of *Mycosphaerella* species infecting banana and plantain in Nigeria. *Plant Pathology*, 62, 298–308.

Published by Burleigh Dodds Science Publishing Limited, 2019.

Chapter 10

Identifying and managing plant health risks for key African crops: legumes

Manuele Tamò, *International Institute of Tropical Agriculture (IITA), Bénin;* **Léonard Afouda**, *Université de Parakou, Bénin;* **Ranajit Bandyopadhyay**, *International Institute of Tropical Agriculture (IITA), Nigeria;* **Harry Bottenberg**, *United States Agency for International Development, USA;* **Laura Cortada-Gonzales**, *International Institute of Tropical Agriculture (IITA), Kenya;* **Harun Murithi**, *International Institute of Tropical Agriculture (IITA), Tanzania;* **Alejandro Ortega-Beltran**, *International Institute of Tropical Agriculture (IITA), Nigeria;* **Barry Pittendrigh**, *Michigan State University, USA;* **Rachidatou Sikirou**, *Institut National des Recherches Agricoles au Bénin, Bénin;* **Abou Togola**, *International Institute of Tropical Agriculture (IITA), Nigeria; and* **Kerstin D. Wydra**, *Erfurt University of Applied Sciences, Germany (corresponding author: m.tamo@cgiar.org)*

1 Introduction

This chapter deals with plant health challenges affecting two grain legumes that have been part of the research portfolio of IITA since its establishment: cowpea and soybean.

Although grain legumes were included in the IITA mandate 'from the outset of its establishment in July 1967' as it reads in the first Grain Legumes Improvement Program (GLIP) report covering 1970–3, the history of plant health in grain legumes at IITA started effectively with the establishment of GLIP mid-1971. That two of the first scientists to be appointed in GLIP were a pathologist and entomologist underscores the focus on plant health from the very beginning. Earlier work in plant pathology included screening for foliar and viral diseases in both soybean and cowpea, transmission studies to boost the screening protocols, and assessing effective seed treatments.

http://dx.doi.org/10.19103/AS.2018.0043.11
Published by Burleigh Dodds Science Publishing Limited, 2019.

Initial efforts in entomology, mostly in cowpea, were mainly directed at testing seed treatments and low toxicity insecticides, complemented by screening for host plant resistance.

This research focus continued throughout the 1970s, and culminated with the production of a cowpea pest and disease manual and a soybean production manual which included plant health aspects as well. In the 1980s, the focus shifted from resistance screening to studying the biology of key pests and epidemiology/aetiology of major disease, in order to better understand the underlying mechanisms of resistance.

With the establishment of the IITA research divisions in 1991, GLIP continued as a breeding programme under the Crop Improvement Division, whereas the main activities in plant health were carried out under the Plant Health Management Division within its three research orientations, Host Plant Resistance, Biological Control and Habitat Management. The research divisions were subsequently re-organized into projects, so the entire cowpea and soybean plant health activities were included into the Integrated Management of Legume Pests Project, and this project continued as a stand-alone entity even after IITA management abandoned the divisional structure.

More recently, with the advent of the CGIAR Research Programs (CRPs) in 2011–12, all grain legume plant health activities were integrated in the CRP on Grain Legumes led by the sister institution ICRISAT.

Parallel to these internal developments at IITA, grain legume plant health researchers had also begun establishing collaborative networks with NARS institutions. One of the first of these networks was RENACO (Reseau de Recherche sur le Niébé pour l'Afrique de l'Ouest et du Centre), established during phase II of the Semi-Arid Food Grains Research and Development Project in 1986, which was followed by the project Protection Ecologiquement Durable du Niébé (PEDUNE), a project focusing entirely on IPM for cowpea, with a strong Farmer Field School component. Later on, in 2000 RENACO and PEDUNE were merged into the Projet Niébé pour l'Afrique (PRONAF), which ended in 2006. In 2007, the Bean-Cowpea Collaborative Research Support Program, which was initially set up to link African NARS scientists with US land-grant universities, was expanded to include CG centres such as IITA and ICRISAT. This opened a new era of collaborative research, which has continued with success until today under the Feed the Future Legume Innovation Lab.

This chapter has self-contained sections on major biotic stresses attacking cowpea and soybean, the two current mandate legume crops of IITA, with detailed descriptions of the target organisms and how they affect cowpea production, and the impact of the research interventions by IITA and collaborators towards their sustainable management using an integrated plant health approach.

Published by Burleigh Dodds Science Publishing Limited, 2019.

2 Cowpea

Cowpea (*Vigna unguiculata*) Walp. (Leguminosae) is the most important grain legume crop in sub-Saharan Africa (SSA), grown by tens of millions of smallholders. West Africa, with over 10 M ha and a production of close to 6 M tons, holds 85% of the world production (FAOSTAT 2016). Cowpea is a protein-rich grain that complements staple starchy crops, and also provides fodder for livestock, soil improvement benefits through nitrogen fixation and household benefits in the form of cash and income diversity. However, on-farm grain yields in SSA reach only 10–30% of their biological yield potential, due primarily to biotic constraints such as insects and diseases (Ehlers and Hall 1997).

2.1 Foliar diseases

2.1.1 Anthracnose

A major foliar disease affecting cowpea crops is anthracnose caused by *Colletotrichum gloeosporioides* Penz. and other *Colletotrichum* spp. However, relatively minor emphasis has been given to this disease in our research programmes. Earlier studies have demonstrated the influence of maize pollen on both conidial germination of *Colletotrichum* spp. and higher fungal virulence when cowpea was intercropped with maize (Allen and Skipp 1982). Methods to evaluate disease resistance under field conditions were developed in the early 1990s (Adebitan et al. 1992). In parallel, the optimal vegetative stage to screen for resistance to anthracnose and the influences of genotypes and cropping practices on development of the disease were determined (Adebitan et al. 1993). Resistant varieties have been developed at IITA, for example TVx 3236, but, given the high variability of the pathogen (Emechebe and Florini 1997), other low-input control methods can be combined for efficient control, such as aqueous applications of papaya leaf extracts (Masangwa et al. 2013).

Further research is needed to better understand the interaction between isolates and genotypes of both anthracnose and cowpea bacterial blight (CoBB, see next section). There is the need to elucidate the role of insects as a pathway of anthracnose transmission, as discussed by Zandjanakou-Tachin et al. (2007) for CoBB. Opportunities for biological control as a component for a sustainable cowpea crop management should be also explored. Last but not least, cowpea land races have remained largely unexplored and provide valuable germplasm with important traits such as resistance to major diseases. We have identified several land race accessions with superior resistance to CoBB (A. Ortega-Beltran, *unpublished results*) and shall evaluate their susceptibilities to anthracnose and other diseases.

Published by Burleigh Dodds Science Publishing Limited, 2019.

2.1.2 Leaf blight and leaf spot

Other foliar diseases that affect cowpea are caused by *Ascochyta phaseolorum* Sacc., and *Septoria* leaf spot caused by *Septoria vignae* Henn. and *S. vignicola* V.G. Rao. There are few reports on these diseases from research programmes. However, our research has identified sources of resistance to *Septoria* leaf spot in cowpea genotypes, including genes Rsv 1 and Rsv 2 which are important for resistance to *S. vignae* (Abadassi et al. 1987), and elucidated the epidemiology of these diseases (Emechebe and Florini 1997). However, while multiple genotypes have been screened for other cowpea diseases (Williams 1977), there is scope for similar research to be conducted for *Ascochyta* leaf blight and *Septoria*, including newly developed cowpea genotypes, and the assessment of compatible control methods which can be applied by resource-poor farmers who can rarely afford the use of fungicides.

2.2 Soil-borne diseases

2.2.1 Charcoal rot

Macrophomina phaseolina (Tassi) Goid, the causative agent of charcoal rot of cowpea and many other crops, is a devastating fungal pathogen in many regions worldwide. Epidemics and yield losses occur preferably under dry conditions, when plants already suffer from abiotic stress. Single control measures are ineffective or not feasible under farmers' conditions because of the seed-, soil-borne and polyphagous nature of the fungus (Singh et al. 1990a). Integrated approaches to control the fungus include its accurate detection and characterization as well as biological control.

Nowadays, molecular methods are intensively used for the detection and characterization of *M. phaseolina* (Azarmanesh et al. 2011), but enzymatic and serological methods have also proved effective for its detection and quantification in plant tissues (Afouda et al. 2009). Thus, the pathogen was demonstrated as a latent infection in symptomless plants, causing epidemics under favourable (dry) conditions. Biological control has been promoted as a component of an integrated management approach especially under arid conditions. Bacterial antagonists isolated from soil samples collected from the rhizosphere of healthy as well as *M. phaseolina*-infected cowpea inhibited the growth and microsclerotia production by *M. phaseolina* on artificial media (Afouda et al. 2001). The most promising of these antagonists were identified as *Bacillus subtilis* Cohn, *Lysinibacillus sphaericus* Ahmed and *Paenibacillus polymyxa* Ash. They were able to reduce the incidence of *M. phaseolina* by 89% in greenhouse experiments. The antagonist-treated plants exhibited lower *M. phaseolina* colonization as confirmed by DAS-ELISA tests, and the antagonist *B. subtilis* A11 was rhizosphere competent with high population densities (Afouda et al. 2012). Therefore, seed coating with *B. subtilis* A11 in combination with other cultural practices such as irrigation and host plant resistance is recommended for the management of *M. phaseolina* under arid conditions (Afouda et al. 2012). More recent studies have confirmed the need for integrating antagonistic mycoparasites (e.g. *Clonostachys rosea* (Link) Schroers, Samuels, Seifert & Gams) with soil amendments such as compost, for effectively reducing *M. phaseolina* inoculum in the soil (Ndiaye et al. 2010).

2.2.2 Bacterial blight

Bacterial blight of cowpea (CoBB), caused by *Xanthomonas citri* pv. *vignicola* (Burkholder) Constantin (*Xcv*) (Constantin et al. 2016), former *X. axonopodis* pv. *vignicola* (Burkholder) Dye, is widespread in most areas where cowpea is cultivated. It causes heavy losses in susceptible genotypes when the infection occurs in an early stage of crop growth. Infected seeds cause pre-emergence and plant mortality up to 67% and 80%, respectively, and yield losses range from 3% to 92% according to cultivar and stage of infection (Kishun 1989).

A first research effort by IITA scientists was directed at clarifying the origin of CoBB in East Africa. In fact, at a certain point it was believed that the disease was introduced to Tanzania after IITA started exchanging seeds for research programmes in 1968. However, further investigations demonstrated than the disease was known in Tanzania prior to the establishment of IITA (Allen 1981). Other research efforts were geared towards developing inoculation methods to evaluate germplasm susceptibilities to *Xcv*, identify sources of resistance to CoBB (Allen et al. 1981; Okechukwu and Ekpo 2004; Sikirou et al. 2001), and detect genetic markers associated with resistance to *Xcv* (Agbicodo et al. 2010). Studies establishing relationships between seed to plant transmission of *Xcv* confirmed seed as a major inoculum source (Okechukwu et al. 2010). In parallel, integrated strategies to control CoBB that include treating seeds with bactericides, intercropping with maize or cassava, sowing pathogen-free seeds and cultivation of resistant germplasm were developed (Sikirou and Wydra 2008; Wydra et al. 2001).

Xcv strains are variable in their pathogenicity, virulence, multiplication and distribution in the plant and in seed infection. Most virulent strains induce systemic symptoms including wilting of non-inoculated leaves, and stem symptoms. In susceptible genotypes *Xcv* moves through the whole stem, while in less susceptible genotype it moves predominantly upwards from the inoculation point (Sikirou and Wydra 2011). Genotypes with high or low disease severity were able to produce non-contaminated seeds (Sikirou and Wydra 2011).

Xcv survived for up to 60 days when placed on the soil surface and up to 45 days when buried at depths of 10 and 20 cm. It survived at least 30 days in the field after spray inoculation on the weed species *Euphorbia heterophylla*

L. (Euphorbiaceae), *Digitaria horizontalis* Willd. (Poaceae) and *Synedrella nodiflora* (L.) Gaertn. (Asteraceae); 20 days on *Panicum subalbidum* Kunth (Poaceae); 10 days on *Euphorbia hirta* L. (Euphorbiaceae) and 5 days on *Talinum triangulare* (Jacq.) Willd. (Talinaceae). When it was leaf-infiltrated under glasshouse conditions, the pathogen was detected up to 90 days in the same weeds. Among 12 legume species tested as alternative hosts of *Xcv*, only *Sphenostylis stenocarpa* (African yam bean) showed typical symptoms of CoBB in a glasshouse experiment following artificial inoculation (Sikirou and Wydra 2004). The fact that *Xcv* was detected in the faeces, the intestines, and on the legs and mandibles of *Zonocerus variegatus* Linnaeus (Orthoptera: Pyrgomorphidae), from organs and faeces of two grasshopper species, bees and three leaf-feeding beetle species collected in bacterial blight-infected cowpea fields suggests possible insect transmission of *Xcv* (Zandjanakou-Tachin et al. 2007). This was confirmed when pathogen contaminated *Z. variegatus* initiated an epiphytic population of *Xcv* on healthy cowpea leaves sufficient to evoke symptoms.

CoBB can be managed by associating different methods including varietal resistance, sanitation and agronomic measures – crop association and sowing time (Wydra et al. 2001). Combining results of spray inoculation on 38 cowpea varieties under field conditions, in different ecozones and different years and leaf infiltration in the glasshouse, some 16 IITA and 6 local varieties from Benin and Togo were identified as resistant to *Xcv*, 9 as medium resistant genotypes and 7 as susceptible genotypes (Sikirou and Wydra 2011).

Intercropping cowpea with cassava (*Manihot esculenta* Crantz) or maize (*Zea mays* L.) as a measure to reduce CoBB showed that a moderate effect in reduction of disease incidence and severity was observed using alternate-row inter-planting, compared to monoculture at high density, though with variable results across years.

Compared with monoculture, cowpea yield was reduced by about 60% in intercropping with cassava and about 40% with maize. Although no stable effect of intercropping on disease reduction was observed, intercropping cowpea with maize or cassava 'in alternate rows' reduced bacterial blight in some cases and generally gives the farmer more yield security in case of failure of one crop. Therefore, intercropping as one element of an integrated control strategy for CoBB is recommended (Sikirou and Wydra 2008).

When the effect of sowing time was evaluated on CoBB development in the Forest Savanna Transition zone for three sowing times, sowing in May reduced disease severity by up to 77% depending on onset of the epidemic (at early inoculation time) and by 38% at late sowing time and additionally increased seed weight. When the disease appeared later in the season, late sowing induced a higher (57%) severity reduction compared to mid-sowing time (Table 1).

Table 1 Effect of sowing time on cowpea bacterial blight and yield of cowpea genotype IT84D-449 in the Forest Savanna Transition zone in 1997

		Sowing time		
Inoculation	dpi	Early (30 May)	Mid (15 June)	Late (30 June)
Early	15	10.02 a***	0.36 c	0.16 b
	30	0.08 a	0.51 c	0.27 b
	45	0.26 a	0.71 b	0.54 b
		0.12*	0.52	0.32
Seed weight (g)		24.22 b**	21.57 b	10.23 a
Late	15	0.29 b	0.22 b	0.05 a
	30	0.56 b	0.89 c	0.32 a
	45	0.90 b	1.19 c	0.64 a
		0.58*	0.76	0.33
Seed weight (g)		23.44 b**	23.64 b	8.82 a

* Mean severity over the test period.
** Mean seed weight of 40 plants.
*** log10 means of 80 plants.
dpi = days post inoculation.
Mean followed by the same letter in each line (dpi) are not significantly different according to *P* value of the difference of means by SAS Mixed Model procedure.

With regard to seed treatment, bacterial cells in artificially inoculated seeds were eradicated by hot water at 60°C for 30 min and 70°C for 10 min, and hot air at 65°C for 120 h and 70°C for 96 h, without negative effect on seed germination, while ethanol at 70% for 30 min did not eradicate the bacteria (R. Sikirou, *unpublished results*).

In conclusion, both *M. phaseolina* and CoBB are difficult to control due to their soil-borne nature. Only integrated measures comprising host plant resistance as the most important measure specifically for CoBB, use of healthy seeds, intercropping in a time pattern adapted to the ecozone and further elaboration of biological control options for charcoal rot can reduce disease pressure.

2.3 Nematodes

More than 51 species and 23 genera of nematodes have been reported to reproduce in cowpea (Sawadogo et al. 2009; Adegbite and Amusa 2010), although traditionally the main genera identified with yield losses have been *Meloidogyne*, *Pratylenchus* and *Scutellonema*. Among these, the root-knot nematode species *M. incognita* Chitwood, *M. javanica* Chitwood and *M. arenaria* Chitwood have been singled out as the most damaging ones causing yield losses in SSA that range from 25% to 94%, depending on the level of nematode inoculum in the soil (Caveness 1992). Nematode-infested plants are prone to suffer secondary diseases caused by fungi or bacteria (Caveness 1992) that penetrate into the damaged roots, causing multi-disease complex systems that are difficult to manage by farmers.

The most profitable, effective and sustainable method to manage root-knot nematodes overall is the use of resistant cultivars (Starr and Mercer 2009). A number of cowpea lines showed resistance (Adegbite et al. 2005; Caveness 1992; Fatokun et al. 2002; Swanson and van Gundy 1984), some with cross-resistance to additional pests (Singh et al. 1997). Resistance to root-knot nematodes lies in the *Rk* locus (*Rk*, *Rk2* and *rK3* genes) that has been used in conventional breeding programmes (Williamson and Roberts 2009). Contrary to other root-knot nematode resistance genes, the *Rk* locus is characterized by a late resistance response allowing nematodes to penetrate into the root tissue of the plant and establish a feeding site, although they will fail to maintain it and subsequently die (Das et al. 2008). A quantitative trait locus (QTL) co-localized with the genome region associated with *Rk* (*QRk-vu11.1*) has been recently identified (Huynh et al. 2016), representing a major improvement for future marker-assisted breeding programmes of cowpea. In spite of its effectiveness, continuous monocropping of root-knot nematode-resistant cowpea lines should be avoided, for it could lead to the selection of root-knot nematode virulent populations as observed with other root-knot nematode-resistant crops carrying *R*-genes (Verdejo-Lucas et al. 2009).

The impact of using cowpea itself as a rotational crop to manage plant parasitic nematodes has been mainly assessed in the Sahelian cereal-legume farming systems, but the results presented in literature are not conclusive and vary widely among studies (Franke et al. 2018). This disparity could have been caused either by the different responses of the cowpea lines studied against the nematodes (e.g. susceptibility, tolerance, resistance), or due to the different nematode species present in the soil. Throughout, the use of root-knot nematode-susceptible cowpea cultivars in such systems should be avoided (Riekert and Henshaw 1998). Other technologies that have been used to control root-knot nematodes in cowpea are crop residues, green amendments with marigold, *Tagetes erecta* L. (Asteraceae) and sunn hemp, *Crotalaria juncea* L. (Leguminosae) (Adekunle 2011). Yard-waste compost amendments, however, showed limited effect over *M. incognita* and other plant parasitic nematodes infecting cowpea after one cropping season (McSorley and Gallaher 1996). *In vitro* application of foliar sprays of salicylic acid has been reported to induce cowpea's resistance against root-knot nematode (Nandi et al. 2003), and the use of *Bacillus subtilis* Cohn, *B. thuringiensis* Berliner and *B. cereus* Frankland & Frankland applied as a seed coat in cowpea significantly reduced the root galling caused by *M. javanica* (Dawar et al. 2008).

Published by Burleigh Dodds Science Publishing Limited, 2019.

In conclusion, the deployment and access to root-knot nematode-resistant varieties will be crucial to minimize yield loses and reduce the use of synthetic pesticides. Awareness must also be created among rural communities and national extension services, regarding plant parasitic nematodes and the importance of sustainably managing them. Future research efforts should aim at developing IPM packages for cowpea growers, where root-knot nematode-resistant varieties should be combined and integrated with additional cultural and biological methods for root-knot nematode suppression.

2.4 Insect pests

In the savannas of West Africa, where most of the cowpea is grown, insect pests alone can cause up to 80% yield loss (Singh et al. 1990b). Among them, the most economically important ones are the cowpea or groundnut aphid, *Aphis craccivora* Koch (Homoptera: Aphididae); the flower thrips, *Megalurothrips sjostedti* Trybom (Thysanoptera: Thripidae); the flower and pod borer, *Maruca vitrata* Fabricius (Lepidoptera: Crambidae); and a complex of pod sucking bugs dominated by *Clavigralla tomentosicollis* Stål (Heteroptera: Coreidae) (Singh et al. 1990b). Bruchids, particularly *Callosobruchus maculatus* Fabricius (Coleoptera: Bruchidae) have for a long time been the most damaging post-harvest insect pests in stored cowpea, for which nowadays a simple non-chemical solution exists based on hermetic storage (Purdue Improved Cowpea Storage, PICS-bags), which is currently being implemented in most of Africa. Adoption rates of up to 64% have been reported from Northern Nigeria (Abdoulaye et al. 2012). This success has been widely documented elsewhere (Moussa et al. 2011; Baoua et al. 2012) and will not be reported in detail in this chapter.

In spite of a wealth of indigenous knowledge to diagnose and manage insect pests in the cowpea cropping systems (e.g. Kitch et al. 1997), farmers mostly rely on synthetic insecticides for their control. Synthetic pesticides have the advantage of showing an immediate response and plant health improvement which can be easily appreciated by the low-literate farmer (Singh et al. 1990b). However, adverse side effects from the indiscriminate use of synthetic pesticides have started to emerge even in rural areas of West Africa (e.g. Diabate et al. 2002). Among them, those impacting on human health are the most pressing and critical. The use of insecticides for cowpea pest control is certainly one of the most critical issues for implementing IPM approaches, particularly because spraying early in the cropping season can wipe out the very fragile hymenopteran parasitoids which further prevented from impacting on pest populations later during the season, leading to pest outbreaks during the yield formation process, as is well documented in other cropping systems (Heong et al. 1992).

As a response to these constraints, throughout the years IITA has advocated a holistic plant health approach (Jackai and Adalla 1997), starting with the prevention of yield losses through host plant resistance and conservation of biological control. However, in most cases further interventions are required to keep the insect damage below the economic threshold, such as improved resistance (including the use of GMOs), habitat management, the use of bio-pesticides or the introduction of more efficient natural enemies. In the subsequent sections, we highlight critical aspects of managing individual pests, while our ultimate target should be the deployment of a comprehensive pest management package addressing all key pest species, thereby assuring complementarity and avoiding negative interferences when single options are combined.

2.4.1 Aphids

Aphis craccivora Koch (Homoptera: Aphididae) is among the most widespread and destructive insect pests of cowpea. Aphids are particularly active during dry spells, and feed on young leaves, stem tissues, growing tips and petioles resulting in stunting, leaf distortion, premature defoliation and death of seedlings (Abate and Ampofo 1996; Togola et al. 2017). They can also attack flowers and fresh pods. Severe damage comes from the transmission of cowpea aphid-borne mosaic virus (Ofuya 1997). Yield loss due to aphid attack reaches 50% and more when high infestations of cowpea aphids are not controlled or in cases of transmission and spread of virus even at low population densities (Obopile and Ositile 2010; Togola et al. 2017). This aspect, however, is presented in Chapter 5: Legg et al. dealing with viruses.

Research undertaken at IITA to manage aphids focused mainly on host plant resistance, an environmentally-friendly, cost-effective and sustainable pest management option for minimizing the pests' incidence and severity (Togola et al. 2017). Many accessions with moderate to good levels of resistance were disseminated for farmers' use, TVu3000 being the most promising (Singh et al. 1982). Breeding efforts in GLIP and national programmes during the 1990s resulted in additional resistant lines such as IT84S-2246-4 and KVx-426-1 (Jackai and Adalla 1997; Ofuya 1993, 1997).

The genetic basis of the resistance in these lines was established to be monogenic conferred by single dominant genes (Bata et al. 1987; Pathak 1988; van Emden 1991; Ofuya 1997). In addition, some antibiosis factors in cowpea were identified as responsible for aphid resistance (Ombakho et al. 1987; Bata et al. 1987; Pathak 1988) and plant compounds such as phenols and flavonoids, namely quercetin, kaempferol and isorhamnetin, induced a deterrent effect on feeding by *A. craccivora* (MacFoy and Dabrowski 1984; Lattanzio et al. 2006).

Published by Burleigh Dodds Science Publishing Limited, 2019.

However, as it often happens with single gene resistance, it was soon overcome by new biotypes of the pest. For instance the single dominant gene in TVu-3000 became no longer effective (Bata et al. 1987; Pathak 1988). Also, concerted attempts to cross wild *Vigna* species, for example *Vigna vexillata* (L.) A. Rich and *V. oblongifolia* (L.) A. Rich (Leguminosae), having high levels of resistance to various insect pests with cultivated cowpea remained largely unsuccessful (Fatokun et al. 1997; Jackai et al. 2001; Fatokun 2002). IITA therefore pursued research in host plant resistance in order to identify genotypes with a more robust basis of resistance to different aphid biotypes. Effective phenotyping protocols were developed under semi-field conditions (screen cages and screen-houses) where several hundred accessions could be evaluated in a short time (Souleymane et al. 2013). Moreover, new breeding approaches such as marker-assisted recurrent selection were deployed to enable faster and more targeted progress.

For many years, the only biocontrol agents observed attacking aphids were predators, mainly coccinellids and syrphids (Ofuya 1997). No hymenopteran parasitoids had ever been recorded from aphids in West Africa until two of them started to appear serendipitously in most countries around 2008. *Aphidius colemanii* Viereck (Hymenoptera: Aphidiidae) was the first to appear, with spectacular parasitism rates, but its activity was very soon disrupted by the local hyperparasitoid *Syrphophagus africanus* Gahan (Hymenoptera: Encyrtidae) and subsequently displaced by the second parasitoid *Lysiphlebus testaceipes* Cresson (Hymenoptera: Aphidiidae) (Saethre et al. 2011; Tepa-Yotto et al. 2013), which was less prone to hyperparasitism (Ganyo et al. 2012). The extent to which *L. testaceipes* is effectively controlling *A. craccivora* has never been assessed, but should be the object of further investigations in the different agro-ecologies where aphids still are a recurrent problem. Because aphids are seldom the target of insecticidal sprays if appearing alone early during the cropping season, it is reasonable to assume that combining better sources of resistance/tolerance with biological control by *L. testaceipes* will provide substantial reduction of yield losses.

2.4.2 Thrips

The key thrips species attacking cowpea is the flower bud thrips, *Megalurothrips sjostedti* Trybom (Thysanoptera: Thripidae), which can cause total yield loss in case of high populations attacking plant reproductive organs such as flower buds and flowers, causing their distortion and abortion (Omo-Ikerodah et al. 2009). Other flower thrips species encountered on cowpea include *Frankliniella schultzei* Trybom, but they have never been reported causing substantial damage. However, leaf thrips *Hydatothrips adolfifriderici* Karny (syn. *Sericothrips occipitalis* Hood) can induce severe leaf damage leading to defoliation on off-season cowpea (e.g. seed multiplication plots) and during long dry spells (Ezueh 1981).

For many years, control options against the flower thrips consisted of host plant resistance and insecticidal applications. The first cowpea cultivar with moderate thrips resistance was TVu 1509, which was used as a donor parent in breeding programmes for several years (Singh 1977; Salifu et al. 1988). It is only in the late 1990s during the PEDUNE regional project that other thrips-resistant lines were identified from local accessions, particularly Sanzisabinli from Ghana and Sewe from Benin. The main conclusion from these studies was that these local resistant lines were displaying high levels of terpenoids and flavonoids in their floral organs (flower buds, peduncle cushion and flowers). These compounds were associated with high thrips larval mortalities, suggesting that these findings could be further exploited for enhancing resistance to flower thrips (Alabi et al. 2004, 2011). Recently, new sources of resistance to *M. sjostedti* were found in some IITA cowpea mini-core genotypes under field conditions (Togola et al., 2018). So far, no QTLs for resistance to flower thrips have become available. However, three QTLs for resistance to foliar thrips were reported by Muchero et al. (2010) and Lucas et al. (2012).

With the invention of the Electrodyn sprayer in the 1970s, and the use of its custom-made 'bozzle' containing Cymbush Super ED (Cypermethrin and Dimethoate), it was possible to keep cowpea fields free of insect pests including thrips, with substantial yield gains and minimal environmental disruption (Tamò and Baumgärtner 1993). However, because of its peculiar business model failing to take into consideration local socio-economic factors (high cost of purchasing the equipment and the sealed, non-refillable 'bottle-and-nozzle' pesticide container meaning that it could only be sourced from the original company), the Electrodyn sprayer was never adopted and was soon phased out while conventional pesticide applications are still being practised now. On the other hand, national programmes involved in the PEDUNE project had started to substitute synthetic pesticides with botanical bio-pesticides, mainly using aqueous formulations of neem grains, neem leaves and papaya leaves, the latter being particularly active against thrips (Dabiré et al. 2012).

It was only in the late 1980s that flower thrips started to be the object of comprehensive studies involving the whole ecosystem, including modelling approaches (Tamò and Baumgärtner 1993; Tamò et al. 1993a). The same studies also investigated the natural enemy complex of *M. sjostedti*, which led to the discovery of hymenopteran egg and larval parasitoids, hitherto unreported in West Africa (Tamò et al. 1993b). Unfortunately, none of these newly discovered parasitoids was found to be able to control thrips populations because of the poor adaptation to the target host. This evidence spurred the search for better adapted parasitoid species known to attack other *Megalurothrips* species worldwide, which culminated in the discovery of an *M. sjostedti*-specific strain of *Ceranisus femoratus* Gahan (Hymenoptera: Eulophidae) in Cameroon (Tamò et al. 2003). After obtaining release permits from the regulatory authorities, this strain was released on nursery plots of *Tephrosia candida* (Roxb.) (Leguminosae) in Benin, Ghana and Nigeria. It became established and started to colonize other key *M. sjostedti* wild host plants, all Leguminosae, such as *Lonchocarpus sericeus* (Poir.) DC., *Philenoptera cyanescens* (Schumach. & Thonn.) Roberty, *Pterocarpus santalinoides* DC. and *Millettia thonningii* (Schumach.) Baker. Until today, *C. femoratus* continues to keep flower thrips populations under control (Fig. 1), albeit confined to areas close to patches of these wild host plants, mostly in hydromorphic areas and along water bodies (Tamò et al. 2012).

Among the critical issues in implementing management options against flower thrips remain the interferences of chemical pesticides used to control other insect pests attacking cowpea flower buds and flowers such as the legume pod borer *Maruca vitrata* (see next subsection). Also, we would need to be able to combine the available, moderately resistant varieties with a biological control approach enhanced by habitat management, for example for improving the dispersal range of *C. femoratus* beyond the hydromorphic areas. This could be achieved by establishing and/or protecting wild host plant habitats serving as refugia for thrips parasitoids. In case of high pest population outbreaks, host plant resistance and biological control could be complemented by punctual applications of bio-pesticides (e.g. emulsifiable neem oil mixture, Sokame et al. 2015).

With regard to measuring the impact of thrips management options, because of the peculiarity of the cowpea insect pest diversity, farm-level impact of control interventions cannot be assessed for an individual pest, but would

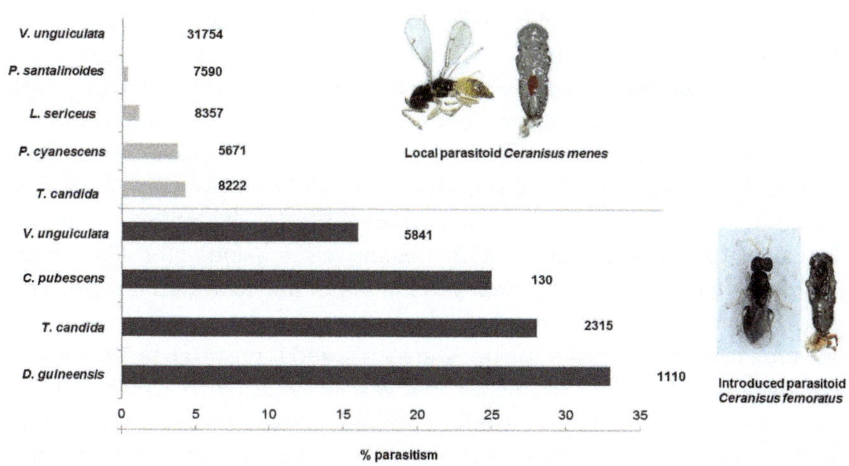

Figure 1 Percentage parasitism of flower thrips (*Megalurothrips sjostedti*) larvae inflicted by the local parasitoid *Ceranisus menes* (light grey bars) compared to that of the introduced parasitoid *Ceranisus femoratus* (dark grey bars). The numbers close to the bars represent the total of *M. sjostedti* larvae reared from the flowers of each host plant.

Published by Burleigh Dodds Science Publishing Limited, 2019.

need to be validated through sophisticated exclusion experiments under controlled conditions. Unfortunately, these studies are very costly and the donor community largely prefers to invest in scaling out technologies rather than validating the ones coming out of the development pipeline. This is also true for all other key cowpea insect pests and this will be valid throughout the entire cowpea insect section. Hence, impact of control measures will have to be assessed in the context of IPM for the whole range of pests attacking cowpea (Togola et al. 2017; Agunbiade et al. 2018), see also Chapter 15: Coyne et al.

2.4.3 Maruca *pod borer*

At the onset of flowering, cowpea reproductive organs (flower buds, flowers and pods) are attacked by caterpillars of lepidopteran pests, of which the legume pod borer *Maruca vitrata* Fabricius (Lepidoptera: Crambidae) is the most economically important throughout tropical Africa. Other, minor or sporadic lepidopteran attacking flowers and pods include two lycaenids, but they are seldom reported to cause economic damage. If cowpea crops are left unprotected in areas of high population pressure by *M. vitrata*, yield losses can reach over 80% (Singh et al. 1990b). In these areas, cowpea farmers very often resort to use of various chemical insecticides to control *M. vitrata*. However, apart from the negative unintended effects described in the introduction, the inappropriate application of insecticides has already led to the appearance of insecticide resistance in *M. vitrata* in Nigeria (Ekesi 1999).

Given its devastating pest status across West Africa, finding resistant varieties to the pod borer became one of the major endeavours of the late 1970s. Two varieties exhibiting field resistance/tolerance to the pod borer were identified as TVu 946 and TVu 4557. Both varieties had peduncles elongating above the canopy, and the pods were attached at a wide angle, preventing *M. vitrata* caterpillars from boring into the pods (Oghiakhe et al. 1991a, 1992a). Other studies elucidated the impact of physical resistance to *M. vitrata* caterpillar feeding (Oghiakhe et al. 1991b). However, damage on flower buds and flowers could not be prevented by this type of tolerance (Singh et al. 1990b). Unfortunately, after extensive screening of the IITA cowpea germplasm, it appeared quite clearly that none of the cultivated cowpea lines would possess any kind of useful resistance to the pod borer. However, some hope arose when wild *Vigna* species were screened for resistance (Jackai et al. 1996; Bottenberg et al. 1998), including *V. vexillata* which displayed high levels of resistance to several pests including *M. vitrata* (Jackai and Oghiakhe 1989; Oghiakhe et al. 1992b; Fatokun et al. 1997; Oigiangbe et al. 2001). Sadly, sustained efforts to cross some of these promising wild species with cultivated cowpea remained largely unsuccessful (Fatokun 2002). Later on, scientists at the Commonwealth Scientific and Industrial Research Organisation succeeded in transforming cowpea with a gene construct encoding Cry1Ab protein from *Bacillus thuringiensis*, which was highly toxic to *M. vitrata* (Popelka et al. 2006). This initiative involved mainly scientists from national programmes in Burkina Faso, Ghana, Malawi and Nigeria, but virtually no direct participation by IITA. However, IITA recently became involved in a couple of studies addressing environmental risk assessment of *Bt*-cowpea (Huesing et al. 2011) and insect resistance management (Onstad et al. 2012). Once available to farmers, *Bt*-cowpea will be a powerful tool to fight the pod borer in the field, although there remain issues to be solved with respect to the logistics of producing and distributing a huge quantity of seeds, getting the farmer to buy them (against the cultural notion that 'a good farmer is the one that can store her/his own seeds') and taking up the challenge of implementing a proper resistance management plan easily understandable by farmers.

As soon as it became obvious that conventional host plant resistance alone would not be able to solve the pod borer problem in cowpea, investigations towards the use of bio-pesticides were intensified, starting with botanical extracts. Neem (*Azadirachta indica*) oil (Jackai and Oyediran 1991; Tanzubil 2000) and leaf extracts (Bottenberg and Singh 1996) were reported to effectively control *M. vitrata* (Jackai and Oyediran 1991; Tanzubil 2000), also by reducing *M. vitrata* feeding on cowpea flowers and pods (Jackai et al. 1992). Moreover, farmers had indicated that aqueous preparations of the African bush mint *Hyptis suaveolens* would protect cowpea from insect pests, including *M. vitrata* (Adetonah et al. 2005). While cowpea farmers valued botanical bio-pesticides for being safe and cheap, they mostly complained about the tedious preparation process, which consumes precious time during the cropping season. They would be most happy if they could buy them off-the-shelf, even at a price higher than chemical pesticides (Agyekum et al., *unpublished results*). Thus, the availability of bio-pesticides remains an issue which can only be solved by engaging community-based organizations and/or the private sector in their commercial production. This is also true for another category of bio-pesticides, based on entomopathogenic microorganisms such as fungi and viruses. Among them, the fungus *Beauveria bassiana* (Balsamo) Vuillemin was tested with some success in Benin, and proved to be as efficient as standard pesticides (Mehinto et al. 2014). Another interesting biopesticide was the specific baculovirus *Maruca vitrata* Multi-Nucleopolyhedrovirus (*Mavi*MNPV) discovered by AVRDC (now WorldVeg) in Taiwan (Lee et al. 2007), and which was introduced into Benin for confined testing on *M. vitrata*. Afterwards, it was dispatched for multi-locational field testing in several West African countries, with encouraging results and mortality rates reaching >95% on young larval instars (Tamò et al. 2012). As observed by Sokame et al. (2015), mixing *Mavi*MNPV with emulsifiable neem oil resulted in suppression of *M. vitrata* populations and subsequent yield increases comparable to standard chemical insecticides.

Published by Burleigh Dodds Science Publishing Limited, 2019.

But perhaps the most fascinating and critical aspect of research carried out to control *M. vitrata* was related to its pest status, particularly with regard to its true origin, and the associated natural enemies biodiversity. For many years, *M. vitrata* has been implicitly labelled as an indigenous pest in Africa, simply because it was already there and causing damage before scientists started to work on it. However, as we started to examine its ecology in more detail, it appeared that the locally available hymenopteran parasitoids attacking its different life stages in West Africa were quite unspecific and therefore inflicting low mortality rates, including in stable ecosystems such as natural host plant habitats (Arodokoun et al. 2006). This fact was quite surprising for organisms thought to have coevolved for ages even before cowpea was domesticated in West Africa, and prompted the idea of investigating the natural enemies' biodiversity in the suspected area of origin of *M. vitrata*, tropical Asia, as earlier suggested by Tamò et al. (1997) based on observations by Kalshoven and van der Vecht (1950). Indeed, a biodiversity assessment in Taiwan allowed the identification of a first set of hymenopteran parasitoids, of which *Apanteles taragamae* Viereck (Hymenoptera: Braconidae) looked to be the most promising candidate (Huang et al. 2003). In Taiwan, this parasitoid was able to inflict up to 60% parasitism on *M. vitrata* caterpillars feeding on leaves of a common green manure legume, *Sesbania cannabina* (Retz.) (Srinivasan et al. 2015). However, in spite of encouraging parasitism rates obtained in the laboratory and screenhouse assays in Benin, small-scale releases of this parasitoid in Benin, Burkina Faso, Ghana and Niger (after obtainment of import and release permits by the respective national regulatory authorities) did not result in substantial establishment. Unfortunately, *A. taragamae* seemed to be poorly adapted to major wild host plants for *M. vitrata* such as *Lonchocarpus* spp., *Pterocarpus santalinoides* and *Milletthia thonningii* (Dannon et al. 2012). In the meantime, population genetic studies comparing worldwide strains of *M. vitrata* (Periasamy et al. 2015) confirmed earlier assumptions that South East Asia was indeed the putative area of origin of *M. vitrata*, which justified renewed efforts to study natural enemies of *M. vitrata* beyond Taiwan. Two of the natural enemies already observed by Kalshoven and van der Vecht (1950) controlling *M. vitrata*, *Phanerotoma philippinensis* (Ashmead) and *Bassus janvanus* (Bhat & Gupta) (Hymenoptera: Braconidae) could easily be recovered from continental Asia, albeit under a revised taxonomy, namely as *Phanerotoma syleptae* and *Therophilus javanus* (Srinivasan et al. 2015). This time, bio-assays and experimental releases (again after obtaining of import and release permits by the respective national regulatory authorities) involving a total of 123 900 parasitoids (Table 2) indicated good ecological adaptation to the native host plant habitat in West Africa. The releases were carried out together with farming communities and local authorities (Fig. 2), and were preceded by a wide sensitization campaign (Fig. 3).

Table 2 Total number of exotic parasitoids (*Therophilus javanus* and *Phanerotoma syleptae*) released in Benin, Burkina Faso and Niger in 2016 as biocontrol agents against the legume pod borer *Maruca vitrata*

Country	Parasitoid	# released
Benin	*Therophilus javanus*	41 600
Benin	*Phanerotoma syleptae*	36 300
Burkina Faso	*Therophilus javanus*	23 000
Burkina Faso	*Phanerotoma syleptae*	18 500
Niger	*Therophilus javanus*	3000
Niger	*Phanerotoma syleptae*	1500
Total		123 900

Both parasitoid species were successfully established in Benin and Burkina Faso (Tamò et al. 2017). While this is still work in progress and will need a proper impact assessment to compare previous *M. vitrata* population levels with those observed after releases, light trap catches of *M. vitrata* adults in an area of very

Figure 2 Former Director General of INRAB (Benin National Agricultural Research Institute), Dr David Arodokoun, releasing parasitoids of the legume pod borer *Maruca vitrata* using a newly developed collapsible cage. The collapsible cage contains up to 1500 parasitoids, which are allowed to fly away immediately after opening all walls of the cage (as illustrated in the detailed picture on the lower right corner).

Published by Burleigh Dodds Science Publishing Limited, 2019.

Figure 3 Sensitization campaign in rural communities of Mono Department (Benin) using pictorials done by local artists, and mounted on plywood boards, to be displayed like 'wooden PowerPoint slides' and explained in local language.

high infestation (Pénéssoulou, Central Benin) have been consistently much lower for the past two years after initial releases (B. Datinon, *unpublished data*), indicating a possible reduction of *M. vitrata* populations. This initial success in biological control against the pod borer needs to be seen in the light of a comprehensive IPM approach, which can be extended to similar pests in other cropping systems. Biological control agents will lower the overall population pressure on alternative host plants, both during the off season when cowpea is not planted, and also on cowpea and surrounding host plants during the cropping season. As mentioned earlier in this section, farmers might soon have access to *Bt*-cowpea which will protect the crop directly in the field. Farmers not having access to *Bt*-cowpea will still have the choice of applying biorational pesticides such as combinations of emulsifiable neem oil with pod borer-specific entomopathogens, which both are not likely to interfere severely with biological control by the Asian parasitoids. To empower low-literacy farmers to make informed decisions about pod borer control, a Farmer Interface App (FIA) has been developed in collaboration with Michigan State University and its Scientific Animations Without Borders programme. The prototype FIA has been field-tested in Benin and proved robust enough to be used on simple smartphones (Agunbiade et al. 2018). The challenge remains to develop this

FIA into a fully operational App, by providing real-time access to climate and phenology data and models as developed by the VIPS platform (https://www.vips-landbruk.no/information/1/) of the Norwegian Institute of Bioeconomy Research (NIBIO).

2.4.4 Pod sucking bugs

Pod sucking bugs are post-flowering insect pests of cowpea in tropical Africa of which *Clavigralla tomentosicollis* Stål., *Anoplocnemis curvipes* Fab. (Heteroptera: Coreidae) and *Riptortus dentipes* Fab. (Heteroptera: Alydidae) are the most devastating species (Jackai and Daoust 1986; Jackai and Singh 1988; Pitan and Odebiyi 2001; Oparaeke et al. 2005). Nymphs and adult bugs damage the seeds by injecting digestive enzymes (Soyelu et al. 2007) and sucking nutrient from the developing seeds, which lose their germination ability (Aina 1975; Akingbohungbe 1982; Togola et al. 2017). The uncontrolled infestation of these insects on cowpea causes losses in the magnitude of 80% and 95% of pod and seed yield, respectively (Egho 2011).

The use of synthetic insecticides such as dimethoate and cypermethrin is the most effective option for controlling the pod sucking bugs complex (Jackai and Ndlovu 1988; Asante et al. 2001). However, the use of synthetic pesticides is subject to high environmental and human health concerns and undesired socio-economic impact (Tamò et al. 2003), which call for a more sustainable, integrated plant health approach.

Host plant resistance to pod sucking bugs was the first option to be embraced at IITA. To date, some cultivated varieties such as TVu-1890 (Jackai and Singh 1988) and IT86D-716 (Dabiré-Binso et al. 2010) were found with operational levels of resistance to pod sucking bug attacks. Also, some wild cowpea relatives such as TVNu-151 displayed high levels of resistance (Koona et al. 2002). The mechanism of resistance to pod sucking bugs relies essentially on biochemical compounds such as cyanogenic heterosides, flavonoids, tannins and trypsin inhibitors (Dabiré-Binso et al. 2010). Similarly, MacFoy et al. (1983) found that cyanogenic heterosides and flavonoids displayed larvicidal activity against *C. tomentosicollis*. Additionally, some plant biophysical characteristics associated with non-preference proprieties were found to be involved in resistance against the pod sucking bugs complex. For instance, the dense hairs found on different parts of the wild cowpea *V. vexillata* (Oghiakhe et al. 1992b), and the dense and long trichomes on some cowpea cultivars were found to increase their resistance to pod sucking bugs damage (Oigiangbe et al. 2002, 2006; Togola et al. 2017).

Nowadays, 373 mini-core accessions from the IITA cowpea collection are under field evaluation for resistance to *C. tomentosicollis* (A. Togola et al., *pers. comm.*). Resistance in identified lines will be confirmed in screen cage and laboratory screening and genetic studies will be initiated to determine if there is allelism with earlier reported resistant lines. Data generated on the cowpea mini-core lines would be used in genome-wide association studies to identify QTLs associated with pod sucking bugs resistance. Transfer of these genes to improved cowpea lines remain a promising prospect for IITA and its partners.

On the other hand, some potential protein candidates such as lectin extracted from African yam bean (Omitogun et al. 1999; Okeola et al. 2002), *para*-aminophenylalanine (PAPA) extracted from *V. vexillata* and cysteine protease inhibitor (E-64) (Soyelu et al. 2007) are known to inhibit the development of *C. tomentosicollis* and can be explored using transgenic methods for developing resistant cowpea lines against this pest.

Among the natural enemies attacking *C. tomentosicollis*, the egg parasitoid *Gryon fulviventris* (Crawford) (Hymenoptera: Scelionidae) seems to be the most interesting one. In fact, field parasitism of egg masses can often reach over 90%. However, there seems to be a disconnect between the population dynamics of the pest and its parasitoid. In fact, Asante et al. (2000) described a slow response of the parasitoid to the migrating population of *C. tomentosicollis* in cowpea fields, which was confirmed by recent field studies in Burkina Faso. Field parasitism data suggest that the parasitoids only start to become active on egg masses oviposited by *C. tomentosicollis* females of the second generation, that is they cannot efficiently recognize egg masses laid by the first invading wave which is mainly composed of migrating gravid females (C. Dabiré et al., *unpublished data*). The nymphs and adults evolved from the first generation escaping egg parasitism will thus feed in high numbers on developing pods, causing devastating damage. The reasons why this first-generation eggs are able to escape parasitism can be explained by the host-finding mechanism. As known from other studies involving pod sucking bugs, egg parasitoids have been observed exploiting cues from male aggregation pheromones to locate fresh egg masses (Lim and Mainali 2013), so we have good reason to hypothesize that this is also the case for *C. tomentosicollis*. Preliminary olfactometric studies have shown that *G. fulviventris* is indeed attracted by caged *C. tomentosicollis* males, and these results have now been confirmed by more detailed chemical ecology studies (H. Kpongbe et al., *unpublished results*). Hence, because the migrating, invasive *C. tomentisicollis* population at the onset of podding in cowpea is essentially composed of gravid females, there are no males around to produce any aggregation pheromones. The critical research question here is whether we can engineer a system whereby the egg parasitoids are attracted earlier in the season by the synthetic male aggregation pheromone and attack first-generation egg masses. Coupled with the existing levels of field tolerance, and hopefully with novel resistance genes, this approach could easily be integrated with control strategies proposed for the other insect pests in the previous sections.

3 Soybean

Soybean (*Glycine max*) Merr. (Leguminosae) cultivation in SSA started to boom in the 1970s with the advent of adapted varieties and the use of Rhizobia inoculation. Today, it is grown on an area of 1.8 M ha producing close to 2.5 M tons. Unlike in other areas of the tropics, in Africa soybean is rarely attacked by insect pests, but can suffer heavy losses from plant pathogens (Wrather et al. 1997). In this chapter, we will focus on one major foliar disease affecting soybean, soybean rust.

3.1 Foliar diseases

3.1.1 Soybean rust

The Asian soybean rust (ASR), *Phakopsora pachyrhizi* (Uredinales: Phakosporaceae) can cause yield losses of up to 80% (Levy et al. 2005) due to premature defoliation, early maturation of plants and reduced seed number and size (Kumudini et al. 2008). ASR was first identified in Japan in 1902 and has so far spread to almost all soybean growing regions in the world due to the ease by which the spores are carried by wind currents. Although ASR was observed in Africa as early as 1978 (Javaid and Ashraf 1978), the first confirmed report was in Uganda in 1996, followed by Nigeria in 2001 (Akinsanmi et al. 2001), Ghana (Bandyopadhyay et al. 2007) and the Democratic Republic of Congo (Ojiambo et al. 2007), and so far, it has been reported in more than ten African countries (Murithi et al. 2016) (Fig. 4).

The rapid spread of *P. pachyrhizi* and its potential for severe yield losses and complete crop destruction makes ASR the most destructive foliar disease of soybean. In areas where ASR is endemic, several fungicide sprays are required to manage the disease. The pathogen is highly variable and different races have been reported. There have been instances of resistance being overcome by new ASR races in India, Australia, Brazil and Taiwan thus making breeding for resistance a continuous process to stay ahead of population shifts in the pathogen (Hartman et al. 2011).

As a rapid response to the importance of ASR when the disease was first reported in Uganda and anticipating its spread to other African nations, more than 300 IITA soybean breeding lines from West Africa were evaluated in Uganda and nearly all were susceptible to the disease. In Nigeria, all released varieties and breeding lines succumbed to ASR after its invasion. In 2003, IITA began an extensive West Africa-focused research programme in partnership with the US Department of Agriculture – Agricultural Research Service (USDA-ARS) to understand the disease risk, elucidate ASR epidemiology, identify sources of resistance and determine the virulence and genetic diversity of the pathogen in Nigeria. Studies to characterize pathogenic variability and molecular diversity

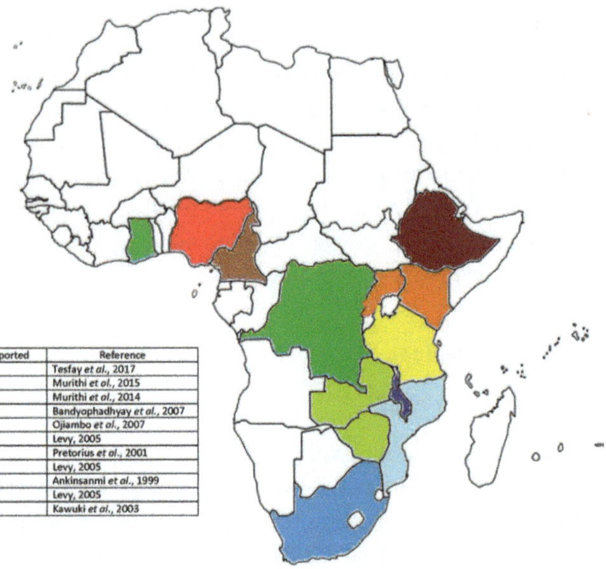

	Country	Year Reported	Reference
	Ethiopia	2017	Tesfay *et al.*, 2017
	Malawi	2014	Murithi *et al.*, 2015
	Tanzania	2014	Murithi *et al.*, 2014
	Ghana	2007	Bandyophadhyay *et al.*, 2007
	DR Congo	2007	Ojiambo *et al.*, 2007
	Cameroon	2003	Levy, 2005
	South Africa	2001	Pretorius *et al.*, 2001
	Mozambique	2000	Levy, 2005
	Nigeria	1999	Ankinsanmi *et al.*, 1999
	Zambia, Zimbabwe	1998	Levy, 2005
	Kenya, Rwanda, Uganda	1996	Kawuki *et al.*, 2003

Figure 4 Occurrence of soybean rust (*Phakopsora pachyrhizi*) in Africa.

of *P. pachyrhizi* in Nigeria revealed the occurrence of seven pathogenic groups with one being more virulent than others, and the presence of a more diverse pathogen population in the Derived Savanna zone (Twizeyimana et al. 2009, 2011b). Our epidemiological studies showed that temperature, duration of leaf wetness and solar radiation affected severity of ASR. These studies further demonstrated that early planting during the rainy season helped susceptible cultivars escape the disease and could hence be included in an integrated disease management approach (Twizeyimana et al. 2011a).

One of the most common ASR control practices in intensive soybean cultivation (e.g. the United States and Brazil) is the application of fungicides. However, their use significantly increases the cost of production, poses health and environmental risks and can result in fungicide resistance due to application of multiple sprays of only one group of fungicides. Fungicides may also not be easily accessible to many smallholder farmers in an African farming context. Despite this, fungicides are widely used by large-scale farmers especially in the southern Africa region. Fungicide resistance has been reported in South America and efforts are now directed towards developing broad-spectrum fungicide (Godoy et al. 2016). There is limited information on fungicide resistance to soybean rust in Africa except in South Africa, therefore future studies should focus on evaluating the efficacy of the available fungicides in order to guide growers in selecting the right combinations to avoid the risk of fungicide insensitivity.

However, recognizing that the use of resistant cultivars was the most cost-effective method of ASR control for smallholder farmers, major efforts were directed to the development and release of ASR-resistant cultivars. At first, techniques to evaluate disease resistance in the laboratory (detached leaf method), greenhouse and field were developed (Twizeyimana et al. 2007). Subsequently, 178 IITA soybean breeding lines and 101 putatively resistant accessions from the USDA Soybean Germplasm Collection (Urbana, IL, USA) and the National Agriculture Research Organization (Uganda) were evaluated to identify promising resistant lines (Twizeyimana et al. 2008). One breeding line (TGx 1835-10E; released by Makerere University as MAKSOY-1 for general cultivation in Uganda) and three accessions had high levels of resistance and were considered as useful sources of ASR resistance genes (Twizeyimana et al. 2008). IITA breeding lines were crossed with ASR-resistant sources to develop ASR-resistant cultivars that are now widely used in various national breeding programmes in Africa (Tukamuhabwa and Maphosa 2012; Kawuki et al. 2004). Through this initiative, seven ASR-resistant cultivars have also been released by the National Plant Variety Release Committee in Nigeria and are available for farmers' use. Also, as a reverse flow of technology, three IITA-bred lines were released publicly in the United States for their ASR resistance and high yield potential traits (Paul et al. 2010).

In the meantime, owing to the high virulence of ASR, even more virulent pathotypes had evolved and rendered most of the resistant genes ineffective (Godoy et al. 2016). For a successful breeding programme, however, it remains vital to understand and monitor virulence changes in the field and to continuously identify new sources of stable resistance, while at the same time understanding the genetic structure of ASR to improve the durability of resistance. Hence, with the shift of focus of IITA rust research to East Africa, some initial 11 rust pathotypes were reported (Murithi et al. 2017, Twizeyimana et al. 2009) confirming that ASR-resistant genes are pathotype specific. Current IITA studies have led to the discovery of four additional pathotypes, with a South African pathotype being most virulent, and four resistant genes effective against local rust populations (Murithi et al. 2017). Efforts are also underway to understand the genetic diversity of ASR isolates from Africa using next-generation sequencing platforms. A combination of three ASR-resistant genes remarkably conferred resistance to isolates with different geographical backgrounds (Yamanaka et al. 2015). Future efforts will be geared towards stacking ASR-resistant genes that could help provide durable resistance to rust (Hartman et al. 2005). Furthermore, new techniques like field pathogenomics using nanopore technology will be deployed for surveillance to rapidly identify changes in virulence of ASR populations in Africa and guide timely interventions.

4 Conclusions and future trends

During the past 50 years, IITA plant health scientists and collaborators have learnt quite a number of lessons with regard to tackling grain legume pests and diseases. Scientific contributions to the grain legume research agenda moved away from providing mere screening services for the breeding programmes, to evolve into proper scientific investigations, with detailed observations and descriptions of resistance mechanisms, epidemiology and population dynamics, and further into ecological studies encompassing alternative host plants and biodiversity of natural enemies.

Parallel to this scientific evolution, while it became clear that host plant resistance alone - mostly against insect pests - would not provide sufficient protection, the different plant health disciplines started to interact more closely for developing integrated approaches, which were subsequently adapted and further validated through regional networks. This has also been the case for most soil-borne diseases, where only an integrated approach including soil health management (e.g. through the application of beneficial rhizosphere organism and soil amendments such as compost), and resistant varieties would be able to provide sufficient suppression of the pathogen.

Conversely, foliar diseases, and soybean rust in particular, are quite an exception, because they can easily be managed by resistant varieties alone. However, given the tight race between pathogen variability - very common for rust fungi - and resistance genes, a proper resistance management plan needs to be developed which includes detailed and real-time monitoring of pathogen variability across regions.

Synthetic pesticides remain, in certain cases, the main control approach implemented by farmers, whereas our IPM strategy considers them of value in cases where a crop would be totally lost because no other options exist. Also, their negative interference with biological control approaches remains one of the most critical issues in insect pest management. However, with the progress of plant health research and IPM implementation described in this chapter, we infer that their use will gradually be reduced as farmers become more confident with truly integrated approaches which can provide the same yields without all the unintended consequences on human and environmental health. Our final goal should be a 'precision-IPM' approach, where even low-literacy farmers can use simple smartphone apps in their own language to scout for a given pest problem, upload field data and receive guidance on how to manage the problem in real-time. Therefore, the FIA might hopefully be expanded to include all cowpea pests and diseases highlighted in this chapter.

5 Where to look for further information

Cowpea and soybean are part of the research portfolio of the CGIAR Research Program on Grain Legumes and Dryland Cereals http://gldc.cgiar.org/

IITA (iita.org) has the world mandate for research on cowpea, and the mandate for improving soybean in Africa. Details of both crops can be found at http://www.iita.org/crops/cowpea/ and http://www.iita.org/crops/soybean-3/, respectively.

A recent review of cowpea pests and their control has been provided by Agunbiade et al. (2018), while for soybean the most recent compendium can be found at http://biblio.iita.org/documents/U14BkMurithiIntegratedNothomNodev.pdf-ac9019985b7a26be9b7e890e423110c6.pdf

6 References

Abadassi, J. A., Singh, B. B., Ladeinde, T. O. A., Shoyinka, S. A. and Emechebe, A. M. 1987. Inheritance of resistance to brown blotch, *Septoria* leaf spot and scab in cowpea (*Vigna unguiculata* (L.) Walp.). *Indian Journal of Genetics* 47: 299–303.

Abate, T. and Ampofo, J. K. O. 1996. Insect pests of beans in Africa: Their ecology and management. *Annual Review of Entomology* 41: 45–73.

Abdoulaye, T., Ayedun, B., Musa, S. A., Lowenberg-Deboer, J., Baributsa, D., Yakubu, S. A. and Idris, A. A. 2012. Adoption of PICS bags in Northern Nigeria: A case study of Jigawa, Kano, and Katsina States. In: Boukar, O., Coulibaly, O., Fatokun, C., Lopez, K. and Tamò M. (Eds), *Enhancing Cowpea Value Chains through Research Advances*. Proceedings of the 5th World Cowpea Research Conference, 26 September - 1 October 2010, Saly, Senegal, pp. 396–409.

Adebitan, S., Ikotun, T., Dashiell, K. and Singh, S. 1992. Use of three inoculation methods in screening cowpea genotypes for resistance to two *Colletotrichum* species. *Plant Disease* 76: 1025–8.

Adebitan, S., Ikotun, T. and Dashiell, K. 1993. Determination of growth stage to screen cowpea genotypes for resistance to two *Colleototrichum* species. *Fitopatologia Brasileira* 18: 51.

Adegbite, A. A. and Amusa, N. A. 2010. The major economic field diseases of cowpea in the humid agro-ecologies of south-western Nigeria. *Archives of Phytopathology and Plant Protection* 43: 1608–18.

Adegbite, A. A., Amusa, N. A., Agbaje, G. O. and Taiwo, L. B. 2005. Screening of cowpea varieties for resistance to *Meloidogyne incognita* under field conditions. *Nematropica* 35: 155–60.

Adekunle, O. K. 2011. Amendment of soil with African marigold and sunn hemp for management of Meloidogyne incognita in selected legumes. *Crop Protection* 30: 1392–5.

Adetonah, S., Atachi, P., Coulibaly, O. and Tamò, M. 2005. Perceptions paysannes et protection de l'environnement: Gestion integrée de lutte contre le foreur des fleurs et gousses du niébé *Maruca vitrata* au Bénin. *Annales des Sciences Agronomiques du Bénin* 7: 139–57.

Afouda, L., Wydra, K., Schulz, D. and Wolf, G. 2001. Studies on the biological control of charcoal rot of cowpea (*Vigna unguiculata)*, caused by *Macrophomina phaseolina*, with bacterial antagonists. *Tri-trophic Interactions in the Rhizosphere, IOBC/wprs Bulletin* 24: 1–4.

Afouda, L., Wolf, G. and Wydra, K. 2009. Development of a sensitive serological method for specific detection of latent infection of *Macrophomina phaseolina* in cowpea. *Journal of Phytopathology* 157: 15–23.

Afouda, L., Dietmar D., Wolf G. and Wydra, K. 2012. Biological control of *Macrophomina phaseolina* on cowpea (*Vigna unguiculata*) under dry conditions by bacterial antagonists. *Journal of Biological and Chemical Sciences* 6: 5068–77.

Agbicodo, E., Fatokun, C., Bandyopadhyay, R., Wydra, K., Diop, N., Muchero, W., Ehlers, J., Roberts, P., Close, T. and Visser, R. 2010. Identification of markers associated with bacterial blight resistance loci in cowpea [*Vigna unguiculata* (L.) Walp.]. *Euphytica* 175: 215–26.

Agunbiade, T. A., Sun, W., Coates, B. S., Traore, F., Ojo, J. A., Lutomia, A. N., Bello-Bravo, J., Miresmailli, S., Huesing, J. E., Agyekum, M., Tamò, M. and Pittendrigh, B. R. 2018. Insect pests and integrated pest management techniques in grain legume cultivation. In: Sivasankar, S., Bergvinson, D., Gaur, P., Agrawal, S., Beebe, S. and Tamò, M. (Eds), *Achieving Sustainable Cultivation of Grain Legumes Volume 1: Advances in Breeding and Cultivation Techniques*. Burleigh Dodds Science Publishing, Cambridge, UK.

Aina, J. O. 1975. The life history of *Riptortus dentipes* F. (Alydidae, Heteroptera): A pest of growing cowpea pods. *Journal of Natural History* 9: 589–96.

Akingbohungbe, A. E. 1982. Seasonal variation in cowpea crop performance at Ile-Ife, Nigeria, and the relationship to insect damage. *Insect Science and its Applications* 3:287–96.

Akinsanmi, O. A., Ladipo, J. L. and Oyekan, P. O. 2001. First report of soybean rust (*Phakopsora pachyrhizi*) in Nigeria. *Plant Disease* 85: 97.

Alabi, O. Y., Odebiyi, J. A., Tamò, M. and Jackai, L. E. N. 2004. Effect of host plant resistance on growth and developmental parameters of the flower bud thrips, *Megalurothrips sjostedti* (Trybom) on some resistant cowpea (*Vigna unguiculata* {L.} Walp.) cultivars. *Crop Protection* 23: 83–8.

Alabi, O., Odebiyi, J. A., Tamò, M. and Omoloye, A. 2011. The roles of plant secondary metabolites from cowpea floral structures in resistance to the flower bud thrips. *Journal of Agricultural Science and Technology* 1: 262-9.

Allen, D. 1981. 'Two bacterial diseases of cowpea in East Africa': A reply. *International Journal of Pest Management* 27: 144.

Allen, D. and Skipp, R. 1982. Maize pollen alters the reaction of cowpea to pathogens. *Field Crops Research* 5: 265–9.

Allen, D., Nebane, C. and Raji, J. 1981. Screening for resistance to bacterial blight in cowpea. *International Journal of Pest Management* 27: 218–24.

Arodokoun, D. Y., Tamò M., Cloutier C. and Brodeur, J. 2006. Larval parasitoids occurring on *Maruca vitrata* Fabricius (Lepidoptera: Pyralidae) in Benin, West Africa. *Agriculture, Ecosystems and Environment* 113: 320–5.

Asante, S. K., Jackai, L. E. N. and Tamò, M. 2000. Efficiency of *Gryon fulviventris* (Hymenoptera: Scelionidae) as an egg parasitoid of *Clavigralla tomentosicollis* (Hemiptera: Coreidae) in northern Nigeria. *Environmental Entomology* 29: 815–21.

Asante, S. K., Tamò, M. and Jackai, L. E. N. 2001. Integrated management of cowpea insect pests using elite cultivars, date of planting and minimum insecticide application. *African Crop Science Journal* 9: 655–65.

Azarmanesh, N., Bond, J. P., Vick, A., Mengistu, A., Fakhoury, A. M. 2011. A qPCR assay to detect and quantify *Macrophomina phaseolina* in soybean roots. *Phytopathology* 101: S11.

Bandyopadhyay, R., Ojiambo, P. S., Twizeyimana, M., Asafo-Adjei, B., Frederick, R. D., Pedley, K. F., Stone, C. L. and Hartman, G. L. 2007. First report of soybean rust caused by *Phakopsora pachyrhizi* in Ghana. *Plant Disease* 91: 1057.

Baoua, I. B., Margam, V., Amadou, L. and Murdock, L. L. 2012. Performance of triple bagging hermetic technology for postharvest storage of cowpea grain in Niger. *Journal of Stored Products Research* 51: 81–5.

Bata, H. D., Singh, B. B., Singh, S. R. and Ladeinde, T. A. O. 1987. Inheritance of resistance to aphid in cowpea. *Crop Science* 27: 892–4.

Bottenberg, H. and Singh B. B. 1996. Effect of neem leaf extract applied using the 'broom' method, on cowpea pest and yield. *International Journal of Pest Management* 42: 207–9.

Bottenberg, H., Tamò, M. and Singh, B. B. 1998. Occurrence of phytophagous insects on wild *Vigna* sp. and cultivated cowpea: Comparing the relative importance of host-plant resistance and millet intercropping. *Agriculture, Ecosystems and Environment* 70: 217–29.

Caveness, F. E. 1992. Nematological research at IITA 1969–1988. A summary of investigations conducted by Fields E. Caveness. Plant Health Management Research Monograph 2. Lowe, J. (Ed.). IITA, Ibadan, Nigeria, 52pp.

Constantin, E. C., Cleenwerck, I., Maes, M., Baeyen, S., Van Malderghem, C., De Vos, P. and Cottyn, B. 2016. Genetic characterization of strains named as *Xanthomonas axonopodis* pv. *dieffenbachiae* leads to a taxonomic revision of the *X. axonopodis* species complex. *Plant Pathology* 65:792–806.

Dabiré, C., Tignegre, J. B., Ba, M. N., Tamò, M., Sanon, A., Ouedraogo, T. J., Neya, B., Muleba, N., Hammond, W. and Coulibaly, O. 2012. An historical review of progress to control key cowpea biotic constraints in Burkina Faso. In: Boukar, O., Coulibaly, O., Fatokun, C., Lopez, K. and Tamò M. (Eds), *Enhancing Cowpea Value Chains through Research Advances*. Proceedings of the 5th World Cowpea Research Conference, 26 September – 1 October 2010, Saly, Senegal, pp. 273–86.

Dabiré-Binso, C. L., Ba, N. M., Sanon, A., Drabo, I. and Bi, K. F. 2010. Resistance mechanism to the pod-sucking bug *Clavigralla tomentosicollis* (Hemiptera: Coreidae) in the cowpea IT86D-716 variety. *International Journal of Tropical Insect Science* 30: 192–9.

Dannon, E. A, Tamò, M., Agboton, C., van Huis, A. and Dicke, M. 2012. Effect of *Maruca vitrata* (Lepidoptera: Crambidae) host plants on life-history parameters of the parasitoid *Apanteles taragamae* (Hymenoptera: Braconidae). *Insect Science* 19: 518–28.

Das, S., DeMason, D. A., Ehlers, J. D., Close, T. J. and Roberts, P. A. 2008. Histological characterization of root-knot nematode resistance in cowpea and its relation to reactive oxygen species modulation. *Journal of Experimental Botany* 59: 1305–13.

Dawar, S., Tariq, M. and Zaki, M. J. 2008. Application of *Bacillus* species in control of *Meloidogyne javanica* (Treub) Chitwood on cowpea and mash bean. *Pakistan Journal of Botany* 40: 439–44.

Diabate, A., Baldet, T., Chandre, F., Akogbeto, M., Guiguemde, T. R., Darriet, F., Brengues, C., Guillet, P., Hemingway, J., Small, G. J. and Hougard, J. M. 2002. The role of agricultural use of insecticides in resistance to pyrethroids in *Anopheles gambiae* s.l. in Burkina Faso. *American Journal of Tropical Medicine and Hygiene* 67: 617–22.

Published by Burleigh Dodds Science Publishing Limited, 2019.

Egho, E. O. 2011. Effects of two agro-ecological zones on insect species of cowpea (*Vigna unguiculata* L.) Walp. during the late cropping season, Delta State, Southern Nigeria. *Agriculture and Biology Journal of North America* 2: 448–53.

Ehlers, J. D. and Hall, A. E. 1997. Cowpea (*Vigna unguiculata* (L.) Walp.). *Field Crops Research* 53: 187–204.

Ekesi, S. 1999. Insecticide resistance in field populations of the legume pod-borer, *Maruca vitrata* Fabricius (Lepidoptera: Pyralidae), on cowpea, *Vigna unguiculata* (L.), Walp. in Nigeria. *International Journal of Pest Management* 45: 57–9.

Emechebe, A. M. and Florini, D. A. 1997. Shoot and pod diseases of cowpea induced by fungi and bacteria. In: Singh, B. B., Mohan Raj, D. R., Dashiell, K. E. and Jackai, L. E. N. (Eds), *Advances in Cowpea Research*. Co-publication of International Institute of Agriculture (IITA) and Japan International Center for Agricultural Sciences (JIRCAS). IITA, Ibadan, Nigeria, pp. 176–92.

Ezueh, M. I. 1981. Nature and significance of preflowering damage by thrips to cowpea. *Entomologia Experimentalis et Applicata* 29: 305–12.

Fatokun, C. A. 2002. Breeding cowpea for resistance to insect pests: Attempted crosses between cowpea and *Vigna vexillata*. In: Fatokun, C. A., Tarawali, S. A., Singh, B. B., Kormawa, P. M. and Tamò, M. (Eds), *Challenges and Opportunities for Enhancing Sustainable Cowpea Production*. Proceedings of the 3rd World Cowpea Conference held at the International Institute of Tropical Agriculture (IITA), Ibadan, Nigeria, 4–8 September 2000. IITA, Ibadan, Nigeria, pp. 52–61.

Fatokun, C. A., Perrino, P. and Ng, N. Q. 1997. Wide crossing in African *Vigna* species. In: Singh, B. B., Mohan Raj, D. R., Dashiell, K. E. and Jackai, L. E. N. (Eds), *Advances in Cowpea Research*. Co- publication of International Institute of Agriculture (IITA) and Japan International Center for Agricultural Sciences (JIRCAS). IITA, Ibadan, Nigeria, pp. 50–7.

Food and Agriculture Organization of the United Nations. 2016. FAOSTAT statistics database. FAO, Rome.

Franke, A. C., van den Brand, G. J., Vanlauwe, B. and Giller, K. E. 2018. Sustainable intensification through rotations with grain legumes in Sub-Saharan Africa: A review. *Agriculture, Ecosystems and Environment* 261:172–85.

Ganyo, K. K., Tounou, A. K., Agboton, C., Dannon, E. A., Pittendrigh, B. R. and Tamò, M. 2012. Interactions between the aphid parasitoid *Lysiphlebus testaceipes* (Hymenoptera: Aphidiidae) and its hyperparasitoid *Syrphophagus africanus* (Hymenoptera: Encyrtidae). *International Journal of Tropical Insect Science* 32: 45–55.

Godoy, C. V., Seixas, C. D. S., Soares, R. M., Marcelino-Guimarães, F. C., Meyer, M. C. and Costamilan, L. M. 2016. Asian soybean rust in Brazil: Past, present, and future, *Pesquisa Agropecuaria Brasileira* 51: 407–21.

Hartman, G. L., Miles, M. R. and Frederick, R. D. 2005. Breeding for resistance to soybean rust. *Plant Disease* 89: 664–6. doi:10.1094/PD890664.

Hartman, G. L., Hill, C. B., Twizeyimana, M., Miles, M. R. and Bandyopadhyay, R. 2011. Interaction of soybean and *Phakopsora pachyrhizi*, the cause of soybean rust. *CAB Review Perspectives in Agriculture Veterinary Science Nutrition and Natural Resources* 6: 13.

Heong, K. L., Aquino, G. B. and Barrion, A. T. 1992. Population dynamics of plant- and leafhoppers and their natural enemies in rice ecosystems in the Philippines. *Crop Protection* 11: 371–9.

Huang, C. C., Peng, W. K. and Talekar, N. S. 2003. Characteristics of infestation by the bean pod borer, *Maruca vitrata* (Lepidoptera: Pyralidae) on *Sesbania cannabina*. *Formosan Entomology* 23: 1–11.

Huesing, J., Romeis, J., Ellstrand, N., Raybould, A., Hellmich, R., Wolt, J., Ehlers, J., Dabiré, C., Fatokun, C., Hokanson, K., Ishiyaku, M. F., Margam, V., Obokoh, N., Mignouna, J., Nangayo, F., Ouedraogo, J., Pasquet, R., Pittendrigh, B., Schaal, B., Stein, J., Tamò, M. and Murdock, L. 2011. Regulatory considerations surrounding the deployment of Bt-expressing cowpea in Africa: Report of the deliberations of an expert panel. *GM Crops* 2: 1–14.

Huynh, B. L., Matthews, W. C., Ehlers, J. D., Lucas, M. R., Santos, J. R. P., Ndeve, A., Close, T. J. and Roberts, P. A. 2016. A major QTL corresponding to the Rk locus for resistance to root-knot nematodes in cowpea (*Vigna unguiculata* L. Walp.). *Theoretical and Applied Genetic* 129: 87–95.

Jackai, L. E. N. and Adalla, C. B. 1997. Pest management practices in cowpea: A review. In: Singh, B. B., Mohan Raj, D. R., Dashiell, K. E. and Jackai, L. E. N. (Eds), *Advances in Cowpea Research*. Co- publication of International Institute of Agriculture (IITA) and Japan International Center for Agricultural Sciences (JIRCAS). IITA, Ibadan, Nigeria, pp. 240–53.

Jackai, L. E. N. and Daoust, R. A. 1986. Insect pests of cowpeas. *Annual Review of Entomology* 31: 95–119.

Jackai, L. E. N. and Ndlovu, T. M. 1988. Screen cage and laboratory evaluation of the efficacy of six foliar insecticides on different stages of the cowpea coreid bug, *Clavigralla tomentosicollis*, Stål. (Hemiptera: Coreidae). *International Journal of Tropical Insect Science* 9: 223–8.

Jackai, L. E. N. and Oghiakhe, S. 1989. Pod wall trichomes and resistance of two wild cowpea, *Vigna vexillata*, accessions to *Maruca testualis* (Geyer) (Lepidoptera: Pyralidae) and *Clavigralla tomentosicollis* Stål (Hemiptera: Coreidae). *Bulletin of Entomological Research* 79: 595–605.

Jackai, L. E. N. and Oyediran, I. O. 1991. The potential of neem *Azadirachta indica* (A. Juss.) for controlling post-flowering pests of cowpea *Vigna unguiculata* (L.) Walp. -I. The pod borer *Maruca testulalis*. *Insect Science and its Applications* 12: 103–9.

Jackai, L. E. N. and Singh, S. R. 1988. Screening techniques for hostplant resistance to insect pests of cowpea. *Tropical Grain Legumes Bulletin* 35: 2–18.

Jackai, L. E. N., Inang, E. E. and Nwobi, P. 1992. The potential for controlling post- flowering pests of cowpea, *Vigna unguiculata* Walp. using neem, *Azadirachta indica* A. Juss. *Tropical Pest Management* 38: 56–60.

Jackai, L. E. N., Padulosi, S. and Ng, Q. 1996. Resistance to the legume pod borer, *Maruca vitrata* Fabricius, and the probable modalities involved in wild *Vigna*. *Crop Protection* 15: 753–61.

Jackai, L. E. N., Nokoe, S., Tayo, B. O. and Koona, P. 2001. Inferences on pod wall and seed defenses against the brown cowpea coreid bug, *Clavigralla tomentosicollis* Stål. (Hem., Coreidae) in wild and cultivated *Vigna* species. *Journal of Applied Entomology* 125: 277–86.

Javaid, I. and Ashraf, M. 1978. Some observations on soybean diseases in Zambia and occurrence of *Pyrenochaeta glycines* on certain varieties. *Plant Disease Reports* 62: 46–7.

Kalshoven, L. G. E. and van der Vecht, J. 1950. *De plagen van de culturgewassen in Indonesië*. Van Hoeve, La Haye, 512pp.

Kawuki, R., Adipala, E., Lamo, J. and Tukamuhabwa, P. 2003. Responding to the soybean rust epidemic in sub-Saharan Africa. A Review. *African Crop Science Journal* 11: 301–18

Kawuki, R. S., Tukamuhabwa, P. and Adipala, E. 2004. Soybean rust severity, rate of rust development, and tolerance as influenced by maturity period and season. *Crop Protection* 23: 447–55.

Kishun, R. 1989. Appraisal of loss in yield of cowpea due to *Xanthomonas campestris* pv. *vignicola*. *Indian Phytopathology* 42: 241–6.

Kitch, L. M., Bottenberg, H. and Wolfson, J. L. 1997. Indigenous knowledge and cowpea pest management in sub-Saharan Africa. In: Singh, B. B., Mohan Raj, D. R., Dashiell, K. E. and Jackai, L. E. N. (Eds), *Advances in Cowpea Research*. Copublication of International Institute of Tropical Agriculture (IITA) and Japan International Center for Agricultural Sciences (JIRCAS). IITA, Ibadan, Nigeria, pp. 292–301.

Koona, P., Osisanya, E. O., Jackai, L. E. N., Tamò, M. and Markham, R. H. 2002. Resistance in accessions of cowpea to coreid podbug *Clavigralla tomentosicollis* (Hemiptera: Coreidae). *Journal of Economic Entomology* 95: 1281–8.

Kumudini, S., Godoy, C. V., Board, J. E., Omielan, J. and Tollenaar, M. 2008. Mechanisms involved in soybean rust-induced yield reduction. *Crop Science* 48: 2334–42.

Lattanzio, V., Lattanzio, V. M. and Cardinali, A. 2006. Role of phenolics in the resistance mechanisms of plants against fungal pathogens and insects. *Phytochemistry: Advances in Research* 661: 23–67.

Lee, S. T., Srinivasan, R., Wu, Y. J. and Talekar, N. S. 2007. Occurrence and characterization of a nucleopolyhedrovirus from *Maruca vitrata* (Lepidoptera, Pyralidae) isolated in Taiwan. *BioControl* 52: 801–19.

Levy, C. 2005. Epidemiology and chemical control of soybean rust in Southern Africa. *Plant Disease* 89: 669–74.

Lim, U. T. and Mainali, B. P. 2013. Effect of aggregation pheromone trap of *Riptortus pedestris* (Hemiptera: Alydidae) on the distribution and composition of its egg parasitoids. *Journal of Economic Entomology* 106: 1973–8.

Lucas, M. R., Ehlers, J. D., Roberts P. A. and Close, T. J. 2012. Markers for quantitative resistance to foliar thrips in cowpea. *Crop Science* 52: 2075–81.

MacFoy, C. C. and Dabrowski, Z. T. 1984. Preliminary studies of cowpea resistance to *Aphis craccivora* Koch (Homoptera: Aphididae). *Journal of Applied Entomology* 97: 202–9.

MacFoy, G. A., Dabrowski, Z. T. and Okech, S. 1983. Studies on the legume pod-borer, *Maruca testulalis* (Geyer)–VI. Cowpea resistance to oviposition and larval feeding. *International Journal of Tropical Insect Science* 4: 147–52.

Masangwa, J. I. G., Aveling, T. A. S. and Kritziger, Q. 2013. Screening of plant extracts for antifungal activities against *Colletotrichum* species of common bean (*Phaseolus vulgaris* L.) and cowpea (*Vigna unguiculata* (L.) Walp). *The Journal of Agricultural Science* 151:482–91.

McSorley, R. and Gallaher, R. N. 1996. Effect of yard waste compost on nematode densities and maize yield. *Journal of Nematology* 28: 655–60.

Mehinto, J. T., Atachi, P., Elégbédé, M., Kpindou, O. K. D. and Tamò, M. 2014. Efficacité comparée des insecticides de natures différentes dans la gestion des insectes ravageurs du niébé au Centre du Bénin. *Journal of Applied Biosciences* 84: 7695–706.

Moussa, B., Lowenberg-Deboer, J., Fulton, J. and Boys, K. 2011. The economic impact of cowpea research in West and Central Africa: A regional impact assessment of improved cowpea storage technologies. *Journal of Stored Products Research* 47: 147-56.

Muchero, W., Ehlers, J. D. and Roberts P. A. 2010. QTL analysis for resistance to foliar damage caused by *Thrips tabaci* and *Frankliniella schultzei* (Thysanoptera: Thripidae) feeding in cowpea [*Vigna unguiculata* (L.) Walp.]. *Molecular Breeding* 25: 47–56.

Murithi, H. M., Beed, F., Madata, C., Haundenshield, J. and Hartman, G. L. 2014. First report of *Phakopsora pachyrhizi* on soybean causing rust in Tanzania. *Plant Disease* 98: 1586.

Murithi, H. M., Beed, F., Soko, M., Haundenshield, J. and Hartman, G. L. 2015. First report of Phakopsora pachyrhizi causing rust on soybean in Malawi. *Plant Disease* 99: 420.

Murithi, H. M., Beed, F., Tukamuhabwa, P., Thomma, B. P. H. J. and Joosten, M. H. A. J. 2016. Soybean production in eastern and southern Africa and threat of yield loss due to soybean rust caused by *Phakopsora pachyrhizi*. *Plant Pathology* 65: 176–88.

Murithi, H. M., Haundenshield, J., Beed, F., Mahuku, G., Joosten, M. H. A. J. and Hartman, G. L. 2017. Virulence diversity of *Phakopsora pachyrhizi* isolates from East Africa compared to a geographically diverse collection. *Plant Disease* 101: 1194–200.

Nandi, B., Kundu, K., Banerjee, N. and Babu, S. P. S. 2003. Salicylic acid-induced suppression of *Meloidogyne incognita* infestation of okra and cowpea. *Nematology* 5: 747–52.

Ndiaye, M., Termorshuizen, A. J. and Van Bruggen, A. H. C. 2010. Effects of compost amendment and the biocontrol agent *Clonostachys rosea* on the development of charcoal rot (*Macrophomina phaseolina*) on cowpea. *Journal of Plant Pathology* 92: 173–80.

Obopile, M. and Ositile, B. 2010. Life table and population parameters of cowpea aphid, *Aphis craccivora* Koch (Homoptera: Aphididae) on five cowpea, *Vigna unguiculate* (L. Walp.) varieties. *Journal of Pest Science* 83: 9–14.

Ofuya, T. I. 1993. Evaluation of selected cowpea varieties for resistance to *Aphis craccivora* Koch (Homoptera: Aphididae) at the seedling and podding phase. *Annals of Applied Biology* 123: 19–23.

Ofuya, T. I. 1997. Control of the cowpea aphid, *Aphis craccivora* Koch (Homoptera: Aphididae), in cowpea, *Vigna unguiculata* (L.) Walp. *Integrated Pest Management Reviews* 2: 199–207.

Oghiakhe, S., Jackai, L. and Makanjuola, W. 1991a. Cowpea plant architecture in relation to infestation and damage by the legume pod borer, *Maruca testulalis* Geyer (Lepidoptera: Pyralidae)–I. Effect of Canopy Structure and Pod Position. *International Journal of Tropical Insect Science* 12: 193–9.

Oghiakhe, S., Jackai, L. E. N. and Makanjuola, W. A. 1991b. Anatomical parameters of cowpea, *Vigna unguiculata* (L.) Walp. Stem and pod wall resistance to the legume pod borer *Maruca testulalis* Geyer (Lepidoptera: Pyralidae). *International Journal of Tropical Insect Science* 12: 171–6.

Oghiakhe, S., Jackai, L. and Makanjuola, W. A. 1992a. Cowpea plant architecture in relation to infestation and damage by the legume pod borer, *Maruca testulalis* Geyer (Lepidoptera: Pyralidae) . -II. Effect of pod angle. *Insect Science and its Applications* 13: 339–44.

Oghiakhe, S., Jackai, L. E. N., Makanjuola, W. A. and Hodgson, G. J. 1992b. Morphology, distribution and the role of trichomes in cowpea (*Vigna vexillata*) resistance to the legume borer, *Maruca testulalis* (Lepidoptera: Pyralidae). *Bulletin of Entomological Research* 82:499–505.

Oigiangbe, O. N., Jacakai, L. E. N., Ewete, F. K. and Lajide, L. 2001. The nutritional ecology of *Maruca vitrata* I: Development on flowers of wild and cultivated *Vigna* species. *Insect Science and its Applications* 20: 259–68.

Oigiangbe, O. N., Jackai, L. E. N., Ewete, F. K., Hughes, J. and Lajide, L. 2002. Reduced consumption and use of pods of *Vigna* species (Leguminosae) by *Maruca vitrata* (Lepidoptera: Pyralidae). *African Entomology* 10: 333–40.

Oigiangbe, O. N., Jackai, L. E. N., Ewete, F. K., Lajide, L. and Hughes, J. 2006. Effects of pubescence on the oviposition and feeding behaviour of *M. vitrata* Fabricius on *Vigna* species. *Ghana Journal of Science* 46: 55–65.

Ojiambo, P. S., Bandyopadhyay, R., Twizeyimana, M., Lema, A., Frederick, R. D., Pedley, K. F., Stone, C. L. and Hartman, G. L. 2007. First report of rust caused by *Phakopsora pachyrhizi* on soybean in Democratic Republic of Congo. *Plant Disease* 91: 1204.

Okechukwu, R. and Ekpo, E. 2004. Sources of resistance to cowpea bacterial blight disease in Nigeria. *Journal of Phytopathology* 152: 345–51.

Okechukwu, R., Ekpo, E. and Okechukwu, O. 2010. Seed to plant transmission of *Xanthomonas campestris* pv. *vignicola* isolates in cowpea. *African Journal of Agricultural Research* 5: 431–5.

Okeola, O. G., Machuka, J. and Fasidi, I. O. 2002. Insecticidal activities of the African yam bean seed lectin on the development of the cowpea beetle and the pod-sucking bug. In: Fatokun, C. A. Tarawali, S. A. Singh, B. B. Kormawa P. M. and Tamò M. (Eds), *Challenges and Opportunities for Enhancing Sustainable Cowpea Production*. Proceedings of the 3rd World Cowpea Conference held at the International Institute of Tropical Agriculture (IITA), Ibadan, Nigeria, 4–8 September 2000. IITA, Ibadan, pp. 223–30.

Ombakho, G. A., Tyagi, A. P. and Pathak, R. S. 1987. Inheritance of resistance to the cowpea aphid in cowpea. *Theoretical and Applied Genetics* 74: 817–19.

Omitogun, O. G., Jackai, L. E. N. and Thottappilly, G. 1999. Isolation of insecticidal lectin-enriched extracts from African yam bean (*Sphenostylis stenocarpa*) and other legume species. *Entomologia Experimentalis et Applicata* 90: 301–11.

Omo-Ikerodah, E. E., Fatokun, C. A. and Fawole, I. 2009. Genetic analysis of resistance to flower bud thrips (*Megalurothrips sjostedti*) in cowpea (*Vigna unguiculata* [L.] Walp.), *Euphytica* 165: 145–54.

Onstad, D. W., Kang, J., Ba, N. M., Tamò, M., Jackai, L., Dabire, C. and Pittendrigh, B. R. 2012. Modeling evolution of resistance by *Maruca vitrata* (Lepidoptera: Crambidae) to transgenic insecticidal cowpea in Africa. *Environmental Entomology* 41: 1255–67.

Oparaeke, A. M., Dike, M. C. and Amatobi, C. I. 2005. Botanical pesticide mixtures for insect pest management on cowpea, *Vigna unguiculata* (L.) Walp. plants–2. The pod borer, *Maruca vitrata* Fab. (Lepidoptera: Pyralidae) and the pod sucking bug, *Clavigralla tomentosicollis* Stål. (Heteroptera: Coreidae). *Agricultura Tropica et Subtropica* 38: 1–6.

Pathak, R. S. 1988. Genetics of resistance to aphid in cowpea. *Crop Science* 28: 474–6.

Paul, C., Bowen, C. R., Bandyopadhyay, R., Tefera, H., Adeleke, R., Sikora, E., Pegues, M. D. and Hartman, G. L. 2010. Registration of three soybean germplasm lines resistant to *Phakopsora pachyrhizi* (soybean rust). *Journal of Plant Registration* 4: 244–8.

Periasamy, M., Schafleitner, R., Muthukalingan, K. and Ramasamy, S. 2015. Phylogeographical structure in mitochondrial DNA of legume pod borer (*Maruca vitrata*) population in tropical Asia and sub-Saharan Africa. *PloS one*, 10:e0124057

Pitan, O. O. and Odebiyi, J. A. 2001. The effect of intercropping with maize on the level of infestation and damage by pod-sucking bugs in cowpea. *Crop Protection* 20: 367–72.

Popelka, J. C., Gollasch, S., Moore, A., Molvig, L. and Higgins, T. J. V. 2006. Genetic transformation of cowpea (*Vigna unguiculata* L.) and stable transmission to progeny. *Plant Cell Reports* 25: 304–12.

Pretorius, Z. A., Kloppers, F. J. and Frederick, R. D. 2001. First report of soybean rust in South Africa. *Plant Disease* 85:1288

Riekert, H. F and Henshawi, G. E. 1998. Effect of soybean, cowpea and groundnut rotations on root-knot nematode build-up and infestation of dryland maize. *African Crop Science Journal* 6: 377–83.

Sæthre, M.-G., Godonou, I., Hofsvang, T., Tepa-Yotto, G. T. and James, B. 2011. Aphids and their natural enemies in vegetable agroecosystems in Benin. *International Journal of Tropical Insect Science* 31: 103–17.

Salifu, A. B., Singh, S. R. and Hodgson, C. J. 1988. Mechanism of resistance in cowpea (*Vigna unguiculata* (L.) Walp.) genotype, TVx3236, to the bean flower thrips, *Megalurothrips sjostedti* (Trybom) (Thysanoptera: Thripidae) 2. Nonpreference and antibiosis. Tropical Pest Management 34: 185–8.

Sawadogo, A., Thio, B., Kiemde, S., Drabo, I., Dabire, C., Ouedraogo, J., Mullens, T. R., Ehlers, J. D. and Roberts, P. A. 2009. Distribution and prevalence of parasitic nematodes of cowpea (*Vigna unguiculata*) in Burkina Faso. *Journal of Nematology* 41: 120–7.

Sikirou, R. and Wydra, K. 2004. Persistence of *Xanthomonas axonopodis* pv. *vignicola* in weeds and crop debris and identification of *Sphenostylis sternocarpa* as a potential new host. *European Journal of Plant Pathology* 110: 939–47.

Sikirou, R. and Wydra, K. 2008. Effect of intercropping cowpea with maize or cassava on cowpea bacterial blight and yield. *Journal of Plant Diseases and Protection* 115: 145–51.

Sikirou, R. and Wydra, K. 2011. Multiplication and movement of *Xanthomonas axonopodis* pv. *vignicola* and seed contamination in cowpea (*Vigna unguiculata*) genotypes. *Journal of Plant Diseases and Protection* 118: 93–101.

Sikirou, R., Wydra, K. and Rudolph, K. 2001. Selection of cowpea genotypes resistant to bacterial blight caused by *Xanthomonas campestris* pv. *vignicola*. In: De Boer, S. H. (Ed.), *Plant Pathogenic Bacteria*. Proceedings of the 10th International Conference on Plant Pathogenic Bacteria, Charlottetown, Prince Edward Island, Canada, July 23–7, 2000. Springer, the Netherlands, pp. 309–14.

Singh, S. R. 1977. Cowpea cultivars resistant to insect pests in world germplasm collection. *Tropical Grain Legume Bulletin* 9: 1–7.

Singh, B. B., Singh, S. R. and Jackai, L. E. N. 1982. Cowpea breeding for disease and insect resistance. In FAO, Breeding for durable disease and pest resistance, FAO plant production and protection paper 55: 139.

Singh, S. K., Nene, Y. L. and Reddy, M. V. 1990a. Additions to the host range of *Macrophomina phaseolina*. *Plant Disease* 74: 828.

Singh, S. R., Jackai, L. E. N., Dos Santos, J. H. R., Adalla, C. B. 1990b. Insect pests of cowpea. In: Singh, S. R. (Ed.), *Insect Pests of Tropical Food Legumes*. John Wiley and Sons Ltd., Chichester, U. K., pp. 43–89.

Singh, B. B., Chambliss, O. L and Sharma, B. 1997. Recent advances in cowpea breeding. In: Singh, B. B., Mohan Raj, D. R., Dashiell, K. E. and Jackai, L. E. N. (Eds), *Advances in*

Cowpea Research. Co- publication of International Institute of Agriculture (IITA) and Japan International Center for Agricultural Sciences (JIRCAS). IITA, Ibadan, Nigeria, pp. 30–49.

Sokame, B. M., Tounou, A. K., Datinon, B., Dannon, E. A., Agboton, C., Srinivasan, R., Pittendrigh, B. R. and Tamò, M. 2015. Combined activity of *Maruca vitrata* multi-nucleopolyhedrovirus, *Mavi*MNPV, and oil from neem, *Azadirachta indica* Juss. and *Jatropha curcas* L., for the control of cowpea pests. *Crop Protection* 72: 150–7.

Souleymane, A., Aken'Ova, M. E., Fatokun, C. A. and Alabi, O. Y. 2013. Screening for resistance to cowpea aphid (*Aphis craccivora* Koch) in wild and cultivated cowpea (*Vigna unguiculata* L. Walp.) accessions. *International Journal of Science, Environment and Technology* 2: 611–21.

Soyelu, O. L., Akingbohungbe, A. E. and Okonji, R. E. 2007. Salivary glands and their digestive enzymes in pod-sucking bugs (Hemiptera: Coreoidea) associated with cowpea *Vigna unguiculata* ssp. *unguiculata* in Nigeria. *International Journal of Tropical Insect Science* 27: 40–7.

Srinivasan, R., Yule, S., Lin, M. Y. and Khumsuwan, C. 2015. Recent developments in the biological control of legume pod borer (*Maruca vitrata*) on yard-long bean. *Acta Horticulturae* 1102: 143–50.

Starr, J. L. and Mercer, C. F. 2009. Development of resistant varieties. In: Perry. R. N., Moens, M. and Starr, J. L (Eds), *Root-knot Nematodes*. CABI Publishing, Wallingford, U. K., pp. 326–35.

Swanson, T. A. and Van Gundy, S. D. 1984. Cowpea resistance to root knot caused by *Meloidogyne incognita* and *M. javanica*. *Plant Disease* 68: 961–4.

Tamò, M. and Baumgärtner, J. 1993. Analysis of the cowpea agro-ecosystem in West Africa: I. A demographic model for carbon acquisition and allocation in cowpea. *Ecological Modelling* 65: 95–121.

Tamò, M., Baumgärtner, J. and Gutierrez, A. P. 1993a. Analysis of the cowpea agro-ecosystem in West Africa: II. Modelling the interactions between cowpea and the bean flower thrips *Megalurothrips sjostedti* (Trybom). *Ecological Modelling* 70: 89–113.

Tamò, M., Baumgärtner, J., Delucchi, V. and Herren, H. R. 1993b. Assessment of key factors responsible for the pest status of the bean flower thrips *Megalurothrips sjostedti* (Trybom) (Thysanoptera, Thripidae). *Bulletin of Entomological Research* 83: 251–8.

Tamò, M., Bottenberg H., Arodokoun D. and Adeoti R. 1997. The feasibility of classical biological control of two major cowpea insect pests. In: Singh, B. B., Mohan Raj, D. R., Dashiell, K. E. and Jackai, L. E. N. (Eds), *Advances in Cowpea Research*. Co-publication of International Institute of Agriculture (IITA) and Japan International Center for Agricultural Sciences (JIRCAS). IITA, Ibadan, Nigeria, pp. 259–70.

Tamò, M., Ekesi, S., Maniania, N. K. and Cherry, A. 2003. Biological control, a non-obvious component of integrated pest management for cowpea. In: Neuenschwander, P., Borgemeister, C. and Langewald, J. (Eds), *Biological Control in Integrated Pest Management Systems in Africa*. CABI Publishing, Wallingford, U. K., pp. 295–309.

Tamò, M., Srinivasan, R., Dannon, E., Agboton, C., Datinon, B., Dabiré, C., Baoua, I., Ba, M. N., Haruna, B. and Pittendrigh, B. R. 2012. Biological control: A major component for the long-term cowpea pest management strategy. In: Boukar, O., Coulibaly, O., Fatokun, C., Lopez, K., Tamò, M. (Eds), *Enhancing Cowpea Value Chains through Research Advances*. Proceedings of the 5th World Cowpea Research Conference, 26 September – 1 October 2010, Saly, Senegal, pp. 249–59.

Tamò, M., Datinon, B., Dannon, E., Traoré, F., Dabiré, C., Pittendrigh, B. R. and Srinivasan, R. 2017. Towards successful establishment of exotic parasitoids attacking the pod borer *Maruca vitrata* in west Africa. *Biocontrol News and Information* 38: 12–13.

Tanzubil, P. B. 2000. Field evaluation of neem, *Azadirachta indica* extracts for control of insect pests of cowpea in Northern Ghana. *Journal of Tropical Forest Products* 6: 165–72.

Tepa-Yotto, G., Hofsvang, T., Godonou, I. and Sæthre, M. 2013. Host preference of *Lysiphlebus testaceipes* (Hymenoptera: Braconidae, Aphidiinae), an alien aphid parasitoid in Benin. *International Journal of Tropical Insect Science* 33: 127–35.

Tesfay, A., Kifle, B., Haundenshield, J. S. and Hartman, G. L. 2017. First report of *Phakopsora pachyrhizi* causing rust on soybean in Ethiopia. *Plant Disease* 101: 1049.

Togola, A., Boukar, O., Belko, N., Chamarthi, S. K., Fatokun, C., Tamò, M. and Oigiangbe, N. 2017. Host plant resistance to insect pests of cowpea (*Vigna unguiculata* L. Walp.): Achievements and future prospects. *Euphytica* 213: 239.

Togola, A., Boukar, O., Belko, N., Chamarthi, S., Tamò, M., Oigiangbe, N., Ojo, J., Ibikunle, M. and Fatokun, C. 2018. Resistance to flower bud thrips, *Megalurothrips sjostedti* Trybom (Thysanoptera: Thripidae) in cowpea mini core accessions under field conditions. *Pest Management Science* (in press).

Tukamuhabwa, P. and Maphosa, M. 2012. State of knowledge in breeding for resistance to soybean rust in the developing world. FAO Rome, Italy. FAO Plant Production and Protection Paper 204: 46pp.

Twizeyimana, M., Ojiambo, P. S., Ikotun, T., Paul, C., Hartman, G. L. and Bandyopadhyay, R. 2007. Comparison of field, greenhouse, and detached-leaf evaluations of soybean germplasm for resistance to *Phakopsora pachyrhizi*. *Plant Disease* 91: 1161–9.

Twizeyimana, M., Ojiambo, P. S., Ikotun, T., Ladipo, J. L., Hartman, G. L. and Bandyopadhyay, R. 2008. Evaluation of soybean germplasm for resistance to soybean rust (*Phakopsora pachyrhizi*) in Nigeria. *Plant Disease* 92: 947–52.

Twizeyimana, M., Ojiambo, P. S., Sonder, K., Ikotun, T., Hartman, G. L. and Bandyopadhyay, R. 2009. Pathogenic variation of *Phakopsora pachyrhizi* infecting soybean in Nigeria. *Phytopathology* 99: 353–61.

Twizeyimana, M., Ojiambo, P. S., Hartman, G. L. and Bandyopadhyay, R. 2011a. Dynamics of soybean rust epidemics in sequential plantings of soybean cultivars in Nigeria. *Plant Disease* 95: 43–50.

Twizeyimana, M., Ojiambo, P. S., Haundenshield, J. S., Caetano-Anollés, G., Pedley, K. F., Bandyopadhyay, R. and Hartman, G. L. 2011b. Genetic structure and diversity of *Phakopsora pachyrhizi* isolates from soybean. *Plant Pathology* 60: 719–29.

van Emden, H. F. 1991. The role of host plant resistance in insect pest mis-management. *CAB International. Bulletin of Entomological Research* 81: 123–6.

Verdejo-Lucas, S., Cortada, L., Sorribas, F. J. and Ornat, C. 2009. Selection of virulent populations of *Meloidogyne javanica* by repeated cultivation of *Mi* resistance gene tomato rootstocks under field conditions *Plant Pathology* 58: 990–8.

Williams, R. J. 1977. Identification of multiple disease resistance in cowpea. *Tropical Agriculture Trinidad* 54: 53–9.

Williamson, V. and Roberts, P. A. 2009. Mechanisms and genetics of resistance. In: Perry, R. N., Moens, M. and Starr, J. L. (Eds), *Root-knot Nematodes*. CABI Publishing, Wallingford, U. K., pp. 301–25.

Wrather, J. A., Anderson T. R., Arsyad, D. M., Gai, J., Ploper, L. D., Porta-Puglia, A., Ram, H. H. and Yorinori, J. T. 1997. Soybean disease loss estimates for the top 10 soybean producing countries in 1994, *Plant Disease* 81: 107–10.

Wydra, K., Fanou, A., Sikirou, R., Zandjanakou, M., Zinsou, V. and Rudolph, K. 2001. Integrated control of bacterial diseases of cassava and cowpea in West Africa. In: De Boer, S. H. (Ed.), *Plant Pathogenic Bacteria*. Proceedings of the 10th International Conference on Plant Pathogenic Bacteria, Charlottetown, Prince Edward Island, Canada, 23–27 July 2000. Springer, the Netherlands, pp. 280–7.

Yamanaka, N., Hossain, M. M. and Yamaoka, Y. 2015. Molecular mapping of Asian soybean rust resistance in Chinese and Japanese soybean lines, Xiao Jing Huang, Himeshirazu, and Iyodaizu B. *Euphytica* 205: 311. doi:10.1007/s10681-015-1377-4.

Zandjanakou-Tachin, M., Fanou, A., LeGall, P. and Wydra, K. 2007. Detection, survival and transmission of *Xanthomonas axonopodis* pv. *manihotis* and *X. axonopodis* pv. *vignicola*, causal agents of cassava and cowpea bacterial blight, respectively, in/by insect vectors. *Journal of Phytopathology* 155: 159–69.

Published by Burleigh Dodds Science Publishing Limited, 2019.

Chapter 11

Identifying and managing plant health risks for key African crops: vegetables

Ignace Godonou, *International Institute of Tropical Agriculture (IITA) and Catholic University of West Africa, Benin;* **May-Guri Sæthre**, *International Institute of Tropical Agriculture (IITA), Nigeria;* **Ghislain Tepa-Yotto**, *International Institute of Tropical Agriculture (IITA) and Université Nationale d'Agriculture, Benin;* **Désiré Gnanvossou** *and* **Ouorou Douro-Kpindou**, *International Institute of Tropical Agriculture (IITA), Benin; and* **Danny Coyne**, *International Institute of Tropical Agriculture (IITA), Kenya (corresponding author: ignace.godonou@hotmail.fr)*

1 Introduction

In Africa, the development of urban and peri-urban agriculture, with horticultural production as a key component, is being driven by the increasing food demand generated by the ever-growing urban population (James et al., 2007). Intensive peri-urban agriculture provides informal employment opportunities (Zossou, unpublished results). Its high-value crops are economically important for small- and large-scale growers alike and supply important proteins, vitamins and mineral elements of daily diets (Hallmann and Meressa, 2018).

In West Africa, vegetables are commonly grown in rain-fed upland as well as lowland ecologies (James et al., 2010). In a country like Benin, vegetable crops are grown year-round mainly in small plots in various intercropping and rotation schemes (James et al., 2010). Vegetable production is, however, undermined by a number of limiting factors, such as soil infertility, pests and diseases, weak irrigation infrastructure and poorly developed commercialization (James et al., 2006).

Land scarcity coupled with increased food demand by a growing urban population contribute to the intensification of vegetable production (James

http://dx.doi.org/10.19103/AS.2018.0043.12
Published by Burleigh Dodds Science Publishing Limited, 2019.

et al., 2010; Sæthre et al., 2011), which aggravates pest infestations (Beniest, 1987; d'Arondel de Hayes and Traoré, 1990; Bordat and Goudegnon, 1991). Pest management often receives minimal research inputs and, as a result, farmers' coping strategies against pest damage often consist of frequent and inappropriate use of synthetic pesticides, which results in a largely unsustainable and ineffective vegetable production (James et al., 2006).

Over-reliance on chemical pest control is driven by farmers' perceptions of pest pressure, zero tolerance of pest incidence, poor knowledge of pest management, lack of access to safe alternatives to toxic pesticides and poor public pressure on farmers to produce vegetables under more hygienic conditions.

Arthropod pests (insects and mites), disease-causing organisms (pathogens) and plant parasitic nematodes, mainly root knot nematodes (*Meloidogyne* spp.) are considered key biotic constraints to horticultural production. Many, like plant parasitic nematodes, leaf-feeding mites, moth caterpillars, whiteflies, fruit flies, leaf miner flies, aphids and some fungal pathogens, have a wide host range and cause economic damage to many vegetable crops by feeding on different parts of the plants.

Various authors have given an account of major pests of economic importance attacking vegetables and causing high levels of damage and yield losses in sub-Saharan Africa (Bordat and Goudegnon, 1991; Bordat and Arvanitakis, 2004; James et al., 2010; Ekesi et al., 2011). The IITA project 'Healthy Vegetables through Participatory Integrated Pest Management (IPM) in Peri-Urban Gardens of Southern Benin' established a baseline survey database on common vegetables and related pests and their incidence and economic importance (James et al., 2006). Table 1 lists arthropod pests, plant parasitic nematodes, pathogens with their taxonomic position, associated primary host vegetables and plant parts that they damage.

This chapter reviews key biotic constraints, mainly plant parasitic nematodes, disease-causing pathogens and arthropod pests that undermine vegetable production. The chapter gives an account of the International Institute of Tropical Agriculture's (IITA) contribution for improving crop health by reducing harmful pesticide regimes in vegetable production systems.

2 Biotic constraints to vegetable production

2.1 Plant parasitic nematodes

A range of plant parasitic nematodes attack vegetable crops, with root knot nematodes species as the most important soil-borne pests affecting vegetable crops (James et al., 2010). Being microscopic and not visible to the unaided eye, they are recognized primarily by the symptoms they cause on plant roots, for instance warty (knotted) and disfigured roots and tubers of carrots, beetroot and so on. Infection by *Meloidogyne* spp. also results in unspecific

Published by Burleigh Dodds Science Publishing Limited, 2019.

Table 1 Plant parasitic nematodes, pathogens and arthropod pests attacking horticultural crops in West Africa

Pest type	Pest species	Common vegetable affected	Plant part attack
Nematode	*Root knot nematodes,* Meloidogyne spp. (*Tylenchida: Meloidogynidae)*	African garden eggplant, aubergine, basil, carrot, lettuce, okra, pepper and tomato	Root
Pathogen	Bacterium, *Pectobacterium carotovorum*, (syn. *Erwinia carotovora*)	Beans, cabbage, garlic, ginger, lettuce, onion, peppers and tomato	Leaf, stem, root
	Bacterium*, Ralstonia solanacearum*	Aubergine, beans, ginger, pepper and tomato	Leaf
	Bacterium*, Xanthomonas campestris* pv. *Campestris*	Cabbage, cauliflower, radish	Cotyledons, Leaf
	Fungus, *Fusarium oxysporum* f. sp. *lycopersici*	Tomato	Root
	Fungus*, Phytopththora* spp.	African garden eggplant, amaranth, bean, pepper, potato and tomato	Root, stem, leaf
	Fungus, *Athelia rolfsii* (syn *Sclerotium rolfsii*)	African garden eggplant, bean, cabbage, carrot, ginger, lettuce, okra, pepper, tomato, *Vernonia amygdalina* ('bitter leaf')	Stem, leaf
Insect	Beet webworm, *Hymenia recurvalis* Fabricius (Lepidoptera: Pyralidae)	Amaranth, aubergine, French bean, beetroot, cucurbits (e.g. melon, watermelon)	Leaf
	Leaf caterpillar, *Phycita melongenae* Aina (Lepidoptera: Pyralidae)	African garden eggplant and aubergine	Leaf
	Leaf caterpillar, *Selepa docilis* Butler (Lepidoptera: Noctuidae)	African garden eggplant and aubergine	Leaf
	Leaf caterpillar, *Psara basalis* Walker (Lepidoptera: Pyralidae)	Amaranth, beetroot, cucurbits and radish	Leaf
	Cotton bollworm, *Helicoverpa armigera* Hübner (Lepidoptera: Noctuidae)	Tomato, bell pepper, onion, okra, cruciferous crops	Flower buds, leave, shoots, fruits
	Diamondback moth, *Plutella xylostella* Linnaeus (Lepidoptera: Yponomeutidae)	Brassica crops	Leaf
	Cabbage webworm, *Hellula undalis* Fabricius (Lepidoptera: Pyralidae)	Broccoli, cabbages, radish and sometimes aubergine	Stem, leaf

(Continued)

Table 1 *(Continued)*

Pest type	Pest species	Common vegetable affected	Plant part attack
	Cotton aphid, *Aphis gossypii* Glover (Hemiptera: Aphididae),	African garden eggplant, amaranth, cucumber, common bean, okra, peppers, watermelon, zucchini	Leaf
	Cowpea aphid, *Aphis craccivora* Koch (Hemiptera: Aphididae)	Amaranth, common bean, cowpea, hot pepper, okra, tomato	Leaf
	False cabbage aphid, *Lipaphis erysimi* Kaltenbach (Hemiptera: Aphididae)	Aubergine, cabbage, radish and turnip	Leaf
	Leaf miner fly, Liriomyza *spp. (Diptera: Agromyzidae)*	African garden eggplant, amaranth, cabbage, cucumber, lettuce, tomato and pepper	Leaf
	Whitefly, *Bemisia tabaci* Gennadius (Hemiptera: Aleurodidae)	African garden eggplant, aubergine, beans, cucumber, lettuce, okra, pepper and tomato	Leaf
	Spiralling whitefly, *Aleurodicus dispersus* Russell (Hemiptera: Aleurodidae)	African garden eggplant, pepper, amaranth, tomato, beans, cucurbit, wild ginger, okra	Leaf
	Plant bug, *Helopeltis schoutedeni* Reuter (Hemiptera: Miridae)	African garden eggplant, beans	Leaf
	Fruit borer, *Scrobipalpa ergasima* Meyrick (Lepidoptera: Gelechiidae)	Aubergine	Flower buds, fruit
	Fruit flies (*Bactrocera* spp., *Dacus* spp. and *Zeugodacus cucurbitae* Coquillett (Diptera: Tephrididae)	Cucurbits, solanaceous crops	Fruit
Mite	Broad mite, *Polyphagotarsonemus latus* Banks (Acari: Tarsonemidae)	African garden eggplant, aubergine, cucumber, pepper	Leaf
	Red mites, *Tetranychus ludeni* Zacher and *Tetranychus urticae* Koch (Acari: Tetranychidae)	African garden eggplant, amaranth, aubergine, bean, cucumber, pepper, tomato	Leaf
	Red mite, *Tetranychus evansi* Baker and Pritchard (Acari: Tetranychidae)	African garden eggplant, amaranth, aubergine, pepper, tomato	Leaf
	Aculops lycopersici Tryon (Acari: Eriophyidae)	African garden eggplant, tomato	Leaf, inflorescences, fruits

Published by Burleigh Dodds Science Publishing Limited, 2019.

symptoms, such as stunting, chlorosis, poor vigour, reduced yields and early crop senescence. Additionally, such infections render hosts more susceptible to other pests and diseases (James et al., 2010).

The unspecific nature of symptoms contributes to the underestimation of *Meloidogyne* spp. as serious pests. Estimations of crop losses in the tropics vary considerably, depending on the vegetable crop, *Meloidogyne* spp., level of infection and cropping system, reaching 100% (Hallmann and Meressa, 2018). Across Africa, diagnostic surveys of *Meloidogyne* spp. associated with vegetables have established that *M. javanica* and *M. incognita* are the most commonly occurring species. Interestingly, some species, such as the highly aggressive *M. enterolobii* are very common in West Africa and less so elsewhere, while other *Meloidogyne* species occurred more sporadically (James et al., 2006; Pagan et al., 2015; Janssen et al., 2016).

A number of *Meloidogyne* spp. were observed for the first time, such as *M. izalcoensis* (Jorge Junior et al., 2016), or - mostly in cooler climates - *M. hapla,* or turned out to be new, undescribed species. *Meloidogyne* spp., a formidable pest group, are key pests in both smallholder and commercial vegetable production systems. Their unspecific damage symptoms make diagnosis from field observations difficult. They often occur as multiple species communities, while individual species tend to be highly polyphagous, able to infect an astonishing range of crop and plant species. This creates immense difficulties for their management, especially where resistance to one species is not conferred to another species.

2.2 Arthropod pests

Arthropod pests are known to be among the main biotic constraints for vegetable production. Those regularly encountered and causing substantial damage on various vegetables crops in West Africa are listed in Table 1. The beet webworm, *Hymenia recurvalis,* and the leaf caterpillar, *Psara basalis,* are the most economically important pests regularly feeding on amaranth. The leaf caterpillar, *Selepa docilis*; the flower bud caterpillar, *S. ergasima;* and the broad mite, *Polyphagotarsonemus latus,* are economically important pests that undermine the production of Africa garden eggplant (Bordat and Goudegnon, 1991; Bordat and Arvanitakis, 2004; James et al., 2010).

The broad mite, *P. latus* attacks a wide range of vegetables (Table 1) and many other non-vegetable crops. It is the most economically important mite pest on solanaceous species and particularly on the African garden eggplant in Benin. This tiny tarsonemid mite occurs throughout the world and has been reported on more than 60 plant species (Adango et al., 2007). By sucking sap from leaves, it causes tiny pale spots and leaves narrowing, tapering and yellowish marks leading to severe food and economic losses (James et al.,

2010). Its damage symptoms, leaf curling, can easily be mistaken for virus infection (Smith Meyer, 1996).

The problem of spider mites in vegetable crops has recently been exacerbated by the accidental introduction of the tomato red spider mite *Tetranychus evansi,* a highly destructive invasive alien species attacking solanaceous crops in Africa. Originating from Brazil, it was first encountered in Southern Africa in the 1980s, from where it moved to Eastern and Central Africa within two decades. In 2008, outbreaks of *T. evansi* were found causing severe damage on tomato and eggplant in horticultural settings at Sémé-Kpodji in Southern Benin, with yield losses estimated at 65% for tomato, 56% for African garden eggplant and 25% for amaranth (Azandémè-Hounmalon et al., 2015). The same study reported increased and inappropriate pesticide use to control this pest, often using doses higher than those recommended in the belief that these will give better control. Foreign explorations in Southern Brazil meanwhile identified a population of the phytoseiid predator *Phytoseiulus longipes* Evans (Acari: Phytoseiidae), which was subsequently introduced into Kenya for the successful control of the pest (Fiaboe et al., 2006; Furtado et al., 2006). It is currently under evaluation for its prospective introduction into Benin from Kenya for mass production and experimental releases in a classical biological control approach.

The African bollworm, *Helicoverpa armigera,* is a pest of major importance in most areas where it occurs and damages a wide range of food, fibre, oilseed, fodder and horticultural crops. Its considerable significance is based on the peculiarities of its biology: its mobility, polyphagy, high reproductive rate and diapause, which make it particularly well adapted to exploit transient habitats such as man-made agro-ecosystems (Greathead and Girling, 1989).

Among vegetable crops, *H. amigera* mainly damages tomato, where, young caterpillars feed on flowers, and bore into and feed on tomato, which rot later. Before pupating, a single caterpillar can destroy many tomato fruits. Yield losses range from 20% to 50% (Kibata, 2002). Biological control options for the control of *H. armigera* in Africa have been discussed by Cherry et al. (2003) and will require situation-specific combinations of, for example conservation biocontrol and the application of microbial pesticides in order to develop and improve IPM strategies in vegetable agro-ecosystems against this pest.

The cabbage webworm *Hellula undalis*, diamondback moth *Plutella xylostella* and the false cabbage aphid *Lipaphis erysimi* are important insect pests that limit cabbage production in West Africa. If cabbage seedlings or young plants are attacked by *H. undalis*, the plants wilt and die quickly, causing severe losses in cabbage nurseries and newly transplanted cabbage fields. Mature plants respond by producing many growing points forming several small heads, which are unsuitable for consumption or sale, resulting in severe economic losses (Bordat and Arvanitakis, 2004; James et al., 2010).

Published by Burleigh Dodds Science Publishing Limited, 2019.

Plutella xylostella is an economically important pest of cruciferous crops, particularly cabbage in urban and peri-urban areas of West Africa (Bordat and Goudegnon, 1991), and globally (Talekar and Shelton, 1993). On cabbage, its caterpillars voraciously eat chlorophyll from the undersides of leaves, growing buds and cabbage heads. The damage leaves a papery epidermis of the upper leaf surface intact and translucent windows in the leaf blades. The 'windows' later break down and appear as irregular holes in the leaves, resulting in complete loss of the crop (James et al., 2010). Chemical control is a common coping strategy of vegetable producers against this pest. On a global scale, diamondback moth control is estimated to cost about US$4–5 billion annually (Zalucki et al., 2012). Diamondback moth is known to have developed resistance to a wide range of insecticides including *Bacillus thuringiensis* Berliner, a widely used biological agent (Liu et al., 1995). New biological control approaches are described under Section 4.1.

Aphids are recognized worldwide as economically important pests (Remaudière and Autrique, 1985). Faunistic studies carried out by Bioforsk/IITA in Benin recorded aphid species and associated natural enemies in vegetable production systems (Sæthre et al., 2011) culminating in new biological control programmes (see Section 4.4). In vegetable production systems in Benin and most West African countries, *Lipaphis erysimi* and *Aphis gossypii* are the main economically important aphids. *L. erysimi* inflicts substantial direct damage to cruciferous crops by sucking sap from the leaves and can also be a vector of viruses that cause diseases in plants. *A. gossypii* causes feeding damage attacking particularly solanaceous and cucurbit crops, where it also transmits, for example watermelon mosaic virus, cucumber mosaic virus and zucchini yellow mosaic virus as discussed in Part 1. Additional losses may be incurred by possible attraction of other pests to honeydew-contaminated surfaces (Slosser et al., 1989).

Leaf miner flies pose enormous problems to both small- and large-scale horticultural producers seeking to respond to the new market opportunities around the world. They are listed as quarantine pests in a number of countries, and therefore also affect trade. The damage inflicted on their host plants is similar for all species. The larval stages feed within the leaves of the host plants and, at high densities, this feeding can severely reduce yields and/or kill the plants (Spencer, 1973). Economic losses to vegetables and cut flowers are substantial (Waterhouse and Norris, 1987), but of minor importance in Benin (James et al., 2010).

Farmers often respond to leaf miner attack by heavy applications of pesticides. Insecticide-resistant populations are therefore fast developing (Murphy and LaSalle, 1999) and the concurrent loss of natural enemies resulting from indiscriminate synthetic pesticide applications exacerbates the situation (Ewell et al., 1990; Hofsvang et al., 2005). Indigenous natural enemy communities of *Liriomyza* spp., particularly parasitoids, are diverse within their native ranges and there is evidence that in pesticide-free areas, together with polyphagous arthropod predators, they can control leaf miners. Since parasitoids of agromyzid are polyphagous, they can readily exploit alien invasive leaf miners also (Neuenschwander et al., 1987; Murphy and LaSalle, 1999).

Whitefly, *Bemisia tabaci,* belongs to a small group of insects that could reasonably be considered the most globally destructive pests (Legg et al., 2003). It is an important pest of many vegetable crops and often reported as vector of a range of leaf curl disease-inducing geminiviruses on tomato and therefore an increasingly severe constraint to tomato production (Pietersen et al., 2000). In Benin, the main damage to the crop is caused by viruses transmitted by *B. tabaci*, for example Tomato Yellow Leaf Curl Virus (Bordat and Arvanitakis, 2004). In severely affected tomato plants, flowers abort and fruit set fails. Whiteflies also secrete honeydew that encourages the growth of black sooty mould (fungus) on plant leaves and stems, which reduces the plant's ability to absorb light (Bordat and Arvanitakis, 2004; Legg et al., 2003).

Tephritid fruit flies are a highly important threat to horticultural production systems. They are highly polyphagous attacking several important fruits and vegetables (Roessler, 1989; White and Elson-Harris, 1992; Ekesi and Billah, 2007; Goergen et al., 2011). Fruit flies damage crops in two ways. Firstly, females cause direct damage by puncturing the surface of fruits to lay eggs underneath. In the process, the insects introduce bacteria which cause tissues of the fruits to rot (James et al., 2010). Fruit flies cause both direct damage to the fruits and indirect losses due to their quarantine pest status. Their mere presence in a particular country therefore already restricts free trade and export of fresh horticultural produce to large lucrative markets abroad (Ekesi et al., 2011).

In Benin and West Africa generally, several species of tephritid fruit flies are recognized as the most economically important group of arthropods attacking cucurbit and other fruit vegetables (Vayssières and Carel, 1999; Goergen et al., 2011; Gnanvossou et al., 2017). Flower and fruit losses range from 10% to 100% depending on cucurbit species and season (Gupta and Verma, 1992; Dhillon et al., 2005; Sakpota et al., 2010; Deguine et al., 2012).

Farmers' coping strategy to control fruit flies has been mainly limited to the use of synthetic chemical insecticides, as natural enemies are rarely perceived as being efficient in controlling fruit fly populations to avoid crop damage (White and Elson-Harris, 1992). A more recent review on control measures and techniques to control fruit flies by Ekesi et al. (2006) concluded that an integrated approach to the management of fruit flies offers better prospects for the horticulture industry than single control options. Components include bait sprays, soil inoculation with entomopathogenic organisms, orchard sanitation and biological control. Fruit fly control on mango and citrus sometimes involves the same species as on vegetables and is discussed in more detail in the accompanying chapter on fruit and other tree crops in Part 2.

2.3 Pathogens

The key plant pathogenic organisms observed to negatively impact horticultural production are presented in Table 1. However, IITA has primarily been involved with collaborators in the investigation of the bacterium *Ralstonia solanacerum*. This pathogen constitutes a serious obstacle to the cultivation of many solanaceous plants in both tropical and temperate regions (review by Tahat and Sijam, 2010). *R. solanacearum* causes rapid wilting of the entire plant. Damage symptoms begin on the youngest leaves, which become flaccid, followed by rapid wilting of the entire plant without yellowing of the leaves. The greatest economic damage by *R. solanacearum* has been reported on potato, tomato and other, non-vegetable crops in various countries around the world with average losses of 15% on tomato and 10% on aubergine and pepper (Elphinstone, 2005). Disease severity mostly increases if *R. solanacearum* is associated with plant parasitic nematodes, especially *Meloidogyne* spp., as demonstrated in India, where the combined pathogenic effects of *R. solanacearum* and *Meloidogyne javanica* were greater than the independent effects of either (Sitaramaiah and Sinha, 1984).

Collaborative studies led by the Benin National Institute of Agricultural Research (INRAB) have identified *R. solanacearum* as the causal agent of tomato bacterial wilt for the first time in Benin in 2009 (Sikirou et al., 2009). Subsequent studies have allowed assessing the distribution and incidence of this disease in the different agro-ecological zones of Benin (Sikirou et al., 2017). *R. solanacearum* was present in Southern and Central Benin, and in two localities growing solanum potato in the far North. The biochemical characterization of these isolates revealed that they belong to pathovar I and III (race 1), both being highly virulent to tomato plants. Because this is a relatively recent discovery, control options so far have been limited to abandoning tomato cultivation. However, there exist several control options, which have been summarized by Nion and Toyota (2015). These include the use of soil amendments, biological control agents such as beneficial microorganisms and grafting onto resistant rootstock, which hold promise for providing tomato farmers with an integrated package against this destructive disease.

The other pathogens affecting vegetable production are briefly summarized from James et al. (2010). The bacterium *Pectobacterium carotovorum* occurs in soil, stems and leaves which are in close contact with the soil. *P. carotovorum* damages foliage of a wide range of vegetables for example beans, cabbage, garlic, ginger, lettuce, onion, peppers and tomato. The bacterium *Xanthomonas campestris* pv. *campestris* causes black rot disease. In infected young plants of cabbage, the cotyledons are discoloured and fall prematurely. In older plants, the damage symptoms appear as lesions along the edges of the lower leaves. As the disease advances, veins darken, leaves wilt and drop. Worldwide, black

Published by Burleigh Dodds Science Publishing Limited, 2019.

rot is the most important disease of crucifers, particularly on cabbage. The fungus *Fusarium oxysporum* f. sp. *lycopersici* damages roots of tomato, making this fungus an economically important pathogen. It is recommended to avoid planting tomatoes in fields with a known history of *Fusarium* wilt, though quality control efforts will probably miss low levels of infection on seeds. *Phytophthora* species mainly occur on roots, stems and foliage of plants. These fungi damage roots and foliage of vegetables and are particularly important on amaranth and tomato. Their dry rot kills seedlings more easily than mature plants, while rapid blighting of leaves occurs mostly during warm moist periods causing 100% crop loss. The fungus *Athelia rolfsii* occurs in the soil and on plant debris. Infected plants show progressive yellowing and wilting of the leaves. The pathogen is best recognized by a white mat of fungal mycelia at the bases of infected stems or just below the soil surface as well as on leaf debris. *A. rolfsii* is an economically important pathogen on tomato and pepper.

3 Problems in pesticide use

Field diagnoses carried out by IITA in vegetable production systems identified severe pest problems and revealed alarming pesticide practices in Southern Benin (James et al., 2010). Farmers frequently applied broad-spectrum foliar pyrethroids, highly toxic organophosphates and carbamates, which are mostly imported for other cash cropping systems, principally cotton. Vegetable producers applied at least 15 different chemical pesticides and many bought and used chemical pesticides without prior training in pesticide safe use practices (James et al., 2006).

Cabbage, tomato and African garden eggplant stood out as indicator crops of harmful pesticide regimes. In attempts to control *Plutella xylostella,* cabbage producers applied up to 46.8 L pesticide concentrate per ha on the crop in 19 applications. Talstar (bifenthrin) was applied 12 times and Decis (deltametrine) 7 times on each cabbage bed within 3 months prior to harvest. Similarly, tomato received 14 applications of pesticide concentrates amounting to 16.4 L of endosulfan and Topsin-M (thiophanate-methyl) and 28.8 g of fenpropathrin per ha within 10 weeks of crop growth for the control of the eriophyid mite *Aculops* sp., *Helicoverpa armigera*, *Meloidogyne* spp. and whiteflies. On African garden eggplants, farmers applied 14 L of concentrates of Talstar and Decis, and 27.8 g of furadan (carbofuran) and maneb (which includes the related active ingredients mancozeb and metiram) per ha in 12 applications within 10 weeks to control the broad mite and root rot nematodes.

Biologically based alternatives were rarely used by vegetable growers. They included aqueous leaf extracts of tobacco, *Nicotiana tabacum* L. (Solanaceae), neem, *Azadirachta indica* A. Juss. (Meliaceae), and papaya, *Carica papaya* L. (Caricaceae), against foliage feeders. Where botanical products were used, farmer experimentation and training in seed and leaf extract preparation, dosage and application frequency were still required (James et al., 2006, 2010).

Subsequent studies by Sæthre et al. (2012) in Benin detected worrying levels of pesticide residues involving 41 different pesticides and/or metabolites. Leaves of African garden eggplant contained the highest number of residues and also the highest concentrations of pesticides. One sample had a very high content of dimethoate and omethoate (sum 8.1 mg/kg). Consumption of such leaves indeed represents a health risk, as both the acceptable daily intake and acute reference dose were far exceeded. Similar conclusions were drawn from a thorough study in Ghana (Donkor et al., 2016).

4 Biological alternatives to pesticides for pest control

4.1 Diamondback moth Plutella xylostella

Biological options in an integrated pest management (IPM) approach offer a solution to sustainable control of *P. xylostella*. In West Africa, farmers use botanical pesticides, mostly aqueous seed extracts of the neem tree against *P. xylostella* and a wide range of other arthropod pests (Goudegnon et al., 2000). Inoculative releases of the parasitoids *Cotesia plutellae* Kurdyumov (Hymenoptera: Braconidae), *Diadromus collaris* Gravenhorst (Hymenoptera: Ichneumnidae), *Oomyzus sokolowskii* Kurdjumov (Hymenoptera: Braconidae) and *Microplitis plutellae* Muesebeck (Hymenoptera: Braconidae) have been attempted against *P. xylostella* in Africa, with success in a few cases (Yaseen, 1978; Kfir, 2003). Microbial pesticides, for example, the granulovirus (PlxyGVNay01) endemic to Kenya, have shown success against *P. xylostella* in field trials (Ogutu et al., 2004).

In Brazil five isolates of the entomopathogenic fungi *Beauveria bassiana* (Bals.-Criv.) Vuill. and *Metarhizium anisopliae* (Metchnikoff) Sorokin (both (Clavicipitaceae) tested against *P. xylostella* caused larval mortalities ranging from 26% to 96% (Valda et al., 2003). *B. bassiana* and *M. anisopliae* are ubiquitous in nature, specific to target pests, persist in the environment and are easy to mass produce. They are therefore increasingly gaining attention as commercial microbial pesticides (Wraight et al., 2001; SP-IPM, 2006). IITA maintains isolates of indigenous *B. bassiana* and *M. anisopliae* in the institute's microbial collection at its station in Benin. The isolates were collected from the insect pests *Sesamia calamistis* Hamp. (Lepidoptera: Noctuidae) and *Eldana saccharina* Walker (Lepidoptera: Pyralidae) (Godonou et al., 2009) and eight of them were screened in the laboratory for their virulence against *P. xylostella*. Mortality of *P. xylostella* caused by Bba5653 was 92% at two days after treatment (DAT) and 94% at 6 DAT, and was significantly higher than that for the other isolates tested (Godonou et al., 2009). Thirty-one percent DBM larvae killed by Bba5653 developed signs of mycosis (mycelial growth or sporulation), significantly higher than the 14% achieved for Ma182, the next best isolate. In farmer participatory field trials with 28 community organizers at seven farmer field school locations, Bba5653 showed the highest virulence to control *P. xylostella* larvae and pupae.

Under farmers' field conditions, Bba5653 was more efficient in *P. xylostella* control than the insecticide Talstar (bifenthrin). Across sites, *P. xylostella* population densities were consistently higher in plots treated with the chemical insecticide Talstar and in control plots than in those treated with Bba5653. Additionally, cabbage yield was at least 2.5 times higher in plots treated with the biopesticide than in plots treated with Talstar or left untreated, indicating that *P. xylostella* had become resistant to this insecticide (Godonou et al., 2009). However, because of the very high dose of Bba5653 used to obtain the above results (1000 g of conidia powder/ha) trials were initiated to test potential dose reductions. They demonstrated that the application of as low as 12.5 g of conidia powder/ha was able to control *P. xylostella* similar to conventional applications of deltamethrin (O. Douro Kpindou, unpublished results). More recent studies by K. Agboyi et al. (unpublished results) indicated that an optimal dosage of 53 g conidia powder of Bba5653 was able to significantly suppress the population of *P. xylostella*.

4.2 Beet webworm, **Hymenia recurvalis** *and leaf caterpillar* **Psara basalis**

In another screening test, *B. bassiana* isolates for virulence to Lepidoptera pests of amaranth, *B. bassiana* Bba5648 isolate proved more virulent to *H. recurvalis* than any other isolate tested. Bba5648 caused 100% mortality of *H. recurvalis* larvae 5 days after inoculation and 83% of the dead larvae showed signs of mycosis (fungal sporulation). Similarly, the isolate Bba5644 caused 100% mortality of *P. basalis* larvae with 83% dead larvae showing fungal sporulation (James et al., 2007). Evidently, Bba5648 and Bba5644 are potential biological candidates to control *H. recurvalis* and *P. basalis,* respectively.

4.3 Root rot nematodes **Meloidogyne** *spp.*

The incorporation of large volumes of organic material into the soil, in general, reduces *Meloidogyne* spp. damage (Viaene et al., 2013; Hallman and Meressa, 2018). Various sources of organic material, such as oil cake, coffee husks, neem, marigold leaves, crustacean chitin waste, sawdust, chicken manure and bagasse have been used with varying success (Muller and Gouch, 1982; Stirling, 1991; Sikora, 1992). Farmer participatory field trials showed that cassava epidermal peels, fresh or in dried powdered form, have nematicidal properties and can be promising soil amendments in the control of *Meloidogyne* spp. (James et al., 2006, 2010).

The addition of organic matter to soil helps in protecting crop roots either through chemical metabolites or through the improvement of the soil environment, encouraging greater microbial activity that helps suppress nematodes. However, one setback to this approach is access to sufficient quantities, or balancing the costs of transport of sufficient quantities (James et al., 2010). Of critical importance is the use of healthy seedlings as a standard practice and creating a mind-set with farmers to embrace this approach. Initially, farmers need to understand the benefits of investing greater resources in more costly healthy seedlings, compared with traditional, field cultivated seedlings.

In studies by IITA and partners in East Africa, the use of healthy seedlings, reared in seedling trays under protected screen houses, led to higher yields of better quality tomato fruits, compared with farmer seedlings (D. Coyne, unpublished results). Furthermore, healthier seedlings in general lowered incidence of pests and diseases, and when combined with improved IPM practices reduced reliance on chemical pesticides by approximately 70%.

4.4 Aphids

Field surveys carried out by the Bioforsk/IITA project revealed six aphid species and seven natural enemies of aphids occurring in vegetable production systems in Benin (Sæthre et al., 2011). Aphid natural enemies recorded included three coccinellid and one syrphid predator, two primary parasitoids *Aphelinus* sp. near *ficusae* Prinsloo and Neser (Hymenoptera: Aphelinidae) and *Lysiphlebus testaceipes* Cresson (Hymenoptera: Braconidae), and one entomopathogen *Neozygites* sp. The surveys indicated that three aphid predators were widely distributed in all agro-ecological zones of Benin (Sæthre et al., 2011), among them, *Cheilomenes sulphurea* Olivier (Coleoptera: Coccinellidae), which has already been identified as one of the key natural enemies of *Aphis craccivora* in Uganda (Davies, 1972) and has been imported to France (Hodek et al., 1984).

The surveys also provided the first records of the aphid parasitoid *L. testaceipes* and *Aphis spiraecola* Patch from West Africa (Sæthre et al., 2011). The parasitoid *L. testaceipes* had already been introduced to Kenya in 1911 for the biological control of *Schizaphis graminum* Rondani (Starý et al., 1985; Greathead, 2003) and has long been established in South Africa (Starý et al., 1985). It is unclear how *L. testaceipes* appeared in Benin, but we speculate that it was slowly spreading from the original establishments in Kenya, as there have never been any targeted introductions in West Africa.

To assess the potential of *L. testaceipes* as a biocontrol candidate against aphids in horticultural crops, field trials were conducted from April to July 2011 on the IITA-Benin station to examine the responses of naturally occurring *L.*

testaceipes on sweet pepper plots initially infested with different densities of *A. gossypii*. These studies revealed that *L. testaceipes* was attacked by the hyperparasitoid *Syrphophagus africanus* (Gahan) (Hymenoptera: Encyrtidae). Three predatory species of aphids were recorded: *Cheilomenes propinqua* (Mulsant), *C. sulphurea* (Olivier) (Coleoptera: Coccinellidae) and *Ischiodon aegyptius* (Wiedemann) (Diptera: Syrphidae). The results showed that *L. testaceipes* responded more quickly to increasing aphid densities than the predators (Fig. 1; G. Tepa-Yotto et al., unpublished results). This supports earlier findings by Fernandes et al. (1998) indicating that *L. testaceipes* can maintain aphid densities below the economic threshold, if introduced early enough as augmentative biocontrol agent. Furthermore, these field studies established that *L. testaceipes* and the three aphid predators can be found simultaneously on *A. gossypii*. These findings corroborate observations by Rutledge et al. (2004) that *L. testaceipes* is essential for the sustainable management of aphids.

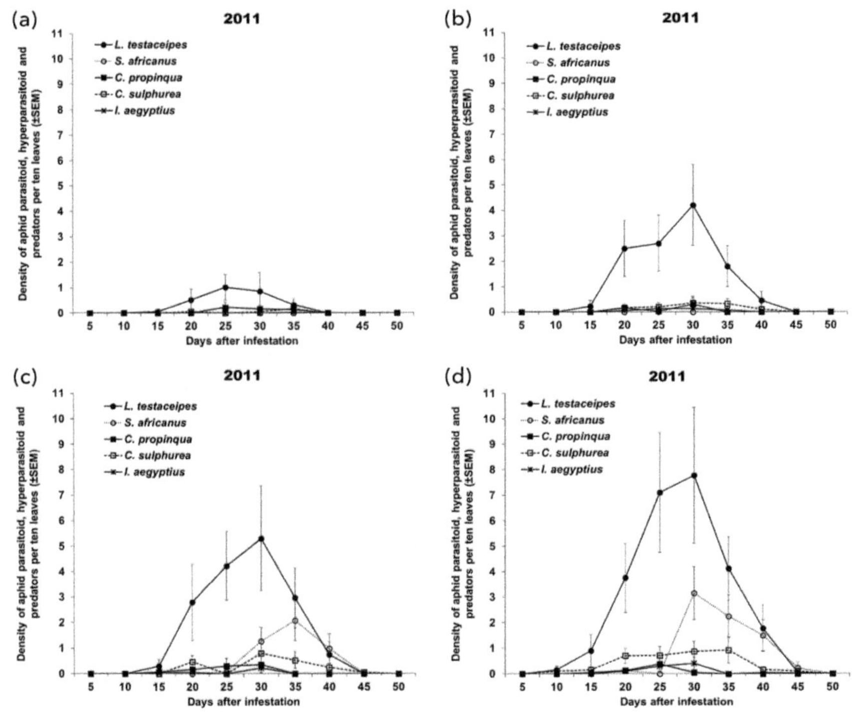

Figure 1 Field occurrence of the aphid parasitoid *Lisyphlebus testaceipes*, its hyperparasitoid *Syrphophagus africanus*, competing coccinelids *Cheilomenes propinqua* and *Cheilomenes sulphurea* and hoverfly *Ischiodon aegyptius* on sweet pepper plants initially infested with varying densities of *Aphis gossypii* [zero (a), ten (b), fifty (c) and one hundred (d) aphids].

4.5 Mites

Diagnostic surveys of major mite pests and associated natural enemies on two indigenous leafy vegetables, amaranth and African garden eggplant, carried out by the IITA/SDC vegetable project revealed broad mite, *Polyphagotarsonemus latus*, on African garden eggplant and, in addition, *Tetranychus ludeni* and *T. urticae* on amaranth (Adango et al., 2007). The surveys also identified potential indigenous phytoseiid species associated with the mite pests and demonstrated that these predatory phytoseiid mites could control *P. latus* in the vegetable production systems when integrated in an environment free of chemical pesticides.

5 Conclusions

In sub-Saharan Africa, intensive vegetable production in urban and peri-urban areas has become the fastest growing agricultural subsector contributing significantly to food security, national incomes, livelihoods of people and informal employment opportunities. Under pressure from land scarcity and increasing food demands for escalating urban populations, abuse of synthetic chemical pesticides against the numerous vegetable pests and diseases has become the norm.

Field surveys carried out by IITA indicated severe health and environmental impacts similar to those that followed the agricultural intensification in Asia, where intensified agricultural production paralleled the intensification of input use, with the result that the health of marginal farmers was undermined by exposure to highly toxic pesticides (Mrema et al., 2017; Schreinemachers et al., 2017).

These insecticides destroyed indigenous parasitoids and predators thereby creating new pests such as many leaf miners and tetranychid mites. Elsewhere in Africa, urban and peri-urban areas followed a similar pattern (Williamson et al., 2008).

There is an urgent need to develop IPM strategies with focus on the use of biocontrol agents and biopesticides, combined with host resistance and sound and sustainable cultural practices, such as adequate crop rotation, intercropping, crop association and incorporation of organic manure. IITA has established strong baseline information on pest diversity, distribution and importance, as well as impacts and consequences of pesticide use in intensive vegetable systems. This information should now be used to test and implement IPM strategies elsewhere in Africa, where vegetable production and pest problems are similar. Consequently, intensive training of farmers and extension agents and increased public pressure on farmers to produce vegetables under more hygienic conditions are needed.

Published by Burleigh Dodds Science Publishing Limited, 2019.

Apart from the described key biotic constraints, other limiting factors undermine plant health such as soil infertility and weak irrigation infrastructure. Vegetable production moreover suffers from poorly developed commercialization including labelling 'organic' produce. As vegetables are perishable, a well-developed road and transport system coupled with efficient storage and cold-chain capacities is needed to avoid post-harvest losses and to improve farmers' income and livelihood.

6 References

Adango, E., Onzo, A., Hanna, R., Atachi, P. and James, B. (2007). Mite pests of major importance on indigenous leafy vegetables in Benin: Search for appropriate control strategies. In: Chadha, M. L., Kuo, G. and Cowda, C. L. L. (Eds), *Proceedings of the 1st International Conference on Indigenous Vegetables and Legumes Prospectus for Fighting Poverty, Hunger and Malnutrition*. Hyderabad, India, 12–15 December 2006. *Acta Horticulturae* 752, pp. 311–17.

Azandémè-Hounmalon, G. Y., Affognon, H. D., Komlan, F. A., Tamò, M., Fiaboe, K. K., Kreiter, S. and Martin, T. (2015). Farmers' control practices against the invasive red spider mite, *Tetranychus evansi* Baker & Pritchard in Benin. *Crop Protection* 76, 53–8.

Beniest, J. (1987). *Guide pratique du maraîchage au Sénégal. Centre de Développement de l'Horticulture*. Collection cahier d'information no. 1, Clairafrique, Dakar, 144pp.

Bordat, D. and Arvanitakis, L. (2004). *Arthropodes des cultures légumières d'Afrique de l'Ouest, Centrale, Mayotte et Réunion*. Martine Duportal, CIAD, Montpellier 5, 291pp.

Bordat, D. and Goudegnon, E. (1991). *Catalogue des principaux ravageurs des cultures maraîchères au Bénin*. CIRAD-IRAT, Montpellier, France, 15pp.

Cherry, A., Cock, M., van den Berg, H. and Kfir, R. (2003). Biological control of *Helicoverpa armigera* in Africa. In: Neuenschwander, P., Borgemeister, C. and Langewald, J. (Eds), *Biological Control in IPM Systems in Africa*. CAB International, Wallingford, UK, pp. 329–46.

d'Arondel de Hayes, J. and Traoré, G. (1990). *Cultures maraîchères en zone Soudano-Sahéliennes; recueil de fiches techniques*. CNRST/CIRAD/IRAT/Ministère des enseignements secondaires, supérieurs et de la recherche scientifique, Ouagadougou, 79pp.

Davies, J. C. (1972). Studies on the ecology of *Aphis craccivora* Koch (Hem.: Aphididae), the vector of rosette disease of groundnuts in Uganda. *Bulletin of Entomological Research* 62, 169–82.

Deguine, J.-P., Atiama-Nurbel, T., Douraguia, E., Chiroleu, F. and Quilici, S. (2012). Species diversity within a community of the cucurbit fruit flies *Bactrocera cucurbitae, Dacus ciliatus*, and *Dacus demmerezi* roosting in corn borders near cucurbit production areas of Reunion Island. *Journal of Insect Science*, Published online, 1 March 2012. doi:10.1673/031.012.3201.

Dhillon, M. K., Singh, R., Naresh, J. S. and Sharma, H. C. (2005). The melon fruit fly, *Bactrocera cucurbitae*: A review of its biology and management. *Journal of Insect Science* 5, 40.

Donkor, A., Osei-Fosu, P., Dubey, B., Kingsford-Adaboh, R., Ziwu, C. and Asante, I. (2016). Pesticide residues in fruits and vegetables in Ghana: A review. *Environmental Science and Pollution Research* 23, 18966–87.

Ekesi, S. and Billah, M. K. (2007). *A Field Guide to the Management of Economically Important Tephritid Fruit Flies in Africa*. ICIPE Science Press, Nairobi, Kenya, 98pp.

Ekesi, S., Mohamed, S. A., Hanna, R., Lux, S. A., Gnanvossou, D. and Bokonon-Ganta, A. (2006). Fruit fly suppression – purpose, tools and methodology. In: Ekesi, S. and Billah, M. K. (Eds), *A Field Guide to the Management of Economically Important Tephritid Fruit Flies in Africa*, 2nd edition. The International Centre of Insect Physiology and Ecology (ICIPE) Science Press, Nairobi, Kenya, pp. D1–D15.

Ekesi, S., Chabi-Olaye, S., Subramanian, S. and Borgemeister, C. (2011). Horticultural pest management and the African economy: Successes, challenges and opportunities in a changing global environment. In: Wesonga, J. and Kahane, R. (Eds), *Proceedings of the First all African Horticultural Congress*. Nairobi, Kenya, 31 August to 3 September 2009. *Acta Horticulturae* 991, 165–83.

Elphinstone, J. G. (2005). The current bacterial wilt situation: A global overview. In: Allen, C., Prior, P. and Hayward, A. C. (Eds), *Bacterial Wilt Disease and the Ralstonia solanacearum Species Complex*. American Phytopathological Society Press, St Paul, MN, pp. 9–28.

Ewell, P., Fano, H., Raman, K. V., Alcazar, J., Palacios, M. and Carhuamaca, J. (1990). *The leaf miner fly in Cañete: A pesticide treadmill. Farmer management of potato insects in Peru*. Food Systems Research Series 6, International Potato Center, Lima, Peru, pp. 37–45.

Fernandes, O. A., Wright, R. J. and Mayo, Z. B. (1998). Parasitism of greenbugs (Homoptera: Aphididae) by *Lysiphlebus testaceipes* (Hymenoptera: Braconidae) in grain sorghum: Implications for augmentative biological control. *Journal of Economic Entomology* 91, 1315–19.

Fiaboe, K. K. M., Fonseca, R. L., de Moraes, G. J., Ogol, C. K. P. O. and Knapp, M. (2006). Identification of priority areas in South America for exploration of natural enemies for classical biological control of *Tetranychus evansi* (Acari: Tetranychidae) in Africa. *Biological Control* 38, 373–9.

Furtado, I. P., de Moraes, G. J., Kreiter, S. and Knapp, M. (2006). Search for effective natural enemies of *Tetranychus evansi* in south and southeast Brazil. *Experimental and Applied Acarology* 40, 157–74.

Gnanvossou, D., Hanna, R., Goergen, G., Salifu, D., Tanga, C. M., Mohamed, S. A. and Ekesi, S. (2017). Diversity and seasonal abundance of tephritid fruit flies in three agro-ecosystems in Benin, West Africa. *Journal of Applied Entomology* 141, 798–809.

Godonou, I., James, B., Atcha-Ahowé, C., Vodouhè, S., Kooyman, C., Ahanchédé, A. and Korie, S. (2009). Potential of *Beauveria bassiana* and *Metarhizium anisopliae* isolates from Benin to control *Plutella xylostella* L. (Lepidoptera: Plutellidae). *Crop Protection* 28, 220–4.

Goergen, G., Vayssières, J-F., Gnanvossou, D. and Tindo, M. (2011). *Bactrocera invadens*, a new fruit fly problem in the Afrotropics: Distribution and host-plant range in West and Central Africa. *Environmental Entomology* 40, 844–54.

Goudegnon, E. A., Kirk, A. A., Schiffers, B. and Bordat, D. (2000). Comparative effects of deltamethrin and Neem kernel to *Plutella xylostella* and *Cotesia plutellae* populations in Cotonou periurban area (Benin). *Journal of Applied Entomology* 124, 141–4.

Greathead, D. J. (2003). Historical overview of biological control in Africa. In: Neuenschwander, P., Borgemeister, C. and Langewald, J. (Eds), *Biological Control in IPM Systems in Africa*. CABI, Wallingford, UK, pp. 1-26.

Greathead, D. J. and Girling, D. J. (1989). Distribution and economic importance of *Heliothis* and of their natural enemies and host plants in southern and eastern Africa. In: King, E. G. and Jackson, R. D. (Eds), *Proceedings of the Workshop on Biological Control of Heliothis: Increasing the Effectiveness of Natural Enemies*. 11-15 November 1985, New Delhi, India. Far Eastern Regional Research Office, United States Department of Agriculture, New Delhi, India, pp. 329-45.

Gupta, D. and Verma, A. K. (1992). Population fluctuations of the maggots of fruit flies *Dacus cucurbitae* Coquillett and *D. tau* (Walker) infesting cucurbitaceous crops. *Advances in Plant Science* 5, 518-23.

Hallmann, J. and Meressa, B. (2018). Nematode parasites of vegetables. In: Sikora, R. A., Coyne, D., Hallmann, J. and Timper, P. (Eds), *Plant Parasitic Nematodes in Subtropical and Tropical Agriculture*, 3rd edition. CAB International, Wallingford, UK, pp. 346-410.

Hodek, I., Chakrabarti, S. and Rejmanek, M. (1984). The effect of prey density on food intake by adult *Cheilomenes sulphurea* (Col.: Coccinellidae). *BioControl* 29, 179-84.

Hofsvang, T., Snøan, B., Andersen, A., Heggen, H. and Anh, L. N. 2005. *Liriomyza sativae* (Diptera: Agromyzidae), an invasive species in South-East Asia: Studies on its biology in northern Vietnam. *International Journal of Pest Management* 51, 71-80.

James, B., Godonou, I., Atcha-Ahowé, C., Baimey, H., Adango, E., Boulg, J., Goudégnon, E. and Zannou, E. (2006). *Healthy Vegetables through Participatory IPM in Peri-Urban Areas of Benin. Summary of Activities and Achievements, 2003-2006*. International Institute of Tropical Agriculture (IITA), Ibadan, Nigeria, pp. 134.

James, B., Godonou, I., Atcha-Ahowé, C., Glitho, I., Vodouhe, S. and Ahanchede, A. (2007). Extending integrated pest management to indigenous vegetables. In: Chadha, M. L., Kuo, G. and Gowda, C. L. L. (Eds), *Proceedings of the 1st International Conference on Indigenous Vegetables and Legumes Prospectus for Fighting Poverty, Hunger and Malnutrition*. Hyderabad, India, 12-15 December 2006. *Acta Horticulturae* 752, pp. 89-93.

James, B., Atcha-Ahowé, C., Godonou, I., Baimey, H., Goergen, H., Sikirou, R. and Toko, M. (2010). *Integrated Pest Management in Vegetable Production: A Guide for Extension Workers in West Africa*. International Institute of Tropical Agriculture (IITA), Ibadan, Nigeria, 120pp.

Janssen, T., Karssen, G., Verhaeven, M., Coyne, D. and Bert, W. (2016). Mitochondrial coding genome analysis of tropical root-knot nematodes (*Meloidogyne*) supports haplotype based diagnostics and reveals evidence of recent reticulate evolution. *Scientific Reports* 6, 22591. doi:10.1038/srep22591.

Jorge Junior, A. S., Cares, J. E., Almeida, M. R. A., Mattos, V. S., Carneiro, R. M. D. G. and Coyne, D. (2016). First report of *Meloidogyne izalcoensis* on coffee and other crops in Africa. *Plant Disease* 100, 2173.

Kfir, R. (2003). Biological control of the diamondback moth *Plutella xylostella* in Africa. In: Neuenschwander, P., Borgemeister, C. and Langewald, J. (Eds), *Biological Control in IPM Systems in Africa*. CAB International, Wallingford, UK, pp. 363-75.

Kibata, G. N. (2002). The status of the African bollworm, *Helicoverpa* (*Heliothis*) *armigera* Hübner (Lepidoptera: Noctuidae) and future research needs for its management in

Kenya. In: Sithanantham, S., Kariuki, C. and Baya, J. (Eds), *Helicoverpa Management in Kenya. Research Status and Needs. Proceedings of the KARI-ICIPE Workshop on Biocontrol Based IPM of the African Bollworm in Kenya*, pp. 11–15.

Legg, J., Girling, D. and Neuenschwander, P. (2003). Biological control of whiteflies in Sub-Saharan Africa. In: Neuenschwander, P., Borgemeister, C. and Langewald, J. (Eds), *Biological Control in Integrated Pest Management Systems in Africa*. CABI Publishing, Wallingford, UK. pp. 87–100.

Liu, Y. B., Tabashnik, B. E. and Johnson, M. W. (1995). Larval age affects resistance to *Bacillus thuringiensis* in diamondback moth (Lepidoptera: Plutellidae). *Journal of Economic Entomology* 88, 788–92.

Mrema, E. J., Ngowi, A. V., Kishinhi, S. S. and Mamuya, S. H. (2017). Pesticide exposure and health problems among female horticulture workers in Tanzania. *Environmental Health Insights* 11, doi: 1178630217715237.

Muller, R. and Gouch, P. S. (1982). Organic amendments in nematode control. An examination of the literature. *Nematropica* 12, 319–26.

Murphy, S. T and LaSalle, J. (1999). Balancing biological control strategies in the IPM of New World invasive *Liriomyza* leafminers in field vegetable crops. *Biocontrol News and Information* 20, 91N–104N.

Neuenschwander, P., Murphy, S. T. and Coly, E. V. (1987). Introduction of exotic parasitic wasps for the control of *Liriomyza trifolii* (Dipt., Agromyzidae) in Senegal. *Tropical Pest Management* 33, 290–7.

Nion, Y. A. and Toyota, K. (2015). Recent trends in control methods for bacterial wilt diseases caused by *Ralstonia solanacearum. Microbes and Environments* 30, 1–11.

Ogutu, W. O., Oduor, G. I., Panel, M., Mano, D. W., Ogol, C. K. P. O., Grywacz, D. and Poole, J. (2004). Evaluation of a naturally occurring baculovirus for the management of diamondback moth, *Plutella xylostella* L., in Kenya. In: Kirk, A. A. and Bordat, D. (Eds), *Improving Biocontrol of Plutella xylostella, Proceedings of the International Symposium*, Montpellier, France, 21–24 October 2002, pp. 222–8.

Pagan, C., Coyne, D., Carneiro, R., Kariuki, G., Luambano, N., Affokpon, A. and Williamson, V. M. (2015). Mitochondrial haplotype-based identification of ethanol-preserved root-knot nematodes from Africa. *Phytopathology* 105, 350–7.

Pietersen, G., Idris, A. M., Kruger, K. and Brown, J. K. (2000). Tomato curl stunt virus a new begomovirus within the TYLCV-Is cluster, causing a severe disease of tomato in South Africa. *Plant Disease* 84, 810.

Remaudière, G. and Autrique, A. (1985). Introduction. In: Remaudière, G., Autrique, A., Eastop, V. F., Stary, P., Aymonin, G., Kafurera, J. and Dedonder, R. (Eds), *Contribution à l'écologie des Aphides Africains*. FAO, Rome, pp. 7–9.

Roessler, Y. (1989). Insecticidal bait and cover sprays. In: Robinson, A. S. and Hooper, G. (Eds), *World Crop Pests – Fruit Flies. Their Biology, Natural Enemies and Control*. Elsevier Science Publishers B. V., pp. 329–36.

Rutledge, C. E., O'Neil, R. J., Fox, T. B. and Landis, D. A. (2004). Soybean aphid predators and their use in integrated pest management. *Annals of the Entomological Society of America* 97, 240–8.

Sæthre, M.-G., Godonou, I., Hofsvang, T., Tepa-Yotto, G. T. and James, B. (2011). Aphids and their natural enemies in vegetable agroecosystems in Benin. *Internal Journal of Tropical Insect Science* 31, 103–17.

Sæthre, M.-G., Assogba-Komlan, F., Oseth-Svendsen, N., Holen, B. and Godonou, I. (2012). Pesticide residues analysis of three vegetable crops for urban consumers in

Benin - human and environmental consequences of abuse and misuse of synthetic pesticides. *Acta Horticulturae* 1007, 393–401.

Sapkota, R., Dahal, K. C. and Thapa, R. B. (2010). Damage assessment and management of cucurbit fruit flies in spring-summer squash. *Journal of Entomology and Nematology* 2, 7–12.

Schreinemachers, P., Chen, H., Nguyen, T. T. C., Buntong, B., Bouapao, L., Gautam, S., Le, N. T., Pinn, T., Vilaysone, P. and Srinivasan, R. (2017). Too much to handle? Pesticide dependence of smallholder vegetable farmers in Southeast Asia. *Science of the Total Environment* 593–594(1), 470–7. doi:10.1016/j.scitotenv.2017.03.181.

Sikirou, R., Beed, F., Ezin, V., Gbèhounou, G., Miller, S. and Wydra, K. (2009). First report of bacterial wilt of tomato (*Solanum lycopersicum*) caused by *Ralstonia solanacearum* in Benin. *Plant Disease* 93, 549.

Sikirou, R., Beed, F., Ezin, V., Hoteigni, J. and Miller, S. (2017). Distribution, pathological and biochemical characterization of *Ralstonia solanacearum* in Benin. *Annals of Agricultural Sciences* 62, 83–8.

Sikora, R. A. (1992). Management of the antagonistic potential in agricultural ecosystems for the biological control of plant parasitic nematodes. *Annual Review of Phytopathology* 30, 245–70.

Sitaramaiah, K. and Sinha, S. K. (1984). Interaction between *Meloidogyne javanica* and *Pseudomonas solanacearum* on brinjal. *Indian Journal of Nematology* 14, 1–5.

Slosser, J. E., Pinchak, W. E. and Rummel, D. R. (1989). A review of known and potential factors affecting the population dynamics of the cotton aphid. *Southwestern Entomologist* 14, 302–13.

Smith Meyer, M. K. P. (1996). *Mite Pests and their Predators on Cultivated Plants in Southern Africa: Vegetable and Berries*. Biosystematics Division, ARC-Plant Protection Research Institute, Pretoria, 90pp.

SP-IPM (2006). *Biological Alternatives to Harmful Chemical Pesticides*. IPM Research Brief No.4. SP-IPM Secretariat, International Institute of Tropical Agriculture (IITA), Cotonou, Benin, 24pp.

Spencer, K. A. (1973). *Agromyzidae (Diptera) of Economic Importance*. Dr W. Junk, The Hague, The Netherlands, p. 418.

Starý, P., Remaudière, G. and Autrique, A. (1985). Les aphidiides parasites de pucerons en région Ethiopienne. In: Remaudière, G., Autrique, A., Eastop, V. F., Stary´, P., Aymonin, G., Kafurera, J. and Dedonder, R. (Eds), *A l'écologie des aphides africains*. Etude FAO production végétale et protection des plantes 64. FAO, Rome, pp. 95–102.

Stirling, G. R. (1991). *Biological Control of Plant Parasitic Nematodes*. CAB international, Wallingford, UK, 282pp.

Tahat, M. M. and Sijam, K. (2010). *Ralstonia solanacearum*: The bacterial wilt causal agent. *Asian Journal of Plant Sciences* 9, 385–93.

Talekar, N. S. and Shelton, A. M. (1993). Biology, ecology, and management of the diamondback moth. *Annual Review of Entomology* 38, 275–301.

Valda, C. A. S., Reginaldo, B., Edmilson, J. M. and Jorge, B. T. (2003). Susceptibility of *Plutella xylostella* (L) (Lepidoptera: Plutellidae) to the fungi *Beauveria bassiana* (Bals.) Vuill. and *Metarhizium anisopliae* (Metsch.) Sorok. *Neotropical Entomology* 32(4), 653–8.

Vayssières, J.-F. and Carel, Y. (1999). Les Dacini (Diptera: Tephritidae) inféodés aux Cucurbitaceae à La Réunion: Gamme de plantes hôtes et stades phénologiques préférentiels des fruits au moment de la piqûre pour des espèces cultivées. *Annales de la Société Entomologique de France* 35, 197–202.

Viaene, N., Coyne, D. L. and Davies, K. (2013). Biological and cultural control. In: Perry, R. and Moens, M. (Eds), *Plant Nematology*, 2nd edition. CAB International, Wallingford, UK, pp. 384–410.

Waterhouse, D. F. and Norris, K. R. (1987). *Biological Control: Pacific Prospects*. Inkata Press, Melbourne, Australia, 644pp.

White, I. M. and Elson-Harris, M. M. (1992). *Fruit Flies of Economic Significance: Their Identification and Bionomics*. CAB International, Wallingford, UK, 601pp.

Williamson, S., Ball, A. and Pretty, J. (2008). Trends in pesticide use and drivers for safer pest management in four African countries. *Crop Protection* 27, 1327–34.

Wraight, S. P., Jackson, M. A. and de Kock, S. L. (2001). Production, stabilization and formulation of fungal biocontrol agents. In: Butt, T. M., Jackson, C. W. and Magan, N. (Eds), *Fungi as Biocontrol Agents Progress, Problems and Potential*. CAB International, Wallingford, UK, pp. 253–87.

Yaseen, M. (1978). The establishment of two parasites of the diamond-back moth *Plutella xylostella* (Lep.: Plutellidae) in Trinidad, W1. *Entomophaga* 23, 111–14.

Zalucki, M. P., Shabbir, A., Silva, R., Adamson, D., Shu-Sheng, L. and Furlong, M. J. (2012). Estimating the economic cost of one of the world's major insect pests, *Plutella xylostella* (Lepidoptera: Plutellidae): Just how long is a piece of string? *Journal of Economic Entomology* 105, 1115–29.

Chapter 12

Identifying and managing plant health risks for key African crops: fruit and other tree crops

Peter Neuenschwander *and* **Désiré Gnanvossou**, *International Institute of Tropical Agriculture (IITA), Benin;* **Stefan Hauser**, *International Institute of Tropical Agriculture (IITA), Nigeria;* **Georg Goergen**, *International Institute of Tropical Agriculture (IITA), Benin;* **Rachid Hanna**, *International Institute of Tropical Agriculture (IITA), Cameroon;* **Lindsey Norgrove**, *Bern University of Applied Sciences, Switzerland; and* **Koffi Negloh** *and* **Cyriaque Agboton**, *International Institute of Tropical Agriculture (IITA), Benin (corresponding author: p.neuenschwander@cgiar.org)*

1 Introduction

The present chapter covers commodities that do not belong to the so-called mandated crops of IITA, that is those crops for which IITA has primary responsibility for genetic improvement and varietal selection within the Consultative Group for International Agricultural Research (CGIAR), in addition to improved farm management and plant protection options. In most African countries, mango, papaya and coconut are not extensively traded internationally, but are part of smallholder farms, where they provide important nutritional and small cash contributions. Cacao and cashew, by contrast, are internationally traded plantation crops, which in most parts of tropical Africa are grown by smallholders. IITA's involvement with these crops

http://dx.doi.org/10.19103/AS.2018.0043.13
Published by Burleigh Dodds Science Publishing Limited, 2019.

is based on its proven capacity to implement biological control and IPM solutions on the one hand and the importance of these crops in smallholder farms on the other hand.

Mango (*Mangifera indica* L., Anacardiaceae) is of Indian origin and was introduced into Africa only in the middle of the nineteenth century, where it mostly replaced the fruits of bush-mango *Irvingia gabonensis* [(Aubry-Lecomte ex O'Rorke) Baill., Irvingiaceae], the indigenous tree and 'royal fruit' of the rainforest zone of Africa. Today, mango is cultivated across the continent, in the humid south in the forest environment as huge trees with small, fibrous fruits, as well as in the savannah zone in plantations and gardens with numerous, often grafted varieties with commercialized fruits. There is hardly a school yard, where this tree is not grown, providing free delicious fruits for children. A few pests of African origin have adapted to this new resource, but generally up to the 1970s, mango trees were free of serious insect pests and diseases. This changed when in the 1980s the mealybug *Rastrococcus invadens* Williams (Hemiptera: Pseudococcidae) was accidentally introduced into West Africa (review: Moore 2004). With the advent of a devastating fruit fly, which was later identified as *Bactrocera dorsalis* (Hendel) (Diptera: Tephritidae) of Pacific origin (Section 3) (Drew and Hancock 1994; Drew et al. 2005; Schutze et al. 2014), the previously well-established and successful biological control of mango mealybug (Neuenschwander et al. 1994) was disrupted by intense insecticide applications and, in commercial plantations, necessitated the development of an IPM programme against *B. dorsalis* as described below (Ekesi 2016). Add to this the mango fruit weevil *Sternochetus mangiferae* (Fabricius) (Coleoptera: Curculionidae), which can be controlled by means of weaver ants (Peng and Christian 2007).

Similar to mango, papaya (*Carica papaya* L., Caricaceae) of neotropical origin was introduced into Africa in the seventeenth century by the Portuguese (Ligon 1673). It spread rapidly, and is today a mainstay of gardens and farms across all of Africa. Because of its rather delicate fruits, export to other continents is rare and most fruits are traded and consumed locally. Varieties used in traditional farming in Africa are mostly dioecious and cannot be reliably sexed until flowering. As a result, papaya orchards are planted at high tree densities to allow a weeding out of unwanted male plants. With a wider availability of hermaphrodite and high performing varieties, this constraint has now been alleviated. Again, up until recently, few important pests were noted; but, in the 1980s, the highly polyphagous spiralling whitefly *Aleurodicus dispersus* Russell (Hemiptera: Aleyrodidae) was accidentally introduced from Central America to Asia (Ramani et al. 2002) and from there to Africa (Manzano et al. 1995). It became a pest on papaya, mango, cassava, vegetables and other crops before biological control by parasitoids that had followed its spread took over. Crop losses of papaya, cassava, mango, due to this whitefly, ceased immediately (Neuenschwander 1994; D'Almeida et al. 1998).

Published by Burleigh Dodds Science Publishing Limited, 2019.

A few years later, a much more serious pest, the papaya mealybug *Paracoccus marginatus* Williams and Granara de Willink (Hemiptera: Pseudococcidae) of Mexican origin was accidentally introduced into West Africa and spread across the continent. A successful biological control programme had been implemented in Asia (Muniappan et al. 2006; Meyerdirk et al. 2004) and was quickly followed up first in Ghana, then across West Africa out of Benin (Goergen et al. 2011a).

Cacao (*Theobroma cacao* L., Malvaceae) of Brazilian origin has been cultivated in the humid zone of Africa since the nineteenth century (Bartley 2005). As cocoa is an export product, its local use and transformation has only been developed recently, but still absorbs only a marginal fraction of the total production.

Côte d'Ivoire is by far the world's largest producer (1.69 million tons in 2015/16), followed by Ghana, Cameroon and Nigeria (Cacaoweb 2018). Cacao also is attacked by mealybugs (Hemiptera: Pseudococcidae) and true bugs (Hemiptera: Miridae) as well as diseases. It is further threatened by a fungal disease caused by *Phytophthora megakarya* Brasier and Griffin, which – compared to all other cocoa pests and disease – is responsible for the largest yield losses (Bowers et al. 2001). Invasive species such as cocoa witches' broom, *Moniliophthora perniciosa* (Stahel) from Brazil (Aime and Phillips-Mora 2005; Meinhardt et al. 2008), up to now, have not been able to take hold in Africa, thanks to the successful implementation of severe quarantine measures. In a workshop in Ghana in 2000, for the first time, CABI, IITA and the chocolate industry teamed up to find solutions to plant health problems in cacao (Vos and Neuenschwander 2002). In West Africa, genetic improvement of cacao and replacement of older stocks have practically been confined to Ghana. The few new varieties developed in the past decades in Cameroon, for instance, display good yield traits, but are extremely poor with regard to pest and disease tolerance or resistance. Only in the last 10 years were some hybrids developed that had good yield characteristics and tolerated *Phytophthora megakarya* (S. Hauser unpublished results). Since mass propagation of such materials is slow and cost-intensive, these hybrids have not been introduced at large scale. Cameroonian cacao farmers therefore maintain plantations 70 years and older, often called 'German cacao' varieties, which produce low, yet, reliable yields year by year. The newer varieties across West and Central Africa all suffer greatly from *Phytophthora megakarya*, which is far more detrimental than *Phytophthora palmivora* (E. J. Butler), which it is replacing in West and Central Africa. *Phytophthora megakarya* has spread into West Africa, causing major production and income losses and increasing, drastically, production costs due to the need to spray more frequently and with more expensive systemic fungicides containing metalaxyl in addition to the usual copper oxide or copper hydroxide-based contact formulations. In addition, cacao will suffer from climate change-related uncertainties in rainfall, specifically the length of the rainy phases, changes that are likely to bring along new pests and diseases.

Coconut (*Cocos nucifera* L., Palmaceae) is the basis for several industries around the world (oil, coconut, juice, flour, flakes, potting medium, etc.) (De Taffin 1993; Punchihewa 1995). It remains a rural crop found in small plantations along coastal areas and in villages. The most economically important pest is the coconut mite, *Aceria guerreronis* Keifer (Prostigmata: Eriophyidae) with Brazil as its most likely original home (Navia et al. 2005). In Africa, it has been recorded since 1967, first in Benin and later in other West African countries (Mariau 1977), and in Tanzania – the largest producer of coconut in Africa – in the 1980s (Seguni 2002). The coconut mite spread to Sri Lanka in 1997 (Fernando et al. 2003). The pest is likely to continue spreading from India eastwards to countries in South and East Asia, the Pacific Islands and Oceania, where nearly 80% of the world's coconuts are produced.

Cashew, *Anacardium occidentale* L. (Anacardiaceae), a native to Brazil, was introduced into West Africa by Portuguese explorers in the fifteenth century as an ornamental plant and for fixing dunes (Goujon et al. 1973). In Benin, nut production doubled in the last decade, reaching an estimated 130 000 tons in 2016 and covering 200 000 ha (CashewInfo.com 2016). Cashew can be intercropped with food crops or cotton, and is also used for its pharmacological properties (roots, leaves, bark, etc.) and wood. Cashew nuts contribute about 8% of national exports and 28% of total agricultural exports. Cashew from Benin is appreciated for its high quality, good taste and the absence of pesticide residues (A. Tandjiékpon pers. comm.). In recent years, organizational constraints concerning planting material and transformation as well as biotic constraints, particularly the wood borer *Apate terebrans* Pallas (Coleoptera: Bostrichidae), have been interfering with the good prospects of this crop.

In sub-Saharan Africa, horticulture – fruit and vegetable production – is an important source of income, employment and nutrition for both rural and urban populations. Production, sales and foreign exports of vegetables and tree fruits are, however, hampered by damage caused by several tephritid fruit flies, mostly *Ceratitis* species (Lux et al. 2003; Vayssières et al. 2009; Ekesi et al. 2016; Gnanvossou et al. 2017). Infestation rates vary among countries, seasons and varieties, ranging from 5% to 100% (Goergen et al. 2011b; Ekesi et al. 2016). In 2003, species composition shifted considerably with the invasion of Africa by *B. dorsalis*, which became the most important tephritid pest of tree fruits (Clarke et al. 2005; Goergen et al. 2011b). Its polyphagous nature and widespread invasion throughout sub-Saharan Africa created a crisis for tree fruit production. IITA has collaborated closely with the International Centre for Agricultural Research for Development (CIRAD), the International Centre of Insect Physiology and Ecology (*icipe*) in Kenya, and several national agricultural research organizations to develop options for managing fruit flies.

2 Mango mealybug

A mealybug that was accidently introduced into West Africa in the 1980s was identified as a new species, *Rastrococcus invadens* of Indian origin (Williams 1986). Although reported on 45 species of crop plants, ornamentals and wild plants (Agounké et al. 1988; Ivbijaro et al. 1992) its main attack was on mango and, to a lesser extent, on citrus *(Citrus* spp., Rutaceae), frangipani *(Plumeria* spp., Apocynaceae) and *Ficus* spp. (Moraceae). Owing to the heavy accumulation of honeydew and the resulting sooty mould, growth, flowering and fruiting of attacked trees were arrested. This often led farmers to cut the affected trees down.

R. invadens and its natural enemies were investigated in India (Narasimham and Chako 1988) and two promising parasitoids, *Gyranusoidea tebygi* Noyes and *Anagyrus mangicola* Noyes (Hymenoptera: Encyrtidae), were sent to quarantine by the International Institute of Biological Control (IIBC) for further study (Noyes 1988, 1990; Willink and Moore 1988; Agricola et al. 1990; Cross and Moore 1992). A biological control project was started by national programmes in Togo and Benin and eventually continued by IITA as a regional programme. *G. tebygi's* efficiency was assessed (Boavida and Neuenschwander 1995; Boavida et al. 1995b; Bokonon-Ganta and Neuenschwander 1995). *G. tebygi* was released in seven West and Central African countries at 42 sites and *Anagyrus mangicola* in six countries at 22 sites (Fig. 1). Today, *G. tebygi* is established wherever *R. invadens* occurs in Africa, while *Anagyrus mangicola* is confined to big towns (Neuenschwander et al. 1994). A sister-species, *R. iceryoides* (Green), has spread to East Africa from India (Tanga et al. 2016); but up until now, no biological control programme has been initiated. By contrast, the recent invasion by *R. invadens* into French Guiana (Germain et al. 2015)

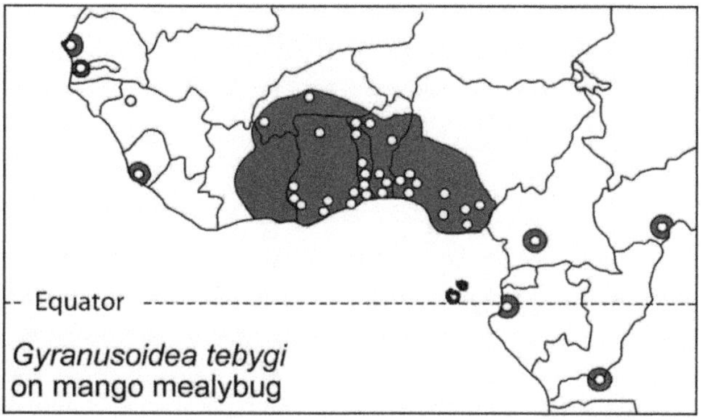

Figure 1 Distribution of *Rastrococcus invadens* in West Africa. White dots: Release sites of *Gyranusoidea tebygi.* By the 1990s, the parasitoid has reached all infested areas. Source: Neuenschwander et al. (1994).

Published by Burleigh Dodds Science Publishing Limited, 2019.

offers the possibility of extending the African biological control programme without great cost to South America.

The parasitoids almost totally eliminated mealybug populations on monitored trees (Fig. 2; Boavida and Neuenschwander 1995). Their efficacy was also confirmed in sleeve cage experiments (Boavida et al. 1995b). Both parasitoids coexist on the same trees because they have different niches: *G. tebygi* accepts second instar hosts and prefers third instars; *Anagyrus mangicola* does not accept second instars, but thrives on fourth instars (Moore and Cross 1992; Boavida et al. 1995a; Bokonon-Ganta et al. 1996). A socio-economic study in southern Benin concluded that the benefits accruing from this biological control project for Benin alone were a staggering US$531 million over a period of 20 years with a benefit: cost ratio of 145:1 (Bokonon-Ganta et al. 2002).

Surveys in Benin showed that mealybug populations remained higher in towns than in rural areas, and particularly where trees were constantly exposed to exhaust fumes from engines, for instance at taxi stands (Neuenschwander et al. 1994). These observations were later detailed in Nigeria (Pitan et al. 2002; Pitan 2008).

Spread of *G. tebygi* to the north was relatively slow and the persistence of damaging mealybug populations on mango planted in plantations in the

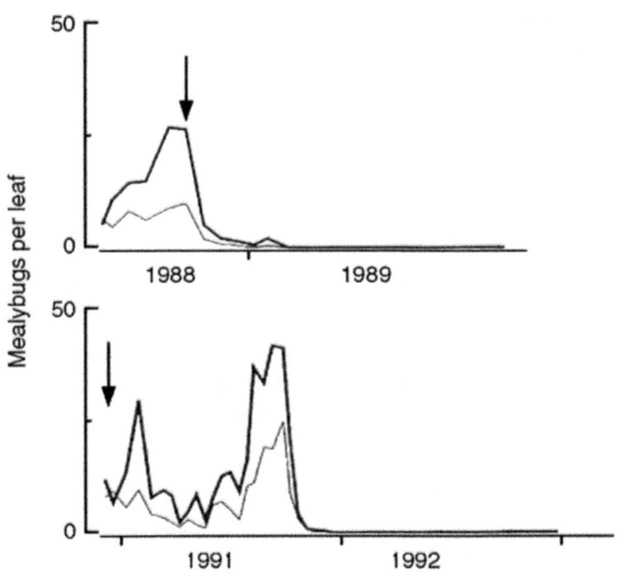

Figure 2 Collapse of the mango mealybug population (all stages) on monitored mango trees following the introduction of *Gyranusoidea tebygi*. Thick line – on young leaves; thin line – on old leaves. Note the almost complete extinction of the mealybug at the level of an orchard. Source: Boavida and Neuenschwander (1995).

Published by Burleigh Dodds Science Publishing Limited, 2019.

north, where *Anagyrus mangicola* is mostly absent, has led to claims in Ghana and Côte d'Ivoire of mealybug 'resurgence', that is a diminishing control capacity by the parasitoids. Since the possibility always exists that another of the many South-East Asian mealybug species (Galanihe and Watson 2012) had been introduced inadvertently, the mealybugs in the resurgence spots were compared in taxonomic and genomic studies with those under evident biological control as well as Indian species. Although minor differences were observed, the claim of a new introduction was refuted (G. Goergen and G. Watson unpublished results). Recent studies in affected orchards showed, however, that the parasitoids were still present and that intensive insecticide treatments against the newly invading fruit fly *B. dorsalis* and other pests were the likely explanations for this resurgence (G. Goergen and P. Neuenschwander pers. comm.). In addition, in some orchards, weaver ants are likely to interfere with biological control (Tanga et al. 2015).

3 Fruit flies

The discovery of *Batrocera dorsalis* in Africa in 2003 triggered numerous studies, to track its distribution, host range and seasonality in several regions in sub-Saharan Africa with collateral gains concerning the biology and ecology of native fruit flies and their associated natural enemies. By 2008, *B. dorsalis* was reported from 42 countries in sub-Saharan Africa with widespread distribution in at least 35 countries (De Meyer et al. 2010; Goergen et al. 2011b). Host range studies in Africa identified at least 46 plant species belonging to 23 families. *B. dorsalis* largely replaced native tephritid species, particularly in areas that are optimal – geographically and seasonally – for *B. dorsalis* and sub-optimal for the native species (Ekesi et al. 2009; Goergen et al. 2011b).

Multi-year trapping studies in mango orchards using parapheromones showed that species composition, abundance and seasonality of fruit flies varied more among agro-ecologies than from one year to the next (Gnanvossou et al. 2017). Among the 12 fruit fly species found in traps on mango trees, *B. dorsalis* and *Ceratitis cosyra* Walker dominated, exhibiting temporal and spatial differences in their abundance as influenced by wild fruits (Ekesi et al. 2006; Goergen et al. 2011b; Vayssières et al. 2015). In citrus orchards in Benin, *B. dorsalis* was by far the dominant fruit fly species, while *Ceratitis cosyra* was rare. Generally, infestations were relatively low (up to 13%), but were much higher in orchards (overall means of 20–32%) where sweet orange, *Citrus sinensis*, was present. Infestations occurred preferentially on mature fruits while immature fruits were never attacked. In contrast to the situation in North Africa, Europe, Central America and western Asia, *Ceratitis capitata* (Wiedemann) – though caught in traps – was never found infesting citrus in Benin or elsewhere in West and Central Africa (Hanna et al. 2009).

Spot sprays of GF-120 (protein attractant plus Spinosad) were used for population control. In northern Benin, such spot sprays on mango reduced fruit fly infestations from 48% to 8% (Vayssières et al. 2009). Similarly in citrus orchards, application of GF-120 reduced fruit infestations from 20% to 3% (Hanna et al. 2010). When parapheromone attractants for males and chemical mixes were deployed in traps at high densities (about 40 units/ha) for population suppression in mango orchards in Benin, the percentage of infested fruits dropped from 50% in control orchards to 25% in treated orchards (Hanna et al. 2008). Since these results corroborate previous studies in Kenya (Ndlela et al. 2016) we recommend such spot sprays or traps to be adopted as IPM tools for mango fruit fly management.

Prior to the invasion of *B. dorsalis*, information on parasitoids associated with fruit flies in West and Central Africa was fragmented and scarce, while a diverse assemblage of parasitoid species is known to be associated with native tephritid fruit flies in other regions (Wharton 1999). Several native parasitoid species are now known to be associated with *Ceratitis* species in Africa. Their impact on populations of native hosts is, however, low because parasitism rarely exceeds 20% (Mohamed et al. 2010; Vayssières et al. 2011). The lack or poor capacity of native parasitoids to establish new associations with *B. dorsalis* led to a campaign of classical biological control, first with foreign exploration in Sri Lanka, the putative origin of the *B. dorsalis*. Unfortunately, the parasitoids found associated with *B. dorsalis* in Sri Lanka could not be shipped due to difficulties in obtaining approval for export to Africa. Subsequently, a population of the egg-larval parasitoid *Fopius arisanus* (Sonan) (Hymenoptera: Braconidae), the best candidate against *Bactrocera* species (Vargas et al. 1993, 2007; Rousse et al. 2005; Mohamed et al. 2010), was imported from Hawaii to Kenya (*icipe*) and from there to IITA in Benin and Cameroon for multiplication and releases in West and Central Africa.

Under no-choice and choice conditions, the effect of different wild host fruit species on parasitism of *B. dorsalis* by *F. arisanus* was assessed. Parasitoids were most attracted to guava and mango, followed by the wild bush-mango (*I. gabonensis*). Guava and mango were therefore used for mass rearing *F. arisanus*. Starting November 2008, adult *F. arisanus* from the IITA-Benin insectary were repeatedly released across several agro-ecosystems in Benin and subsequently in Cameroon (Gnanvossou et al. 2016). In both countries, the first three years of releases targeted mango orchards. The parasitoid was consistently recovered, but initially failed to persist through the dry season. Subsequently, bush mangos were targeted. In Cameroon in mango orchards surrounded by alternate host plants including *I. gabonensis*, parasitism of *B. dorsalis* eggs by *F. arisanus* varied according to host fruit species. Per cent parasitism on mango (up to 2%) and other host fruits (up to 10%) was generally lower than that on bush mango (up to 21%) (Gnanvossou et al. 2016). In south-western

Benin, *B. dorsalis* populations were reduced by 33 and 65% in 2011 and 2012, respectively, because of *F. arisanus* population increases. In Cameroon, the parasitoid failed to pass through the dry season until a breakthrough occurred in 2017 when *F. arisanus* was recovered in a bush mango orchard in the central region of Cameroon, two years after the last *F. arisanus* release in the same location, suggesting that *F. arisanus* was effectively controlling *B. dorsalis* (D. Gnanvossou unpublished results).

Fruit samples from bush mango orchards and surrounding vegetation from 2012 to 2014 showed that *F. arisanus* had spread at least 8 km from the release orchards (Gnanvossou et al. 2016). While this rate of spread is much slower than the one of other exotic parasitoids, we consider *F. arisanus* to be firmly established in Benin and Cameroon. It is likely to continue to spread in both countries with the potential for substantial impact on *B. dorsalis*. Additional efforts are now needed to use the acclimatized populations for further releases in other countries in sub-Saharan Africa.

The African weaver ant *Oecophylla longinoda* (Latreille) (Hymenoptera: Formicidae) has been the subject of intensive research in West Africa to determine its role in the management of tephritid fruit flies in mango orchards (Van Mele et al. 2007). Several studies in Benin have shown that orchards with high levels of *O. longinoda* have extremely low levels of mango infestations compared with orchards with low or absent *O. longinoda* (Sinzogan et al. 2008; Adandonon et al. 2009; Vayssières et al. 2016). Subsequent field observations in Central and Northern Benin and Northern Cameroon have, however, failed to show any significant impact by *O. longinoda* on fruit fly infestations on mango, possibly due to less than optimal abundance of *O. longinoda* in those orchards (R. Hanna unpublished data). A potentially negative effect of *O. longinoda* on fruit fly parasitoids has been observed under confined conditions (Appiah et al. 2014; Migani et al. 2017), but never in open field situations (J.-F. Vayssières unpublished results). Similarly, Peng and Christian (2013) showed that the Asian *O. smaragdina* (Fabricius) in the field either benefited or had no impact on natural enemy diversity and abundance. At present, interactions between ants and other natural enemies such as the introduced parasitoid *F. arisanus* are being studied in order to improve the efficiency of weaver ants in mango orchards. Furthermore, farmers would have to be trained on how to handle and manage this highly aggressive ant (Sinzogan et al. 2008).

4 Spiralling whitefly

Aleurodicus dispersus was described from old museum specimens (Russell 1965) before it spread, presumably out of Central America across the Pacific into India (Ramani et al. 2002), the Canary Islands (Manzano et al. 1995), and finally the African continent. In Nigeria, this polyphagous pest was observed for the first time also on cassava (Akinlosotu et al. 1993). Before parasitoids that had been deployed in the Pacific region (Kumashiro et al. 1983) could be imported into Benin, both species, namely *Encarsia,* cited as sp. near *haitiensis* Dozier, which was later newly described as *Encarsia dispersa* Polaszek, Manzari and Quicke (Polaszek et al. 2004) as well as *E. guadeloupae* Viggiani (Hymenoptera: Aphelinidae) turned up serendipitously (Neuenschwander 1994). In Benin, they spread north together with *Aleurodicus dispersus* and reduced host populations levels by 80% (Fig. 3; D'Almeida et al. 1998). This made the already initiated introduction of the predator *Nephaspis oculatus* Blatchley (Coleoptera: Coccinellidae) obsolete. Similar spontaneous introductions of aphelinid parasitoids of *Aleurodicus dispersus* had been observed before in the Pacific region (Waterhouse and Norris 1989).

In Tanzania, the pest was reported for the first time in 2002. The two parasitoids were introduced into Tanzania in 2009 out of IITA-Benin. By 2014, parasitized hosts were recorded in 69 out of 100 sites over a spread perimeter of more than 3000 km, but with large gaps between agro-ecologies. Parasitism was highest in areas with high humidity. Spiralling whitefly population densities dropped from 113 per leaf in 2010 to 14 per leaf in 2012. Moreover, the presence of the parasitoids in border regions suggests that the natural enemies may have already reached Kenya, Rwanda, Burundi, Malawi, Zambia and Mozambique, and this without any investments or interventions from farmers and from the neighbouring countries (B. Pallangyo and R. Hanna unpublished results).

5 Papaya mealybug

Originating from Mexico and Central America *Paracoccus marginatus* (Hemiptera: Pseudococcidae) became an invasive species around the mid-1990s when it started to spread in the Caribbean archipelago. From there it expanded within only two decades to most tropical regions of the globe. The pest was first detected on the African continent by the end of 2009 when outbreaks were causing severe damage in papaya orchards near Accra, Ghana (Goergen et al. 2011a). *Paracoccus marginatus* is highly polyphagous and reported to have a host range of more than 80 host plants in 33 botanical families. Beside its preferred host plant papaya, it attacks other economically important tropical crops such as cassava, sweet potato, yams, mango, avocado, pineapple, cotton, citrus and cashew, as well as ornamentals, shrubs and weeds. Infestations are found on all plant parts and observed as dense mealybug colonies, dusted with white wax and often bearing cottony egg masses. Immature and adult mealybugs typically produce ample honeydew resulting in sooty mould, reduced photosynthesis as well as lowered yield and quality of crops. Strong pest attacks may cause leaf distortion, defoliation, premature fruit dropping and ultimately death of the host plant.

The absence of natural enemies outside its native range led immediately to explosive proliferations of mealybug populations in the main papaya production areas of Ghana. This soon was reflected in serious yield reductions, leading to shortfalls in local marketing and exports, as well as unemployment of 1700 people in the sector (Goergen et al. 2011a). This alarming situation prompted the Ghanaian government and FAO to initiate an emergency technical cooperation programme involving national plant protection partners of Ghana and IITA. Thus, in a classical biological control approach, first introductions and releases of the endoparasitic wasp *Acerophagus papayae* Noyes and Schauff (Hymenoptera: Encyrtidae), a specific and highly efficient natural enemy of the papaya mealybug, were made in Ghana in 2010. Meanwhile, the steady expansion of papaya mealybug infestations in West and Central Africa necessitated a scaling-up of the successful FAO pilot project. To address this growing menace, a new regional project bringing together national partners in Ghana, Togo, Benin, Nigeria, Cameroon and Gabon was launched by the end of 2012 (Goergen et al. 2014).

In 2013, prior to parasitoid releases, country-wide surveys revealed a particularly high incidence and attack severity under humid tropic conditions, such as in coastal regions. Population levels averaging more than 50 mealybugs per leaf in southern coastal regions of Benin, Nigeria and Togo gradually decreased towards northern regions of these countries. Strong papaya mealybug infestations were more frequent in urban areas than in rural environments. Following inoculative field releases since 2010, *Acerophagus papayae* successfully established in all six project countries and by 2015 it had naturally spread under various climatic and agro-ecological conditions to an estimated area of 0.6 million km^2.

Country-wide surveys carried out in Benin over three consecutive years until 2015 showed that *Acerophagus papayae* successfully controlled papaya mealybug in all affected agro-ecozones. Data from 450 papaya trees surveyed at 20-km interval in 90 sites throughout the country demonstrated that despite decreasing pest incidence *Acerophagus papayae* was recovered on 94% (2014) and 90% (2015) of plants with papaya mealybug. At the same time the proportion of papaya trees that became entirely free of pest attacks increased from 66% in 2014 to 83% in 2015 (Fig. 3).

Data from all experimental sites and project countries similarly confirmed that papaya mealybug populations decreased sharply shortly after the inoculative releases of *Acerophagus papayae* and were thereafter kept low over long periods (typically below 10% when compared to the initial situation) (Fig. 4). This control was achieved despite the presence of five species of native hyperparasitoids. However, depending on the sites, countries and age of the orchards, temporary pest resurgences, peaking at moderate densities, were observed especially when equilibrium levels of mealybug populations

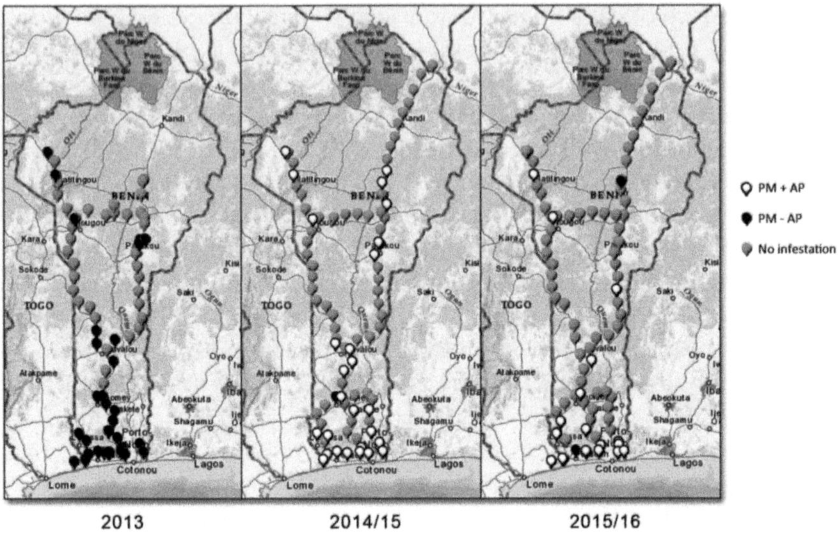

Figure 3 Spread of the papaya mealybug (black dots) mainly in the south and jumping to big towns in the north. Two years later, the parasitoid *Acerophagus papayae* has caught up almost everywhere (white dots) to be found in almost all sites again the next year. Most orchards in central and northern Benin remain uninfected (grey dots). Source: Goergen et al. (2014).

were particularly low while at the same time rainfall increased (Goergen et al. 2014). The successful control of *Acerophagus papayae* in West and Central Africa parallels similar biological control actions in the Caribbean, Pacific islands and elsewhere (Meyerdirk et al. 2004; Muniappan et al. 2006; Sankar et al. 2012).

These results were corroborated by positive economic trends and resulted in a substantial impact for the papaya sector in surveyed project countries.

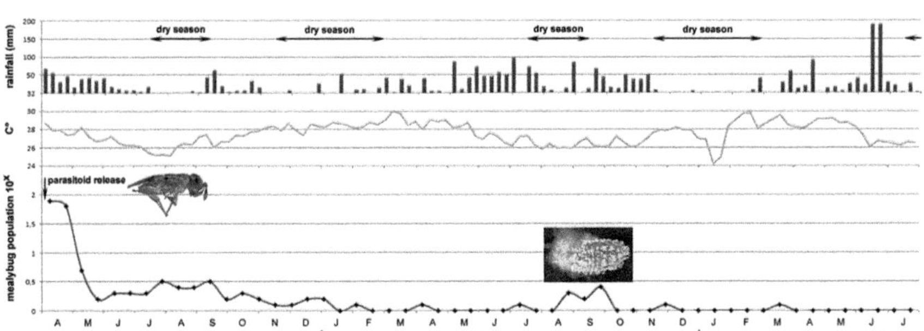

Figure 4 Following the release of *Acerophagus papayae* (arrow) papaya mealybug populations crash and stay low irrespective of the season.

Thus, between 2012 and 2015, the total surface area under papaya cultivation increased by 45% in Benin, 48% in Togo, and 81% in Ghana. At the same time papaya production increased by 76%, 43% and 157%, respectively. The reduction in pesticide use achieved through effective biological control resulted in a decrease in production costs by 3% in Benin and even 19% in Ghana, where papaya is produced on a large scale mainly for export. Profitability indicators show that the benefit-cost ratio of using biological control to manage the papaya mealybug ranges between 21:1 in 2020 and 78:1 in 2030 for Ghana and between 12:1 and 37:1 for Benin (R. Adeoti and G. Goergen unpublished results). Corresponding effects on export volumes, income generation, job creation and food security were observed (S. Tovignan and S. Ferrigno unpublished results). By extending the measured economic impact without additional costs to other crops that were similarly affected by papaya mealybug, returns become even more lucrative (Myrick et al. 2014).

6 *Phytophthora megakarya*

Traditionally, cacao plantations were established under thinned forest cover or after forest clearing, in which trees other than cacao were either planted or allowed to grow to provide shade to the cacao saplings. Although cacao in its original environment lives under forest shade, there is scant evidence that the adult tree requires shade, although shaded cacao systems can retain soil faunal diversity and activity better, which might improve nutrient cycling (Norgrove et al. 2009) and also conserve carbon (Norgrove and Hauser 2013). However, the heavily shaded systems common in Cameroon and Nigeria appear to suffer heavily from the fungal disease *Phytophthora megakarya*. This 'new-encounter' pathogen of cacao was first identified in Nigeria (Brasier and Griffin 1979) and recognized as a separate species from the pantropical *Phytophthora palmivora*. Globally, there are seven *Phytophthora* species affecting cacao, causing estimated annual losses of 700 000 tonnes, which makes it the disease causing the highest yield loss (Ploetz 2016).

Phytophthora megakarya has moved from an African host, possibly *I. gabonensis*, the bush mango, to cacao (Holmes et al. 2003). Currently in an invasive phase, *Phytophthora megakarya* has been spreading west from Ghana through Côte d'Ivoire (Kebe et al. 2002). There are possibly two groups of this pathogen, a Western group spreading west from Nigeria and a Central group, separated biogeographically near the Cameroon/Nigeria border (Nyassé et al. 1999). These diseases cause pod rot and are promoted by the higher humidity and longer lasting water films on leaves and fruits in shaded plantations.

Methods to reduce losses include incorporating resistance, biological control, fungicide application and integrated pest management (Ploetz 2016). However, incorporating resistance into varieties is slow. Amelonado-type accessions appear to be less susceptible than Trinitario and Criollo types (Ploetz 2016). An alternative is to find healthy individual trees among the genotypic diversity on farms under high disease conditions.

Biological control measures have been researched by IRAD (Cameroon) and partners with some success, identifying several *Trichoderma* spp. (Hypocreaceae) as effective antagonists to *Phytophthora megakarya* (Tondje et al. 2007). *Trichoderma* endophytes are plant symbionts, which can protect their hosts from diseases through various mechanisms: competitive exclusion, antibiosis, induced resistance and mycoparasitism (Ploetz 2016). CABI and USDA have been screening *Trichoderma* spp. that exhibit these properties and colonize cacao tissue for their potential as biocontrol agents (Bailey et al. 2008). However, today there is still no commercially available product based on biological agents to combat the disease.

Chemical fungicides are widely recommended; however, implementation is yield- and price-sensitive (Ploetz 2016). Protectant sprays of copper-based fungicides, together with the systemic fungicide metalaxyl, applied at 4-week intervals, greatly increase yield (Norgrove 2007). Fungicides are applied with a knapsack sprayer. More effective control methods, including rational fungicide use, improved sprayers and spraying techniques (e.g. motorized mist blowers), need to be recommended to farmers (Bateman 2004). By comparison with other continents, the use of fungicides and fertilizer in cacao in West Africa has been minimal.

Cultural control methods have been investigated, specifically the removal of young infected pods (Ndoumbe-Nkeng et al. 2004). However, most research on pod removal was conducted in combination with fungicide application and none of the studies could recommend only pod removal as an effective measure. IITA research within the Sustainable Tree Crops Programme showed that pod removal alone was not suitable to control pod rot or to increase yields when no fungicide was applied (Norgrove 2007). Removing diseased fruits and pruning can minimize yield losses from *Phytophthora palmivora*, yet have a negligible effect in *Phytophthora megakarya*-infected areas (Opoku et al. 2000; Norgrove 2007).

The aim of integrated pest management is to eliminate *Phytophthora megakarya* inoculum and to prevent its movement from the soil (Ploetz 2016). This fungus is inhibited by light and has a lower temperature optimum than *Phytophthora palmivora* (Brasier and Griffin 1979). Shade and canopy management practices that increase light transmission to the cacao canopy and airflow (appropriate spacing, pruning, weed control) can thus reduce *Phytophthora megakarya*. Leaf litter mulches and ground covers might also reduce the spread of the disease by being a barrier to rain splash and by promoting the microbial decomposition of *Phytophthora*-infected debris.

In Cameroon, with the end of both governmental control over the cocoa trade and provision of fungicides and organized insecticide application, coupled

Published by Burleigh Dodds Science Publishing Limited, 2019.

with low cocoa bean prices, many plantations were abandoned and were not weeded, sprayed nor pruned for several years. Instead, farmers felled and sold valuable timber trees to compensate for lost cocoa income. When in 2007, cocoa prices increased, many farmers tried to recover production, yet faced the high cost of fungicides to control *Phytophthora megakarya*. At this moment a unique opportunity to develop integrated cacao production systems offered itself. Reduction in shade tree densities, as had already been started with timber tree removal, plus yield monitoring with consequent removal of unproductive trees and therefore focusing fungicide use on the remaining productive trees would have been the first steps to recover production. Replacing unproductive trees with new disease-tolerant cacao hybrids or diversifying by planting fruit trees would have been the next steps towards systems that can improve livelihoods and preserve some forest trees with accompanying biodiversity albeit with the use of agrochemicals. But, unfortunately, the donors' priority on labelling cacao production as 'rainforest friendly' was the probable cause of this missed opportunity.

7 Coconut mite

The search for natural enemies against *Aceria guerreronis* focused on South America. Baseline surveys highlighted the high value of coconut for household income (40–60%) in both Brazil and Benin and the widespread perception of damage caused by coconut mite on coconut production and farmers' income. The mite is widely distributed in Brazil, particularly in the northern and north-eastern parts where potential natural enemies could be found by IITA in collaboration with the University of São Paulo, the Institute of Plant Protection in Vienna, Austria and the University of Amsterdam in the Netherlands. Coconut mite abundance and damage to coconut were far lower in Brazil than in Africa and the predator fauna associated with coconut mite in Brazil richer than elsewhere (Lawson et al. 2008a,b; Negloh et al. 2011).

The three most common predatory mites found in association with coconut mite in Brazil were introduced into IITA-Benin for side-by-side comparisons with African populations of the same species. The biological characteristics of four predatory mite species were studied (Lawson-Balagbo et al. 2007a; Negloh et al. 2008). Geographic populations of the common *Neoseiulus paspalivorus* (De Leon) and *Neoseiulus baraki* Athias-Henriot (Acari: Phytoseiidae) showed considerable variation in morphology and biology (Lawson Balagbo et al. 2007b; Famah Sourassou et al. 2011, 2012, 2013, 2014; Negloh et al. 2010, 2012). Of particular interest was the Brazilian population of *Neoseiulus baraki*, which was more voracious than *Neoseiulus paspalivorus* (Negloh et al. 2012). Molecular and morphological characterization showed that the so-called Brazilian *Neoseiulus baraki* was different from the African population, making it a promising candidate for experimental field evaluations in Africa (Famah Sourassou et al. 2012). On this basis, Brazilian *Neoseiulus baraki*, yet to be described as a new species, was multiplied at the IITA station in Benin and released in six orchards in Benin starting in 2011. It established and spread to neighbouring plantations without affecting the abundance of endemic predatory mites. Within three years, total predatory mite abundance increased three-fold with average 1.17-fold declines in coconut mite densities (K. Negloh unpublished results). Further releases are planned.

8 Cashew wood borer

The indigenous *Apate terebrans* was identified as the most important pest, followed by mirid bugs (Agboton et al. 2014) and a leafminer, which has a rich coterie of indigenous parasitoids (Agboton et al. 2018). Collecting and burning all dead twigs and branches attacked by *Apate terebrans* early in the season proved to be the most appropriate measure to substantially reduce later infestations (Agboton et al. 2017; Onzo et al. 2018). Where weaver ants, *O. longinoda,* were present, yields and quality of cashew nuts were significantly improved, which was attributed to their effect on mirids, as well as on thrips (Anato et al. 2015). Cashew trees were, however, less well adapted to the development of weaver ant colonies than mango trees (Wargui et al. 2015).

Arbuscular mycorrhizal fungi were experimentally applied against several diseases such as red rust, anthracnose and bacterial leaf and nut spots. They improved seedling growth and cashew tree productivity (Afouda et al. 2013; I. Balogoun unpublished results); but how these beneficial fungi can be applied by producers remains to be investigated.

9 Conclusions

The CGIAR was founded to improve the standard of living of farmers in the tropics, particularly subsistence farmers. In the wake of the Green Revolution, its institutions, IITA included, concentrated on genetic improvement of so-called mandated crops. Yet, there are probably no farms in Africa where only these crops are grown. Fruit trees are present in all backyards and fields and contribute to the local nutrition and small commerce. Since the 1980s, IITA therefore started to apply its know-how gained from biological control programmes on cassava also to tree crops. In an opportunistic manner, we dealt with new pest problems, against which classical biological control was deemed to have a good chance of success. Indeed, several such programmes came to fruition, favoured by the stable presence and climate on trees as compared to field crops, where host plants, phytophagous pests and their parasitoids are often absent every year over long periods with harsh weather. All the above-mentioned activities were financed by special projects but profited from a stable infrastructure and know-how available at the institute and its partners. Most benefits of these projects went directly and without additional costs to all farmers of the affected regions.

All biological control programmes concerning Homopteran pests were of relatively long duration (sometimes without additional funding). They resulted in establishment and spread of the exotic parasitoids and measurable impact on plants and, in the case of mango and papaya pests, on yield and farm income. Biological control programmes involving an exotic phytoseiid mite on coconut and an exotic parasitoid on fruit flies also led to their establishment and a modest spread. All these projects had impacts that will continue beyond the project duration.

In the case of fruit fly management with different attractants and insecticides, and management of a fungus on cacao, promising methods were developed and, as a 'proof of concept', their impact was demonstrated (though sometimes not yet published in peer-reviewed journals). Similarly, for the control of various indigenous cashew pests, cultural control options were developed. Whether these methods will be implemented beyond their projects is, however, not assured.

Sometimes, honest conflicts of interest emerged as in the case of cacao, where the industry wanted the rainforest label, while opening the canopies would lower the use of fungicides by small-scale farmers. While international initiatives driven by the chocolate industry, interested in securing raw material supplies, seem to take care of the crop, it remains questionable if such measures taken at regional level can benefit smallholder farmers and plantation operators alike. In fact, Gockowski and Sonwa (2011) argued that the main driver of deforestation in West and Central Africa has been the expansion of extensive smallholder agriculture and non-adoption of intensification measures. Had fertilizer and fungicides been applied to cacao in the 1960s, much deforestation would have been avoided.

10 References

Adandonon, A., Vayssieres, J.-F., Sinzogan, A. and Van Mele, P. 2009. Density of pheromone sources of the weaver ant *Oecophylla longinoda* affects oviposition behaviour and damage by mango fruit flies (Diptera: Tephritidae). *International Journal of Pest Management* 55, 285–92.

Afouda, L. C. A., Zinsou, V., Balogoun, R. K., Onzo, A. and Ahohuendo, B. C. 2013. Inventaire des agents pathogènes de l'anacardier (*Anacardium occidentale* L.) au Bénin. *Bulletin de la Recherche Agronomique du Bénin (BRAB)* 73, 13–19.

Agboton, C., Onzo, A., Ouessou, F. I., Goergen, G., Vidal, S. and Tamò, M. 2014. Insect fauna associated with *Anacardium occidentale* (Sapindales: Anacardiaceae) in Benin, West Africa. *Journal of Insect Science* 14(229), 1–11.

Agboton, C., Onzo, A., Korie, S., Tamò, M. and Vidal, S. 2017. Spatial and temporal infestation rates of Apate terebrans (Coleoptera: Bostrichidae) in cashew orchards in Benin, West Africa. *African Entomology* 25, 24–36.

Agboton, C., Onzo, A., Akohou, A. H., Goergen, G., Vidal, S. and Tamò, M. 2018. Population dynamics of the cashew leafminer, *Eteoryctis gemoniella* (Lepidoptera: Gracillariidae), and inventory of its parasitoids in cashew orchards of Northern Benin, West Africa. *International Journal of Tropical Insect Science* 38, 132–44.

Agounké, D., Agricola, U. and Bokonon-Ganta, H. A. 1988. *Rastrococcus invadens* Williams (Hemiptera: Pseudococcidae), a serious exotic pest of fruit trees and other plants in West Africa. *Bulletin of Entomological Research* 78, 695–702.

Agricola, U., Agounké, D., Fischer, H. U. and Moore, D. 1990. Biology of the mealybug *Rastrococcus invadens* Williams (Hemiptera: Pseudococcidae) and its control with *Gyranusoidea tebygi* Noyes (Hymenoptera: Encyrtidae) in Togo. *Mitteilungen der Deutschen Gesellschaft für Allgemeine und Angewandte Entomologie* 7, 647–52.

Aime, M. C. and Phillips-Mora, W. 2005. The causal agents of witches' broom and frosty pod rot of cacao (Chocolate, *Theobroma cacao*) form a new lineage of Marasmiaceae. *Mycologia* 97, 1012–22.

Akinlosotu, T. A., Jackai, L. E. N., Ntonifor, N. N., Hassan, A. T., Agyakwa, C. W., Odebiyi, I. A., Akingbohungbe, A. E. and Rossell, H. W. 1993. Spiralling whitefly, *Aleurodicus dispersus*, in Nigeria. FAO Plant Protection Bulletin 41, pp. 127–9.

Anato, F. M., Wargui, R. B., Sinzogan, A. A. C., Offenberg, J., Adandonon, A., Vayssières, J. F. and Kossou, D. K. 2015. Reducing losses inflicted by insect pests on cashew, using weaver ants as a biological control agent. *Agricultural and Forest Entomology* 17, 285–91.

Appiah, E. F., Ekesi, S., Afreh-Nuamah, K., Obeng-Ofori, D. and Mohamed, S. A. 2014. African weaver ant-produced semiochemicals impact on foraging behaviour and parasitism by the Opiine parasitoid, *Fopius arisanus* on *Bactrocera invadens* (Diptera: Tephritidae). *Biological Control* 79, 49–57.

Bailey, B. A., Bae, H., Strem, M. D., Crozier, J., Thomas, S. E., Samuels, G. J., Vinyard, B. T. and Holmes, K. A. 2008. Antibiosis, mycoparasitism, and colonization success for endophytic *Trichoderma* isolates with biological control potential in *Theobroma cacao*. *Biological Control* 46, 24–35.

Bartley, B. G. D. 2005. *The Genetic Diversity of Cacao and Its Utilization*. CABI, Wallingford, UK, 341pp.

Bateman, R. 2004. The use of narrow-angle cone nozzles to spray disperses pods and other slender biological targets. *Crop Protection* 23, 989–99.

Boavida, C. and Neuenschwander, P. 1995. Population dynamics and life tables of the mango mealybug, *Rastrococcus invadens* Williams and its introduced natural enemy, *Gyranusoidea tebygi* Noyes in Benin. *Biocontrol Science and Technology* 5, 489–508.

Boavida, C., Ahounou, M., Vos, M., Neuenschwander, P. and van Alphen, J. J. M. 1995a. Host stage selection and sex allocation by *Gyranusoidea tebygi* (Hymenoptera: Encyrtidae), a parasitoid of the mango mealybug, *Rastrococcus invadens* (Homoptera: Pseudococcidae). *Biological Control* 5, 487–96.

Boavida, C., Neuenschwander, P. and Herren, H. R. 1995b. Experimental assessment of the impact of the introduced parasitoid *Gyranusoidea tebygi* Noyes on the mango mealybug *Rastrococcus invadens* Williams by a physical exclusion method. *Biological Control* 5, 99–103.

Bokonon-Ganta, A. H. and Neuenschwander, P. 1995. The impact of the biological control agent *Gyranusoidea tebygi* Noyes (Hymenoptera: Encyrtidae), on the mango mealybug, *Rastrococcus invadens* Williams (Homoptera: Pseudococcidae) in Benin. *Biocontrol Science and Technology* 5, 95–107.

Bokonon-Ganta, A. H., van Alphen, J. J. M. and Neuenschwander, P. 1996. Competition between *Gyranusoidea tebygi* and *Anagyrus mangicola*, parasitoids of the mango mealybug, *Rastrococcus invadens*: Interspecific host discrimination and larval competition. *Entomologia Experimentalis et Applicata* 79, 179–85.

Bokonon-Ganta, A. H., De Groote, H. and Neuenschwander, P. 2002. Socioeconomic impact of biological control of mango mealybug in Benin. *Agriculture, Ecosystems and Environment* 93, 367–78.

Bowers, J. H., Bailey, B. A., Hebbar, P. K., Sanogo, S. and Lumsden, R. D. 2001. The impact of plant diseases on world chocolate production. *Plant Health Progress*. Online. doi:10.1094/PHP-2001-0709-01-RV.

Brasier, C. M. and Griffin, M. J. 1979. Taxonomy of Phytophthora palmivora on cocoa. *Transactions of the British Mycological Society* 72, 111–43.

Cacaoweb. 2018. http://www.cacaoweb.net/countries.html (accessed on 24 January 2018).

CashewInfo.com. 2016. Cashew week. An initiative of cashewinfo.com. Special edition to celebrate *3rd World Cashew Convention*, 09–11 February, Singapore, 18(4), 1–61.

Clarke, A. R., Armstrong, K. F., Carmichael, A. E., Milne, J. R., Raghu, S., Roderick, G. K. and Yeates, D. K. 2005. Invasive phytophagous pests arising through a recent tropical evolutionary radiation: The *Bactrocera dorsalis* complex of fruit flies. *Annual Review of Entomology* 50, 293–319.

Cross, A. E. and Moore, D. 1992. Developmental studies on *Anagyrus mangicola* (Hymenoptera: Encyrtidae), a parasitoid of the mealybug *Rastrococcus invadens* (Homoptera: Pseudococcidae). *Bulletin of Entomological Research* 82, 307–12.

D'Almeida, Y. A., Lys, J. A., Neuenschwander, P. and Ajuonu, O. 1998. Impact of two accidentally introduced *Encarsia* Species (Hymenoptera: Aphelinidae) and other biotic and abiotic factors on the Spiralling Whitefly *Aleurodicus dispersus* (Russell) (Homoptera: Aleyrodidae), in Benin, West Africa. *Biocontrol Science and Technology* 8, 163–73.

De Meyer, M., Robertson, M. P., Mansell, M. W., Ekesi, S., Tsuruta, K., Mwaiko, W. and Peterson, A. T. 2010. Ecological niche and potential geographic distribution of the invasive fruit fly *Bactrocera invadens* (Diptera, Tephritidae). *Bulletin of Entomological Research* 100, 35–48.

De Taffin, G. 1993. Le Cocotier. Le Technicien d'Agriculture Tropicale, Agence, CIRAD et CTA, 166pp.

Drew, R. A. I. and Hancock, D. L. 1994. The *Bactrocera dorsalis* complex of fruit flies (Diptera: Tephritidae: Dacinae) in Asia. *Bulletin of Entomological Research* 2, 1–68.

Drew, R. A. I., Tsuruta, K. and White, I. M. 2005. A new species of pest fruit fly (Diptera: Tephritidae: Dacinae) from Sri Lanka and Africa. *African Entomology* 13, 149–54.

Ekesi, S. 2016. Baiting and male annihilation techniques for fruit fly suppression in Africa. In: Ekesi, S., Mohamed, S. A. and De Meyer, M. (Eds), *Fruit Fly Research and Development in Africa - Towards a Sustainable Management Strategy to Improve Horticulture*, Springer Nature, Switzerland, Basel, pp. 275–92. ISBN: 978-3-319-43226-7.

Ekesi, S., Nderitu, P. W. and Rwomushana, I. 2006. Field infestation, life history and demographic parameters of the fruit fly *Bactrocera invadens* (Diptera: Tephritidae) in Africa. *Bulletin of Entomological Research* 96, 379–86.

Ekesi, S., Billah, M. K., Nderitu, P. W. and Rwomushana, I. 2009. Evidence for competitive displacement of *Ceratitis cosyra* by the invasive fruit fly *Bactrocera invadens* (Diptera: Tephritidae) on the mango and mechanisms contributing to the displacement. *Journal of Economic Entomology* 102, 981–91.

Ekesi, S., De Meyer, M., Mohamed, S. A., Virgillo, M. and Borgemeister, C. 2016. Taxonomy, ecology and management of native and exotic fruit fly species in Africa. *Annual Review of Entomology* 61, 210–38.

Famah Sourassou, N., Hanna, R., Zannou, I., de Moraes, G. J., Negloh, K. and Sabelis, M. W. 2011. Morphological variation and reproductive incompatibility of three coconut-mite-associated populations of predatory mites identified as *Neoseiulus paspalivorus* (Acari: Phytoseiidae). *Experimental and Applied Acarology* 53, 323–38.

Famah Sourassou, N., Hanna, R., Zannou, I. D., Breeuwer, H., de Moraes, G. J. and Sabelis, M. W. 2012. Morphological, cross-breeding and molecular analysis of the coconut mite associated predatory mites identified as *Neoseiulus baraki:* Evidence for cryptic species. *Experimental and Applied Acarology* 57, 15–36.

Famah Sourassou, N., Hanna, R., Negloh, K., Breeuwer, J. A. J. and Sabelis, M. W. 2013. Females as intraguild predators of males in cross-pairing experiments with phytoseiid mites. *Experimental and Applied Acarology* 61, 173–82.

Famah Sourassou, N., Hanna, R., Breeuwer, J. A. J., Moraes, G. J. and Sabelis, M. W. 2014. The endosymbiotic bacteria *Wolbachia* and *Cardinium* and their effects in three populations of the predatory mite *Neoseiulus paspalivorus*. *Experimental and Applied Acarology* 64, 207–21.

Fernando, L. C. P., Aratchige, N. S. and Peiris, S. G. 2003. Distribution patterns of the coconut mite *Aceria guerreronis* Keiffer and its predator *Neoseiulus paspalivorus* in coconut palms. *Experimental and Applied Acarology* 31, 71–8.

Galanihe, L. D. and Watson, G. W. 2012. Identification of *Rastrococcus rubellus* Williams (Hemiptera: Pseudococcidae): A new record to Sri Lanka. *Tropical Agricultural Research and Extension* 15, 7–10.

Germain, J. F., Laplace, D., Devarieux, A. and Boavida, C. 2015. First records of the mealybug *Rastrococcus invadens* Williams (Hemiptera: Pseudococcidae) in French Guiana and the Americas. *Zootaxa* 3905, 447–50.

Gnanvossou, D., Hanna, R., Bokonon-Ganta, A. H., Ekesi, S. and Mohamed, S. A. 2016. Release, establishment and spread of the natural enemy *Fopius arisanus* (Hymenoptera: Braconidae) for control of the invasive oriental fruit fly *Bactrocera dorsalis* (Diptera: Tephritidae) in Benin, West Africa. In: Ekesi, S., Mohamed, S. A. and De Meyer, M. (Eds), *Fruit Fly Research and Development in Africa - Towards a Sustainable Management Strategy to Improve Horticulture*, Springer Nature, Switzerland, Basel, pp. 575–600. ISBN: 978-3-319-43226-7.

Gnanvossou, D., Hanna, R., Goergen, G., Salifu, D., Tanga, C. M., Mohamed, S. A. and Ekesi, S. 2017. Diversity and seasonal abundance of tephritid fruit flies in three agro-ecosystems in Benin, West Africa. *Journal of Applied Entomology* 141, 798–809.

Gockowski, J. and Sonwa, D. 2011. Cocoa intensification scenarios and their predicted impact on CO_2 emissions, biodiversity conservation, and rural livelihoods in the Guinea rain forest of West Africa. *Environmental Management* 48, 307–21.

Goergen, G., Tamò, M., Kyofa-Boamah, M. E., Bokonon-Ganta, A. and Neuenschwander, P. 2011a. Papaya mealybug, *Paracoccus marginatus* (Hemiptera: Pseudococcidae), a new invading pest in West Africa. *Biocontrol News and Information* 32, 9N–10N.

Goergen, G., Vayssières, J.-F., Gnanvossou, D. and Tindo, M. 2011b. *Bactrocera invadens*, a new fruit fly problem in the Afrotropics: Distribution and host-plant range in West and Central Africa. *Environmental Entomology* 40, 844–54.

Goergen, G., Ajuonu, O., Kyofa-Boamah, M. E., Umeh, V., Bokonon-Ganta, A., Tamò, M. and Neuenschwander, P. 2014. Classical biological control of papaya mealybug in West Africa. *Biocontrol News and Information* 35, 5N–6N.

Goujon, P., Lefèbvre, A., Leturcq, P., Marcellesi, A. P. and Praloran, J. 1973. Etudes sur l'anacardier. *Revue Bois et Forêts des Tropiques* 151, 27–53.

Hanna, R., Gnanvossou, D. and Grout, T. 2008. La lutte régionale contre les mouches des fruits en Afrique de l'Ouest. *Technique d'annihilation des mâles pour la suppression de B. invadens au Bénin. COLEACP/CIRAD. Lettre d'information no.* 7, p. 3.

Hanna, R., Gnanvossou, D. and Grout, T. 2009. Infestations d'agrumes par des mouches des fruits Tephritidae en Afrique subsaharienne: Les premiers résultats au Bénin montrent que la mouche du Sri Lanka *Bactrocera invadens* prédomine. *COLEACP/CIRAD. Lettre d'information no.* 2, p. 3.

Hanna, R., Gnanvossou, D., Bokonon-Ganta, A. and Ekesi, S. 2010. Fruit flies species composition and seasonality, and relative efficiency of four female attractants in mango orchards in Benin. In: Sabater Munoz, B., Navarro Llopis, V. and Urbaneja Garcia, A. (Eds), *Book of Abstracts of the 8th International Symposium on Fruit Flies of Economic Importance*. Editorial Universitat Politècnica de València, Spain, 312pp. ISBN: 978-84-693-4446-0.

Holmes, K. A., Evans, H. C., Wayne, S. and Smith, J. 2003. *Irvingia*, a forest host of the cocoa black-pod pathogen, *Phytophthora megakarya*, in Cameroon. *Plant Pathology* 52, 486–90. https://doi.org/10.1017/S1742758417000339.

Ivbijaro, M. F., Udensis, N., Ukwela, U. M. and Anno-Nyako, F. V. 1992. Geographical distribution and host range in Nigeria of the mango mealy bug, *Rastrococcus invadens* Williams, a serious exotic pest of horticulture and other crops. *Insect Science and its Application* 13, 411–16.

Kebe, I. B., N'Guessan, F. K., Keli, J. Z. and Bekon, A. K. 2002. Cocoa IPM research and implementation in Côte d'Ivoire. In: Vos J., And Neuenschwander, P. (Eds), *Proceedings of the West African Regional Cocoa IPM Workshop*. CPL Press, Newbury, UK, pp. 45–52.

Kumashiro, B. R., Lai, P. Y., Funasaki, G. Y. and Teramoto, K. K. 1983. Efficacy of *Nephaspis amnicola* and *Encarsia haitiensis* in controlling *Aleurodicus dispersus* in Hawaii. *Proceedings of the Hawaiian Entomological Society* 24, 261–70.

Lawson-Balagbo, L. M., Gondim, Jr. M. G. C., de Moraes, G. J., Hanna, R. and Schausberger, P. 2007a. Food-type effects on life history of the predatory mites *Neoseiulus paspalivorus* and *Proctolaelaps bickleyi*, candidates for biological control of the coconut mite. *Experimental and Applied Acarology* 43, 49–61.

Lawson-Balagbo, L. M., Gondim, Jr. M. G. C., de Moraes, G. J., Hanna, R. and Schausberger, P. 2007b. Refuge use by the coconut mite *Aceria guerreronis*: Fine scale distribution and association with other mites under the coconut perianth. *Biological Control* 43, 102–10.

Lawson-Balagbo, L. M., Gondim, Jr. M. G. C., de Moraes, G. J., Hanna, R. and Schausberger, P. 2008a. Exploration of the acarine fauna on coconut palm in Brazil with emphasis on *Aceria guerreronis* (Acari: Eriophyidae) and its natural enemies. *Bulletin of Entomology Research* 98, 83–96.

Lawson-Balagbo, L. M., Gondim, Jr. M. G. C., de Moraes, G. J., Hanna, R. and Schausberger, P. 2008b. Compatibility of *Neoseiulus paspalivorus* and *Proctolaelaps bickleyi*, candidate biological control agents of the coconut mite *Aceria guerreronis*: Spatial niche use and intraguild interactions. *Experimental and Applied Acarology* 45, 1–13.

Ligon, R. 1673. *A True and Exact History of the Island of Barbados*, 2nd ed. Frank Cass Publishers, London, 136pp.

Lux, S. A., Copeland, R. S., White, I. M., Manrakhan, A. and Billah, M. K. 2003. A new invasive fruit fly species from the *Bactrocera dorsalis* (Hendel) group detected in East Africa. *Insect Science and its Application* 23, 355–61.

Manzano, F., Carnero, A., Pérez Padrón, F. and González, A. 1995. *Aleurodicus dispersus* Russell (Homoptera, Aleurodidae) una 'mosca blanca' de importancia económica en Canarias. *Boletin de Sanidad Vegetal Plagas* 21, 3–9.

Mariau, D. 1977. *Aceria* (*Eriophyes*) *guerreronis*: Un important ravageur des cocoteraies africaines et américaines. *Oléagineux* 3, 101–11.

Meinhardt, L. W., Rincones, J., Bailey, B. A., Aime, M. C., Griffith, G. W., Zhang, D. and Pereira, G. A. G. 2008. *Moniliophthora perniciosa*, the causal agent of witches' broom disease of cacao: What's new from this old foe? *Molecular Plant Pathology* 9, 577–88.

Meyerdirk, D. E., Muniappan, R., Warkentin, R., Bamba, J. and Reddy, G. V. 2004. Biological control of the papaya mealybug, *Paracoccus marginatus* (Hemiptera: Pseudococcidae) in Guam. *Plant Protection Quarterly* 19, 110–14.

Migani, V., Ekesi, S., Merkel, K. and Hoffmeister, T. 2017. At lunch with a killer: The effect of weaver ants on host-parasitoid interactions on mango. *PLoS One* 12(2): e0170101/doi: 10.1371

Mohamed, S. A., Ekesi, S. and Hanna, R. 2010. Old and new host-parasitoid associations: Parasitization of the invasive fruit fly *Bactrocera invadens* (Diptera: Tephritidae) and five other African fruit fly species by *Fopius arisanus*, an Asian opiine parasitoid. *Biocontrol Science and Technology* 20, 183–96.

Moore, D. 2004. Biological control of *Rastrococcus invadens*. *Biocontrol News and Information* 25, 17N–27N.

Moore, D. and Cross, A. E. 1992. Competition between two primary parasitoids, *Gyranusoidea tebygi, Encarsia meritoria* species-complex (Hymenoptera, Aphelinidae), parasitoids of whiteflies (Hemiptera, Aleyrodidae) of economic importance. *Zoologica Scripta* 33, 403–21.

Muniappan, R., Meyerdirk, F. M., Sengebau, D. D., Berringer, G. V. P. and Reddy, G. V. 2006. Classical biological control of the papaya mealybug, *Paracoccus marginatus* (Hemiptera: Pseudococcidae) in the Republic of Palau. *Florida Entomologist* 89, 212–17.

Myrick, S., Norton, G. W., Selvaraj, K. N., Natarajan, K. and Muniappan, R. 2014. Economic impact of classical biological control of papaya mealybug in India. *Crop Protection* 56, 82–6.

Narasimham, A. U. and Chacko, M. J. 1988. *Rastrococcus* spp. (Hemiptera: Pseudococcidae) and their natural enemies in India as potential biocontrol agents for *R. invadens* Williams. *Bulletin of Entomological Research* 78, 703–8.

Navia, D., de Moraes, G. J., Roderick, G. and Navajas, M. 2005. The invasive coconut mite *Aceria guerreronis* (Acari: Eriophyidae): Origin and invasion sources inferred from mitochondrial (16S) and nuclear (ITS) sequences. *Bulletin of Entomological Research* 95, 505–16.

Ndlela, S., Mohamed, S., Ndegwa, P. N., Ong'Amo, G. O. and Ekesi, S. 2016. Male Annihilation technique using methyl eugenol for field suppression of *Bactrocera dorsalis* (Hendel) (Diptera: Tephritidae) on Mango in Kenya. *African Entomology* 42, 437–47.

Ndoumbe-Nkeng, M., Cilas, C., Nyemb, E., Nyasse, S., Bieysse, D. and Flori, A. 2004. Impact of removing diseased pods on cocoa black pod caused by *Phytophthora megakarya* and on cocoa production in Cameroon. *Crop Protection* 23, 415–24.

Negloh, K., Hanna, R. and Schausberger, P. 2008. Comparative demography and diet breadth of Brazilian and African populations of the predatory mite *Neoseiulus baraki*, a candidate for biological control of coconut mite. *Biological Control* 46, 523–31.

Negloh, K., Hanna, R. and Schausberger, P. 2010. Season- and fruit age-dependent population dynamics of *Aceria guerreronis* and its associated predatory mite *Neoseiulus paspalivorus* on coconut in Benin. *Biological Control* 54, 349–58.

Negloh, K., Hanna, R. and Schausberger, P. 2011. The coconut mite, *Aceria guerreronis*, in Benin and Tanzania: Occurrence, damage and associated acarine fauna. *Experimental and Applied Acarology* 55, 361–74.

Negloh, K., Hanna, R. and Schausberger, P. 2012. Intraguild predation and cannibalism between the predatory mites *Neoseiulus neobaraki* and *N. paspalivorus*, natural enemies of the coconut mite *Aceria guerreronis*. *Experimental and Applied Acarology* 58, 235–46.

Neuenschwander, P. 1994. Spiralling whitefly, *Aleurodicus dispersus*, a recent invader and new cassava pest. *African Crop Science Journal* 2, 419–21.

Neuenschwander, P., Boavida, C., Bokonon-Ganta, A., Gado, A. and Herren, H. R. 1994. Establishment and spread of *Gyranusoidea tebygi* Noyes and *Anagyrus mangicola* Noyes (Hymenoptera: Encyrtidae), two biological control agents released against the mango mealybug *Rastrococcus invadens* Williams (Homoptera: Pseudococcidae) in Africa. *Biocontrol Science and Technology* 4, 61–9.

Norgrove, L. 2007. Effects of different copper fungicide application rates upon earthworm activity and impacts on cocoa yield over four years. *European Journal of Soil Biology* 43, S303–10.

Norgrove, L. and Hauser, S. 2013. Carbon stocks in shaded *Theobroma cacao* farms and adjacent secondary forests of similar age in Cameroon. *Tropical Ecology* 54, 15–22.

Norgrove, L., Csuzdi, C., Forzi, F., Canet, M. and Gounes, J. 2009. Shifts in soil faunal community structure in shaded cacao agroforests and consequences for ecosystem function in Central Africa. *Tropical Ecology* 50, 71–8.

Noyes, J. S. 1988. *Gyranusoidea tebygi* sp. n. (Hymenoptera: Encyrtidae), a parasitoid of *Rastrococcus* (Hemiptera: Pseudococcidae) on mango in India. *Bulletin of Entomological Research* 78, 313–16.

Noyes, J. S. 1990. A new species of *Anagyrus* (Hymenoptera: Encyrtidae) from India attacking *Rastrococcus invadens* (Hemiptera: Pseudococcidae), a pest of mango and citrus in West Africa. *Bulletin of Entomological Research* 80, 203–7.

Nyassé, S., Grivet, L., Risterucci, A. M., Blaha, G., Berry, D., Lanaud, C. and Despréaux, D. 1999. Diversity of *Phytophthora megakarya* in Central and West Africa revealed by isozyme and RAPD markers. *Mycological Research* 103, 1225–34.

Onzo, A, Biaou, J. T. and Agboton, C. 2018. Efficacité du ramassage et de brûlage systématiques du bois mort dans la lutte contre le foreur de bois, *Apate terebrans*, dans les anacarderaies du Nord-Bénin. *Journal of Applied Biosciences* 121, 12168–80.

Opoku, I. Y., Appiah, A. A., Akrofi, A. Y. and Owusu, G. K. 2000. *Phytophthora megakarya*: A potential threat to the cocoa industry in Ghana. *Ghana Journal of Agricultural Science* 33, 237–48.

Peng, R. and Christian, K. 2007. The effect of weaver ant, *Oecophylla smaragdina* (Hymenoptera: Formicidae), on the mango seed weevil, *Sternochetus mangiferae* (Coleoptera: Curculionidae) in mango orchards in the Northern Territory of Australia. *International Journal of Pest Management* 53, 15–24.

Pitan, O. O. R. 2008. Variations in the population of the mango mealybug *Rastrococcus invadens* (Homoptera: Pseudococcidae), and its parasitism, in relation to smoke pollution. *International Journal of Tropical Insect Science* 28, 119–25.

Pitan, O. O. R., Mwansat, G., Akinyem, S. O. S., Adebayo, O. S. and Akinlosotu, T. A. 2002. Effect of mango mealybug and sooty mould attack on mango and the of the released *Gyranusoidea tebygi* Noyes on yield. *Fruits* 57, 105–13.

Ploetz, R. 2016. The impact of diseases on cacao production: A global overview. In: Bailey, B. A. and Meinhardt, L. W. (Eds), *Cacao Diseases: A History of Old Enemies and New Encounters*. Springer Nature, Switzerland, Basel, pp. 33–59. ISBN: 978-3-319-24789-2.

Polaszek, A., Manzari, S. and Quicke, D. L. J. 2004. Morphological and molecular taxonomic analysis of the *Encarsia meritoria* species-complex (Hymenoptera, Aphelinidae), parasitoids of whiteflies (Hemiptera, Aleyrodidae) of economic importance. *Zoologica Scripta* 33, 403–21.

Punchihewa, P. G. 1995. Coconut product diversification. In: Oropeza, C., Howard, F. W. and Ashburner, G. R. (Eds), *Lethal Yellowing: Research and Practical Prospects*. Kluwer Academic Publishers, Dordrecht, pp. 229–47.

Ramani, S., Poorani, J. and Bhumannavar, B. S. 2002. Spiralling whitefly, *Aleurodicus dispersus*, in India. *Biocontrol News and Information* 23(2), 55N–62N.

Rousse, P., Harris, E. J. and Quilici, S. 2005. *Fopius arisanus*, an egg-pupal parasitoid of Tephritidae: Overview. *Biocontrol News Information* 26, 59N–69N.

Russell, L. M. 1965. New species *of Aleurodicus* Douglas and two close relatives (Homoptera: Aleyrodidae). *Florida Entomologist* 48, 47–55.

Sankar, C., Saravanan, P., Kathiravan, J., Marimuthu, R. and Prabhukumar, S. 2012. Classical biological control of papaya mealybug, *Paracoccus marginatus* Williams and Granara de Willink (Hemiptera: Pseudococcidae) in tapioca by an encyrtid parasitoid, *Acerophagus papayae* in Perambalur District of Tamilnadu, India. *Pest Management in Horticultural Ecosystems* 18, 213–16.

Schutze, M. K., Mahmood, K., Pavasovic, A., Bo, W., Newman, J., Clarke, A. R., Krosch, M. N. and Cameron, S. L. 2014. One and the same: Integrative taxonomic evidence that *Bactrocera invadens* (Diptera: Tephritidae) is the same species as the Oriental fruit fly *Bactrocera dorsalis*. *Systematic Entomology* 40, 472–86.

Seguni, Z. 2002. Incidence, distribution and economic importance of the coconut eriophyid mite, *Aceria guerreronis* Keifer in Tanzanian coconut based cropping systems. In: Fernando, L. C. P., Moraes, G. J. and de Wickramananda, I. R. (Eds), *Proceedings of the International Workshop on Coconut Mite*. Coconut Research Institute, Lunuwila, Sri Lanka, pp. 54–7.

Sinzogan, A., Van Mele, P. and Vayssières, J.-F. 2008. Implications of on-farm research for local knowledge related to fruit flies and the weaver ant *Oecophylla longinoda* in mango production. *International Journal of Pest Management* 54, 241–6.

Tanga, C. M., Ekesi, S., Govender, P., Nderity, P. W. and Mohamed, S. A. 2015. Antagonistic Interactions between the African Weaver Ant *Oecophylla longinoda* and the parasitoid *Anagyrus pseudococci* potentially limits suppression of the invasive mealybug *Rastrococcus iceryoides*. *Insects* 7(1), 1. doi:10.3390/insects7010001

Tanga, C. M., Ekesi, S., Govender, P. and Mohamed, S. A. 2016. Host-plant relationships and natural enemies of the invasive mealybug, *Rastrococcus iceryoides* Green in Kenya and Tanzania. *Journal of Applied Entomology* 140, 655–68.

Tondje, P. R., Roberts, D. P., Bon, M. C., Widmer, T., Samuels, G. J., Ismaiel, A., Begoude, A. D., Tchana, T., Nyemb-Tshomb, E., Ndoumbe-Nkeng, M., Bateman, R., Fontem, D. and Hebbar, K. P. 2007. Isolation and identification of mycoparasitic isolates of *Trichoderma asperellum* with potential for suppression of black pod disease of cacao in Cameroon. *Biological Control* 43, 202–12.

Van Mele, P., Vayssières, J.-F., Van Tellingen, E. and Vrolijks, J. 2007. Effects of an African weaver ant, *Oecophylla longinoda*, in controlling mango fruit flies (Diptera: Tephritidae) in Benin. *International Journal of Pest Management* 47, 695–701.

Vargas, R. J., Stark, D. J., Uchida, G. K. and Purcel, M. 1993. Opiine parasitoids (Hymenoptera: Braconidae) of oriental fruit fly (Diptera Tephritidae) on Kauai Island, Hawaii: island-wide relative abundance and parasitism rates in wild and orchard guava habitats. *Environmental Entomology* 22, 246–53.

Vargas, R. J., Leblanc, L., Putoa, R. and Eitam, V. 2007. Impact of introduction of *Bactrocera dorsalis* (Diptera: Tephritidae) and classical biological control releases of *Fopius arisanus* (Hymenoptera: Braconidae) on economically important fruit flies in French Polynesia. *Journal of Economic Entomology* 100, 670–9.

Vayssières, J.-F., Sinzogan, A., Korei, S., Ouagoussounon, I. and Thomas-Odjo, A. 2009. Effectiveness of spinosad bait sprays (GF-120) in controlling mango-infesting fruit flies (Diptera: Tephritidae) in Benin. *Journal of Economic Entomology* 102, 515–21.

Vayssières, J-F., Wharton, R. A., Adendonon, A. and Sinzogon, A. 2011. Preliminary inventory of parasitoids associated with fruit flies in mangoes, guavas, cashew pepper and wild fruit crops in Benin. *Biocontrol* 56, 35–43.

Vayssières, J.-F., De Meyer, M., Ouagoussounon, I., Sinzogan, A., Adandonon, A., Korie, S., Wargui, R., Houngbo, H., Didier, C., De Bon, H. and Goergen, G. 2015. Seasonal abundance of mango fruit flies (Diptera: Tephritidae) and ecological implications for their management in mango and cashew orchards in Benin. *Journal of Economic Entomology* 108, 2213–30.

Vayssières, J.-F., Offenberg, J., Sinzogan, A., Adandonon, A., Wargui, R., Anato, F., Houngbo, H. Y., Ouagoussounon, I., Diamé, L., Quilici, S., Rey, J.-Y., Goergen, G., De Meyer, M. and Van Mele, P. 2016. Chapter 18: The use of weaver ants in the management of fruit flies in Africa. In: Ekesi, S., Mohamed, S. A. and De Meyer, M. (Eds), *Fruit Fly Research and Development in Africa - Towards a Sustainable Management Strategy to Improve Horticulture*, Springer Nature, Switzerland, Basel pp. 389–433. ISBN: 978-3-319-43226-7.

Vos, J. and Neuenschwander, P. (Eds). 2002. *Proceedings of the West African Regional Cocoa IPM Workshop*, Cotonou, Benin, 13–15 November 2001. CPL Press, Newbury, UK, pp. 100–4.

Wargui, R., Offenberg, J., Sinzogan, A., Adandonon, A., Kossou, D. and Vayssières, J. F. (2015). Comparing different methods to assess weaver ant abundance in plantation trees. *Asian Myrmecology* 7, 159–70.

Waterhouse, D. F. and Norris, K. R. 1989. *Aleurodicus dispersus* (Russell). In: *Biological Control, Pacific Prospects*, Supplement 1. ACIAR, Canberra, pp. 13–22.

Wharton, R. A. 1999. A review of the Old World genus *Fopius* Wharton (Hymenoptera: Braconidae: Opiinae), with description of two new species reared from infesting Tephritidae (Diptera). *Journal of Hymenoptera Research* 8, 48–64.

Williams, D. J. 1986. *Rastrococcus invadens* sp. n. (Hemiptera: Pseudococcidae) introduced from the oriental region to West Africa and causing damage to mango, citrus and other trees. *Bulletin of Entomological Research* 76, 695–9.

Willink, E. and Moore, D. 1988. Aspects of the biology of *Rastrococcus invadens* Williams (Hemiptera: Pseudococcidae), a pest of fruit crops in West Africa, and one of its primary parasitoids, *Gyranusoidea tebygi* Noyes (Hymenoptera: Encyrtidae). *Bulletin of Entomological Research* 78, 709–15.

Published by Burleigh Dodds Science Publishing Limited, 2019.

Part 3

Integrated pest management: putting it all together and exchange of knowledge with farmers

Chapter 13

Commercial products promoting plant health in African agriculture

Frederick Schreurs *and* **Ranajit Bandyopadhyay**, *IITA, Nigeria;* **Christiaan Kooyman**, *IITA, Kenya;* **Alejandro Ortega-Beltran** *and* **Adebowale Akande**, *IITA, Nigeria;* **Matieyedou Konlambigue**, *IITA, Ghana; and* **Niels Van den Bosch**, *IITA, Nigeria* *(corresponding author: f.schreurs@cgiar.org)*

1 Introduction

The research institutions of the CGIAR generally produce public goods for their clients, the national programmes. All researches are supposed to be of open access, and in fact up to recently IITA's new crop varieties and biological control agents have been given free of cost to the national programmes. In the case of varieties, the national programmes and seed companies often changed the names also on products that had not been further developed in order to facilitate uptake by farmers. For commercialization, this procedure poses problems as no producer can be assured priority access to the new invention. In the case of Green Muscle®, a product based on an entomopathogenic fungus against locusts and grasshoppers, description of one step in its production was patented and withheld from the public by the collaborating institutes CAB International, U.K., and AGRHYMET, Niger, in order to attract investors. The fate of this product and its derivatives is described in the first section and its application against a local grasshopper is dealt with in detail in Chapter 8.

The second major product, Aflasafe®, was produced by IITA in collaboration with the United States Department of Agriculture – Agricultural Research Service (USDA-ARS) with funding from multiple donors. The product, tailor-made for each country, consists of native atoxigenic strains of *Aspergillus flavus*

http://dx.doi.org/10.19103/AS.2018.0043.14
Published by Burleigh Dodds Science Publishing Limited, 2019.

Link (*Trichocomaceae*) that in its toxigenic form is responsible for the high aflatoxin content in vulnerable grains. The atoxigenic strains thereby replace the dangerous fungus in the soil, an effect that lasts over one to two seasons. The commercialization of Aflasafe® is described in the second part of this chapter and its development and application in maize in Chapter 7.

In the third part, the concept of commercialization and problems of its realization in an international research institution without commercial interests, but funded by outside donors, are explained.

2 Green Muscle

2.1 Introduction

In Africa, augmentative or inundative releases of natural enemies (Van Lenteren, 2012) have been deployed mainly by using pathogens. In collaboration with other institutions, the entomopathogenic fungus *Metarhizium acridum* (Driver & Milner) J.F. Bisch., Rehner & Humbel (*Clavicipitaceae*) was developed and commercialized as Green Muscle® in Africa, where it was used against locusts and Sahelian grasshoppers, especially in Madagascar and Senegal. A similar product was developed for Australia using an Australian strain of the same fungus and commercialized as Green Guard® (Lomer et al., 2001). Since the current licence holder lost interest in these products, a third product, NOVACRID®, was developed in only 2 years thanks to the valuable experience gained during the development of Green Muscle. Similarly, entomopathogenic viruses are being propagated in a rustic manner and applied on vegetables, as described in Chapter 11, and other crops.

2.2 Identifying an efficient isolate

After the devastating locust plague of the 1980s, during which millions of ha were treated with chemical pesticides in North Africa alone, the donor community argued that this should not be repeated. Research institutions were requested to look for less environmentally damaging insecticides. IITA therefore teamed up with the then International Institute of Biological Control (IIBC), part of CAB International in the United Kingdom, and the Département de Formation en Protection des Végétaux (DFPV) of AGRHYMET in Niamey, Niger. Together, in 1989, they executed a research programme, called LUBILOSA (LUtte BIologique contre les LOcustes et les SAuteriaux) that aimed at developing a biological solution for locust control.

Various biological control options were considered and classical biological control, using parasitoids, was dismissed because locusts are too mobile and irregular in occurrence. Biopesticides based on entomopathogenic viruses, bacteria and fungi appeared more promising. The choice eventually fell on

Published by Burleigh Dodds Science Publishing Limited, 2019.

entomopathogenic fungi, because these were known to infect locusts, have conidia that are fairly robust and are relatively easy to produce on artificial substrates. Because few isolates existed in collections at the time the main activities of the first few years of the programme consisted in searching for isolates of such fungi in the field (Shah et al., 1997) and screening them for efficacy against the desert locust, *Schistocerca gregaria* Forsskål (Orthoptera: Acrididae) (Bateman et al., 1996).

About 180 isolates were collected from West Africa and the Arabian Peninsula and sent to CAB International for screening. Most isolates belonged to the genus *Metarhizium*, while others were *Beauveria* or *Synglíocladium*. Screening was carried out under close to natural conditions for the desert locust. Under those conditions, a group of isolates belonging to *Metarhizium* showed good efficacy. Initially, these isolates were identified as *M. flavoviride* Gams & Roszypal, but after a series of subsequent analyses of *Metarhizium* isolates, they were eventually recognized as a new species, *M. acridum* (Driver & Milner) J.F. Bisch., Rehner & Humber. In the end, isolate IMI330189, found infecting the grasshopper *Ornithacris cavroisi* (Finot) (Orthoptera: Acrididae) near Niamey, Niger, was chosen for further development, because it was among the isolates with highest efficacy and had already been studied to some extent by IIBC scientists.

2.3 Developing a new formulation

For many years, entomopathogenic fungi had been applied in water formulations, but efficacy of these formulations was often disappointing. Then Prior et al. (1988) discovered that an oil formulation of *Beauveria bassiana* (Bals.-Criv.) Vuillemont (*Cordycipitaceae*) had a much higher efficacy than a water formulation. This was explained by the fact that conidia of *B. bassiana* are lipophilic. Since *M. acridum* conidia are also lipophilic, it seemed logical that oil formulations would also be better for them. This was demonstrated in laboratory, field cage and 'arena' trials (Bateman et al., 1993, 1998).

The next step was to show that oil formulations of *M. acridum* were truly effective under operational conditions. Though efforts were made to test those formulations against desert locusts, it proved difficult to find opportunities to do so. It was possible, however, to test them against Sahelian grasshoppers and the variegated grasshopper, *Zonocerus variegatus* (L.) (Orthoptera: Pyrgomorphidae) (Douro-Kpindou et al., 1995; Kooyman et al., 1997). These trials confirmed the efficacy of *M. acridum* in the field and were also important in optimizing the oil formulation. The formulating oil chosen at that point was a mixture of 70% groundnut oil and 30% kerosene.

It proved impossible to find an industrial biopesticide producer to produce the required quantities of *M. acridum* for a reasonable price. Permission was therefore requested and granted by the programme donors to convert one of the wings of the IITA research station in Cotonou into a spore-producing unit (Cherry et al., 1999). The facility proved to be an excellent research unit and made it possible to refine realistic technical specifications and to test production, contamination control, spore separation, drying and packaging. A special spore harvesting apparatus was designed based on vacuum cleaner cyclone technology, which later became famous under the name MycoHarvester (http://www.dropdata.net/mycoharvester/).

2.4 Testing a product

Starting from phase 3 of the project (1996), opportunities for trials against other locusts arose: desert locust, brown locust (*Locustana pardalina* Walker), tree locust (*Anacridium melanorhodon* Walker) and red locust (*Nomadacris septemfasciata* Audinet-Serville) (all Orthoptera: Acrididae) (Kooyman and Abdalla, 1998; Kooyman and Godonou, 1997; Langewald et al., 1997; Price et al., 1997). All these field trials taught us that in general, the target insects would become visibly sick from day 4 after application, that mortality would start from day 9 with a peak between days 12 and 15, and that 80–90% control would be achieved within 3 weeks. Where predation pressure was high, mainly by birds, reptiles and insects, the locusts would be controlled much faster. For example during a trial in Mauritania against desert locust nymphs in 2006, all treated hopper bands were eliminated within 10 days (Langewald et al., 1997). It was also found that diesel fuel was a perfect and easily available formulating oil.

Other aspects that were investigated included locust behaviour, (eco) toxicology and mass production techniques. It was discovered that infected locusts would consume less and less food starting from 3 to 4 days after infection. Infected females that made it to adulthood would dispense with the usual fattening period of 10–14 days before developing eggs and lay only one batch before dying. A worrying discovery was that infected locusts would try to increase their body temperature to above the 40°C preferred temperature, dubbed 'behavioural fever', in an attempt to get rid of the fungus. In the field, this means that death is often delayed. Fortunately, *M. acridum* survives these higher temperatures and resumes activity as soon as the body temperature drops below 35°C. During field trials, it was established that roughly 8 h of favourable body temperatures per day was sufficient to achieve more than 80% mortality in less than 3 weeks.

The host range of *M. acridum* was extensively researched. Though some laboratory tests showed infection of non-target insects by the fungus, investigation under more natural conditions showed that the only non-target species likely to be affected by spray operations was the silkworm, *Bombyx mori* (L.) (Lepidoptera: Bombycidae). Otherwise, the host range of the fungus is restricted to species of *Acridomorpha*, that is locusts and grasshoppers.

2.5 Mass production and registration

In the meantime, the parameters for the mass production of *M. acridum* were refined. The production system chosen was a biphasic one with blastospores produced in liquid fermentation followed by the production of aerial conidia on sterilized rice as solid substrate. The Mycoharvester, mentioned before, turned out to be invaluable in obtaining a powder of more than 90% single spores, which was found to be essential for preventing the clogging of filters in spray equipment. It was also discovered that the conidia needed to be dried to <6% humidity in order to be able to store them for up to 5 years at 4°C.

A search for commercial producers of the product, named Green Muscle®, started in the second half of the 1990s. The first company found willing to take the plunge was Biological Control Products (BCP) of South Africa. It first registered the product in South Africa followed by a number of other countries in southern, eastern and northeastern Africa. Though another company was found in France, which managed to obtain a temporary sales permit for the member countries of CILSS in West Africa, this company did not manage to efficiently produce *M. acridum* in their production system. It took until 2004, several years after the termination of the LUBILOSA programme, that a producer was found for West and North Africa. This was the Fondation Agir pour l'Education et la Santé (FAES) founded by the then first lady of Senegal. A purpose-built facility was erected and production started in 2007.

Though both BCP and FAES managed to produce a high-quality product that was successfully tested and registered in several countries, convincing the government agencies responsible for locust control to buy the product turned out to be very hard. Both companies did not want or were unable to join the corrupt practices that were, and often still are, employed by the sellers of chemical insecticides and the procurement officers in the respective countries. Moreover, the competition managed to create a perception among end users that Green Muscle was not efficacious and much too slow to be able to control locusts, no matter that many scientific papers and reports showed the opposite, at least where preventative control of locust nymphs was concerned - and despite a recommendation by FAO. BCP continued to soldier on but only managed to sell significant quantities of Green Muscle to Madagascar, though at least half of this was sourced from FAES, which had a much higher production capacity (50 kg/day). FAES itself was only able to sell products to the Senegalese government for grasshopper control.

2.6 The future

Unfortunately, FAES did not survive political interference, internal wrangles and the loss of the presidential election by Abdoulaye Wade in 2011. Production of Green Muscle ceased in August 2010 and the foundation was closed down some time in 2011. In the meantime, BCP had been bought up by Becker Underwood, which had even earlier acquired the production plant of Green Guard, a similar product developed in Australia. Becker Underwood appeared to be interested in marketing these products, as evidenced by the fact that they submitted a tender for the licence of Green Muscle in West and North Africa in competition with the newly created company Eléphant Vert of Geneva, Switzerland, which the former won. However, things changed when, subsequently, BASF acquired Becker Underwood. Apparently, BASF had mainly been interested in Becker Underwood's expertise in the production of entomogenous nematodes and the *M. acridum* products simply happened to be part of the package. Several years later, there were no signs that BASF was marketing Green Muscle anywhere. The director general of Eléphant Vert then decided that a new product was needed. Because of the expertise acquired during the LUBILOSA programme, this product, NOVACRID®, could be developed within a mere 2 years. By the time of writing, NOVACRID has been field tested against Moroccan locust (*Dociostaurus maroccanus* (Thunberg)), Italian locust (*Calliptamus italicus* (Linnaeus)), migratory locust (*Locusta migratoria* (Linnaeus) (all Orthoptera: Acrididae)), red locust and Sahelian grasshoppers in Central Asia and East and West Africa at doses from 12.5 to 50 g/ha with excellent results. The company is currently concentrating on the Central Asian market, both because a million or more hectares are treated in most years and because there appears to be a serious willingness among the governments to switch to more environmentally friendly products (C. Kooyman, unpublished results).

3 Aflasafe

3.1 Introduction

Aflatoxins are dangerous carcinogens produced by fungi when susceptible crops such as maize and groundnut become naturally infected when growing, during post-harvest and/or storage (Cotty et al., 1994). Most fungi producing aflatoxins belong to *Aspergillus* section Flavi. Aflatoxins suppress the immune system, stunt child growth and development, and sometimes cause rapid death of people and animals (Shephard, 2008; Wu, 2014). In sub-Saharan Africa (SSA), it is estimated that during some seasons and in certain regions, up to 40% of maize is contaminated with aflatoxins. Aflatoxin exposure leads to reduced labour hours owing to poor health (Wu et al., 2014). This leads to compromised productivity and decline of rural economies. Crops with excessive aflatoxin

Published by Burleigh Dodds Science Publishing Limited, 2019.

content are rejected in premium export markets governed by stringent safety regulations (van Egmond et al., 2007).

Aflasafe™ is a biocontrol product that contains four atoxigenic strains of *A. flavus* that decrease crop aflatoxin content in maize and groundnut as active ingredient (Atehnkeng et al., 2014). The product was developed by IITA, the USDA-ARS and other partners. It has been proven to reduce aflatoxins by 80–99% during crop growth and throughout the value chain. The technology has moved beyond proof of concept and is now available for commercial use in Nigeria, Kenya, Senegal, The Gambia, Burkina Faso and Ghana. It will also soon be available in other African nations (Bandyopadhyay et al., 2016). For more details on principles of the technology and product development pathway see Chapter 7.

In this section, we describe our approaches to commercialize Aflasafe in order to make it accessible to a large section of farmers and others affected by aflatoxin. The commercialization process begins with registration of the products for which availability of a product manufacturing facility is a pre-requisite. For scaling-up, it is also necessary to demonstrate that use of Aflasafe can generate profits for farmers and agribusinesses and that they are willing to pay for the product. Finally, we explain our approaches for sustainable scale-up by licensing the manufacturing and distribution responsibilities of Aflasafe to private and public investors.

3.2 Product registration

Laboratory testing, farmer field testing and safety evaluation of Nigeria's candidate biocontrol product Aflasafe was conducted with guidance from the National Agency for Food and Drugs Administration and Control (NAFDAC), the biopesticide regulatory agency of Nigeria. IITA's interactions with NAFDAC personnel involved joint efforts for sensitization on aflatoxin and its management including the biocontrol technology. Additionally, NAFDAC personnel provided input which was incorporated into the research needed to generate data for product registration. This created a synergy to develop the framework to register an atoxigenic biocontrol product in an African nation. Data on efficacy, safety, quality and social/economic value of the product was used to prepare a registration dossier that was approved by NAFDAC. Aflasafe is now a registered biopesticide for application in both maize and groundnut fields throughout Nigeria (Bandyopadhyay et al., 2016). Similar approaches were followed in other nations (i.e. Kenya, Senegal, The Gambia, Ghana and Burkina Faso) where the product has been registered, as well as in other countries where registration should soon be completed. In some African countries where the biopesticides registration framework was underdeveloped, IITA collaborated with USDA-Foreign Agricultural Service to develop the guidelines for biopesticide registration. This eased registration of atoxigenic products in target nations and will facilitate registration of biopesticides in other nations.

3.3 Product manufacturing

Nigeria's candidate Aflasafe product was initially manufactured using a laboratory-scale process adapted from the industrial process used to manufacture the U.S. biocontrol product *Aspergillus flavus* AF36 in Arizona, U.S. (Cotty et al., 2007). The laboratory-scale process was resource and time intensive and allowed production of only relatively small quantities of the product (Atehnkeng et al., 2014) barely sufficient for large-scale product testing and market development research.

As value of the Aflasafe product in reducing aflatoxin was understood by maize and groundnut value chain actors, demand in Nigeria increased. At the same time, testing of tailored candidate products began in other African countries. It was thus necessary to design and construct a manufacturing plant to produce Aflasafe to meet the increasing need for products in various countries. The Bill & Melinda Gates Foundation funded the construction of an Aflasafe manufacturing plant through a grant to the Partnership for Aflatoxin Control in Africa (PACA) of the African Union Commission (AUC). A manufacturing facility was built in IITA-Ibadan in 2013 and is the third aflatoxin biocontrol product factory in the world and first in Africa (Bandyopadhyay et al., 2016). Unlike the AF36 production plant, the Ibadan factory was designed to run on a dry process where the substrate or carrier (sorghum grains) is sterilized using a feed roaster. The substrate is heated on a feed roaster to about 120°C, then cooled using air filtered through an HEPA filter. The cool sterile grains are thereafter coated with spores of the four atoxigenic *A. flavus* strains using a seed treater. A polymer is added to aid sticking of the spores onto the sterile sorghum grain and a food colourant (blue dye) is added to distinguish it from food and feed. The final product is then packed in moisture-proof packets. This was the first plant in the world to run on dry process. At full capacity, the facility can produce 5 tons of Aflasafe products per hour. However, the facility can be modified relatively easily to increase the production capacity. More than 1500 tons of Aflasafe have been produced for both commercial and experimental use in Nigeria, Kenya, Senegal, The Gambia, Burkina Faso, Ghana, Tanzania, Mozambique and Zambia. The demonstration-scale manufacturing plant was constructed envisioning that public, private and public–private sector actors would be motivated to invest in the construction of similar plants.

In order to avoid complicated machineries and complex design the plant in Ibadan was constructed using off-the-shelf material and equipment commonly used by seed, fertilizer and feed industries operating in Africa (Bandyopadhyay et al., 2016). The manufacturing plant jump-started the thrust towards commercialization of Aflasafe and a modular manufacturing plant was built in the Katumani research station of the Kenya Agricultural and Livestock Research Organization in Machakos in 2017. Its current plant capacity is 10 tons/day, and can be scaled to 13 tons/h as demand increases. More than 200 tons of Aflasafe KE01™ will be produced for the 2018 planting season. Another modular manufacturing plant will be operational in Senegal in early 2019. The modular manufacturing plants were designed in a manner such that production capacity can be increased by adding new modules in the plant as demand for the product increases. In both Nigeria and Kenya facilities, manufacturing protocols were developed for approval and certification for commercial production by the national regulators, NAFDAC for Nigeria, PCPB and KEPHIS for Kenya.

3.4 Incentives for large-scale use

Several questions have been raised on the adoption and profitability of biocontrol technology, particularly by smallholder farmers (Ehrlich et al., 2015). Despite Aflasafe's proven efficacy to reduce aflatoxin content in crops, various market barriers including low consumer awareness, access to Aflasafe™, affordability, rootedness to traditional farming practices and inadequate enforcement of regulations on maximum aflatoxin limits have prevented widespread adoption of biocontrol. Through the AgResults Aflasafe™ Pilot ('Pilot'), an innovative approach was developed to overcome the barriers for smallholder farmers to adopt Aflasafe (Bandyopadhyay et al., 2016; Kimberley and Hoffman, 2010). The Pilot is testing a pay-for-results ('pull') model for reducing aflatoxin content and creating sustainable market demand for Aflasafe along the maize value chain in Nigeria. AgResults Aflasafe is a collaborative initiative jointly funded by the Bill & Melinda Gates Foundation, the Department of Foreign Affairs and Trade of Australia, the Department for International Development of the United Kingdom, Global Affairs Canada, USAID and the Commercial Agriculture Development Program of the Government of Nigeria, and technically supported by USDA-ARS.

The Pilot is designed to stimulate the private sector to overcome the market limitations mentioned above, deliver development impact to smallholder farmers and assess influences of pull mechanisms on smallholder adoption of agricultural technologies (Scherer and Yago, 2011). Since its inception in 2012, the Pilot developed and implemented a suite of interventions that reduce aflatoxin contamination from planting, through harvest and all the way until the crop reaches the market. It also boosts socioeconomic benefits. The Pilot offers an incentive payment proportional to the weight of maize verified to have been treated with Aflasafe after proportions of Aflasafe strains associated with the crop have been determined using microbiological and molecular techniques (Bandyopadhyay et al., 2016). The presence of 70% or more of Aflasafe strains on the grain is taken as a proxy for successful Aflasafe application by farmers.

The Pilot relies on private and public sector agricultural enterprises (known as implementers) to sell Aflasafe to smallholder farmers, train them on how to use the product, and aggregate and market the Aflasafe-treated maize produced by the participant smallholder farmers. Another role of the implementer is to stimulate the market by sensitizing farmers and buyers. In addition, implementers facilitate access to technical knowledge and inputs that farmers require to boost crop productivity and move from solely subsistence to commercial production. Implementers are the primary means by which small-scale farmers obtain Aflasafe. The Pilot provides monetary performance-pegged incentives of US$18.75/ton grain to implementers that aggregate maize verified as Aflasafe-treated. Through a shared-value approach, smallholders and implementers benefit from both the performance-pegged incentives and premiums paid by buyers of Aflasafe-treated maize with safe levels of aflatoxins. The primary buyers of Aflasafe maize in Nigeria are poultry producers, feed manufacturers and premium food industries.

During 2013–17, smallholder farmers applied 605 tons of Aflasafe on 60 500 ha of maize to reduce aflatoxin. More than 96% of the maize grain lots harvested from Aflasafe-treated fields showed the mentioned dramatic reduction in aflatoxin content in comparison to non-treated maize. In addition, the yield increased by 50% compared to the national average because of use of improved agronomical practices. On average, 53.6% of the harvested maize from treated fields were aggregated for sale in formal markets, while 22.6% was retained at home, 7% was sold at the farm gate and 16.8% was sold at different times before aggregation or given to family members and friends. Grains retained for home consumption and sold/given to family and friends enhanced safety of food at the rural household level. The smallholder farmers benefitted from premium payments for Aflasafe-treated maize sold to food and feed industries. Grain lots from Aflasafe-treated fields fetched an average 542% return on investment on Aflasafe from sale to anchor buyers. The total gross earnings of all implementers from the premium received from anchor buyers (72.4% of total earning) and incentives from AgResults (27.6% of total earning) were nearly US$150 000. Due to increased income from use of Aflasafe, farmers having prior experience with Aflasafe in Kaduna state were willing to pay US$13.42 $\pm$ 4.19 to treat a hectare of maize field (Ayedun et al., 2017). These results demonstrate that the Pilot contributed not only to upscaling the technology to improve smallholder livelihoods and health, but also how it set the scene for commercialization of innovations and strengthening of local economies.

3.5 Commercialization

Following the successful proof of concept of the AgResult Aflasafe pilot which has demonstrated the commercial use of Aflasafe in Nigeria, IITA and partners are scaling-up Aflasafe in 11 African countries. The scaling-up of Aflasafe in these countries is facilitated by the Aflasafe Technology Transfer and Commercialization (ATTC) initiative. ATTC aims to identify strategic options for partnership with private companies or government entities, execute those partnerships and help to ensure the distinct Aflasafe products reach millions of farmers. Considering the diversity of countries in terms of development of the private sector, the level of awareness about aflatoxins, the existence and the level of implementation of food safety regulations and the scaling-up through commercialization requires distinct strategies for each target country. This has informed the overall approach for commercialization, which starts with the design of a strategy adapted to each country. This strategy provides market projections, analyses manufacturing feasibility and distribution scenarios, investor's landscape as well as enables interventions required to increase the uptake of Aflasafe in the country. This information is packaged to mobilize investors for the manufacturing and distribution during investor forums. After the review of expression of interest and due diligence of interested investors, pre-selected companies submit their business plan. At the end of the process, Technology Transfer and Licensing Agreement (TTLA) are signed with the successful applicants. The TTLA provides an incentive for the private sector to invest, but it also attempts to maintain a balance between profit and public health benefits.

Throughout the execution of the TTLA, ATTC facilitates the transfer of know-how to selected investors. It implements an ambitious awareness-raising strategy combining the design of digital (website, social media loaded with materials) and traditional tools as well as the organization of events focusing on effects of aflatoxins and solutions available. ATTC also provides technical assistance in the development of the right training materials to support the promotion of the products.

To date, IITA licensed two companies for the manufacturing and distribution and signed a distribution agreement with two other companies. With these investors on board, there is leverage of resources to invest in awareness raising to increase the demand, promote the product and organize an efficient supply. As a result, in 2 years, these businesses were able to sell and distribute more than 1000 metric tons of Aflasafe enough to cover 100 000 ha of maize and groundnut farms in Nigeria, Kenya, Senegal, The Gambia and Burkina Faso. The retail price of the product has thereby been reduced from US$1.87 to US$1.5/kg.

Because of the silent nature of aflatoxin contamination the level of awareness about the problem is very limited among the agricultural value chain

Published by Burleigh Dodds Science Publishing Limited, 2019.

actors. In most cases, only 'sophisticated' industrial and institutional buyers and exporters are aflatoxin-conscious due to the strict requirement imposed by the market they are operating in. These large businesses (or premium markets) are currently the strongest drivers for Aflasafe adoption. However, they represent a small part of the addressable demand for the product. In order to unlock adoption levels in different market segments, the development of business cases, regulatory enforcement and awareness campaigns are necessary to complement the investment of the manufacturing and distribution partners.

3.6 Communication and awareness

Across SSA, aflatoxin awareness is predominantly low, which generally contributes to populations consuming maize, either self-produced or obtained from local markets, without questioning its quality (Bandyopadhyay et al., 2016; Unnevehr and Grace, 2013). The need for enhanced awareness is therefore one of the highest priorities (Bandyopadhyay et al., 2008; Cardwell and Miller, 1996).

In the early 2000s, a communication campaign led by IITA, Rotary Clubs and national food standard boards was implemented in Benin, Ghana and Togo to increase public awareness of aflatoxins and associated health risks (James et al., 2007). To increase its effectiveness, the campaign messages incorporated location-specific scientific data on aflatoxin prevalence and management strategies. In addition, promotional materials were adapted to the context of each nation. Although results were positive, the authors concluded that future campaigns should target all value chain actors and introduce the concept of paying premium prices for safer maize. Furthermore, communication strategies should be designed with tailored information for key targets such as farmers, traders, consumers, food/feed industries, scientists, development investors and policy makers/implementers (Johnson et al., 2015).

The first comprehensive communication campaign at a national level was designed in Kenya by IITA and partners in 2017 with the aim of educating farmers on correct usage of Aflasafe KE01. However, for this campaign (and other campaigns across SSA) to be successful, it should be accompanied by continuous investment in extension training, infrastructure and so on to prevent and/or reduce crop aflatoxin contamination throughout the crop value chain (Udomkun et al., 2017). Various communication products such as videos, printable flyers and information sheets about aflatoxin and its management can be found at www.aflasafe.com.

4 Commercialization principles

Developing and producing a technology or product for marketing and commercializing within a research institute poses different challenges.

Commercialization necessitates teamwork between inventors and business professionals. Creating a sustainable commercialization strategy that meets market demands brings up questions of scaling strategies and needs assessments, but also requires cost-price analyses including optimal production costs, start-up costs and so on. Inevitably, such questions require continuous re-strategizing and adapting to always changing (market) circumstances. This proved to be especially challenging in the case of Aflasafe, which entails targeting a specific market segment such as maize and groundnut farmers.

A typical approach used by large-scale agro-suppliers to commercialize (bio)pesticides starts with a clear user benefits analysis. This encompasses analysing how the invention can save costs for the end users (the farmers) and how much it can increase or improve yields. When looking at how an invention can save costs, it is important to include the cost price of the various field applications that it is replacing. Such costs should compare the means used in the absence of the invention such as labour, fuel and tractor costs it replaces. It should be noted that when an invention generates a better quality of the yield, it is often possible to access better market segments and obtain higher prices. Aflasafe is a good example of a biopesticide that improves the quality, rather than the quantity of the yield. In fact, Aflasafe plays an important role for the public health sector similar to the one of vaccines or the chlorination of drinking water. Although difficult to calculate, this in itself creates a higher market value.

User benefits are a determining factor for the adoption of an invention. For a hypothetical (bio)pesticide, for which the commercialization partner invested in product development, registration and initial market development costs (which was not the case for Aflasafe), the following calculation in Table 1 with a relatively high gross gain for the famer exemplifies the necessary calculations for commercialization. If an invention has a saving of US$200/ha and a better/ higher yield adds an additional US$50/ha, the seller would earn US$250/ha. Yet, to assure sustainable commercialization, these earnings should then be weighed against the marketing and promotion costs and the production costs gained by the value chain. Realistically, this should not be more than 90%, leaving 10% of the US$250/ha, which equals in this case US$25/ha. Details are shown in Table 1.

However, the adoption or rejection of an invention is more than just a financial consideration. External socioeconomic factors also play an important role and are often numerous, unpredictable and difficult to control. For instance, security threats, political or economic instability, traditional farming practices, land grabbing, conflicting customary laws, dependency on local authorities and climate change effects are among the factors that may influence the willingness of the end user to take a risk by investing/adopting an invention. This requires a good contextual understanding of the market, a flexible implementation

Table 1 Model calculation for a sustainable supply chain pricing approach

	Margin (%)	Price (US$/kg)	Price at certain position in the supply chain (US$/kg)
Marketing and promotion			
Retail price to farmer		12.50	12.50
Commercial margin local distributor	20	2.50	10
Transport		0.25	9.75
Regional distributor	10	0.975	8.78
Marketing, promotion and awareness	20	2.50	6.28
Transport from factory to regional distributor	9.40	0.50	5.78
Production			
Overhead manufacturer	10	0.48	5.28
Margin manufacturer	20	0.80	4.80
Direct assigned production costs (raw material, staff, water, electricity a.o.)			4
Benefit to farmer by using invention			
Gross commercial gain farmer per ha		250	
Percentage of gain per ha which goes to the supply chain	10	25	
Amount of kg of product used per ha	2	12.50	
Net commercial gain farmer per ha		225	
Return on investment for farmer by using invention		9	

For each US$ invested in the biopesticide the farmer gets US$9 back, that is 2 × US$12.5/US$225.

strategy and a certain amount of risk-taking from the implementers and/or end users.

When introducing inventions like Aflasafe, which supports public health, it is critical to have the support of the government. This is especially the case for Aflasafe, as most of the population is unaware of the dangers of aflatoxins. Although governments have a large outreach and influence, in low- and middle-income countries they often lack the capacity to support mass implementation. It should also be noted that many low- and middle-income countries do not have a large commercial-scale agricultural sector. Most farming is done at subsistence level whereby farmers rely on traditional practices, lack access to new information and have few opportunities to develop their businesses. This makes mass awareness, implementation and legislation efforts through the government difficult. In such cases the private sector is a viable way to successfully introduce an invention and promote its implementation. While the private sector lacks outreach capacity and is unable to push for legislation, it often has a better contextual understanding of the field. Implementation strategies through joint efforts from the government and the private sector therefore offer beneficial and fruitful solutions. The AgResults Pilot and the ATTC project are good examples of how an invention is scaled-up through joint efforts. As mentioned before, the AgResults Pilot created awareness and made Aflasafe available through the private agri-businesses ('the implementers') via a pull mechanism whereby implementers were empowered to stimulate market demand. Simultaneously, ATTC addresses the challenge of scaling-up the invention through both the private sector and the governments. Through AgResults and ATTC's efforts in Nigeria, the government obtained the capacity to implement a national aflatoxin control strategy in their Agricultural Sector Food Security and Nutrition Strategy for 2016–25. In addition, IITA through ATTC partnered with Harvestfield Industries Limited (HIL), a private sector investor, to take over the production and distribution of Aflasafe. This was necessary since IITA's demonstration manufacturing plant was not built for commercial purposes and there was a need to formally transfer and license the technology to the private sector. To assure a successful transition, ATTC supported HIL in setting up a manufacturing plant and laboratory to produce Aflasafe inoculum and enabled it to produce the active ingredient. In addition, as the AgResults Pilot was coming to an end, a better understanding of the challenges faced by the implementers and the opportunities to upscale operations was achieved through a social science approach. This should eventually enable HIL to support and utilize the implementers efficiently. To support and steer these different initiatives seamlessly, IITA set up the Business Incubation Platform (BIP), which supports not only the AgResults Pilot and ATTC, but also other IITA projects and businesses.

5 Conclusion and future trends

For most of its existence, the CGIAR and with it IITA delivered its products free of charge to national programmes and farmers. The costs to develop and promulgate resistant/high-yielding varieties, biocontrol agents and advice to customers was thereby borne by international donors. Here we describe two efforts to commercialize products, which are sold by private enterprises to farmers and governments alike.

Green Muscle, a biorational insecticide against locusts and grasshoppers was developed on request by donors to change the previous insecticide regime with its vast pollution of the environment. Uptake of the newly developed entomopathogen product was difficult because the primary target, migratory locusts develop dangerous population levels only every ten or so years. In its final form, the product could be stored for years, but incentives for commercialization still proved inadequate despite recommendations for its use by FAO. Various grasshoppers that pose economic damage every year offered more promising targets, but given the poverty of the endangered areas and farmers, uptake and subsequently production stalled. Entomopathogen-based products against locusts and grasshoppers, though under other names and based on other local strains of the active ingredient, eventually were profitable only when more promising targets in richer countries like Australia and in Central Asia were found. Throughout the project, the chemical industry successfully competed with Green Muscle or, after having acquired the local producers and their know-how, did not continue production.

Twenty years later and with different scientists at IITA under a changed organogram, Aflasafe was developed without benefitting from experience in licensing gained with Green Muscle. The originally targeted maize farmers were often too near to subsistence level to guarantee successful commercialization of the product that effectively eliminated poisoning of consumers with aflatoxin. Fortunately, a potent donor continued supporting this research. Eventually the breakthrough happened thanks to the commercial interests of the poultry industry in Nigeria. The project developed an impressive succession of arrangements and organizations to foster and advocate Aflasafe in various value chains. In addition, farmers benefitted from advice on other technologies thereby not only producing a healthier crop but also increasing their yields. The process currently followed for transfer of Aflasafe technology and granting commercialization licence to investors is a model for scaling-up for other technologies. Future efforts should focus on developing a fast-track system allowing acceptance of Aflasafe-related evidence from one nation to register biocontrol products in other nations. This would drastically cut time from product development to commercialization.

With both products, effects are slow and circumstantial. Continued advocacy is therefore at the heart of success. Both projects highlight the importance of close collaboration between scientists, engineers, regulators, business professionals and extension services.

6 Where to look for further information

Guidelines for future research are

1. Optimizing production efficiency such as cheaper and better carriers.
2. Improving manufacturing processes such as different scales of production.
3. Improving product efficiency under diverse environmental factors, such as sporulation under drought.
4. Ensuring that strains do not degrade or change.
5. Monitoring product specifications and product performances.

Published by Burleigh Dodds Science Publishing Limited, 2019.

7 References

Atehnkeng, J., Ojiambo, P. S., Cotty, P. J. and Bandyopadhyay, R. (2014), Field efficacy of a mixture of atoxigenic *Aspergillus flavus* Link: Fr vegetative compatibility groups in preventing aflatoxin contamination in maize (*Zea mays* L.), *Biological Control* 72: 62–70.

Ayedun, B., Okpachu, G., Manyong, V., Atehnkeng, J., Akinola, A., Abu, G. A., Bandyopadhyay, R. and Abdoulaye, T. (2017), An assessment of willingness to pay by maize and groundnut farmers for aflatoxin biocontrol product in northern Nigeria, *Journal of Food Protection* 80: 1451–60.

Bandyopadhyay, R., Leslie, J. F. and Frederiksen, R. A. (2008), Nominal group discussion technique: Questions and responses, In: *Mycotoxins: Detection Methods, Management, Public Health, and Agricultural Trade* (J. F. Leslie, R. Bandyopadhyay and A. Visconti, eds), CAB International, Wallingford, UK, pp. 19–25.

Bandyopadhyay, R., Ortega-Beltran, A., Akande, A., Mutegi, C., Atehnkeng, J., Kaptoge, L., Senghor, A. L., Adhikari, B. N. and Cotty, P. J. (2016), Biological control of aflatoxins in Africa: Current status and potential challenges in the face of climate change, *World Mycotoxin Journal*, 9: 771–9.

Bateman, R., Carey, M., Moore, D. and Prior, C. (1993), The enhanced infectivity of *Metarhizium flavoviride* in oil formulations to desert locusts at low humidities, *Annals of Applied Biology* 122: 145–52.

Bateman, R., Carey, M., Batt, D., Prior, C., Abraham, Y., Moore, D., Jenkins, N. and Fenlon, J. (1996), Screening for virulent isolates of entomopathogenic fungi against the desert locust, *Schistocerca gregaria* Forskal, *Biocontrol Science and Technology* 6: 549–60.

Bateman, R., Douro-Kpindou, O., Kooyman, C., Lomer, C. and Ouambama, Z. (1998), Some observations on the dose transfer of mycoinsecticide sprays to desert locusts, *Crop Protection* 17: 151–8.

Cardwell, K. and Miller, J. D. (1996), Mycotoxins in foods in Africa, *Natural Toxins* 4: 103–7.

Cherry, A., Jenkins, N., Heviefo, G., Bateman, R. and Lomer, C. (1999), Operational and economic analysis of a West African pilot scale production plant for aerial conidia of *Metarhizium* spp. for use as a mycoinsecticide against locusts and grasshoppers, *Biocontrol Science and Technology* 9: 35–51.

Cotty, P. J., Bayman, P., Egel, D. S. and Elias, K. S. (1994), Agriculture, aflatoxins and *Aspergillus*, In: *The Genus Aspergillus* (K. Powell, ed.), Plenum Press, New York, NY, pp. 1–27.

Cotty, P. J., Antilla, L. and Wakelyn, P. J. (2007), Competitive exclusion of aflatoxin producers: Farmer driven research and development, In: *Biological Control: A Global Perspective* (C. Vincent, N. Goettel and G. Lazarovits, eds), CAB International, Oxfordshire, UK, pp. 242–53.

Douro-Kpindou, O.-K., Godonou, I., Houssou, A., Lomer, C. and Shah, P. (1995), Control of *Zonocerus variegatus* by ultra-low volume application of an oil formulation of *Metarhizium flavoviride* conidia, *Biocontrol Science and Technology* 5: 131–9.

Ehrlich, K., Moore, G., Mellon, J. and Bhatnagar, D. (2015), Challenges facing the biological control strategy for eliminating aflatoxin contamination, *World Mycotoxin Journal* 8: 225–33.

James, B., Adda, C., Cardwell, K., Annang, D., Hell, K., Korie, S., Edorh, M., Gbeassor, F., Nagatey, K. and Houenou, G. (2007), Public information campaign on aflatoxin contamination of maize grains in market stores in Benin, Ghana and Togo, *Food Additives and Contaminants* 24: 1283–91.

Johnson, N., Atherstone, C. and Grace, D. (2015), The potential of farm-level technologies and practices to contribute to reducing consumer exposure to aflatoxins: A theory of change analysis, IFPRI Discussion Paper 01452, IFRI, Washington DC, pp. 1–27.

Kimberley, A.N. and Hoffman, M. (2010), Pulling agricultural innovation and the market together, CGD Working Paper 215, Center for Global Development, Washington DC, pp. 1-25.

Kooyman, C. and Abdalla, O. (1998), Application of *Metarhizium flavoviride* (Deuteromycotina: Hyphomycetes) spores against the tree locust, *Anacridium melanorhodon* (Orthoptera: Acrididae) in Sudan, *Biocontrol Science and Technology* 8: 215–9.

Kooyman, C. and Godonou, I. (1997), Infection of *Schistocerca gregaria* (Orthoptera: Acrididae) hoppers by *Metarhizium flavoviride* (Deuteromycotina: Hyphomycetes) conidia in an oil formulation applied under desert conditions, *Bulletin of Entomological Research* 87: 105–7.

Kooyman, C., Bateman, R. P., Jürgen, L., Lomer, C. J., Ouambama, Z. and Thomas, M. B. (1997), Operational-scale application of entomopathogenic fungi for control of Sahelian grasshoppers, *Proceedings of the Royal Society London B: Biological Sciences* 264: 541–6.

Langewald, J., Kooyman, C., Douro-Kpindou, O., Lomer, C., Dahmoud, A. and Mohamed, H. (1997), Field treatment of desert locust (*Schistocerca gregaria* Forskal) hoppers in Mauritania using an oil formulation of the entomopathogenic fungus *Metarhizium flavoviride*, *Biocontrol Science and Technology* 7: 603–12.

Lomer, C., Bateman, R., Johnson, D., Langewald, J. and Thomas, M. (2001), Biological control of locusts and grasshoppers, *Annual Review of Entomology* 46: 667–702.

Price, R., Bateman, R., Brown, H., Butler, E. and Müller, E. (1997), Aerial spray trials against brown locust (*Locustana pardalina*, Walker) nymphs in South Africa using oil-based formulations of *Metarhizium flavoviride*, *Crop Protection* 16: 345–51.

Prior, C., Jollands, P. and Le Patourel, G. (1988), Infectivity of oil and water formulations of *Beauveria bassiana* (Deuteromycotina: Hyphomycetes) to the cocoa weevil pest *Pantorhytes plutus* (Coleoptera: Curculionidae), *Journal of Invertebrate Pathology* 52: 66–72.

Scherer, J. and Yago, G. (2011), Paying for outcomes: Protecting human and animal health in sub-Saharan Africa. Financial Innovations Lab™ Report, Milken Institute, Santa Monica, CA, p. 31.

Shah, P., Kooyman, C. and Paraïso, A. (1997), Surveys for fungal pathogens of locusts and grasshoppers in Africa and the Near East, *Memoirs of the Entomological Society of Canada* 129: 27–35.

Shephard, G. S. (2008), Risk assessment of aflatoxins in food in Africa, *Food Additives and Contaminants* 25: 1246–56.

Udomkun, P., Wiredu, A., Nagle, M., Müller, J., Vanlauwe, B. and Bandyopadhyay, R. (2017), Innovative technologies to manage aflatoxins in foods and feeds and the profitability of application – A review, *Food Control* 76: 127–38.

Unnevehr, L. and Grace, D. (2013), Tackling aflatoxins: An overview of challenges and solutions, 2020 vision focus brief 2013, International Food Policy Research Institute, pp. 1–2 van Egmond, H. P., Schothorst, R. C. and Jonker, M. A. (2007), Regulations relating to mycotoxins in food: Perspectives in a global and European context, *Analytica of Bioanalytical Chemistry* 389: 147–57.

Van Lenteren, J. C. (2012), The state of commercial augmentative biological control: Plenty of natural enemies, but a frustrating lack of uptake, *BioControl* 57: 1–20.

Wu, F. (2014), Global impacts of aflatoxin in maize: trade and human health, *World Mycotoxin Journal* 8: 137–42.

Wu, F., Groopman, J. D. and Pestka, J. J. (2014), Public health impacts of foodborne mycotoxins, *Annual Review of Food Science and Technology* 5: 351–72.

Published by Burleigh Dodds Science Publishing Limited, 2019.

Chapter 14

Weeds affecting field crops and water bodies in Africa

David Chikoye, *International Institute of Tropical Agriculture (IITA), Zambia;* **Friday Ekeleme, Stefan Hauser, Abebe Menkir** *and* **Alpha Y. Kamara**, *International Institute of Tropical Agriculture (IITA), Nigeria;* **Peter Neuenschwander** *and* **Obinna Ajuonu**, *International Institute of Tropical Agriculture (IITA), Bénin; and* **Hakeem A. Ajeigbe**, *International Crops Research Institute for the Semi-Arid Tropics (ICRISAT), Nigeria (corresponding author: d.chikoye@cgiar.org)*

1 Introduction

For many agriculture-dependent African countries struggling to feed their booming populations, weeds are bad news. On land, weeds reduce crop yields, increase the cost of crop production, lower crop quality and harbour other crop pests; in water, aquatic weeds interfere with aquaculture and choke waterways used in commercial transportation and recreation. The low investment in R&D and lack of policies to support the development and deployment of efficient and cost-effective weed control options encourage the continued practice of archaic and labour-intensive management methods. The highly ineffective hoe-and-machete-based weed removal system, which still dominates most African smallholdings, takes up to 70% of available farm labour and prevents farmers from expanding their farms, with some even abandoning heavily infested fields. Chemical control, which revolutionized weed management in the developed world, has not been widely adopted in Africa due to the high costs of quality

http://dx.doi.org/10.19103/AS.2018.0043.15
Published by Burleigh Dodds Science Publishing Limited, 2019.

herbicides, poor access to appropriate spraying equipment and the lack of skills to properly and safely apply herbicides. The improper use of chemicals leads to the increase in herbicide resistance in weeds and contamination of the environment, as well as to severe illnesses to farmers due to exposure. Without an effective weed management system that is adapted to the needs of smallholder farmers, agricultural development in Africa will continue to stagnate.

In this chapter, we review the impacts of, and progress towards, developing efficient management options for selected problematic weeds of African crops and waterways, including (1) speargrass, a common weed in the forest–savanna transition zone; (2) *Striga*, a common parasitic weed in cereals and legumes in the savannas; (3) *Chromolaena*, an important weed in the humid zones; and (4) water hyacinth, water lettuce and water fern, all important aquatic weeds.

2 Integrated management of speargrass (*Imperata cylindrica*)

2.1 Introduction

Speargrass, also known as cogongrass [*Imperata cylindrica* (L.) Raeuschel] (Poaceae), is a perennial invasive grass weed that presents a serious problem to agriculture in the tropics and sub-tropics of the world (Terry et al., 1997). It is ranked as the tenth worst weed globally (Willard et al., 1990) and perceived to be the most destructive weed in West Africa (Chikoye et al., 1999). Speargrass is a hardy weed that is tolerant to drought, fire and poor soil fertility, disperses and reproduces by both seeds and rhizomes, and has wide genetic variability. Hubbard et al. (1944) recognize five varieties of *I. cylindrica*: var. *major* (Nees) C. E. Hubbard is found in tropical Asia; var. *africana* (Anderss.) C. E. Hubbard is from Africa; var. *europaea* (Anderss.) Aschers and Graebn. occurs mainly in the Mediterranean region; var. *condensata* (Steud.) Hack. ex Stuckert is a native of the coastal region of Chile and possibly Argentina; and var. *latifolia* (Hook. f.) C. E. Hubbard is found in India. Clayton and Renvoize (1982) claim that only three varieties are commonly recognized, var. *africana, major* and *cylindrica,* and the others are recorded as synonyms.

What encourages the spread of speargrass?

1. *Frequent bush burning*. Burning encourages the spread of the weed as its rhizomes can tolerate fire.
2. *Land tenure systems*. The traditional leasehold system discourages farmers from investing in long-term weed control measures for land they do not own.
3. *Land-use intensification*. Intensive cropping means a shorter fallow period (of less than 2 years), which favours speargrass.

Published by Burleigh Dodds Science Publishing Limited, 2019.

4 *Poor access to improved technologies*. The high costs of inputs such as herbicides deter smallholder farmers from adopting effective control measures against speargrass.
5 *Wide genetic variability*. This trait enables speargrass to survive different environments and control measures.
6 *Ability to reproduce by seeds and rhizomes*. Winds spread speargrass seeds up to 100 m; rhizomes can survive drought and fire.

Speargrass is a serious problem in slash-and-burn agriculture in the savanna-forest transition zone, where it undermines the production of perennial crops (rubber, oil palm and coconuts) and field crops (cereals, legumes and root and tuber crops). Competition from speargrass causes 80% to 100% yield losses in annual crops (Chikoye et al., 1999, 2000, 2001; Lum et al., 2005; Udensi et al., 1999).

Speargrass can invade and quickly overtake disturbed ecosystems, making it almost impossible for other plants to coexist with it. This weed is a huge fire hazard as it burns rapidly even when green, taking with it other native vegetation. With the ability to quickly rebound after a burn, speargrass can displace other ecologically and economically important plant species. Frequent bush burning causes considerable loss of leaf litter and surface organic matter that are vital to maintain soil fertility. The lack of options available to smallholder farmers to effectively control this weed forces them to abandon land in favour of less-infested marginal land (Akobundu et al., 2000).

The International Institute of Tropical Agriculture (IITA), in partnership with National Agricultural Research Organisations in Nigeria and the Republic of Benin, has conducted extensive research to understand the speargrass problem and to develop promising control methods. These partner institutions have tested various mechanical, cultural, chemical and biological alternatives - used either singly or in combination - in comparison with the manual methods used by farmers.

2.2 Manual control

Manual weeding using handheld implements such as hoes and machetes is the most common method of speargrass control in West Africa. Prior to sowing, farmers burn the foliage of speargrass followed by slashing of new shoots (Chikoye et al., 1999, 2000; Udensi et al., 1999). This control method is ineffective as the weed's underground rhizomes remain intact and eventually produce new shoots. In addition, burning tends to stimulate bud sprouting and flowering. To augment this, farmers may perform shallow tillage during the dry season to uncover the rhizomes and expose them to sunlight for drying.

However, during the wet season, shallow tillage could break the dormancy of the rhizomes through fragmentation and result in increased sprouting of speargrass shoots. The control of speargrass through manual means is generally labour-intensive, and time-consuming.

2.3 Mechanical control

Mechanical control is achieved using heavy machinery such as tractor-drawn ploughs and disc harrows. To desiccate the underground rhizomes, tillage operations should be done at least twice to soil depth of no more than 15 cm. The majority of speargrass shoots originates from rhizomes in the 0–15 cm horizon, hence shallow tillage can significantly reduce the rhizomes' biomass (Ivens, 1980). The tillage operations should be done at 2-week intervals, and each operation at right angles to the previous one to maximize rhizome fragmentation and exposure to sunlight. Cultivation should be conducted in the dry season when most of the speargrass biomass is in the rhizomes and sun-drying them is most effective.

Mechanical control can also be used to bury the rhizomes at depths of more than 15 cm. Burying rhizome fragments at 15 cm or deeper makes it difficult for any shoot to emerge as their carbohydrate reserves would have already been depleted before they could break ground and photosynthesize. Mechanical control, however, has its downsides: (1) it is too expensive for smallholder farmers; (2) it causes soil compaction and erosion; and (3) in the long-run, speargrass could re-establish dominance if the land is not immediately planted to crops (Chikoye et al., 2000; Chikoye and Ekeleme, 2003; Ivens, 1980).

2.4 Smothering with cover crops

Cover crops, also referred to as improved fallow systems, are primarily grown to perform multiple agroecosystem functions such as improving soil fertility, reducing soil erosion and controlling weeds. Cover crops suppress weed growth through competition for light, water, nutrients and space. Like other weeds, speargrass does not grow vigorously under shade such as that provided by cover crops. Several green manure crops like *Mucuna* spp.; *Centrosema pubescens* Benth.; *Lablab purpureus* (L.) Sweet; *Pueraria phaseoloides* (Roxb.) Benth.; *Calapogonium mucunoides* Desv.; *Psophocarphus palustris* Auct. non Desv.; *Stylosanthes guianensis* (Aublet) Sw. and *Flemingia* (*Moghania*) *macrophylla* (Willd.) Kuntze (all Leguminosae) as well as food crops like egusi melon (*Citrullus lanatus* [Thunb. Matsum and Nakai]); cowpea (*Vigna unguiculata* (L.) Walp.); pigeon pea (*Cajanus cajan*, (L.) Millspaugh and sweet potato (*Ipomoea batatas* L.) have been evaluated for speargrass suppression in West Africa (Chikoye et al., 2001). *Pueraria* and *Centrosema* spp. were effective, but they have a high likelihood of becoming difficult weeds to control because of their perennial growth cycle and high seed production capacity. The food crops were not very effective in speargrass suppression due to the short duration of approximately 3 months only of their canopy cover. Most of the research on the use of cover crops for speargrass suppression focused on the use of *Mucuna* spp. because of their advantages over other cover crops such as early establishment, faster growth rate, higher biomass production and quick improvements to soil fertility. Several researchers have reported effective control of speargrass by *Mucuna* spp. in West Africa (Akobundu et al., 2000; Chikoye et al., 2001; Udensi et al., 1999). In Nigeria, Udensi et al. (1999) reported that cover cropping with *Mucuna pruriens* var. *utilis* for four consecutive years effectively controlled speargrass. Widespread adoption by farmers with complete elimination of speargrass after 2–3 years of planting *Mucuna* crops has also been reported in southwestern Benin (Versteeg et al., 1998). In fields previously planted with *Mucuna*, cassava and maize yields were higher due to reduced weed competition and improved soil fertility (Chikoye et al., 2001).

However, the widespread adoption of *Mucuna* for speargrass control in West Africa has been constrained by the lack of commercial value of *Mucuna* seed, lack of availability of seeds, reduction in land available for food production, toxicity of *Mucuna* grain when consumed by humans and animals, destruction of accumulated *Mucuna* mulch by dry season fire and inability to inter-crop *Mucuna* with short-stature crops such as cowpea (Akobundu et al., 2000).

2.5 Cultural practices and high-density cropping

As previously described, speargrass can be controlled by good land preparation using mechanical tillage and followed immediately by cultivating the land with crops, especially low-growing ones such as cowpea, egusi melon or sweet potato planted at high-density. Farmers in the derived savanna areas of southwestern Nigeria usually plant melon alone or inter-cropped with maize to suppress speargrass following land tillage (Chikoye et al., 2006). The same practice is observed by farmers in eastern Nigeria who plant cowpea and melon immediately after land preparation (Chikoye et al., 2001). This practice of high-density cropping after mechanical tillage is not only effective in smothering seed-borne speargrass, but it also saves farm labour through one less weeding.

2.6 Alley cropping

Alley cropping is an agroforestry system in which food crops are grown in alleys formed by hedgerows of leguminous trees or shrubs. The trees and shrubs are periodically pruned during the growing season, and the prunings are used to provide nutrient-rich mulch and green manure for the companion crops (Kang et al., 1981). Alley cropping was promoted as an alternative to traditional long bush fallow with a potential to control speargrass.

In the forest–savanna transition zone of southwestern Nigeria, Anoka et al. (1991) observed that hedgerows of *Gliricidia sepium* (Jacq.) Steud and *Leucaena leucocephala* (Lam.) de Wilt reduced speargrass density by 67% and 51%, respectively, after 2 years compared with no hedgerows. Shoot biomass of speargrass also decreased by 81% and 78% between *Gliricidia* and *Leucaena* hedgerows, respectively, while *Gliricidia* and *Leucaena* reduced rhizome biomass by 96% and 90%, respectively. At Fashola in Nigeria, Aken'Ova and Atta-Krah (1986) observed that only few stands of speargrass grew between the alleys of *Gliricidia* compared to plots without trees, which were covered with a vigorous stand of speargrass. They concluded that the dense canopy of uncut *Gliricidia* that persists for 8–9 months in a year provides sufficient shade to discourage the development of speargrass during a relatively short fallow of 2 or 3 years.

Despite the advantages of alley cropping in soil and weed management, the technology was not widely accepted by smallholder farmers in West Africa. Some of the reasons given by farmers include (1) additional labour needed for the establishment, pruning and general management of hedges; (2) reduction of the area of land available for food crops; (3) existing land-use and land tenure system; (4) competition for light, water and nutrients between trees and crops; (5) possible allelopathic effects of the hedgerow plants on crops; and (6) fear that the hedgerow plants might serve as source of pests and diseases for the companion food crops (Bruning and Sander, 1983). The tree species introduced for alley cropping were, however, widely used in compact wood lots.

2.7 Chemical control

Chemical control is probably the cheapest, quickest and least soil-disruptive of the available speargrass management options (Chikoye et al., 2000, 2005). Several compounds including paraquat, glyphosate, imazapyr, fusilade, nicosulfuron, rimsulfuron and many other formulations have been tested and found to be highly effective against speargrass in West Africa (Akobundu and Ekeleme, 2000; Chikoye et al., 2005; Lum et al., 2005; N. Onyia, unpublished results; Udensi et al., 1999). Below, we look at the key characteristics and qualities of some of the more popular herbicides currently being used in Africa:

- *Paraquat* is a broad-spectrum, non-selective contact herbicide that causes shoot desiccation within a short time after application. However, speargrass control is only temporary as regrowth occurs within 1–2 weeks after application. Paraquat is banned in Europe and is restricted for use in the United States because of human health and environmental concerns. Despite the dangers it poses, paraquat is still widely marketed and used in Africa because it is cheaper than hand-weeding. According to the manufacturers, paraquat is a safe and effective herbicide if it is used according to the label. However, the herbicide has been controversial because of its use for suicides and recent link to Parkinson's disease.

- *Glufosinate ammonium* is also a broad-spectrum contact herbicide and is largely considered as an alternative to paraquat. At the rate of 2–2.5 kg a.i./ha, the herbicide causes severe foliar damage to speargrass, but does not effectively prevent regrowth from rhizomes. Higher rates (3–4 kg/ha) have been reported to effectively control speargrass for 5–6 months before regrowth occurs (Brook, 1989). Regrowth of speargrass shoots from rhizomes after treatment is a major limitation to its use. The product is also not readily available on the African market.
- *Glyphosate*, first introduced as Roundup but now available under several other trade names and formulations, is the most commonly used and effective herbicide for speargrass control in West Africa (Chikoye et al., 2000). It is a broad-spectrum, non-selective systemic herbicide usually applied before crop emergence. Glyphosate's popularity can be attributed to several factors including its low mammalian toxicity, absence of phytotoxic residues in the soil and its efficacy against speargrass (Terry et al., 1997). In addition, the efficacy of the herbicide is not dependent upon the carrier volume; hence it can be applied using low volume as well as high volume sprayers. Despite its advantages, glyphosate has severe disadvantages: it is more expensive than other herbicides and has poor rain-fastness (i.e. if rain falls within 6 h after treatment, it needs to be reapplied). The overuse of glyphosate has also been linked to the development of herbicide resistance in 41 weeds. EU nations have been lobbying to ban glyphosate amidst concerns that its use could cause cancer and reduce biodiversity.
- *Imazapyr*, marketed as 'Arsenal' and 'Assault', is a broad-spectrum, systemic and non-selective herbicide registered for use in rights of way and plantations, but has also shown great potential for speargrass control in food crops in West Africa (Udensi et al., 1999). For acceptable control of speargrass, an imazapyr dose of 0.5–1.0 kg a.i./ha is required. In southwestern Nigeria, imazapyr application at the rate of 0.5–1.25 kg a.i./ha provided season-long control (6 months) of speargrass (N. Onyia, *unpublished results*). The herbicide causes more rhizome mortality and prevents the emergence of new shoots better than glyphosate. In Ibadan, Udensi et al. (1999) reported that imazapyr at 0.5 kg a.i./ha was as effective as 1 kg a.i./ha and comparable to glyphosate at 1.8 kg a.i./ha in controlling speargrass. The

major disadvantages of imazapyr are its high cost, slow action in killing the weed and its persistence in the soil, which consequently restricts immediate planting of food crops after application. In studies conducted at IITA in Ibadan, Nigeria, maize could not be planted 2–3 months after imazapyr was applied at 1 kg/ha (N. Onyia, *unpublished results*). The delay in planting of food crops caused by its persistence in the soil shortens the available crop-growing period and, consequently, reduces overall yields. Although effective, this group of herbicides should be used with caution to prevent the evolution of herbicide-resistant weeds.

- *Nicosulfuron and Rimsulfuron* are two selective, post-emergence sulfonylurea herbicides recommended for weed control in maize. Dosages of nicosulfuron between 150 and 200 g/ha applied at 1–2 weeks after planting are effective against speargrass. Maize yield increased by up to 90% compared to untreated fields (Lum et al., 2005). Rimsulfuron at 20 g/ha suppressed speargrass growth and increased maize yield by 50% in the humid tropics of Nigeria (Chikoye et al., 2007). The widespread use of these two herbicides is limited by their high cost and lack of availability on the local market. Sulfonylureas have high-selection pressure, like imazapyr, which may result in rapid evolution of herbicide-resistant weeds.

African smallholder farmers cite the perceived high cost of chemicals and needed application equipment, lack of financial resources, awareness and proper use-skills, the need for skilled labour, and unavailability of herbicides as the major stumbling blocks for the widespread use of chemicals for speargrass control. Moreover, already 250 weed species have shown herbicide resistance (www.weedscience.org).

2.8 Biological control

For sustainable speargrass management, the biocontrol agent should be effective against the weed's foliage, rhizomes or prevent seed production. There are many potential agents (pathogens, insects, nematodes etc.) available for speargrass control (Soerjani, 1970; Evans 1987). IITA-led research has tested three fungi isolates [*Bipolaris sacchari* (E. J. Butler) Shoemaker, Drechslera *gigantea* Heald and F. A. Wolf and *Colletotrichum caudatum* (Sacc.) Peck], two rust pathogens [*Puccinia fragosoana* Beltran and *Puccinia imperatae* (Magnus) G. Poirault] and a head smut fungus [*Sporisorium schweinfurthianum* (Thum.) Vanky] as potential biocontrol agents against speargrass in Benin (A. Breeyen, *unpublished results*). The three fungi were proven effective on speargrass foliage, but did not completely kill the weed. No disease symptoms were observed after inoculation with the two rust pathogens, while the smut fungus reduced only the production of inflorescences. These studies showed that prospects for implementing fungal-based biocontrol programmes as a component of speargrass control were promising and that a more comprehensive survey for virulent pathogens should be carried out.

2.9 Conclusion

Speargrass is a consequence and the sign of intensified farming systems with short fallows under poor land management practices. Currently in Africa, no single management practice provides sustainable control of speargrass in smallholder farms. Farmers need to adopt a combination of several available management options such as slashing, mechanical control, legume cover fallow, intercropping and chemical control in an integrated and well-balanced management programme to reduce the impact of the weed to non-economical levels. Although short-term interventions to suppress speargrass and prevent crop yield losses are vital, long-term strategies to prevent the re-establishment of speargrass are also important. This may be achieved by adopting good agronomic practices that restore or conserve soil fertility and preventing annual bush-fires so that secondary succession can proceed naturally.

3 Integrated management of *Striga hermonthica*

3.1 Introduction

Several parasitic plants pose substantial problems for cereal and legume production in the savanna zones of Africa, the most important being about 41 species of *Striga* (Scrophulariaceae) with greatest diversity in West and Central Africa. The most important species are *Striga hermonthica* (Del.) Benth. on cereals and *S. gesnerioides* (Willd.) Vatke on legumes, each with several races adapted to different wild and cultivated hosts (Berner et al., 2003). Particularly *S. hermonthica* is a serious threat to achieving sustainable and economic food production in Africa (Ejeta and Butler, 1993), often causing yield losses greater than 50% in infested fields (Parker, 1991). Major food and nutrition security crops such as maize, sorghum, pearl millet, upland rice and cowpeas - grown mostly by resource-poor smallholder farmers in the savannas - are especially vulnerable. Also called 'witchweed' and 'vampire weed', *Striga* attaches itself to the roots of the host plant and extracts essential nutrients, leaving its host plant stunted, yielding little or no grain. Nearly 30 million ha are infested with *Striga* annually (Ejeta and Butler, 1993), with annual crop production losses amounting to US$7 billion, affecting over 100 million people in the savanna regions of Africa (M'boob, 1986).

Crops grown under severe moisture deficit and poor soils are especially predisposed to *Striga*. The weed is difficult to control because it produces large quantities of seeds, and its adaptation and dormancy mechanisms permit the seeds to stay alive in the soil for several decades. Farmers have developed a range of coping strategies that include hand-roguing application of inorganic fertilizer, manure and compost, and crop rotation (Emechebe et al., 2004; Debrah et al., 1998). Over the years, there have been numerous studies on *Striga* and a range of technologies have been identified as effective control methods (Parker and Riches, 1993). These include use of *Striga*-resistant maize varieties, and cropping leguminous trap crops that stimulate suicidal germination (Kling et al., 2000). However, researchers agree that *Striga* would best be controlled by combining a range of component technologies into an integrated *Striga* control programme to provide better flexibility and sustainable control over a wide range of biophysical and socio-economic environments. To this end, IITA and its partners have developed component and integrated technologies to control *Striga* particularly in cowpea- and maize-based systems (Berner et al., 2003). This section describes these technologies and limits itself to *S. hermonthica*.

3.2 Maize-legume rotation and related crop management practices

Intercropping is a viable low-cost technology that addresses the interrelated problems of low soil fertility and *Striga* infestation. It is important to identify effective legumes that are compatible with the existing cropping systems and capable of drastically reducing the *Striga* seed bank in the soil. Many crops can be grown in rotation or as inter-crops with susceptible cereals. Roots of cotton and legumes (cowpea, pigeon pea, groundnut and soybean) produce *Striga* germination stimulants, but are not parasitized by *S. hermonthica*. These crops are used as trap crops that cause suicidal germination of *Striga* and, therefore, reduce its seed bank (Parker and Riches, 1993; Schulz et al., 2003). Some soybean, cowpea and groundnut varieties possessing high *Striga* germination-stimulant activity have been identified in West Africa whereby *Striga* seeds collected from different locations respond differently to a range of trap crops. Trials are therefore needed to identify the most effective legume cultivars for a specific ecological area and farming system. IITA has developed and screened soybean and cowpea accessions and selected those that most stimulate *Striga* germination, which are then used in rotation with *Striga*-resistant maize in Nigeria. An example is TGX 1448-2E, a soybean variety that is popular in Nigeria. It produced 33% more grain yield, reduced *Striga* infestation by over 60% and reduced N fertilizer requirement by 50% compared to the local variety (Sanginga et al., 2002). Similarly, two cowpea varieties (IT97K-499-35 and IT90KD-82-2) that are resistant to *S. gesnerioides* and cause suicidal germination of *S. hermonthica* have been released in Nigeria (Kamara et al., 2009). Intercropping of dual-purpose cowpea with cereals can reduce *Striga* infestation by 12–50% each growing season and soybean and

Published by Burleigh Dodds Science Publishing Limited, 2019.

groundnut rotation or intercropping increased grain yields of *Striga*-tolerant or -resistant maize in Nigeria (Franke et al., 2006). The effectiveness of maize-legume rotation in improving maize yields and soil productivity has also been demonstrated in Ghana, Nigeria and Benin Republic, where farmers are now able to obtain good maize yields where there had been crop failures due to *Striga* infestation (Sanginga et al., 2002).

3.3 Application of nitrogen and inorganic fertilizer to stimulate biological control

Applying high rates of nitrogen fertilizer and organic materials can reduce *Striga* damage in maize (Kamara et al., 2008). IITA-led research has validated and promoted various improved soil management technologies (inorganic fertilizer + maize/soybean rotation or/and inorganic fertilizer + manure) in maize-based systems in West Africa (A. Kamara, *pers. comm.*). The adoption of improved soil fertility management technologies has led to (1) increased production and quality of farm products; (2) food diversification and improved nutrition; (3) increased income of farm households; and (4) less land degradation. These technologies were adopted by 49% of all farmers (over 1.5 million farmers) in northern Nigeria (A. Akinola, *unpublished results*). Now, these technologies need to be scaled-out to more beneficiaries in northern Nigeria where over 25 million people are affected by *Striga*.

The mechanism of seed suppression through fertilization is only little understood. It was shown that by improving soil conditions, particularly by adding nitrogen and/or by crop rotation with legumes, growth of biotic agents responsible for observed soil suppressiveness is encouraged (Berner et al., 2003). Thus, ethylene-producing bacteria like *Pseudomonas syringae* van Hall stimulate suicidal germination (Berner et al., 1999), which opens up the potential for seed coating (Ahonsi et al., 2002). Other pathogens like various *Fusarium* spp. directly attack *Striga* seeds and plants even before emergence and can contribute to enhance crop yield (Hess et al., 2002). For large-scale application, microbial delivery systems have been developed (Elzein et al., 2008). A major constraint is given by quarantine regulations, which make delivery across borders difficult and make it necessary that a pathogenic strain of an organism needs to be developed into a mycoherbicide in each country separately (Berner et al., 2003).

3.4 Striga-resistant/tolerant maize and cowpea

The use of *Striga*-resistant/tolerant cereal and cowpea varieties is the most economically feasible and environmentally friendly means available to smallholder farmers for controlling *Striga*. IITA, CIMMYT, ICRISAT and national programmes have developed open-pollinated maize and sorghum varieties (OPVs), hybrids and inbred lines that are resistant to *Striga*. Some of the OPVs give 40% higher yields than farmer varieties and are also tolerant to drought and low soil nitrogen. These varieties are already marketed by seed companies and used by farmers in West Africa. The varieties perform well in areas with low to medium levels of *Striga* infestation (Menkir et al., 2012).

3.5 Herbicide-resistant maize and seed coating with herbicides

Low doses of acetolactase synthase (ALS)-inhibiting herbicides applied as seed coating to maize with ALS herbicide resistance provide effective control of *Striga* (Berner and Kling, 1997; Kanampiu et al., 2001; Menkir et al., 2009; Chikoye et al., 2011). This control option is economical, making it attractive to African subsistence farmers because the herbicides are applied at very low rates and without the need for purchasing and using expensive sprayers. In addition, seed coating with these herbicides controls *Striga* during its subterranean growth stage before it can cause significant reduction in crop yields (Abayo et al., 1998; Kanampiu et al., 2001).

The use of ALS herbicide-resistant maize has been commercialized as an effective *Striga* control strategy and demonstrated through pilot projects in Western Kenya. In West Africa, the IITA breeding programme has incorporated the responsible imidazolinone resistance (IR) gene into maize inbred lines that already have good levels of polygenic resistance to *Striga*. These inbred lines have been used to develop hybrids and synthetic varieties that have been tested on research stations and farmers' fields for the past 5 years in northern Nigeria. Several promising hybrids that combine resistance to the herbicide with natural resistance to *Striga* have been identified. For example, seed coating the three highest yielding IR maize hybrids from the breeding programme provided 650–800 kg/ha more grain yield than plants grown from untreated seed. The numbers of emerged parasites with seed treatment for the ALS herbicide-resistant maize hybrids were 2–17 times lower than those without seed treatment. Such maize hybrids also had 76–85% less emerged parasites and produced 84–169% higher grain yields compared with the *Striga*-tolerant hybrid check without the IR gene (Menkir et al., 2009), demonstrating the high levels of *Striga* resistance of these maize hybrids. Several herbicide-coated and -uncoated hybrids that combined herbicide resistance with *Striga* tolerance were evaluated in three locations in the savanna of Nigeria (Chikoye et al., 2011). They sustained less *Striga* damage symptoms and yield loss under *Striga* infestation condition and supported fewer emerged *Striga* shoots than the uncoated *Striga*-susceptible and -tolerant hybrids, as well as the commercial hybrid check. This effect was possible because the herbicide provided season-long control of the weed, and prevented the formation of the parasite's seeds. To date, research results about ALS herbicide-resistant maize suggest that farmers can achieve an immediate yield increase in the season they plant such as maize and a reduction of the *Striga* seed bank, which will dramatically limit seed return to the soil. On sorghum, coating seeds with herbicides reduced *Striga* infestation, flowering and seed set, but had no effect on the sorghum biomass (D. Ndung`u, *unpublished results*).

Despite the extensive evaluation in farmers' fields and formal registration of IR maize cultivars, the potential of this technology has not yet translated into actual, on-the-ground improvements in maize production. This is due to many technical (seed coating issues) and commercial (quality seed production and supply infrastructure) problems. In addition, there still can be crop damage when rains are sparse after planting or too heavy so that the herbicide may be washed beyond the maize root zone too quickly, allowing *Striga* to emerge and replenish its seed bank. More work is needed to optimize the ALS herbicide-resistant maize technology by improving hybrid adaptability and performance, developing and implementing technology stewardship plans focused on seed agrodealers and maize farmers, facilitating herbicide registration processes with national authorities and improving herbicide formulation and seed coating. This technology is quality dependent and should be marketed by one or two very reliable seed companies. As discussed above, overuse of this family of herbicides can result in the evolution of herbicide-resistant weeds. However, the use of non-dressed seed in rotation with herbicide-dressed seed could potentially slow the evolution of ALS resistance in *Striga* by reducing selection pressure. Although models show that the likely occurrence of resistance in *Striga* is rare when the herbicide is used in seed coating (Gressel et al., 2004), our experience with non-parasitic plants is that resistance to ALS inhibitors should be expected. It is therefore important that farmers using ALS herbicide-resistant maize are trained to use resistance management strategies to ensure a good longevity of the technology.

3.6 Integrated Striga *control*

Since individual technologies have many shortcomings sustained control of *Striga* can only be obtained by the integration of two or more control options. In West Africa, IITA has tested the combined use of *Striga*-resistant/tolerant maize varieties in rotation with legumes that elicit suicidal germination of *Striga* seeds and contribute to improvement in soil fertility. Integrated control methods have been tested by the national programmes in Benin, Burkina Faso, Cameroon, Côte d'Ivoire, Ghana, Mali and Nigeria. Results of 235 on-farm trials conducted in six countries revealed that the combined use of *Striga*-resistant/

tolerant maize varieties in rotation with legumes (e.g. groundnut, cowpea, soybean) trap crops had 44% less emerged *Striga* plants and produced 41% more maize grain as compared to farmers' practices (A. Kamara, *unpublished results*). Extensive on-farm trials were also conducted in northern Nigeria to promote the use of *Striga*-resistant varieties in rotation with legume trap crops that cause suicidal germination of *Striga* seeds (Ellis-Jones et al., 2004; Franke et al., 2006; Kamara et al., 2008). This integrated approach increased crop productivity by an average of 88% (Franke et al., 2006). The use of a resistant maize variety after a legume crop resulted in net benefits of more than 100% over farmers' practice across seasons (Ellis-Jones et al., 2004). Additionally, the use of a *Striga*-resistant variety alone or in rotation with legumes reduced *Striga* seed density in the soil by 29–50% (Franke et al., 2006).

3.7 Conclusions

Despite the impressive results from current efforts to combat *Striga* this weed continues to cause significant crop losses on maize and cowpea farms resulting in food insecurity in many rural communities in sub-Saharan Africa (SSA). Farmers seldom adopt long-term sustainable control measures because their main incentive is to increase crop yields in the short run. Moreover, biophysical and socio-economic factors limit the full utilization of the available control options.

A more sustainable and robust *Striga* threat reduction strategy is therefore needed that identifies and promotes scientifically proven technologies that work on smallholder farmers' fields and have an immediate effect on improving soil fertility, stopping *Striga* emergence, reducing the *Striga* seed bank and improving crop yields.

4 Role and control of *Chromolaena odorata* in the humid tropics

4.1 Introduction

Chromolaena odorata King and Robinson (Asteraceae) arrived in West Africa in the 1930s, and since then has invaded most of humid and sub-humid SSA. It has been present in southern Cameroon for several decades, often dominating the weed flora of crop fields and fallows. It can establish in freshly cleared forest fields. Most farmers perceive *C. odorata* as a noxious weed because it forms woody roots rapidly and sprouts from root stocks even after burning the slashed biomass. Farmers are required to pull out every root stock from the soil before tillage is possible and to prevent immediate infestation. Generally, *C. odorata* is considered a transformer species (Richardson et al., 2000) that causes economic and biodiversity losses (Zachariades et al., 2009).

4.2 Slash and burn

With the currently common land preparation by slash and burn, farmers inadvertently promote the establishment and spread of *C. odorata* because the wind-dispersed seeds fall onto burned soil on which most seeds of potential competitors have been eliminated. Hauser and Mekoa (2009) showed that *C. odorata* biomass production in a plantain field cleared from secondary forest was not different in burned and mulched plots, yet after burning *C. odorata* attained 50% of the total volunteer biomass much earlier than in the mulched plots. In this case, burning caused an earlier dominance of *C. odorata*. In a similar environment but in a plantain plantation under the shade of *Terminalia ivorensis* A. Chev. (Combretaceae) trees, Norgrove et al. (2000) found that *C. odorata* produced more biomass after slash-and-burn at 7 months after planting, and attained between 83% and 90% of the total volunteer biomass at 12 months.

4.3 Biological control

The high labour requirements to clear fields of *C. odorata*, combined with the weed's competitiveness culminated in calls for biological control of the weed. Research on controlling *C. odorata* with classical biocontrol agents started in the 1960s. However, it took more than 30 years (until 1992) before the defoliating moth *Pareuchaetes pseudoinsulata* Rego Barros (Lepidoptera: Arctiidae) was released in Côte d'Ivoire (McFadyen, 1996) to naturally control *C. odorata*. Ghana, through its Crops Research Institute, followed suit, releasing *P. pseudoinsulata* between 1991 and 1997 and monitoring its establishment (Timbilla et al., 2003). A second biocontrol agent, the stem galling fly *Cecidochares connexa* Macquart (Diptera: Tephritidae), was released in Côte d'Ivoire in 2003 and 2004. Today, *P. pseudoinsulata* has reached Nigeria (Uyi, 2011), yet its effects on *C. odorata* remains largely unknown. *C. connexa* has meanwhile reached Benin (G. Goergen and P. Neuenschwander, *unpublished results*).

These biocontrol efforts were challenged by reports of *C. odorata* having strong beneficial effects on soil fertility recovery in fallow phases, shortening fallows in some places by up to 2 years and outcompeting many weeds of the cropping phase (Akobundu and Ekeleme, 1996; De Foresta, 1996; Slaats, 1996; Weise, 1996). Similarly, the benefits of *C. odorata* were described from Asia long before any biocontrol efforts were initiated (Chevalier, 1952; Poilane, 1952). However, more recent research results from West and Central Africa showed no firm conclusion on the benefits or detrimental effects of *C. odorata*. This led to withdrawal of funding for the biocontrol initiatives and abandonment of planned research in some countries (McFadyen, 1996). In the Republic of South Africa efforts to introduce new agents are still continuing because an important negative impact of *C. odorata* in this country is on the biodiversity in national parks, where thickets of this weed can stop even white rhinoceros (C. Zachariades, *pers. comm.*). Interestingly, while the invasion of *C. odorata* across the tropics concerns one biotype, *C. odorata* from RSA belongs to another biotype, which explains partly the failure of some test biological control agents (von Senger et al., 2002) and opens up prospects for finding new agents. Several species are still being tested. *Pareuchaetes insulata* (Walker) was considered the most promising, but eventually showed a rather poor performance and established in only one of its release sites (Uyi et al., 2016).

The untimely withdrawal of support to already-initiated biocontrol programmes in West and Central Africa prevents a scientific follow-up on the spread of the released organisms. While these agents have been tested in quarantine and found to be confined to *C. odorata*, the possible reduction of this weed and the resulting impact on the ecology reverting to the conditions before the introduction, would well have merited a study. And what would this mean for the farmer managing young fallow or the forester rehabilitating a forest patch?

4.4 Conclusion

With the benefits and drawbacks of *C. odorata* not being fully understood and neither ecologically nor economically assessed, it remains impossible to judge if continued biological control would have been the best way forward. The expectation that with biological control of *C. odorata* a return to the previous situation would occur may be unsubstantiated because new invasive species such as *Tithonia diversifolia* (Hemsl.) A. Gray (Asteraceae) would likely replace *C. odorata* as dominant species. As such, it remains in the farmers' hands to combat or accept the presence of *C. odorata* in their fields and fallows and to balance its benefits and disadvantages in each specific situation.

5 Weed control in plantain

5.1 Introduction

There is currently very little concrete information on the response of plantain and banana to weed infestation in smallholder plantations in the humid SSA. While commercial dessert banana plantations maintain high plant density, dense canopy cover and chemical-based weed control, smallholder farms are often at the other end of the spectrum: sparse planting density, scant canopy cover and manual weed control, conditions amenable to high weed infestation. This section discusses the effects of weed control on plantain smallholdings in the humid forest zone of West and Central Africa, with specific examples from southern Cameroon.

Published by Burleigh Dodds Science Publishing Limited, 2019.

5.2 Hand-weeding, herbicide and fertilizer applications

Plantain is a staple in southern Cameroon. It is mostly planted after clearing old bush fallow or secondary forest and burning of the biomass. Ordinarily, this prevents early weed infestation. However, even older patches of forest do not remain weed-free for long as clumps of weeds easily regrow and seeds of weeds are carried and deposited by winds. Weeds such as *C. odorata* and related Asteraceae species sprout soon after crop establishment, with biomass burning often facilitating their rapid germination and growth (Fig. 1). A specific problem after clearing old bush or forest are climbing species, which are common in such vegetation. Although these species often sprout slowly after the clearing, their ability to use plantain as a support renders them particularly detrimental as they may wind around the youngest still furled leaf and arrest the emergence of the following leaves.

In a study of the effects of glyphosate on the vegetation in a plantain plantation after forest clearing, S. Ndumbe (*unpublished results*) found that of about 160 species only 20 were killed and did not reappear after two herbicide applications of 6 L/ha. About 50% of the species suffered damage but did not die and 8–10 species showed no signs of damage. Thus herbicides do not offer a solution in these systems.

These days, the cost of clearing a patch of forest is prohibitively high, while access to forests is becoming increasingly difficult. Thus, farmers try to grow plantain in younger fallows. Except for their soil organic matter content, these fallows are basically identical in chemical composition to newly cleared forest. However, these areas are likely to be infested with weeds from the very start of the crop. An on-farm trial in a previously cropped field at the village

Figure 1 Typical fallow, crop and forest situation in southern Cameroon. *Foreground*: *Chromolaena odorata* invaded the land after forest clearing. *Middle*: Plantain field after forest clearing, with good weed control. *Back*: Secondary growth forest.

Published by Burleigh Dodds Science Publishing Limited, 2019.

of Nkolnda, near Yaoundé, Cameroon, showed that without fertilizer, plantains did not produce fruits unless the field was weeded 8–12 times per year (Fig. 2). When fertilizer was applied, bunch yield increased linearly with increasing weeding frequency. In a secondary forest in the remote village of Mvambot in southern Cameroon, a plantain crop produced higher yields when weeding was combined with fertilizer application (Fig. 3). This showed that under naturally fertile conditions, more frequent weed removal doubled yields, while the addition of fertilizer added only about 30–50% to crop yields (R. Hauser, *unpublished results*).

5.3 Conclusion

Weed control may not be required for the first several months in plantain systems, when established in newly cleared old bush or secondary forest through slash and burn. Today, with plantain production systems moving into less favourable lands and the weed controlling effect of biomass burning becoming less effective due to smaller amounts of biomass, if any, weed infestation appears earlier and has an apparent negative interaction with the soil fertility status. With less fertile soils used for plantain production, weed control will become very important, yet ready to use low-cost and low-labour approaches have not been developed. The low plant density in such plantations allows for easy manual control, yet the cost of such operations is increasing and keeping plantain production in young fallow profitable will require more effort to develop new cost and labour efficient weed control approaches.

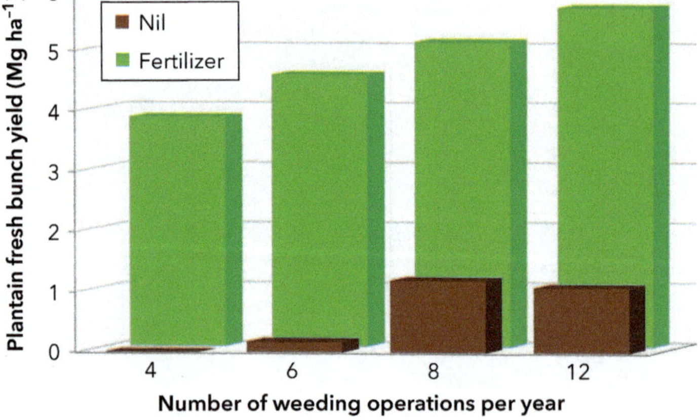

Figure 2 Plantain fresh bunch yield (Mg/ha) from an on-farm trial in a young fallow field at Nkolnda near Yaoundé, southern Cameroon.

Published by Burleigh Dodds Science Publishing Limited, 2019.

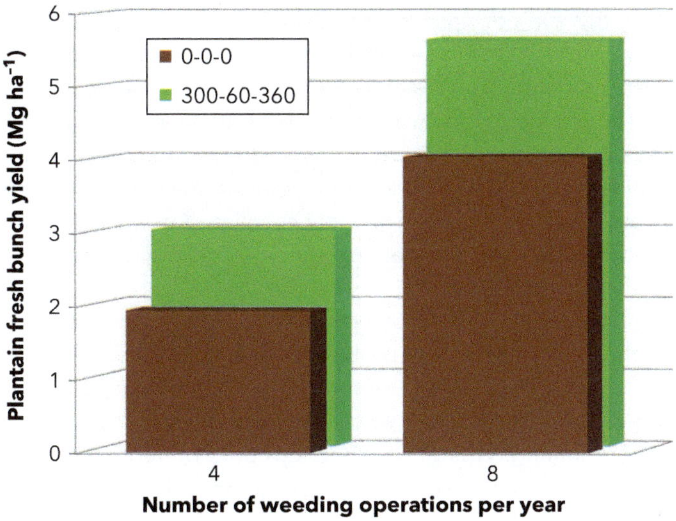

Figure 3 Plantain fresh bunch yield (Mg/ha) in an old secondary growth forest field at Mvambot, South Province, Cameroon.

6 Weed control in cassava systems

6.1 Introduction

Despite the importance of cassava in SSA, there was no considerable effort to develop efficient weed management systems specific to the crop. Infestation by grass (88%) and broad leaf weeds (67%) has been cited by cassava farmers to be the major problems (Agumagu et al., 2008). In Nigeria, cassava root yield losses from weeds range between 25% and 97%, while other SSA countries report losses of 62–90% (Hauser and Ekeleme, 2017). This section will deal specifically with agronomic measures to control weeds in cassava.

6.2 Agronomic measures, herbicides and crop varieties

Recently, IITA embarked on a project that tackled weed control in cassava systems using a combination of agronomy- and variety-based approaches supplemented by mechanical- and herbicide-based weed control (Ekeleme et al., 2016). In on-farm trials across southern Nigeria, the project looked at tillage, cassava variety, cassava density, intercropping (with maize) and fertilizer application against weed biomass production, which was monitored at four weeding operations over the first 24 weeks after planting. Results showed that fertilizer application, as well as intercropping with maize, had no effect on the weed biomass. Increased cassava plant densities ranging from 10 000/ha to 20 000/ha resulted in a 22% drop in weed biomass (at 12 500/ha). Higher planting densities did not produce any significant decrease in weed biomass. However, it was noted that tillage with ridging had the best effect on the weeds, reducing weed biomass by 70% of that found on flat soil across four weeding operations. The choice of cassava variety also had a similarly strong influence on the weeds, with a vigorous cassava mosaic disease-tolerant variety reducing weed biomass by 28% compared with a less vigorous variety. Reduction of weed biomass was only weakly positively correlated with cassava root yields, probably because the various other agronomic interventions masked to a large extend the effects of weeds.

Three or four manual weeding operations using a hand hoe are required to attain highest yields and low probabilities of low yields. Chemical-based control approaches using pre- and post-emergence herbicide applications attained the same root yields as weeding three or four times within the first 24 weeks after planting. Weeding frequency could be reduced under optimal timing. However, there was no concrete information to determine such optimum weeding times (Fig. 4).

6.3 Conclusion

Cassava is one of the most important crops in humid SSA and will gain more importance in the face of climate change due to its ability to survive periods

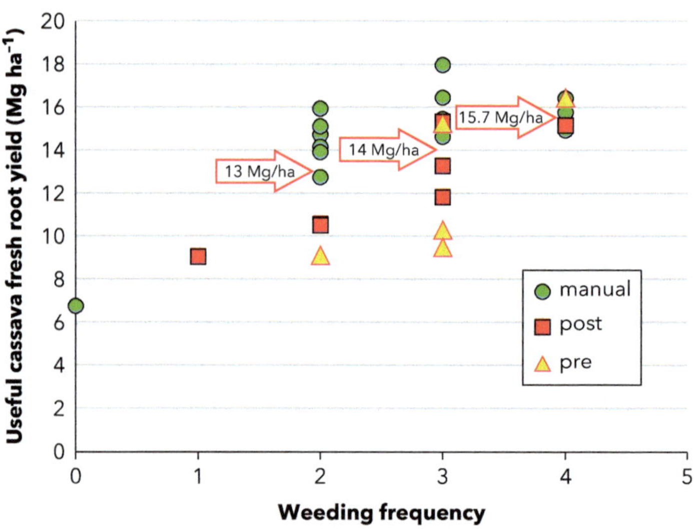

Figure 4 Cassava root yield response to method and frequency of weeding in southwestern Nigeria. Yields in arrows are the average for the respective weeding frequency across weeding methods. Post = post-emergence herbicide application, pre = pre-emergence herbicide application (S. Hauser, *unpublished data*).

of low or erratic rainfall. Weed control under water-limited conditions will become more important as competition will become more severe. Thus, food security in the cassava growing zones will depend to a large extent on the readiness of new and efficient weed control approaches in cassava. Considering the vast areas of land under cassava, collaborative efforts of IITA, federal government, state governments, the private sector, national research institutes, universities and concerned stakeholders along the cassava value chain are required to tackle the weed menace. The common manual weed control is cost-intensive and often insufficient. There is a lack of suitable post-emergence herbicides, and mechanical weed control has not received enough attention to provide suitable machinery. The combination of best agronomic practices, pre-emergence herbicide application and weed density-dependent post-emergence control measures require further research to be developed into decision support tools. The large number of households depending on cassava is calling for tools that would be easily accessible, free of charge and site specific. New initiatives such as the African Cassava Agronomy Initiative have integrated weed control in the research and dissemination framework.

7 Floating aquatic weeds

7.1 Introduction

Most indigenous aquatic plants of Africa, whether submerged, emerging or floating, are of restricted economic importance. However, under particularly favourable conditions, indigenous grasses and sedges become problematic. Since no classical biological control is feasible manual removal is often recommended and used. Manual clearing needs to be repeated several times as the weeds grow back almost immediately, particularly in eutrophic waters.

This scenario similarly applies to the three exotic floating waterweeds common to Africa, which pose far greater economic problems than indigenous plants. These include:

- *Eichhornia crassipes* (Mart.) Solms-Laubach (Pontederiaceae), commonly called water hyacinth, of South American origin, has been described as the world's most important water weed (Winston et al., 2014), more so in Africa (Martin et al., 2007).
- *Pistia stratiotes* L. (Araceae), or water salad, of unclear, probably South American origin, occurs across the continent, but is of lesser importance (Neuenschwander et al., 2007).
- *Salvinia molesta* Mitchell (Salviniaceae), or water fern, of South American origin, occurs in southern and Central Africa, very locally also in West Africa (Côte d'Ivoire-Ghana border) (Julien et al., 2007).

Published by Burleigh Dodds Science Publishing Limited, 2019.

In some cases, these floating waterweeds have led to the abandonment of settlements by totally covering streams and lagoons such as along the Lekki Lagoon in Nigeria and in the Congo on several tributaries of the Ubangi Shari. In Tanzania, important ferries on Lake Victoria have been blocked. In Benin, markets that depend on waterways have been shut down. In addition, floating waterweeds negatively impact fisheries and aquaculture, and, consequently, human health (Cilliers et al., 2003; Ajuonu et al., 2003; Mbati and Neuenschwander, 2005).

Conflicts of interest sometimes arise in the perception of waterweeds. For example, floating waterweeds such as *E. crassipes* are partially recognized as useful for mulching as well as used as raw material for weaving baskets and mats or for producing absorbent powders. Therefore, in such cases, stakeholder discussions are essential to determine whether such weeds are bane or boon.

Traditionally, in most places where floating waterweeds are considered nuisance, chemical and physical methods have been used to control them. Both approaches invariably proved ineffective because of the resulting water pollution, their unsustainability and costs. Several international symposia concluded that water hyacinth was a serious weed, which should be managed through classical biological control using specific insects and pathogens from South America. The use of glass carp and other fish species that typically consume floating waterweeds were not recommended as they were shown to be dangerous especially to local ecosystems.

Successful biocontrol programmes using biological agents from South America were undertaken in Florida, Asia, Australia and South Africa (see below). Research organizations based in these countries undertook the necessary quarantine studies and made available already tested biological control agents, while IITA relied on its vast network of scientists based in national programmes in West, Central and East Africa. Below we discuss work led by IITA on addressing waterweeds in these regions.

7.2 Water hyacinth

Water hyacinth, which was introduced as an ornamental, invaded Benin in 1977, Ghana and Nigeria since 1984 (Akinyemiju, 1987; De Graft Johnson, 1988) and quickly spread across the continent. To control its spread, IITA imported two weevils: *Neochetina eichhorniae* Warner and *Neochetina bruchi* Hustache (Coleoptera: Curculionidae) from Australia and Florida, respectively, via quarantine in England, then mass-cultured and released them in Benin in 1991 (van Thielen et al., 1994). Additionally, IITA provided starter cultures of the agents to Burkina Faso, Côte d'Ivoire, Ghana, Nigeria, Niger Republic, Tanzania, Uganda and Zimbabwe, together with the corresponding training held at IITA-Benin or in the respective countries.

In West Africa, dispersal and impact mainly by *N. eichhorniae* were documented (Ajuonu et al., 2003). Monitoring reports showed that within 5–10 years, water hyacinth coverage was reduced by at least 50%, leaving enough plant material for a new industry, which transforms water hyacinth mulch into an oil-absorbent material (Green-Keeper-Africa, 2018). Economic impact and savings were evaluated from randomized interviews in 24 villages involving 200 000 people. The accumulated savings amounted to around US$260 million, yielding a respectable benefit-cost ratio of 124:1. This ratio included only direct economic effects on the people of Southern Benin and did not consider indirect benefits or the benefits of the project to other countries, where the biological control agents had been successfully introduced. Other effects mentioned, but not measured, included enhanced water quality and human health (De Groote et al., 2003). Releases of the biological agents were then made across West Africa. In 1995, both *Neochetina* spp. were also introduced around Lake Victoria (Kenya Kisumu, Uganda [also Lake Kyoga], Tanzania), in collaboration with RSA scientists. The impact was stunning: total control by *N. bruchi* was achieved within just a few years (Fig. 5) (Wilson et al., 2007).

In Central Africa, releases in the Congo since 1999 led to the establishment mainly of *N. eichhorniae*. The impact was deemed to be satisfactory though local problems persisted (Mbati and Neuenschwander, 2005). Meanwhile, *Eccritotarsus catarinensis* (Carvalho) (Hemiptera: Miridae) and *Niphograpta albiguttalis* (Warren) (Lepidoptera: Crambidae) were released in West Africa, but could not establish widely (Ajuonu et al., 2009). In southern Africa, several more agents such as the mite *Orthogalumna terebrantis* (Acari: Gallumnidae) were released and partially established, but had only limited impact (Martin et al., 2007). Releases of *O. terebrantis* are being planned for West Africa.

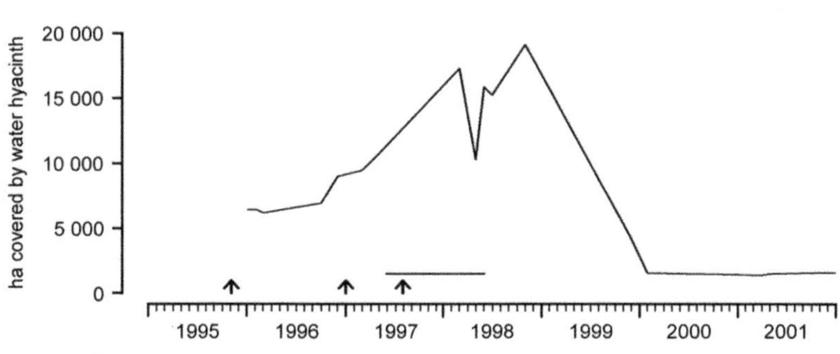

Figure 5 Surface coverage by water hyacinth on Lake Victoria (in ha) as assessed from satellite pictures. Arrows indicate releases of *Neochetina bruchi*, which proved to be the efficient biological control agent in East Africa (Wilson et al., 2007). The horizontal bar indicates an El Niño effect, which was originally claimed to be responsible for the collapse of the water hyacinth population. The low infestation level of 2001 has been maintained by *Neochetina bruchi* up to today.

7.3 Water lettuce

Against this waterweed of lesser importance, *Neohydronomus affinis* Hustache (Coleoptera: Curculionidae) was released in Benin in 1995. It spread 80 km pa and up to the Niger River in Malanville, achieving total control within a few years (Fig. 6), but causing local extinction-reinvasion cycles (Ajuonu and Neuenschwander, 2003). In 1997, *N. affinis* was released in Onne, Nigeria, from where it probably spread to nearby Douala, Cameroon (observed in 2014 by S. Ken, *pers. comm.*). In Central Africa, good control was achieved in the Congo (Mbati and Neuenschwander, 2005).

7.4 Water fern

Because water fern does not exist in Benin, releases of *Cyrtobagous salviniae* Calder and Sands (Coleoptera: Curculionidae) in the Congo were made from quarantine in RSA. The waterweed had spectacularly covered fish ponds and rivers and led to the abandonment of several settlements. The eventual clearing of ponds and rivers of the waterweed was highly celebrated by the local population (Mbati and Neuenschwander, 2005).

7.5 Conclusion

Biological control studies of invasive exotic waterweeds by IITA and by other research organizations in RSA and other countries demonstrated that the approach is generally highly efficient, but remained ineffective in eutrophic

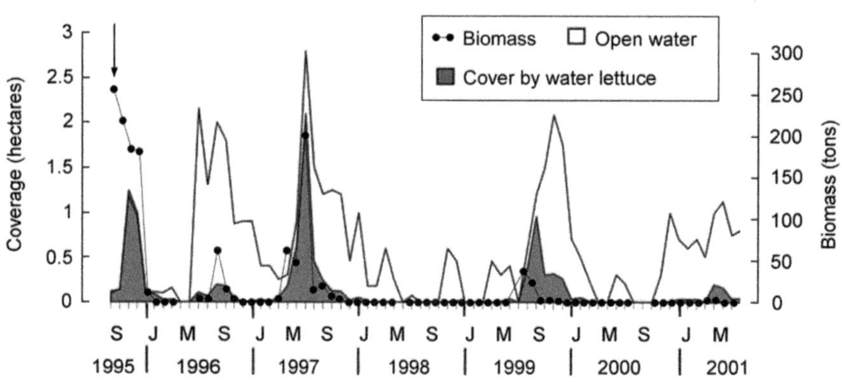

Figure 6 Impact of *Neohydronomus affinis* (release marked by arrow) in Southern Benin. Overall, the water surface varied sharply according to rains. The proportion of open waters (white) increased and the surface covered by water lettuce (grey) decreased until in 2001 almost no water lettuce was found anymore. The wet biomass of water lettuce (dots) gradually declined to zero (Ajuonu and Neuenschwander, 2003).

water that favours plant growth. This means that no satisfactory control could be achieved in urban environments.

Up to now, only a few biological control agents have been successful. Since many more probably exist in South America, it is recommended to test and introduce other agent species, particularly against water hyacinth. These agents have yet to be introduced to many river systems infested with floating waterweeds. However, in Africa, high-cost mechanical control is often favoured over the more cost-effective biological solution for non-scientific reasons (e.g. import of machinery, continuous employment). Except in special situations, such as above hydroelectric dams, such mechanical clearance must be considered a waste of resources and should be stopped.

Meanwhile, an industry exploiting particularly water hyacinth has been developed. While mulching with water hyacinth remains unsatisfactory, particularly the new technique to create oil-absorbent powder has great prospects and should be supported. It will not control the weed, but it will give employment to the local population.

8 Conclusion and future trends

Weeds are 'plants growing in the wrong place', but otherwise some commonalities arise from the above treated weed groups, that is floating water weeds, field weeds and parasitic weeds. Floating water weeds have been controlled, often successfully, by classical biological control mainly by insects with high and sustained impact, but there is a need to distribute these further. The response to terrestrial weeds presented here is integrated weed management, using wherever possible ecologically sound techniques of sustainable management of the soil, its microflora, and accompanying crops and shade trees, and integrating also resistant varieties of crop plants, for example maize or legumes. However adoption of more effective weed management practices is usually low. Where weeds like *Chromolaena odorata* also have positive attributes, conflicts of interest arise. Actions to stimulate widespread adoption and impact of proven weed management options include:

1. Application of integrated approaches that combine a variety of options that are attuned to farmer's agronomic and socio-economic conditions (soil type, climate, costs, local practices and preferences) to obtain synergy and longer-term control.
2. Partners with complementary skills need to work together in a well-coordinated manner to develop and promote effective and long-lasting weed control approaches.
3. More effective ways are needed to deliver research results directly to farmers. This requires the full participation of all stakeholders (i.e.

researchers, extension agents, policy makers, the private sector such as input dealers and farmers) in the design and dissemination of improved agricultural technologies.

4 Adequate and sustained financial support from donor agencies is vital if researchers are to continue developing and promoting weed control options that are well-suited to smallholder-farmer conditions.
5 There is a need to engage private seed companies to improve delivery of improved technologies such as seeds and plant protection products to affected smallholder farmers.

9 Where to look for further information

A good introduction to integrated weed management in smallholder agriculture in SSA can be found in I. O. Akobundu, *Weed Science in the Tropics: Principles and Practices* (see Akobundu, 1987; for full details on the subject).

The best single source of information on biological control of aquatic weeds is P. Neuenschwander, C. Borgemeister and L. Langewald (Eds), *Biological Control in IPM Systems in Africa* (see Cilliers et al., 2003).

A good source of information on *Striga* management is a book by C. Parker and C. R. Riches *Parasitic Weeds of the World: Biology and Control* (see Parker and Riches, 1993) and on speargrass management (see Brook, 1989; Chikoye et al., 2006). The International Institute of Tropical Agriculture is also implementing projects to help smallholder cassava farmers tackle problem weeds in SSA (for more information, please see http://www.cassavaweeds.org) while the Weed Science Society of America is a very useful resource for information on weed management in many continents (http://www.wssa.org).

10 References

Abayo, G. O., English, T., Eplee, R. E., Kanampiu, F., Ransom, J. K. and Gressel, J. 1998. Control of parasitic weeds (*Striga* spp.) on corn (*Zea mays*) resistant to acetolactase synthase inhibitors. *Weed Science* 49: 459–66.

Agumagu, A., Adesope, O. M. and Matthews-Njoku, E. C. 2008. Perception of infestation problems on cassava farms and preference for weed management practices in humid agro-ecological zone of Nigeria. *Agriculturae Conspectus Scientificus* 73: 115–20.

Ahonsi, M. O., Berner, D. K., Emechebe, A. M., Sanginga, N. and Lagoke, S. T. O. 2002. Selection of non-pathogenic ethylene-producing rhizobacteria for accelerated depletion of *Striga hermonthica* seed bank. *African Crop Science Journal* 10: 145–56.

Ajuonu, O. and Neuenschwander, P. 2003. Release, establishment, spread and impact of the weevil *Neohydronomus affinis* (Coleoptera: Curculionidae) on water lettuce (*Pistia stratiotes*) in Benin, West Africa. *African Entomology* 11: 205–11.

Ajuonu, O., Schade, V., Veltman, B., Sedjro, K. and Neuenschwander, P. 2003. Impact of the exotic weevils *Neochetina* spp.: (Coleoptera: Curculionidae) on water hyacinth,

Eichhornia crassipes (Lil: Pontederiaceae) in Benin, West Africa. *African Entomology* 11: 153–61.

Ajuonu, O., Byrne, M., Hill, M., Neuenschwander, P. and Korie, S. 2009. The effect of two biological control agents, the weevil *Neochetina eichhorniae* and the mirid *Eccritotarsus catarinensis* on water hyacinth, *Eichhornia crassipes*, grown in culture with water lettuce, *Pistia stratiotes*. *BioControl* 54: 155–62.

Aken'Ova, M. E. and Atta-Krah, A. N. 1986. Control of speargrass (*Imperata cylindrica* (L.) Beauv.) in an alley cropping fallow. *Nitrogen Fixing Tree Research Reports* 4: 27–8.

Akinyemiju, O. A. 1987. Invasion of Nigerian waters by water hyacinth. *Journal of Aquatic Management* 25: 24–6.

Akobundu, I. O. 1987. *Weed Science in the Tropics: Principles and Practices*. John Wiley Interscience Publication, John Wiley and Sons, New York, NY, 522p.

Akobundu, I. O. and Ekeleme, F. 1996. Potentials for *Chromolaena odorata* (L.) R.M. King and H. Robinson in fallow management in West and Central Africa. In: Prasad, U. K., Muniappan, R., Ferrar, P., Aeschliman, J. P. and de Foresta, H. (Eds), *Proceedings of the Third International Workshop on Biological Control and Management of Chromolaena odorata*. UK Agricultural Experiment Station, University of Guam, Publication 202, pp. 88–98.

Akobundu, I. O. and Ekeleme, F. 2000. Effect of method of *Imperata cylindrica* management on maize grain yield in the derived savanna of south-western Nigeria. *Weed Research* 40: 335–41.

Akobundu, I. O., Udensi, U. E. and Chikoye, D. 2000. Velvetbean (*Mucuna* spp.) suppresses speargrass (*Imperata cylindrica* (L.) Raeuschel) and increases maize yield. *International Journal of Pest Management* 46: 103–8.

Anoka, U. A., Akobundu, I. O. and Okonkwo, S. N. C. 1991. Effects of *Gliricidia sepium* (Jacq.) Steud and *Leucaena leucocephala* (Lam.) de Wit on growth and development of *Imperata cylindrica* (L.) Raeuschel. *Agroforestry Systems* 16: 1–12.

Berner, D. K. and Kling, J. G. 1997. Sustainable control of *S. hermonthica* spp. through a focused integrated pest management programme. In: Badu-Apraku, B., Akoroda, M. O., Ouderaoogo, M. and Quin, F. M. (Eds), *Contributing to Food Self Sufficiency: Maize Research and Development in West and Central Africa*. International Institute of Tropical Agriculture, Ibadan, Nigeria, pp. 1–11.

Berner, D. K., Schaad, N. W. and Völksch, B. 1999. Use of ethylene-producing bacteria for stimulation of *Striga* spp. seed germination. *Biological Control* 15: 274–82.

Berner, D. K., Sauerborn, J., Hess, D. E. and Emechebe, A. 2003. The role of biological control in integrated pest management of *Striga* species in Africa. In: Neuenschwander, P., Borgemeister, C. and Langewald, J. (Eds), *Biological control in IPM systems in Africa*. CABI Publishing, Wallingford, UK, pp. 259–76.

Brook, R. M. 1989. Review of literature on *Imperata cylindrica* (L.) Raueschel with particular reference to South East Asia. *Tropical Pest Management* 35: 12–35.

Bruning, E. F. and Sander, N. 1983. Ecosystem structure and functioning: Some interactions of relevance to agroforestry. In: Huxley, P. A. (Ed.), *Plant Research and Agroforestry*. ICRAF, Nairobi, Kenya, 617pp.

Chevalier, A. 1952. Deux composées permettant de lutter contre *l'Imperata* et empêchant la dégradation des sols tropicaux qu'il faudrait introduire rapidement en Afrique Noire. *Revue Internationale de Botanique Appliquée* 32: 494–6.

Chikoye, D. and Ekeleme, F. 2003. Cover crops for cogongrass (*Imperata cylindrica*) management and effects on subsequent corn yields. *Weed Science* 51: 792–7.

Chikoye, D., Ekeleme, F. and Ambe, J. T. 1999. Survey of distribution and farmer perceptions of speargrass (*Imperata cylindrica*) in cassava based cropping systems in West Africa. *International Journal of Pest Management* 45: 305–12.

Chikoye, D., Manyong, V. M. and Ekeleme, F. 2000. Characteristics of speargrass (*Imperata cylindrica*) dominated fields in West Africa: Crops, soil properties, farmer perceptions and management strategies. *Crop Protection* 19: 481–7.

Chikoye, D., Ekeleme, F. and Udensi, U. E. 2001. Cogongrass suppression by intercropping cover crops in corn/cassava systems. *Weed Science* 49: 658–67.

Chikoye, D., Udensi, U. E. and Ogunyemi, S. 2005. Integrated management of cogongrass [*Imperata cylindrica* (L.) Raueschel] in corn using tillage, glyphosate, row-spacing, cultivar and cover crop. *Agronomy Journal* 97: 1164–71.

Chikoye, D., Ellis-Jones, J., Tarawali, G., Kormawa, P., Nielsen, O. K., Ibana, S. and Avav, T. 2006. Farmers' perceptions of the speargrass (*Imperata cylindrica*) problem and its control in the lowland sub-humid savannah of Nigeria. *Journal of Food, Agriculture and Environment* 4: 118–26.

Chikoye, D., Udensi, U. E., Lum, A. F. and Ekeleme, F. 2007. Rimsulfuron for post-emergence weed control in corn (*Zea mays*) in the humid tropical environment. *Weed Technology* 21: 977–81.

Chikoye, D., Fontem, L. A. and Menkir, A. 2011. Seed coating herbicide tolerant maize hybrids with imazapyr for *Striga hermonthica* (Del) Benth control in the West African Savanna. *International Journal of Food, Agriculture and Environment* 9: 414–21.

Cilliers, C. J., Hill, M. P., Ogwang, A. and Ajuonu, O. 2003. Aquatic weeds in Africa and their control. In: Neuenschwander, P., Borgemeister, C. and Langewald, J. (Eds), *Biological control in IPM systems in Africa*. CABI Publishing, Wallingford, UK, pp. 161–78.

Clayton, W. D. and Renvoize, S. A. 1982. *Flora of Tropical East Africa*. Graminea (Part 3). A.A. Balkema, Rotterdam, The Netherlands, 448pp.

De Foresta, H. 1996. *Chromolaena odorata*: Calamité ou chance pour l'Afrique tropicale. In: Prasad, U. K., Muniappan, R., Ferrar, P., Aeschliman, J. P. and de Foresta, H. (Eds), *Proceedings of the Third International Workshop on Biological Control and Management of Chromolaena odorata*. Agricultural Experiment Station, University of Guam, Publication 202, pp. 45–53.

De Graft Johnson, A. K. K. 1988. Aquatic plant invasion and water resource – the Ghanaian experience. In: Oke, O. L., Imevbore, A. M. A. and Farri, T. A. (Eds), *Water Hyacinth, Menace and Resource*, Proceedings of an International workshop held in Lagos Nigeria, 7–12 August 1988. Federal Ministry of Science and Technology, pp. 32–8.

De Groote, H., Ajuonu, O., Attignon, S., Djessou, R. and Neuenschwander, P. 2003. Economic impact of biological control of water hyacinth in southern Benin. *Ecological Economics* 45: 105–17.

Debrah, S. K., Defoer, T. and Bengaly, M. 1998. Integrating farmers' knowledge, attitude and practice in the development of sustainable *Striga* control interventions in southern Mali. *Netherlands Journal of Agricultural Science* 46: 65–75.

Ejeta, G. and Butler, L. G. 1993. Host plant resistance to *Striga*. In: Buxton, D. R., Shibles, R., Forsberg, R. A., Blad, B. L., Asay, K. H., Paulsen, G. M. and Wilson, R. F. (Eds), *International Crop Science I, 1993*. Crop Science Society of America, Madison, WI, pp. 561-9

Ekeleme, F., Hauser, S., Atser, G., Dixon, A., Weller, S., Olorunmaiye, P., Usman, H., Olojede, A. and Chikoye, D. 2016. Weed management in cassava in Africa: Challenges and opportunities. *Outlooks on Pest Management* 27: 208–12.

Ellis-Jones, J., Schulz, S., Douthwaite, B., Hussaini, M. A., Oyewole, B. D., Olanrewaju, A. S. and White, R. 2004. An assessment of integrated *Striga hermonthica* control and early adoption by farmers in northern Nigeria. *Experimental Agriculture* 40: 353–68.

Elzein, A., Kroschel, J. and Cadisch, G. 2008. Efficacy of Pesta granular formulation of *Striga*-mycoherbicide *Fusarium oxysporum* f. sp. *strigae* Foxy 2 after 5-years of storage. *Journal of Plant Diseases and Protection* 115: 259–62.

Emechebe, A., Ellis-Jones, J., Schulz, S., Chikoye, D., Douthwaite, B., Husseini, M. A., Kureh, I. and Kormawa, P. 2004. Farmers' perception of the *Striga* problem and its control in northern Nigeria. *Experimental Agriculture* 40: 215–32.

Evans, H. C. 1987. Fungal pathogens of some subtropical and tropical weeds and the possibilities for biological control. *Biocontrol News and Information* 8: 7–30.

Franke, A. C., Ellis-Jones, J., Tarawali, G., Schulz, S., Hussaini, M. A., Kureh, I. and Olanrewaju, A. S. 2006. Evaluating and scaling-up integrated *Striga hermonthica* control technologies among farmers in northern Nigeria. *Crop Protection* 25: 868–78.

Green-Keeper-Africa. 2018. https://www.viawater.nl/projects/benin-green-keeper-africa (accessed on 29 January 18).

Gressel, J., Hanafi, A., Head, G., Marasas, W., Obilana, A. B., Ochanda, J. and Tzotzos, G. 2004. Major heretofore intractable biotic constraints to African food security that may be amenable to novel biotechnological solutions. *Crop Protection*, 23: 661-89.

Hauser, S. and Ekeleme, F. 2017. Weed control in cassava cropping systems. In: Hershey, C. (Ed.), *Achieving Sustainable Cultivation of Cassava. Volume 2 Genetics, Breeding, Pests and Diseases*. Burleigh Dodds Series in Agricultural Science 21, Burleigh Dodds Science Publishing, Cambridge, UK, pp. 1–26.

Hauser, S. and Mekoa, C. 2009. Biomass production and nutrient uptake of *Chromolaena odorata* as compared with other weeds in a burned and a mulched secondary forest clearing planted to plantain (*Musa* spp.). *Weed Research* 49: 193–200.

Hess, D. E., Kroschel, J., Traoré, D., Elzein, A. E. M., Marley, P. S., Abbasher, A. A. and Diarra, C. 2002. *Striga*: Biological control strategies for a millennium. In: Leslie, J. F. (Ed.), *Sorghum and Millet Diseases 2000*. Iowa State Press, Ames, IA, pp. 165–70.

Hubbard, C. E., Whyte, R. O., Brown, D. and Gray, A. P. 1944. *Imperata cylindrica*. Taxonomy, distribution, economic significance and control. Imperial Agricultural Bureaux Joint Publication No. 7. Imperial Agricultural Bureaux, Aberystwyth, UK, pp. 5–13.

Ivens, G. W. 1980. *Imperata cylindrica* (L.) Beauv. in West African agriculture. BIOTROP Special Publication No. 5. Indonesia, pp. 149–56.

Julien, M. H., Hill, M. P. and Tipping, P. W. 2007. *Salvinia molesta* D.S. Mitchell (Salviniaceae). In: Muniappan, R., Reddy, G. V. P., Raman, A. and Gandhi, V. P. (Eds), *Weed Biological Control with Arthropods in the Tropics Towards Sustainability*. Cambridge University Press, Cambridge, UK, pp. 378–407.

Kamara, A. Y., Chikoye, D., Ekeleme, F., Omoigui, L. O. and Dugje, I. Y. 2008. Field performance of improved cowpea varieties under conditions of natural infestation by the parasitic weed *Striga gesnerioides*. *International Journal of Pest Management* 54: 189–95.

Kamara, A. Y., Ekeleme, F., Omoigui, L., Menkir, A., Chikoye, D., Dugje, I. Y., Abdoulaye, T. and Amaza, P. 2009. Influence of nitrogen fertilization on the performance of early and late maturing maize varieties under natural infestation with *Striga hermonthica* (Del.) Benth. *Archives of Agronomy and Soil Science* 55: 125–45.

Kanampiu, F., Ransom, J. K. and Gressel, J. 2001. Imazapyr seed dressing for *Striga* control on acetolactase synthase target site resistant maize. *Crop Protection* 20: 885–95.

Kang, B. T., Wilson, G. F. and Sipkens, L. 1981. Alley cropping maize (*Zea mays* L.) and leucaena (*Leucaena leucocephala* Lam.) in Southern Nigeria. *Plant and Soil* 63: 165–79.

Kling, J. G., Fajemisin, J. M., Badu-Apraku, B., Diallo, A., Menkir, A. and Melake-Berhan, A. 2000. *Striga* resistance breeding in maize. In: Haussman, B. I. G., Hess, D. E., Koyama, M. L., Grivet, L., Rattunde, H. F. W. and Geiger, H. H. (Eds), *Breeding for Striga Resistance in Cereals*. Margraf Verlag, Weikersheim, Germany, pp. 273–81.

Lum, A. F., Chikoye, D. and Adesiyan, S. O. 2005. Effect of nicosulfuron dosages and timing on the post emergence control of cogongrass (*Imperata cylindrica*) in corn (*Zea mays*). *Weed Technology* 19: 122–7.

Mbati, G. and Neuenschwander, P. 2005. Biological control of three floating water weeds, *Eichhornia crassipes, Pistia stratiotes*, and *Salvinia molesta* in the Republic of Congo. *BioControl* 50: 635–45.

M'boob, S. S. 1986. A regional program for *Striga* control in West Africa. In: Robson, T. and Broab, H. (Eds), *Improved Striga Management for Africa*. Proc. FAO/OAU All Africa Govt. Consultation on *Striga* Control, Rome, pp. 190–4.

McFadyen, R. C. 1996. Biocontrol of *Chromolaena odorata*: Divided we fail. In: Moran, V. C. and Hoffmann, J. H. (Eds), *Proceedings of the Ninth International Symposium on Biological Control of Weeds*. University of Cape Town Press, Cape Town, pp. 455–9.

Menkir, A., Chikoye, D. and Lum, A. F. 2009. Incorporating an herbicide resistance gene into tropical maize with inherent polygenic resistance to control *Striga hermonthica* (Del.) Benth. *Plant Breeding* 129: 385–92.

Menkir, A., Franco, J., Adepoju, A. and Bossey, B. 2012. Evaluating consistency of resistance reactions of open-pollinated maize cultivars to *Striga hermontica* (Del.) Benth under artificial infestation. *Crop Science* 52: 1051–60.

Neuenschwander, P., Julien, M. H., Center, T. D. and Hill, M. P. 2007. *Pistia stratio*tes L. (Araceae). In: Muniappan, R., Reddy, G. V. P., Raman, A. and Gandhi, V. P. (Eds), *Weed Biological Control with Arthropods in the Tropics Towards Sustainability*. Cambridge University Press, Cambridge, UK, pp. 332–52.

Norgrove, L., Hauser, S. and Weise, S. F. 2000. Response of *Chromolaena odorata* to timber tree densities in an agrisilvicultural system in Cameroon: Aboveground biomass, residue decomposition and nutrient release. *Agriculture, Ecosystems and Environment* 81: 191–207.

Parker, C. 1991. Protection of crops against parasitic weeds. *Crop Protection* 10: 6–22.

Parker, C. and Riches, C. R. 1993. *Parasitic Weeds of the World: Biology and Control*. CAB International, Wallingford, UK, 332pp.

Poilane, E. 1952. L'*Eupatorium odoratum* L. et d'autres plantes de couverture en Indochine. III. Fourrages pour pays tropicaux. *Revue Internationale de Botanique Appliquée et d'Agriculture Tropicale* 32: 496–7.

Richardson, D. M., Pysek, P., Rejmánek, M., Barbour, M. G., Panetta, F. D. and West, C. J. 2000. Naturalization and invasion of alien plants: Concepts and definitions. *Diversity and Distributions* 6: 93–107.

Sanginga, N., Okogun, J., Vanlauwe, B. and Dashiell, K. 2002. The contribution of nitrogen by promiscuous soybeans to maize-based cropping systems in the moist savanna of Nigeria. *Plant and Soil* 241: 223–31.

Schulz, S., Hussaini, M. A., Kling, J. G., Berner, D. and Ikie, F. O. 2003. Evaluation of integrated *Striga hermonthica* control technologies under farmer management. *Experimental Agriculture* 39: 99–108.

Slaats, J. J. P. 1996. The use of *Chromolaena odorata* as fallow in a semi-permanent cropping system in south-west Côte d'Ivoire. In: Prasad, U. K., Muniappan, R., Ferrar, P., Aeschliman, J. P. and de Foresta, H. (Eds), *Proceedings of the Third International Workshop on Biological Control and Management of Chromolaena odorata*. Agricultural Experiment Station, University of Guam, Publication 202, pp. 68–75.

Soerjani, M. 1970. Alang-alang [*Imperata cylindrica* (L.) Beauv.] (1812). Pattern of growth as related to its problem of control. BIOTROP Bulletin, No. 1, Indonesia, 88pp.

Terry, P. J., Adjers, G., Akobundu, I. O., Anoka, A. U., Drilling, M. E., Tjitrosemito, S. and Utomo, M. 1997. Herbicides and mechanical control of *Imperata cylindrica* as a first step in grassland rehabilitation. *Agroforestry Systems* 36: 151–79.

Timbilla, J. A., Zachariades, C. and Braimah, H. 2003. Biological control and management of the alien invasive shrub *Chromolaena odorata* in Africa. In: Neuenschwander, P., Borgemeister, C. and Langewald, J. (Eds), *Biological Control in IPM Systems in Africa*. CABI Publishing, Wallingford, UK, pp. 145–60.

Udensi, E. U., Akobundu, I. O., Ayeni, A. O. and Chikoye, D. 1999. Management of cogon grass (*Imperata cylindrica*) with velveltbean (*Mucuna pruriens* var. *utilis*) and herbicides. *Weed Technology* 13: 201–8.

Uyi, O. 2011. Discovery of *Pareuchaetes pseudoinsulata* in Edo State, Nigeria. *Chromolaena odorata Newsletter* 18: 1–3.

Uyi, O., Zachariades, C., Hill, M. P. and McConnachie, A. J. 2016. Temperature-dependent performance and potential distribution of *Pareuchaetes insulata*, a biological control agent of *Chromolaena odorata* in South Africa. *BioControl* 61: 815–25.

van Thielen, R., Ajuonu, O., Schade, V., Neuenschwander, P., Adité, A. and Lomer, C. J. 1994. Importation, release, and establishment of *Neochetina* spp. (Curculionidae) for the biological control of water hyacinth, *Eichhornia crassipes* (Lil.: Pontederiaceae), in Benin, West Africa. *Entomophaga* 39: 179–88.

Von Senger, I., Barker, N. P. and Zachariades, C. 2002. Preliminary phylogeography of *Chromolaena odorata*: Finding the origin of a South African weed. In: Zachariades, C., Muniappan, R. and Strathie, L. W. (Eds), *Proceedings of the Fifth International Workshop on Biological Control and Management of Chromolaena odorata*. Durban, South Africa, 23–25 October 2000, pp. 90–7.

Versteeg, M. N., Amadji, F., Etéka, A., Gogan, A. and Koudokpon, V. 1998. Farmers' adoptability of *Mucuna* fallowing and agroforestry technologies in the coastal savanna of Benin. *Agricultural Systems* 56: 269–87.

Weise, S. F. 1996. Distribution and significance of *Chromolaena odorata* (L.) R.M. King and H. Robinson across ecological zones in Cameroon. In: Prasad, U. K., Muniappan, R., Ferrar, P., Aeschliman, J. P. and de Foresta, H. (Eds), *Proceedings of the Third International Workshop on Biological Control and Management of Chromolaena odorata*. Agricultural Experiment Station, University of Guam, Publication 202, pp. 29–38.

Willard, T. R., Hall, D. W., Shilling, D. G., Lewis, J. A. and Currey, W. L. 1990. Cogongrass (*Imperata cylindrica*) distribution on Florida Highway Rights-of-Way. *Weed Technology* 4: 658–60.

Wilson, J. R. U., Ajuonu, O., Center, T. D., Hill, M. P., Julien, M. H., Katagira, F. F., Neuenschwander, P., Njoka, S. W., Ogwang, J., Reeder, R. H. and Van, T. 2007.

The decline of water hyacinth on Lake Victoria was due to biological control by *Neochetina* spp. *Aquatic Botany* 87: 90–3.

Winston, R. L., Schwarzlander, M., Hinz, H. L., Day, M. D., Cock, M. J. W. and Julien, M. H. (Eds). 2014. *Biological Control of Weeds: A World Catalogue of Agents and Their Target Weeds*. 5th edition, USDA Forest Service, Forest Health Technology Enterprise Team, Morgantown, WV, 838pp.

Zachariades, C., Day, M. D., Muniappan, R. and Reddy, G. V. P. 2009. *Chromolaena odorata* (L.) King and Robinson (Asteraceae). In: Muniappan, R., Reddy, G. V. P., Raman, A. (Eds), *Biological Control of Tropical Weeds using Arthropods*. Cambridge University Press, Cambridge, UK, pp. 130–62.

Published by Burleigh Dodds Science Publishing Limited, 2019.

Chapter 15

Making integrated pest management (IPM) work in sub-Saharan Africa

Danny Coyne, *IITA, Kenya;* **Michael Abberton**, *IITA, Nigeria;* **Sounkoura Adetonah**, *IITA, Bénin;* **Maria Ayodele**, *IITA, Nigeria;* **Laura Cortada-Gonzales**, *IITA, Kenya;* **Brice Gbaguidi**, *IITA, Bénin;* **Stefan Hauser** *and* **P. Lava Kumar**, *IITA, Nigeria;* **Peter Neuenschwander**, *IITA, Bénin;* **Marc Schut**, *Wageningen University, The Netherlands;* **Manuele Tamò**, *IITA, Bénin; and* **Abou Togola**, *IITA, Nigeria (corresponding author: d.coyne@cgiar.org)*

1 Introduction

Agriculture in sub-Saharan Africa (SSA) is characterized by smallholder subsistence farming involving a multitude of crops, crop cultivars and mixed cropping systems, which may combine perennial crops with semi-perennial and annual crops. With year-round cultivation occurring in places, coupled with the availability of alternative hosts, due to favourable tropical and subtropical climates, the potential for pest build-up, as these systems intensify, is immense. For SSA to feed itself in the coming years, agricultural productivity needs to significantly improve (Hazell and Wood 2008; Vanlauwe et al. 2014), essentially through more intensified cropping systems (Keating et al. 2010). Up to now, most increases in production in SSA come from expansion of cropland, often into marginal and/or biodiversity-rich habitats. This is no longer possible.

http://dx.doi.org/10.19103/AS.2018.0043.16
Published by Burleigh Dodds Science Publishing Limited, 2019.

Losses to pests are still amongst the most important factors affecting productivity in SSA (Savary et al. 2012), and their mitigation should be achieved by implementing integrated pest management (IPM). Since its original definition, IPM has become a frequently used, even misused, term (Kogan 1998). Initially, IPM tended to be entomology biased. It is often still viewed as a form of insecticide management, advocating the 'judicious' use of pesticides (in addition to resistant varieties, agronomic practices and biocontrol). Evidently, this recommendation is not heeded: despite a clear increase in pesticide use, crop losses due to pests have not significantly decreased since the 1960s (Oerke 2006). Consequently, the justified and safe use of appropriate synthetic pesticides within the IPM concept is still considered acceptable, *not as the first option but as the last resort*, provided they are as selective as possible, available and accessible to farmers (Jones 2014).

The concept was further expanded to take into account the complexity of ecological factors that affect crop growth and development constraining agricultural production (Bajwa and Kogan 2002). To achieve the full potential of such a decision-based support system for the selection and use of various control options, farmers need to become both experimenters and generators of technology (Van Huis and Meerman 1997). Consequently, improved farmer access to decision making, in addition to science-based ecological control, including biocontrol and biopesticides, resistant crop varieties and good soil and natural resource management, have become embedded within IPM considerations. This also includes considering the all-important, and previously neglected, aspects of gender, which have been integrated into the farmer exchange concept of Farmer Field Schools (FFS) and Farmer Field Fora (FFF).

IITA's contribution to pest management was initially based on its breeding programmes aiming to improve the six IITA mandate crops (cassava, yams, maize, cowpea, soybean, banana/plantain), developing new cultivars with resistance against key biotic threats as well as other traits. Breeding cultivars with resistance against key biotic threats therefore formed the basis of IITA's IPM strategy. Later, IITA implemented biological control-based projects across the continent, and thus became instrumental in the development and implementation of IPM programmes across SSA. Within the CGIAR system IITA is unrivalled through its systematic attention to plant health, and by maintaining a diverse multidisciplinary team of plant health specialists over the years as described in Chapter 2. While some achievements were noteworthy and highly publicized, others passed without significant notice: evidently, preventing catastrophe is less impressive than treating one.

In this chapter, rather than reviewing individual IPM programmes, we describe the key challenges in implementing IPM in Africa, IITA's role in promoting IPM in SSA and how IPM can contribute in creating an enabling environment towards improving the sustainability of cropping systems. The following cross-cutting issues, which are dealt with only cursorily in many chapters, are critical to make IPM work.

2 Healthy planting material

Healthy planting material is essential to arrest the dissemination of pests, especially through infected, vegetatively propagated material, such as yam, banana, cassava and so on. The development of rapid multiplication mechanisms of healthy, tissue-culture-derived material in state-of-the-art laboratories has received much attention at IITA, such as for cassava (e.g. Legg et al. 2015), bananas (e.g. Dubois et al. 2013) and yam (Aighewi et al. 2015). Eradication of virus-infected plants and replacement with virus-free planting material has been used as a major strategy for managing banana bunchy top disease (Kumar et al. 2015) and cassava mosaic and brown streak diseases (Legg et al. 2015).

Simple disinfection techniques suitable for smallholder farmers have also been developed, such as boiling water therapy for nematode and weevil control in bananas (Coyne et al. 2010; Hauser and Coyne 2010). While effective, the conventional hot water technique was a challenge and needed careful monitoring of temperatures, which made it unattractive to farmers (Elsen et al. 2004). Instead, by dipping suckers into boiling water for 30 s the technique was simplified and farmers became much more receptive to adopting it (Hauser 2007; Hauser and Messiga 2010), see also Chapter 9.

The Germplasm Health Unit thereby synchronized IITA activities with National Plant Protection and Quarantine Organizations by providing diagnostic information and services and building capacity to develop diagnostic skills in pathogen isolation and identification, field inspection and symptom identification, pest risk analysis, surveillance and import/export certification, thus preventing international distribution of infected/infested germplasm material as detailed in Chapter 3. It also worked closely with the Genetic Resources Center (Genebank), providing a constant check on the health of the thousands of accessions and cultivars, which it maintains and dispatches to partners.

3 The role of pesticides

Although synthetic pesticides enabled farmers to increase crop productivity across the globe, their use has become significantly restricted, with many products removed from the market, due to the threat they pose to environmental and human health. Consequently, any agricultural revolution in Africa will undoubtedly have a different signature than previously witnessed in the Green Revolution in Asia and elsewhere as outlined in Chapter 2. A search on the FAO database (FAOSTAT 2018) on the use of pesticides in Africa (Eastern, Western, Central, Southern and Northern) indicates a substantial decline in the use of insecticides between 2005 and 2014, across regions (Fig. 1). Of interest however, is that despite this apparent reduction in the use of insecticides, overall import values of total pesticides have increased annually until 2014 to US$ 21 billion (Fig. 2). Data for West Africa show that herbicides comprise, by far, the greatest portion of these imports (62%), followed by insecticides (26%) and fungicides, growth regulators, rodenticides and nematicides (12%) (Diarra and Haggblade 2017). Consequently, a rapid market growth for herbicides, rather than for insecticides, appears to be taking place, in part driven by the rise in generic products, the emerging low-cost Asian suppliers and falling global pesticide prices (Diarra and Haggblade 2017). In Mali and Ethiopia, for example, the associated costs for the use of herbicides was about half the amount of hiring labour for weeding (Haggblade et al. 2017a), translating into a rapid rise in the use of herbicides in accessible rural areas (Haggblade et al. 2017b). However, caution is advised in the interpretation of much of these data. While FAOSTAT is likely the most reliable and accessible source of such information, data are often not available and missing, providing a distorted vision of the situation. For example, Morocco reported the highest pesticide usage among all African countries in 2005 and 2006, for which there are no further data for Morocco. In Central Africa, only data for 2009 and 2010 are available for the Democratic Republic of Congo, while for countries such as Kenya or Nigeria, where better

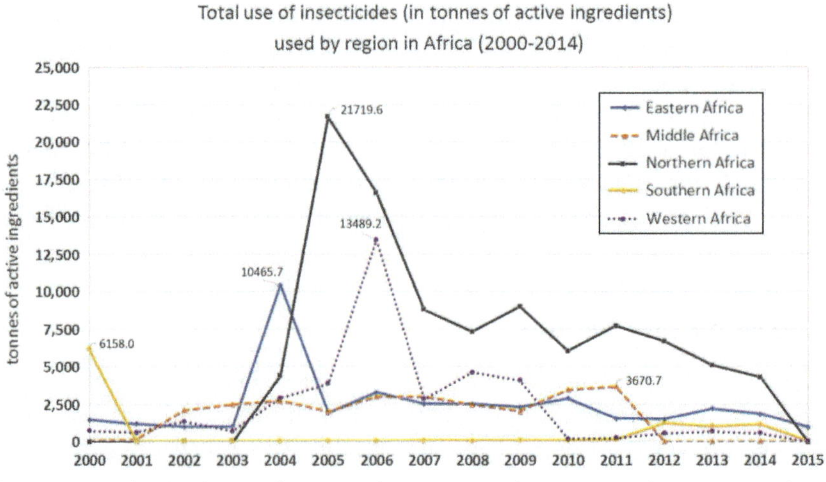

Figure 1 Total annual use of insecticides (expressed as tonnes of active ingredients) used in Africa from 2000 to 2015 by region. Data sourced from FAOSTAT (2018) related to pesticide use filtered by: (1) Region; (2) Items: 'Insecticides + (Total)'; (3) Elements: 'Use'; and (4) Years: '2000 to 2014'.

phytosanitary policies and a more regulated pesticide market exist, data are also missing. Evidently, no data does not transcribe into no pesticide use.

In SSA, pesticides have not generally formed the mainstay of pest management, as they have elsewhere, except for use in certain systems, such as vegetables, cowpeas and export-oriented production systems. This, by default, effectively equates to organic farming for most other cropping systems, although it is not referred to as such. The question therefore, is how to increase yields without losing this advantage. The option of relying on cheap synthetic

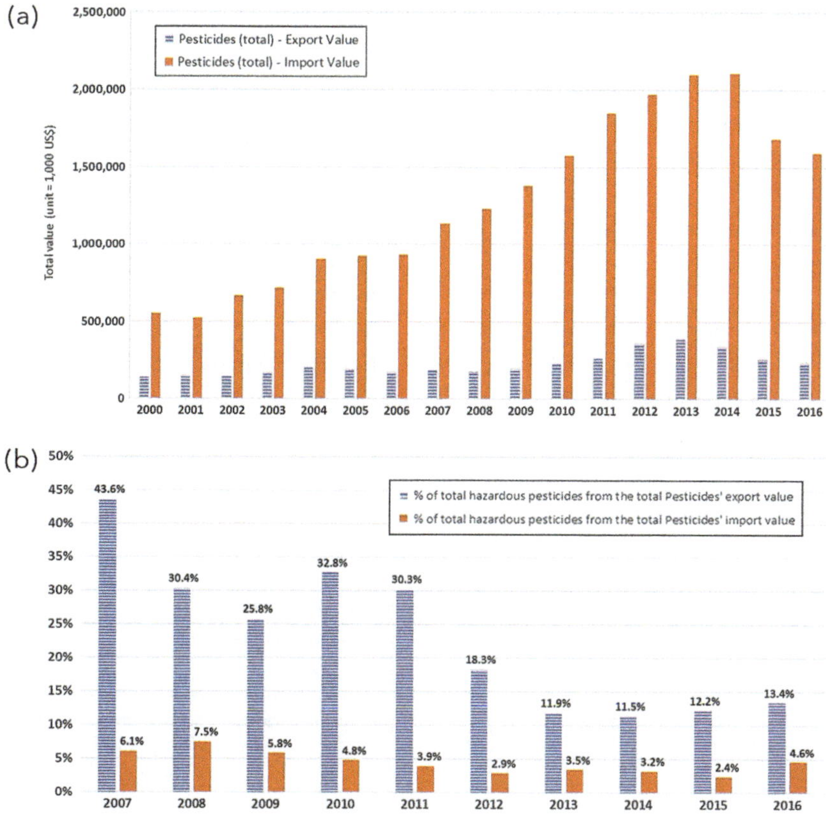

Figure 2 (a) Total export (blue-pattern bars) and import (orange-solid bars) and value (in 1000 US$) of pesticide trade (including hazardous pesticides, insecticides, fungicides, herbicides, disinfectants and other) in Africa from 2000 to 2016; (b) Percentage (%) of export (blue-pattern bars) and import (orange-solid bars) value of hazardous pesticides only from the total pesticide trade for Africa from 2007 to 2016. Search was conducted on the FAOSTAT (2018) related to pesticide trade, filtered by: (1) Region: 'Africa > (List)'; (2) Items: 'Pesticides (Total)' and 'Hazardous pesticides'; (3) Elements: 'Import Value' and 'Export Value'; and (4) Years: '2000 to 2016'. The pesticides (Total) traded for retail sale or as preparations or articles are those classified under code 38.08 in the Harmonized System Nomenclature (HS).

Table 1 List of hazardous pesticides (by active ingredient) recorded for Nigeria

No	Compound	Use
1	Aldrin (TN)	Insecticide
2	Binapacryl	Acaricide (miticide) fungicide
3	Captafol	Fungicide
4	Chlordane	Insecticide
5	Chlordimeform	Acaricide (miticide)
6	Chlorobenzilate	Acaricide
7	DDT	Insecticide, acaricide
8	Delta HCH	Insecticide, fungicide
9	Dieldrin	Insecticide
10	Di-nitro-ortho-cresol - DNOC	Herbicide, insecticide and fungicide
11	Dinoseb	Herbicide
12	Endosulfan	Insecticide
13	Endrin	Insecticide, acaricide and rodenticide
14	Ethylene dibromide (EDB)	Insecticide
15	Ethylene oxide	Fumigant
16	Fluoroacetamide	Rodenticide, insecticide
17	Heptachlor	Insecticide
18	Lindane	Insecticide, fungicide
19	Methamidophos	Systemic insecticide
20	Methyl bromide	Herbicide
21	Methyl parathion or Parathion methyl	Insecticide, acaricide
22	Monocrotophos	Systemic insecticide, acaricide
23	Paraquat dichloride	Herbicide
24	Parathion	Insecticide, acaricide
25	Parathion-ethyl	Insecticide, acaricide
26	Phosphamidon	Insecticide, acaricide
27	Toxaphene	Insecticide

Source: The National Agency for Food and Drug Administration and Control (NAFDAC); IUPAC = International Union of Pure and Applied Chemistry; Pan International Consolidated Lists (2018).

pesticides evidently does not work, as shown by the fact that the smallholder farmers, who use them, usually misuse or overuse them, as demonstrated in Chapter 11. This is due to various reasons, including a limited understanding of, or ability to identify, causal pests or their symptoms by farmers and agricultural staff (Abang et al. 2014). Irregular and erratic availability of products, especially quality products, worsens the situation, while available products often are adulterated, outdated or ineffective. They are also often not available in suitable-sized packages, at the right time, or have poor labelling and instructions (Williamson et al. 2008; Dey 2010). Civil unrest can create an additional vulnerability to the unregulated arrival of obsolete and hazardous stocks of pesticides. Unpublished survey data (L. Cortada, *pers. comm.*) for Northern Somalia (Hargeisa Province) obtained in 2012 from agro-input dealers showed that most insecticides available for use on food crops were based on nine active ingredients, two of which are specified for veterinary use (Malathion, Triazophos) and four that are banned under the EU 1107/2009 regulation (Dichlorovos, Hexaconazole, Oxytetracyclin, Triazophos). The smuggling of chemicals across porous borders further hampers the adequate control over hazardous and/or illegal pesticides (Keri 2009; Diallo and Tasie 2017). Though there is evidence of newer pesticide formulations being adopted (Mengistie et al. 2015), 'older' (established), non-patented (generic), more acutely toxic and environmentally persistent agents, which are less expensive, are still widely used in some countries (see Table 1). The situation is further aggravated by the zealous salesmanship and the widespread presence of unlicensed retailers, where advice to farmers is often provided with only minimal or zero knowledge (Mengistie et al. 2014). Such conditions increase the health and environmental hazards of pesticide use (Belfroid et al. 1998).

However, there are options for integrating chemical pesticides in IPM schemes, such as the use of selective or narrow-spectrum pesticides that can efficiently complement biological control (Waterhouse and Norris 1987). On cowpea, effective pest management was achieved by combining resistant cultivars with improved cultural practices, biological control and minimum pesticide applications (Jackai and Daoust 1986; Oghiakhe 1995, see also Chapter 10). When assessing options to create a source of healthy yam planting material, of all the various options tested, a simple insecticide plus fungicide dip proved the most efficient and cost effective, which was preferred by farmers (Claudius-Cole et al. 2017; see also Chapter 8). Used in moderation, there is therefore scope to include the use of pesticides as an important component of IPM programmes because of their potential role in reducing pest populations where other tactics are unsuccessful. For an effective use, archaic but common application techniques with brooms of leaves are, however, no longer acceptable. Reducing application rates, better targeting or using carriers that extend the longevity of pesticides are sought-after options for reducing their environmental impact while maintaining their effectiveness (Matthews et al. 2014). Unfortunately, benefits of improved techniques for applying pesticides, such as the Electrodyn sprayers, have not been widely accepted because of relatively high costs, as detailed in Chapter 10.

With the removal of numerous Class I pesticides from the market, the search for new, environmentally sensitive alternatives, with lower mammalian toxicity, has accelerated. Biopesticides, that is plant-derived or microbial-based products, are increasingly considered as alternative options also in SSA, as detailed in Chapter 13. The pest management toolbox has thus expanded to include more environmentally friendly, targeted and efficient products, allowing beneficial flora and fauna to prosper. A good example is Kenya, the lead country in East Africa in the use of biological control products for the horticulture and floriculture industry, which is driving the upsurge in regional biopesticide use. A Kenyan company produces products locally (www.realipm.com/), including predatory mites and a local isolate of *Trichoderma asperellum* (isolate TR900), which is one of very few *Trichoderma*-based products commercialized worldwide for the control of plant parasitic nematodes. Other companies import biological control agents and products into the region (www.koppert.com/company/subsidiaries/kenya/; www.dudutech.com). In contrast to the usual commercial systems, some companies also provide an extensive network of advisory services to smallholder farmers, which requires an entirely different approach than for large-scale commercial systems (H. Wainwright, *pers. comm.*). However, information about the use of biopesticide and biologically based products remains scarce, with data in FAOSTAT (2018) available for only nine countries from Africa, with consumption representing just 0.29% (541.22 tonnes of a.i.) of all insecticides used for crop protection in Africa since 2000. Genetically engineered crops, designed to produce their own insecticides or exhibit resistance to broad-spectrum herbicides are also becoming increasingly available. Unfortunately, GMOs are not readily available, approved or licensed in much of Africa. At present, Bt-maize is still the only commercialized GM food crop and is cultivated in Burkina Faso, Egypt, Republic of South Africa and Sudan only (Gathigah 2017). Initially introduced for the management of stem borers, such as *Busseola fusca* Fuller (Lepidoptera: Noctuidae) and *Chilo partellus* Swinhoe (Lepidoptera: Crambidae), the current threat of the fall armyworm *Spodoptera frugiperda* (J.E. Smith) (Lepidoptera: Noctuidae) is now challenging the decision on the use of Bt-maize. For example, contained field trials are currently underway in Kenya, Mozambique and Uganda as a potential option against this invasive pest (AATF 2017). Not only do Bt crops reduce losses to lepidopteran pests, but they also contribute towards reducing the need for and application of synthetic insecticides (Zilberman et al. 2007). Cowpea has also been transformed to express the Bt protein, and is expected to be introduced within the next couple of years into Burkina Faso, Ghana, Malawi and Nigeria, as detailed in Chapter 10. The increasing occurrence of resistance by insect pests against Bt (Tabashnik 2015), however, will limit the desirability of such GMOs.

4 Gender access to IPM

Gender analysis explores and highlights the relationships between different societal groups (e.g. men and women of different ages), and the inequalities in those relationships. It does so by asking: Who does what? Who has what? Who decides? How? Who gains? Who loses? When we pose these questions, we also ask: Which men? Which women? Of which age? Gender analysis looks at how power relations within the household interrelate with those at the international, state, market and community level. A study in Northern Ghana showed, for instance, that both men and women farmers are aware of new crop varieties, but that women often do not have access to these inputs (Adetonah, *unpublished results*). Until recently, little attention has been paid to the fact that men and women of different age have different access to IPM innovations, that is that gender is relevant to IPM (Georgina 2012), as shown in a study on communities affected by banana bunchy top disease in SSA (Ajambo et al. 2018).

IPM packages are designed, or selected, to eliminate or mitigate economic and health damage caused by pests, minimize the use of pesticides and the risk to human and environmental health associated with pesticide use (Beyond Pesticides 2017). In a study in West Africa, women were found to be significantly less aware of pesticide dangers and IPM techniques than men and were exposed to pesticides through other pathways than men (Christie et al. 2015). This is especially true, where pesticide use is likely to rise, or is already high, such as for vegetable systems.

In many parts of Africa, due to gender divisions of labour, women's agricultural work is concentrated around certain types of tasks, many of which are directly linked to pest management. For example, women and girls are often responsible for weeding and cleaning, sorting and storage of agricultural produce, activities that are highly relevant for pest control, while soil preparation, tillage and insecticide application are the reserve of adult men. Since technological needs and priorities of women in IPM differ from those of men, research and development of IPM technologies need to be adjusted accordingly.

From the beginning, women in farming households, as direct beneficiaries of IPM, should therefore be included in the study groups that develop and test IPM innovations (Erbaugh et al. 2003). Reducing gender inequality and recognizing the contribution of women to agriculture is critical to achieving global food security (USAID 2016). When developing sustainable IPM programmes, the concepts of FFS and FFF (see below) will help to alleviate this discrepancy.

5 Improved exchange with farmers

Participating in the generation, analysis and ownership of knowledge opens the door for creativity and inclusion of marginalized groups (Chambers 1994). Therefore, key to the success of any IPM programme is the early involvement of farmers, towards ensuring both farmer understanding and suitability of any proposed interventions. FFS are a well-known approach to human development in agriculture, particularly in IPM and integrated crop management training for environmentally sound practices (Gallagher 2003). Less well known are FFF, which draw strongly on the FFS approach to promote, refine and reorient the experimentations and *in situ* conservation of genetic resources (FAO 2005). FFS and FFF became recognized platforms for the implementation of IPM (Van den Berg and Jiggins 2007), including at IITA. The original methods have been streamlined and rendered less expensive in order to respond to often voiced criticism. Numerous studies have since confirmed the effectiveness of FFF, for example for yam pest and disease management in Ghana (Opare-Atakora et al. 2014). The success of the FFF was further accentuated through its adaptation to Farmer Welfare Fora (FWF). The FWF diffused information on crop technologies, which built capacity for choices to sustain and improve livelihoods of people living with HIV/AIDS. The same concept was recently adapted for rice farmers to promote production practices that reduce malaria vectors (Gbaguidi and Coulibaly 2013).

In FFF, the exchange of experiences and knowledge between farmers, scientists and extension agents is emphasized (Navarro 2008) in a non-formal education process (Knowles et al. 2015). Discovery-based learning is related to agro-ecological principles in a participatory learning process throughout a crop cycle. Consequently, thanks to such participatory learning platforms, farmers become adept at solving their own, specific problems, within the agro-ecological context of their own farms. They are able to innovate, plan and conduct their own experiments towards improving productivity (Gbaguidi and Coulibaly 2013). The FFF is both a technical education and capacity-building approach to empower farmers, who are both users and owners of the knowledge, to improve their productivity and livelihoods. Furthermore, the use of an agro-ecosystem analysis is also used as a decision-making tool for farming or post-farming activities or operations (http://ffs.ipm-info.org/). This pedagogical approach allows producers to make their own decisions and experimentations based on their field monitoring of diseases, pests and their natural enemies. Through agro-ecosystem analysis, farmers are encouraged to examine their own crop problems, evaluate available options and make appropriate crop management decisions to promote natural defences.

> 'Before all insects are pests for me. But now I know that in my field there are natural enemies, pests and period for pest damage'
>
> *Athanase Alidjinou Atchakpa, Benin, 2002.*
>
> 'I was trained on cowpea IPM but now I'm applying the FFF knowledge on tomato, cotton, maize . . .'
>
> *Alexis Atchakpa, Benin 2002.*

Between 2003 and 2011 over 10000 farmers in nine West African countries benefited from FFF through the Projet Niébé pour l'Afrique [cowpea project for Africa] (PRONAF) project at IITA, as detailed in Chapter 10. The number of producers trained by facilitators in this novel, inclusive format continues to grow, covering agriculture, livestock, fisheries, health and the environment.

6 Improved collaboration

IITA has been especially visible in respect to biological control programmes as detailed in Chapters 6, 12 and 14. All of these programmes have had a regional perspective, transcending national boundaries, which present administrative and political obstacles. Similarly, all projects have, of course, been conducted jointly with local national partners, together with numerous international partners and funding, and the many farmers involved. Across the CGIAR the System-wide Program for IPM (SP-IPM) (www.spipm.cgiar.org) extended and facilitated plant health research and the IPM approach until it was discontinued (Makkouk and Emechebe 2003; Hoeschle-Zeledon et al. 2013).

7 Future trends and conclusion

IPM will be critical to providing a support system for decision-based selection and use of pest control options that are suitable for the predominantly smallholder-based systems in SSA, as well as for more progressive, larger-scale operations. Approaches to implementing suitable IPM strategies, however, will also require ever-greater inclusion of how farmers, particularly women, and communities adopt IPM. Furthermore, governmental acceptance and support of new IPM technologies is crucial. Current policies tend to discourage investors from registering new products, such as biologically based pest management products, in part due to official reluctance to endorse such products because of a lack of understanding, fear of getting it wrong or conflicts with other commercial interests. However, there appears to be increasing and promising acceptance from both national and international companies to bring new biologically based products onto the market in SSA. For farmers to have access to new, effective products or new pest-resistant cultivars, investments are necessary for their registration in each country, and further, to make them available at the right time and the right quantity, before it is possible to incorporate them into IPM programmes. The procrastination surrounding biosafety regulations for the approval of GMOs, such as traditional cultivars modified for pest resistance, further demonstrates institutional reluctance to change policies, even where strong scientific supporting evidence exists for both their safety and their benefit. For risk-based security measures to be effective, supportive policies need to be in place, while careful attention to deployment of surveillance activities are necessary to ensure optimal cost-effectiveness (Kroschel et al. 2013).

The impact of climate change will also need to be considered, including its influence on pest dynamics and creating more favourable conditions for new pests. A shift to more commercial production systems will be necessary to feed the African continent, but meantime, the complex nature of smallholder farming systems will strongly influence pest management strategies in the near future.

Published by Burleigh Dodds Science Publishing Limited, 2019.

Systems will need to adapt and adopt techniques and tactics according to these changes, including the development of policies that support this changing agricultural landscape. Phytosanitary services will need to better intercept pests at national boundaries, an area where informal trade and exchange of plant material can be substantial. Such services will, in turn, require support and more up-to-date information on important pest species. And in turn, extension services will need to be effectively coordinated with such services towards the efficient and speedy delivery of information to farmers. Embracing new ICT tools, developing apps and taking advantage of the expanding mobile telephone network will ultimately be essential to reaching people efficiently. Of course, for this to be effective, sound diagnostics are essential to accurately detect and differentiate species (or strains) of danger or benefit, and to ensure optimum preservation of biodiversity. Meanwhile there appears to be great scope for the exploitation of chemical communication within the food web (Pickett et al. 2014), either through host repellence of pests or non-host attraction (e.g. Kihika et al. 2017). While biological control in its classical application may be expensive and time consuming to develop, the potential returns that have been demonstrated by notable examples in this volume make it an extremely attractive option in numerous situations for SSA, which must not be discounted for less-expensive 'quick fixes' (Neuenschwander 2004). This may be especially pertinent given the current knowledge surrounding the dangers of synthetic pesticide use, while the absence of numerous Class I pesticides that were previously relied upon necessitates that we seek and pursue safer alternative options.

8 Where to look for further information

In the near future crop protection in SSA will largely be determined by smallholder farming systems, which dominate the agricultural landscape. A gradual shift to more intensive commercial systems is likely however, which in turn, will change the *modus operandi* for pest and disease management. Research focus therefore needs to prioritize smallholder conditions, while remaining cognizant of the likely shift in farming systems. In either case, however, research needs to focus on identifying management options that negate or limit the use of synthetic pesticides towards safer and healthier cropping systems that help to maintain biodiversity. More detailed information is available through a number of channels or networks, such as the landmark texts by Maredia et al. (2003) and Neuenschwander et al. (2003). Useful links for information, advice and networks include those of the International Association for the Plant Protection Sciences (IAPPS) (https://www.plantprotection.org), the International Organization for Biological Control (IOBC) (http://www.iocbc.com/) and Croplife International (https://www.croplife.com/), amongst others, while a notable endeavour, supported by USAID under the Feed the Future programme, includes the Innovation Lab for IPM (https://www.oired.vt.edu/ipmcrsp/).

9 References

AATF. 2017. African states urged to adopt Bt maize to control armyworm. Partnerships: A newsletter of the African Agricultural Technology Foundation 19, 2-6. Available at https://www.aatf-africa.org/files/Partnerships-Issue-19-Jan-Jun-2017.pdf

Abang A. F., Kouame C. M., Abang M., Hanna R. and Fotso A. K. 2014. Assessing vegetable farmer knowledge of diseases and insect pests of vegetable and management practices under tropical conditions. *International Journal of Vegetable Science* 20, 240–53.

Aighewi B. A., Asiedu R., Maroya N. and Balogun M. 2015. Improved propagation methods to raise the productivity of yam (*Dioscorea rotundata* Poir.). *Food Security* 7, 823–34.

Ajambo S., Rietveld A., Nkengla L. W., Niyongere C., Dhed'a D. B., Olaosebikan D. O., Nitunga E., Toengaho J., Kumar P. L., Hanna R., Kankeu R. S. and Omondi A. 2018. Recovering banana production in bunchy top-affected areas in sub-Saharan Africa: Developing gender-responsive approaches. *Acta Horticulturae* 1196, 219–28.

Bajwa W. I. and Kogan M. 2002. Compendium of IPM definitions – What is IPM and how is it defined in the worldwide literature? IPPC publication no. 998. Oregon State University, Corvallis, OR, pp. 1–15.

Belfroid A. C., Van Drunen M., Beek M. A., Schrap S. M., Van Gestel C. A. M. and Van Hattum B. 1998. Relative risks of transformation products of pesticides for aquatic ecosystems. *Science of the Total Environment* 222, 167–83.

Beyond Pesticides. 2017. 35 years reflections on beyond pesticides: Vision for the future. Beyond Pesticides, Washington DC. https://www.beyondpesticides.org/about/our-mission (last accessed: 7 June 2018).

Chambers, R. 1994. Participatory rural appraisal (PRS): Analysis of experience. *World Development* 22, 1253–68.

Christie M. E., Van Houweling E. and Zseleczky L. 2015. Mapping gendered pest management knowledge, practices, and pesticide exposure pathways in Ghana and Mali. *Agriculture and Human Values* 32, 761–75.

Claudius-Cole A. O., Kenyon L. and Coyne D. L. 2017. Effect of pre-plant treatments of yam (*Dioscorea rotundata*) setts on the production of healthy seed yam, seed yam storage and consecutive ware tuber production. *Journal of Agriculture and Rural Development in the Tropics and Subtropics* 118, 297–306.

Coyne D. L., Wasukira A., Dusabe J., Rotifa I. and Dubois T. 2010. Boiling water treatment: A simple, rapid and effective technique for producing healthy banana and plantain (*Musa* spp.) planting material. *Crop Protection* 29, 1478–82.

Dey N. C. 2010. Use of pesticides in vegetable farms and its impact on health of farmers and environment. *Environmental Science and Technology* 11, 134–40.

Diallo B. and Tasie O. 2017. National implementation of regional pesticide policies in west Africa: The Gambia case study report. Department of Agricultural, Food, and Resource Economics, Michigan State University, East Lansing, MI. http://foodsecuritypolicy.msu.edu/uploads/resources/FSP_Research_Paper_69.pdf

Diarra A. and Haggblade S. 2017. Implementing regional pesticide policies in West Africa during a period of rapid market growth: Achievements, Challenges and

Opportunities. Innovation Lab for Food Security Policy. Presented to the USAID, Washington DC. http://foodsecuritypolicy.msu.edu/uploads/resources/regional_pesticide_policy_implementation.pdf

Dubois T., Dusabe Y., Lule L., Van Asten P., Coyne D., Hobayo J.-C., Nkurunziza S., Ouma E., Kabunga N., Qaim M., Kahangi E., Mwirigi P., Mwaura P., Kisii D., Kizito H. and Mugisha J. 2013. Tissue culture banana (*Musa* spp.) for smallholder farmers: Lessons learnt from East Africa. *Acta Horticulturae* 986, 51–60.

Elsen A., Goossens B., Belpaire B., Neyens A., Speijer P. R. and De Waele D. 2004. Recolonisation by nematodes of hot water treated cooking banana planting material in Uganda. *Nematology* 6, 215–21.

Erbaugh M. J., Donnermeyer J., Amujal M. and Kyamanywa S. 2003. The role of women in pest management decision making in Eastern Uganda. *Journal of International Agricultural and Extension Education* 10, 71–81.

FAO. 2005. Champs de Diversité: Définition, principes, déroulement général et outils pour la préparation, l'animation et la réalisation des CD, Tome 1. Guide Pratique FAO pour les Champs de Diversité, version juin 2005. FAO, Rome, Italy, pp. 16.

FAOSTAT. 2018. Food and agriculture data. http://www.fao.org/faostat/en/(last accessed: 2 July 2018).

Gallagher K. 2003. Fundamental elements of a farmer field school. *LEISA (Low External Input and Sustainable Agriculture)* 19, 5–6.

Gathigah M. 2017. Biotechnology part of the solution to Africa's food insecurity, scientists say. *Reliefweb*. https://reliefweb.int/report/world/biotechnology-part-solution-africa-s-food-insecurity-scientists-say (last accessed: 15 June 2018).

Gbaguidi B. and Coulibaly O. 2013. PRONAF's Farmer Field Fora. MEAS Case Study Series on Human Resource Development in Agricultural Extension. MEAS Case Study Series #2. USAID, Washington DC, p. 6. https://agrilinks.org/sites/default/files/resource/files/MEAS%20HRD%20CS%202%20-%20Farmer%20Field%20Fora%20-%20Jan%202013.pdf delete, or cite differently, at least add (last accessed: 10 July 2018).

Georgina A. 2012. Genre et changement climatique. Boîte à outils. Orient Longman, Hyderabad, India, 44p.

Haggblade S., Smale M., Kergna A., Theriault V. and Assima A. 2017a. Causes and consequences of increasing herbicide use in Mali. *The European Journal of Development Research* 29, 648–74.

Haggblade S., Minten B., Pray C., Reardon T. and Zilberman D. 2017b. The herbicide revolution in developing countries: Patterns, causes and implications. *The European Journal of Development Research* 29, 533–59.

Hauser S. 2007. Plantain (*Musa* spp. AAB) bunch yield and root health response to combinations of mechanical, thermal and chemical nematode control measures on suckers. *African Plant Protection* 13, 1–15.

Hauser S. and Coyne D. 2010. SP-IPM Technical Innovation Brief 10. Boiling water treatment of banana and plantain: A hot bath cleans all. IITA, Ibadan, Nigeria, 2pp.

Hauser S. and Messiga F. N. K. 2010. Nematode control on plantain suckers (*Musa* spp. AAB genome) through submergence in boiling water: Emergence rates, early growth, bunch yield and root health. *Acta Horticulturae* 879, 323–31.

Hazell P. and Wood S. 2008. Drivers of change in global agriculture. *Philosophical Transactions of the Royal Society B* 363, 495–515.

Published by Burleigh Dodds Science Publishing Limited, 2019.

Hoeschle-Zeledon I., Neuenschwander P. and Kumar L. 2013. Regulatory challenges for biological control. SP-IPM Secretariat. IITA, Ibadan, Nigeria, 43pp.

Jackai L. E. N. and Daoust R. A. 1986. Insect pests of cowpeas. *Annual Review of Entomology*, 31, 95–119.

Jones K. 2014. Promotion of integrated pest management by the plant science industry: Activities and outcomes. In: Pimentel D. and Peshin R. (Eds), *Integrated Pest Management*. Springer, Dordrecht, The Netherlands, pp. 394–408.

Keating B. A., Carberry P. S., Bindraban P. S., Asseng S., Meinke H. and Dixon J. 2010. Eco-efficient agriculture: Concepts, challenges, and opportunities. *Crop Science* 50, 109–19.

Keri J. H. 2009. Nigeria's status on pesticide registration and maximum residue levels. *Workshop on Pesticide Maximum Residue Levels*. Alexandria, Egypt, 30th March – 2nd April. Available at http://www.gmup.org/MRLWorkshop/NIGERIA%E2%80%99S%20STATUS%20ON%20PESTICIDE%20REGISTRATION%20AND%20MAXIMUM%20RESID.pdf

Kihika R., Murungi L. K., Coyne D., Ng'ang'a M., Hassanali A., Teal P. E. A. and Torto B. 2017. Parasitic nematode *Meloidogyne incognita* interactions with different *Capsicum annum* cultivars reveal the chemical constituents modulating root herbivory. *Scientific Reports* 27, 2903.

Knowles M. S., Holton E. F. and Swanson R. A. 2015. *The Adult Learner: The Definitive Classic in Adult Education and Human Development*, 8th ed. Gulf Publishing Company, Houston, TX, 402pp.

Kogan M. 1998. Integrated pest management: Historical perspectives and contemporary development. *Annual Review of Entomology* 43, 243–70.

Kroschel J., Beed F., Garrett K., Coyne D., van Etten J., Forbes G., Herrera B., Kreuze J., Parsa S., Sparks A., Tamò M., Subramanian S. and Tonnang H. 2013. Management of critical pests and diseases through enhanced risk assessment and surveillance and understanding climate impacts through enhanced modelling. CCAFS and CRP-RTB Workshop Report. Copenhagen, Denmark; Lima, Peru. www.ccafs.cgiar.org.

Kumar P. L., Selvarajan R., Iskra-Caruana M.-L., Chabannes M. and Hanna R. 2015. Biology, etiology and control of virus diseases of banana and plantain. *Advances in Virus Research* 91, 229–69.

Legg J. P., Kumar L., Makeshkumar T., Ferguson M., Kanju E., Ntawuruhunga P., Tripathi L. and Cuellar W. 2015. Cassava virus diseases: Biology, epidemiology and management. *Advances in Virus Research* 91, 85–142.

Makkouk K. and Emechebe A. 2003. Bridging the gap with the CGIAR systemwide program on integrated management. In: Maredia K., Dakouo D. and Mota-Sanchez D. (Eds), *Integrated Pest Management in the Global Arena*. CAB International, Wallingford, UK, pp. 419–34.

Maredia K., Dakouo D. and Mota-Sanchez D. (Eds). 2003. *Integrated Pest Management in the Global Arena*. CAB International, Wallingford, UK, 560pp.

Matthews G., Bateman R. and Miller P. 2014. *Pesticide Application Methods*, 4th ed. Wiley Blackwell, Chichester, UK, 536pp.

Mengistie B. T., Mol A. P. J., Oosterveer P. and Simane B. 2014. Information, motivation and resources: The missing elements in agricultural pesticide policy implementation in Ethiopia. *International Journal of Agricultural Sustainability* 13, 240–56.

Mengistie B. T., Mol A. P. J. and Oosterveer P. 2015. Pesticide use practices among smallholder vegetable farmers in Ethiopian Central Rift Valley. *Environmental and Development Sustainability* 13, 240–56.

Navarro M. 2008. On the path to sustainable agricultural development: Enhancing extension agents' contribution. *The International Journal of Environmental, Cultural, Economic and Social Sustainability* 4, 71–8.

Neuenschwander P. 2004. Harnessing nature in Africa. Biological control can benefit the pocket, health and the environment. *Nature* 432, 801–2.

Neuenschwander P., Borgemeister C. and Langewald J. (Eds). 2003. *Biological Control in IPM Systems in Africa*. CAB International, Wallingford, UK, 448pp.

Oerke E.-C. 2006. Centenary review: Crop losses to pests. *Journal of Agricultural Science* 144, 31–43.

Oghiakhe, S. 1995. Progress and future prospects in utilisation of host plant resistance in the control of insect pests of cowpea, *Vigna unguiculata* in Africa. *International Journal of Tropical Insect Science* 16, 123–43.

Opare-Atakora D. Y., Donkoh S. A. and Alhassan A. 2014. Farmer Field Fora and adoption of yam integrated pest and disease management technologies in Northern Ghana. *Journal of Agricultural Extension and Rural Development* 6, 143–52.

Pan International Consolidated. 2018. http://pan-international.org/pan-international-consolidated-list-of-banned-pesticides/(last accessed: 15 June 2018).

Pickett J. A., Woodcock C. M., Midega C. A. O. and Khan Z. R. 2014. Push-pull farming systems. *Current Opinion in Biotechnology* 26, 125–32.

Savary S., Ficke A., Aubertot J.-N. and Hollier C. 2012. Crop losses due to diseases and their implications for global food production losses and food security. *Food Security*. doi:10.1007/s12571-012-0200-5.

Tabashnik B. E. 2015. ABCs of insect resistance to Bt. *PLoS Genetics* 11(11), e1005646. https://doi.org/10.1371/journal.pgen.1005646

USAID. 2016. A Food-Secure 2030: A global vision and call to action. Unprecedented progress and momentum. Feed the Future. USAID, Washington DC. https://www.usaid.gov/what-we-do/agriculture-and-food-security/a-food-secure-2030 (last accessed: 5 June 2018).

Van den Berg H. and Jiggins J. 2007. Investing in farmers – The impacts of farmer field schools in relation to integrated pest management. *World Development* 35, 663–86.

Van Huis A. and Meerman F. 1997. Can we make IPM work for resource-poor farmers in sub-Saharan Africa? *International Journal of Pest Management* 43, 313–20.

Vanlauwe B., Coyne D., Gockowski J., Hauser S., Huising J., Masso, C., Nziguheba, G., Schut, M. and Van Asten, P. 2014. Sustainable intensification and the African smallholder farmer. *Current Opinion in Environmental Sustainability* 8, 15–22.

Waterhouse D. F. and Norris K. R. 1987. *Biological Control Pacific Prospects*. Inkata Press, Melbourne, Australia, 454pp.

Williamson S., Ball A. and Pretty J. 2008. Trends in pesticide use and drivers for safer pest management in four African countries. *Crop Protection* 27, 1327–34.

Zilberman D., Ameden H. and Qaim M. 2007. The impact of agricultural biotechnology on yields, risks, and biodiversity in low-income countries. *Journal Development Studies* 43, 63–78.

Published by Burleigh Dodds Science Publishing Limited, 2019.

Part 4

Conclusions and future challenges

Chapter 16

Improving plant health in sub-Saharan Africa: conclusions and future challenges

Peter Neuenschwander *and* **Manuele Tamò**, *International Institute of Tropical Agriculture (IITA), Benin; and* **May-Guri Sæthre**, *International Institute of Tropical Agriculture (IITA), Nigeria (corresponding author: p.neuenschwander@cgiar.org)*

1 Introduction

IITA has experienced tremendous growth since its establishment 50 years ago and especially after 2012 when staff and budget increased three-fold in about 5 years and the institute's tagline was revised from *'Research to Nourish Africa'* to *'Transforming African Agriculture'* with the corresponding changes in perspective. When IITA received the 2018 Africa Food Prize the Director General noted that

> 'A shift is under way from agriculture being seen as a series of commodities alone, toward being seen as a whole ecosystem that has people at its centre. And that by creating job opportunities for rural youth and improving livelihoods, we can ensure food security and nutrition for Africans, and provide a powerful antidote to threats posed by political instability and extremism in the continent' (https://www.africafoodprize.org/2018-africa-food-prize-awarded/)

Africa's problems indeed are enormous. Because its population, estimated at 650 million in 1990, is projected to reach 1.5 billion people by 2025, food demand will more than double within a time frame of less than four decades.

http://dx.doi.org/10.19103/AS.2018.0043.17
Published by Burleigh Dodds Science Publishing Limited, 2019.

The key challenge therefore is to increase agricultural production sustainably by preserving biodiversity and reducing the pressure on ecosystems and natural resources.

IITA grew from a few dozen internationally employed staff in 1967 with a budget of about US$2.5 million p.a. to 224 scientists in 2018 and a mostly project-based budget of over US$100 million p.a. Yet, the number of staff of the entire CGIAR with its 15 institutes across the world only roughly equals the one of corresponding French institutions and is dwarfed by Brazil and India with their national agricultural research institutions with thousands of scientists. Despite of its modest size, IITA is the biggest coordinating research institution within its mandate area of tropical Africa. It is now represented by its own stations in 16 African countries, including those where IITA is hosted by other institutions (Fig. 1).

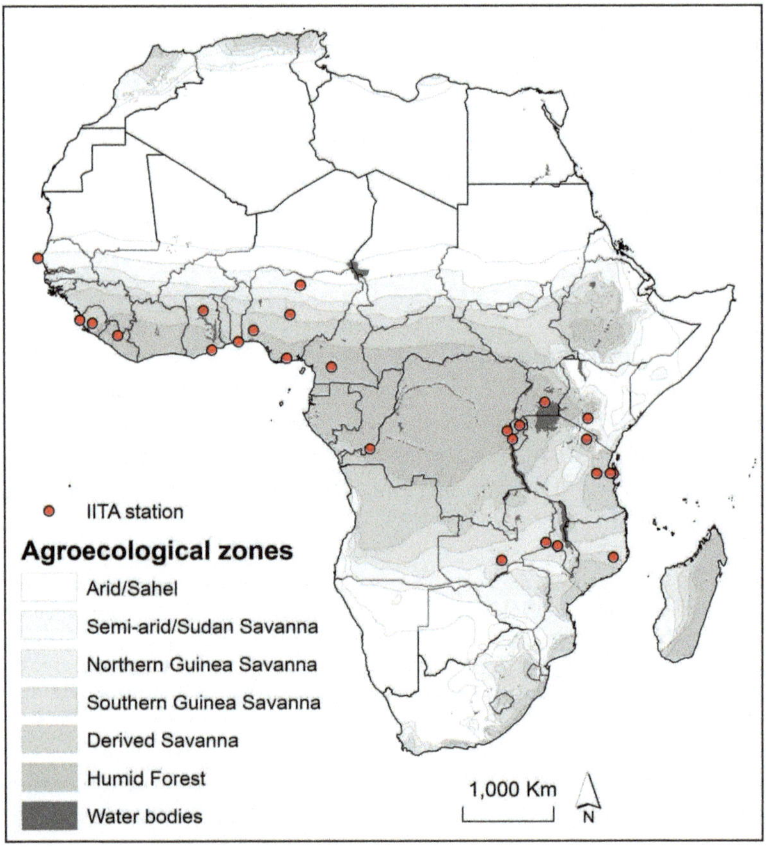

Figure 1 Agroecological zones of Africa with IITA stations as dots.

Published by Burleigh Dodds Science Publishing Limited, 2019.

IITA research to transform Africa puts it on a knife's edge between the activities of big implementing non-government agencies and universities – or in the position of Louis Pasteur in the Pasteur's quadrant (Stokes, 1997). On the one hand, scientific research aims not only to describe but to understand mechanisms (though not all scientists are equally interested in this aspect); on the other hand, the institute implements and documents how its results are being adopted and adapted in order to increase production, sustainability and the well-being of farmers. In the early days, confirming yield increases was the main goal. Today, ideally, we aim to achieve and document higher income and well-being, but also question how this is achieved, with the understanding that this is the only way science can advance. In order to find a holistic solution for all stakeholders, poor versus rich, women and youth, we also address environmental and ecological challenges in close collaboration with our first clients, the various national programmes.

Today, we face increased pressure from donors to stress 'transformation', that is implementation of results. Instant successes at farm level within shortest time periods are thereby expected. Yet, research means that not everything is already known, means accepting failures and accepting also negative results. This requires much patience and confidence by all involved. With the present book on plant health research for development (R4D) we sketch the history and problems in African agriculture (Chapter 2) and pause for a moment in the relentless year-to-year hunting for funds that often makes it difficult to keep the larger and long-term view. The aim is to consolidate and extend the successes of the past 50 years into the future. The cited chapters are listed at the end of the chapter in the section 'Where to look for further information'.

2 The views of IITA clients and peers

In a first step, we asked former collaborators for their opinion about IITA in order to listen to their advice and complaints (Chapter 1). Generally, there seems to be good agreement between requests by national institutions, universities and donors, on the one hand, and IITA's output, on the other hand. In fact, often these requests had already been discussed with our colleagues in the national programmes even before they surfaced as official demands.

A special case is the CGIAR System-wide Program for Integrated Pest Management (SP-IPM), which was very well accepted in Africa, but less so on other continents. It could have served as a model for the CGIAR Research Programmes (CRPs), but was dismissed by donors – in our view a missed chance. A list of CGIAR Research Programs in which IITA is involved is provided in Table 1.

Generally, a holistic IPM approach, coupled with robust phytosanitary arrangements, is given importance by our interviewees. One dissenting voice,

Table 1 CGIAR Research Programs in which IITA is involved

CGIAR Research Program on Climate Change, Agriculture and Food Security	CCAFS generates evidence and supports adoption of climate-smart agricultural policies, practices and services that alleviate poverty, increase gender equity and support sustainable landscapes. CCAFS also plays an integrating role across all of the CRPs.
CGIAR Research Program on Agriculture for Nutrition and Health	A4NH is built on the notion that agriculture can do much more than reduce hunger and poverty - it has an enormous potential to significantly improve the nutrition and health of people around the world. The links between nutrition and climate change are a focus area.
CGIAR Research Program on Policies, Institutions, and Markets	PIM is an action-oriented research to provide support for policies that help poor farmers, both men and women; improve their lives; produce nutritious and affordable foods; and protect the soil, water and biodiversity in rural landscapes.
CGIAR Research Program on Grain Legumes and Dryland Cereals	GLDC aims to increase the productivity, profitability, resilience and marketability of critical and nutritious grain legumes and cereals within the semi-arid and sub-humid dryland agro-ecologies of sub-Saharan Africa and South Asia. In these agro-ecologies poverty, malnutrition, climate change and soil degradation are among the most acute globally.
CGIAR Research Program on Maize	MAIZE is an international collaboration between more than 300 partners that seeks to mobilize global resources in maize research and development to achieve a greater strategic impact on maize-based farming systems in Africa, South Asia and Latin America.
CGIAR Research Program on Roots, Tubers and Bananas	RTB is working globally to harness the untapped potential of those crops in order to improve food security, nutrition, income, climate change resilience and gender equity of smallholders.
CGIAR Genebank Platform	Development of trait-based subsets for climate resilience, conservation and characterization of diversity for both staple and underutilized crops for future climate options.
CGIAR Platform for Big Data in Agriculture	Agriculture to harness the capabilities of big data to accelerate and enhance the impact of international agricultural research.

however, complains about too little reliance on systems models, too weak a fight for organic agriculture (without mineral fertilizers, synthetic pesticides, or GMOs) and requests a total change in IITA's approach to plant health management. This viewpoint will challenge us throughout this chapter. All colleagues stressed the importance of partnerships, which in the end will have an equal if not higher impact than the direct research results. Capacity building must therefore be considered as one of the most important contributions IITA has made.

3 Comparing various IITA projects and their impact

IITA works on demand by the national programmes, but is also heavily influenced by donor requests, which have become ever more detailed. Today, IITA has become a project-based institute, which differs radically from the prevailing reliance on core funding during the first half of its existence. Therefore our analysis follows the fate of projects (and their follow-ups) (Table 2).

Comparing projects with different actors, in different areas and at different times, against different targets, is like comparing 'apples and oranges'. A total of 58 targets that were combatted through sometimes consecutively funded projects or continuously through core funding (up to about 2017) are presented in Table 2 and arranged according to crop plant or environment, pest group and target pest. In order to estimate the importance that IITA gave these projects we characterized them by a confounded criterion from (1) to (5), taking into account the numbers of scientists, collaborating institutions, approximate years of duration, countries where the project had activities and number of donors involved. Evidently not all projects fell clearly into one category and occasionally one of the defining criteria was not met. Achievements were ranked from type of publication to proof of concept to use and adaptation of the results. The contribution of different techniques to the desired project outcome was roughly estimated, namely for (1) synthetic chemicals; (2) so-called biorational techniques involving the use of botanicals, *Bacillus thuringiensis* (Bt), fungi, viruses and so on; (3) biological control with insects or mites, by introducing agents or favouring existing agents, as well as competitive exclusion; (4) resistant varieties; and (5) cultural practices including phyto-sanitation and use of virus-free planting material. Projects were characterized according to which of these techniques contributed to more than half of the impact. Where no technique contributed more than half, the project was labelled as 'mixed'.

The present discussion will be presented according to main techniques (>50% contribution to impact) responsible in each project. We start with biological control because this is the one intervention that farmers can control only little. It is therefore considered as the basis for all plant health management. This will be followed by resistant varieties, which was the main contribution of IITA in the first half of this period, then the use of biorational methods and the control by synthetic chemicals, which was the traditional intervention in early IPM, then cultural control, that is good farming practices that are at the root of all other interventions. We shall compare these interventions for different crops and environments, followed by their combination in IPM.

Table 2 Summary of 58 IITA projects

					6 - Percentage contribution						
1- Crop etc.	2 - Pest group	3 - Pest	4 - Size of project	5 - Achievements	Synth. chem.	Biorational	Res. var.	Biocontrol	Cultural	>50%	7 - Chapter citation
Maize	Insect	Fall army worm	5	*publ*	10	30	10	40	10	*mixed*	4
Maize	Fungus	Downy mildew	3	*use**			100			*res*	7
Maize	Insect	Maize stemborers	4	*proof*			15	15	70	*cult*	7
Maize	Insect	Larger grain borer	5	*proof*				20	80	*cult*	7
Maize	Fungus	Mycotoxins	5	*use**		60	20		20	*biorational*	7; 13
Maize	Virus	Maize viruses	5	*use**			80		20	*res*	3; 5
Cassava	Virus	Cassava viruses	5	*use**			80		20	*res*	3; 5
Cassava	Insect	Cassava mealybug	5	*adapt**			5	90	5	*biocon*	6
Cassava	Mite	Cassava greenmite	5	*use**			15	85		*biocon*	6
Cassava	Insect	*Zonocerus*	2	*test*		90			10	*biorational*	6
Cassava	Bacterium	Cassava bacterial blight	3	*proof*			55		45	*res*	6
Cassava	Fungus	Cassava anthracnose	2	*proof*			100			*res*	6
Cassava	Insect	ARTS	2	*proof*					100	*cult*	6
Cassava	Bacterium	Cassava root diseases	1	*publ*				10	90	*cult*	6
Cassava	Plant	Various weeds	2	*unpubl*			10		90	*cult*	14
Yams	Virus	Yam viruses	4	*use*			55		45	*res*	3; 5; 8

Published by Burleigh Dodds Science Publishing Limited, 2019.

Yams	Nematode	Root-knot nematodes	4	*publ*	20	20	20		40	*mixed*	4; 8
Yams	Fungus	Fungal pathogens	4	*publ*		20	10		70	*cult*	8
Cowpea	Virus	Cowpea viruses	1	*publ*			100			*res*	3; 5
Cowpea	Fungus	Anthracnose	1	*proof*	10		80		10	*res*	10
Cowpea	Fungus	Leaf spot	1	*publ*			60		40	*res*	10
Cowpea	Fungus	Charcoal rot	1	*publ*			20	40	40	*mixed*	10
Cowpea	Bacterium	*Xanthomonas*	1	*publ*			60	10	30	*res*	10
Cowpea	Nematode	Root-knot nematodes	1	*publ*			60	20	20	*res*	10
Cowpea	Insect	Cowpea aphids	2	*proof*			40	60		*biocon*	10
Cowpea	Insect	Thrips	3	*proof*	20		40	40		*mixed*	10
Cowpea	Insect	*Maruca* pod borer	4	*use*		20	10	70		*biocon*	10
Cowpea	Insect	*Clavigralla*	2	*proof*	20	10	30	40		*mixed*	10
Soybean	Fungus	Soybean rust	1	*proof*			100			*res*	10
Soybean	Virus	Soybean viruses	1	*publ*			100			*res*	3; 5
Banana	Virus	Banana streak virus	3	*use*			20		80	*cult*	3; 5; 9
Banana	Virus	Banana bunchy top virus	3	*use*			20		80	*cult*	3; 5; 9
Banana	Fungus	*Fusarium* wilt	1	*unpubl*					100	*cult*	9
Banana	Fungus	Black sigatoka	1	*test*	10		20		70	*cult*	9; 14
Banana	Bacterium	*Xanthomonas* wilt	1	*proof*			100			*res*	9

(*Continued*)

Table 2 (*Continued*)

1- Crop etc.	2 - Pest group	3 - Pest	4 - Size of project	5 - Achievements	6 - Percentage contribution: Synth. chem.	Biorational	Res. var.	Biocontrol	Cultural	>50%	7 - Chapter citation
Banana	Nematode	Nematodes	2	*use*					100	*cult*	9
Banana	Insect	Banana weevil	3	*use*	10	20		10	60	*cult*	9
Plantain	Plant	*Chromolaena*	1	*unpubl*	10				90	*cult*	14
Vegetables	Nematode	*Meloidogyne*	2	*use*					100	*cult*	11
Vegetables	Bacteria	*Ralstonia*	2	*proof*			20	20	60	*cult*	11
Vegetables	Insect	Diamond-back moth	2	*use*	10	60		30		*biorational*	11
Vegetables	Insect	Beet web-worm	2	*proof*		100				*biorational*	11
Vegetables	Mite	Broad mite	1	*publ*				100		*biocon*	11
Mango	Insect	Mango mealybug	5	*adapt**				90	10	*biocon*	12
Mango	Insect	*Bactrocera dorsalis*	5	*test*	20			40	40	*mixed*	12
Mango	Insect	Spiralling whitefly	1	*use**				100		*biocon*	12
Mango	Insect	Papaya mealybug	5	*use**				100		*biocon*	12
Cacao	Fungus	*Phytophthora megakarya*	2	*proof*	20		20	20	40	*mixed*	12
Coconut	Mite	Coconut mite	1	*unpubl*				100		*biocon*	12
Cashew	Insect	Cashew wood borer	1	*publ*				10	90	*cult*	12

Published by Burleigh Dodds Science Publishing Limited, 2019.

Sahel	Insect	Locusts	5	*use**		100				*biorational*	13
Savanna	Plant	Speargrass	4	*use**	20			5	75	*cult*	14
Savanna	Plant	*Striga*	5	*use**	5	5	20	10	60	*cult*	14
Savanna/ forest	Plant	*Chromolaena*	*1*	*test*	5			25	70	*cult*	14
Open water	Plant	Water hyacinth	4	*use**		10		80	10	*biocon*	14
Open water	Plant	Water fern	2	*use**				100		*biocon*	14
Open water	Plant	Water lettuce	3	*use**				100		*biocon*	14
Human health	Insect	Mosquitoes	2	*publ*	20				80	*cult*	16

1 – Crop.
2 – Pest group.
3 – Main target pest(s) in this core activity or project, including follow-up projects with similar goal.
4 – Importance of project: (1) small project with one student, part of supervisor time share, participation of one university, of few years duration, in one African country, supported by one donor; (2) small project with several students, part of supervisor time share, more than one university, of a few years duration, in one to two African countries, often supported by several donors; (3) part of a larger project with a few IITA international staff, several students at various universities, active in a few countries, with support from several donors and, in the past, continuous core funding; (4) large self-standing project with several collaborating supervisors from different CGIAR centres and professors of several universities active in several countries over up to five years, often funded by several donors; (5) large self-standing projects with numerous international staff of different CGIAR centres, other international organizations and universities, with numerous students, active across the continent, funded by an organized donor consortium (eventually planned) over five or more years.
5 – Achievement: *unpubl* – unpublished results concerning implementation; *publ* – scientific results published in peer-reviewed journal; *proof* – proof-of-concept documenting impact under experimental or restricted conditions; *test* – tested under farmers' field conditions; *use* – results ('goodies') and impact are documented and used by governments/NGO/farmers; *adapt* – results from IITA were adapted to new circumstances, extended to other continents or crops; * – with socio-economic analysis; *fail* – activity has later been acknowledged as a failure/failed investment.
6 – Contributing factors to success in % (very rough estimates) after the project has been implemented: % control by synthetic chemicals, % biorational control (including botanicals, Bt, fungi, viruses etc.), % biological control with insects, mites (introducing agents or favouring existing agents) including competitive exclusion, % resistant varieties, % cultural practices, including phytosanitation and use of virus-free planting material. Contribution >50%, if all <50% = mixed.
7 – Citation: Chapter number in this volume.

3.1 The impact of IITA projects: biological control

3.1.1 Comparing results

Biological control is mentioned in 30 projects, and in 12 of them it was the most important (>50%), in six even the only component. Averaging the confounded criterion for importance and size of the project with values from column 4 (Table 2), over all 12 projects dominated by biological control interventions, yielded a score of 3.2. All these projects collaborated with other institutions exchanging material, relying on an international network and fraternity of biological control scientists. Nine projects (30%) remained academic studies, nine (30%) reached proof-of-concept stage or were already tested in the field and 12 were in use by farmers (40%). Among the 12 projects with predominantly biological control input, six had size 4 or 5 and reached the stage of *use* or even *adaptation*. In two of these cases, IITA provided material to other continents (Wyckhuys et al., 2018a), repeating the impact of biological control often without additional financial support. Three projects with size 1 or 2 did not go beyond *proof* of concept, indicating that bigger projects generally brought higher rewards. Yet, three projects had high impact (*use* or *adapt*) with minimal input (1 or 2). Those were mostly projects without proper funding, which we could execute thanks to personal relations ('old-boys' network') by receiving already tested agents for free. Most of our biological control successes concerned using specific hymenopterous parasitoids against Hemiptera, slightly polyphagous phytoseiid mites against other mites, or specific curculionid beetles against plant pests, thereby confirming rather old international statistics (https://www.cabi.org/isc).

Interestingly, classical measures of efficiency of natural enemies (the intrinsic rate of natural increase r_m for instance) did not well predict the actual impact of some natural enemies in the field (Chapter 6). *Anagyrus lopezi*, the parasitoid that brought cassava mealybug under control, had a low r_m, but achieved its success thanks to its exceptional host-finding and strong dispersal capacities. Similarly, the most promising phytoseiid mite predator of cassava green mite, *Typhlodromalus aripo,* had a low reproductive capacity, but excelled because of its habitat choice, which corresponded to the one of its prey, and its capability to remain on the host plant throughout the year thanks to its utilization of alternative food sources. While assessing the reproductive capacity under laboratory conditions gives useful data for population modelling, the above cases highlight that these studies should be followed up by ecological and behavioural studies in order to understand and better exploit natural enemies.

In most projects, taxonomic problems were encountered (Chapter 4), be it at the level of the host plant (e.g. *Chromolaena,* Chapter 14), the pest (e.g. *Maruca,* Chapter 10; *Aspergillus*, Chapter 8) or its natural enemies

Published by Burleigh Dodds Science Publishing Limited, 2019.

(*Anagyrus, Neoseiulus,* Chapter 6; *Phanerotoma,* Chapter 10). This underlines the importance of having a taxonomic capacity as realized in the staff of the biodiversity collection at IITA-Benin. In close collaboration with this biodiversity centre, most biocontrol projects invested in studies of food webs, indicating which hosts were attacked by which natural enemies. Importantly, none of the often-criticized non-target effects (Lynch and Thomas, 2000; Roy et al., 2016) were reported. These studies were, however, hampered by the often still weak taxonomic knowledge about indigenous insects in tropical Africa.

A high degree of biological control has been achieved and documented mainly for exotic pests: cassava mealybug, cassava green mite (Chapter 6), mango and papaya mealybugs, spiralling whitefly (Chapter 12) and floating water weeds (Chapter 14). On mandated crops, these biological control projects always revealed that some existing varieties were less susceptible to the invader than others and were worthy to be recommended, thereby demonstrating that biological control interventions were generally fully compatible with existing resistant varieties (Thomas and Waage, 1996). A better consideration of the effect of the host plant on natural enemies, as for instance selecting for 'domatia' – shelters for *T. aripo* on cassava (Chapter 6) – or hairiness of leaves on the performance of small parasitoids, could open up new prospects for IPM (Wang et al., 2009; Pickett et al., 2014; Stenberg et al., 2015).

In addition, several studies revealed incipient successes by providing 'proof of concept' for the effect of natural enemies against several exotic pest species. Four projects demonstrated unexploited potential for biological control based on improved taxonomic studies.

In the case of *Maruca* pod borer (Chapter 10), which was generally assumed to be of African origin, taxonomic studies indicated an Asian origin (with new complications mentioned in Chapter 4), thereby opening up the possibility for classical biological control, which is starting to show an incipient success.

Similarly, better control of larger grain borer might still be achieved across Africa by exploiting different sources of predators (Chapter 7). The same is true for the control of different biotypes of *Chromolaena* of different origins (Chapter 14). Finally, inner-African differences among stem borers and their parasitoids were exploited for classical biological control, as shown in first feasibility studies (Chapter 7).

In the case of the new fruit fly attacking mangoes and citrus, later identified as *Bactrocera orientalis,* a parasitoid has now been established and is slowly spreading (Chapter 12). Though biological control of fruit flies has famous precedents (van den Bosch et al., 1951), it has in recent cases proven to be difficult and requiring an IPM approach including augmentative releases of parasitoids and sterile male techniques (Rousse and Quilici, 2009).

In all these cases, further studies that go well beyond the funding periods of the corresponding projects are necessary to measure and perhaps explain the slowness of impact of released natural enemies – or arrive at the conclusion that the equilibrium levels achieved by the released and established agents are unacceptably high.

Despite some abuse of pesticides, often in confined conditions as in vegetables (Chapter 11), but also in cowpea (Chapter 10), cocoa (Chapter 12) and other cash crops, African small-scale agriculture is still relatively free of pesticides and could be called 'organic'. Considering the huge numbers of insect species present in these tropical environments, indigenous pest species are indeed astonishingly few (Prinsloo and Uys, 2015). Natural biological control, though mostly undocumented, evidently plays an important role. It is often only recognized when indigenous natural enemies are being killed by insecticide treatments that make pests out of many spider mites, leaf-mining flies and others. This control is often achieved through non-specific natural enemies (Chapter 11). Similarly, abusive insecticide applications against fruit flies endangered the well-established biological control on mango mealybug, requiring now the development of a holistic IPM approach encompassing all pests on mango (Chapter 12).

Yet, difficult-to-tackle problems with indigenous pest species remain. Most stem borers on maize, sorghum and sugar cane (Chapter 7), cowpea thrips (Chapter 10), witch weeds on maize and legumes (Chapter 14), and all indigenous field and submersed weeds (Chapter 14) remain problematic. Classical biological control against them is no option, but inundative releases of natural enemies can be envisaged. Because of the problematic logistics of large-scale insect rearing this approach has been difficult in tropical Africa, but favouring parasitoids through changes in the environment like shelters or alternative host plants has given good results (Chapter 10; Guerci et al., 2018). Generally, control against indigenous pest species requires an array of IPM interventions and, in the case of parasitic weeds like *Striga* (Chapter 14), will concern mainly the management of soil fertility.

Increasingly, IITA's R4D is influenced and guided by international treaties. Since its first biological control projects, IITA was involved in the development of guidelines for the import and release of natural enemies (FAO, 1996) and has since meticulously followed the FAO recommendations. More recently, the Intergovernmental Panel on Climate Change (IPCC, 2014) gives directions to reduce CO_2 emissions, to prepare breeding programmes for higher temperatures and lower rainfall, and to study the influence of a changed climate on food webs, particularly the relationship between natural enemies, their phytophagous hosts and their host plants. The corresponding and much younger Intergovernmental Science-Policy Platform on Biodiversity and Ecosystem Services with its focus on the epic disappearance of plant and animal populations still struggles to include players also in developing countries and foster economic and sociological analyses to reach the same acceptance as the IPCC (Masood, 2018).

The best estimates about the health of the biosphere come from studies of birds, a particularly well-known and popular taxon. These studies show that among the 1469 globally threatened species, 74% are disappearing because of agriculture, a far higher number than the one attributed to any other activity like logging or hunting (Birdlife, 2018). IITA with its goal to further agricultural production for people therefore rightly focusses on the sustainability of its technologies. Thus, ecosystem services are discussed in many projects and IITA participated in worldwide projects on classical biological control for the protection of natural ecosystems (van Driesche et al., 2010; Chapter 14). Nature's contribution to people is increasingly taken into account in research and led to fruitful collaboration with non-academic activists for nature protection (Chapter 4) as suggested by McNeely and Scherr (2001) and recently stated more generally by Kremen and Merenlender (2018).

3.1.2 Challenges

Alien exotic species continue to reach Africa and force us to find long-lasting solutions. In eastern Africa, five major invasive alien species threaten smallholder mixed maize farming in Ethiopia, Kenya, Malawi, Rwanda, Tanzania and Uganda (Pratt et al., 2017). Thus, spotted stem borer (*Chilo partellus*), maize lethal necrosis disease, the weed *Parthenium hysterophorus*, the *Liriomyza* leaf-miner complex and the South American tomato leaf-miner *Tuta absoluta* caused annual losses of US$0.9–1.1 billion on mixed maize smallholders in these six countries alone. Losses are likely to increase as the weed and pests continue to spread.

Another example concerns the environmental and economic costs of invasive alien plants in South Africa (van Wilgen and De Lange, 2011). Costs for lost water, diminished grazing and negative impacts on biodiversity are estimated at US$450 million p.a., and could rise to >5% of GDP if invasive plants are allowed to reach their full potential.

Finally, the estimated losses caused by the recent invasion of fall army worm (FAW) correspond to 21–53% of the annual maize production in Africa, or between US$2.5 and US$6.2 billion annually. Other food crops are also likely to suffer losses from FAW (Prasanna et al., 2018).

Biological control against such invasive exotic pests is the method of choice and in the past yielded benefits that equalled whole breeding programmes (Neuenschwander, 2004). Yet, classical biological control programmes face opposition (starting with Howarth, 1991), because of the widespread abhorrence of anything exotic fuelled by scientists in North America and Europe and transferred to Africa by some NGOs. Fortunately, viewing the spread of organisms as inherently natural – in opposition to the dogmatic view of anything exotic as being 'unnatural' that is needing extirpation – is gaining importance (Guiaşu, 2016). IITA's practical view is to interfere only on request by clients and only under severe damage. We respond to cries of alarm. Therefore, not every new insect or pathogen becomes a target. Ideally, projects are coupled with a cost:benefit analysis before launching a classical biological control programme. Such programmes generally are an initially expensive, government-controlled action that does not need much involvement by farmers because introduced agents spread on their own.

Under these premises, one recurrent practical problem concerns the timing of the intervention. Cassava mealybug, for instance, was allowed to spread and do damage for years before the logistics of foreign exploration, external quarantine and natural enemy rearing could be tackled satisfactorily (Chapter 6). When, after many years, an efficient parasitoid had been found and reared and all import permits and so on were ready, damage was huge and subsequent control spectacular. In the case of papaya mealybug (Chapter 12), educated personnel, facilities and knowledge at the level of quarantine authorities were ready, control was equally spectacular, even faster than against cassava mealybug, but publicity and corresponding recognition were much lower, if not absent. Consequently, project funding was limited and did not allow as much basic research, which advances the science of biological control, as in the case of cassava mealybug. For the control of spiralling whitefly (Chapter 12), procedures for importing the proven parasitoids were well advanced when these turned up serendipitously.

With some pest groups, as for instance leaf-miner flies, whose parasitoids do not attack the larvae directly but lay their eggs into the mine, whereupon the parasitoid larva searches its prey by itself, indigenous parasitoids have been shown to control these exotic invaders (Neuenschwander et al., 1987). Such control of exotic invaders by indigenous natural enemies very much depends on the systems involved and remains controversial (Levine et al., 2004). While already present natural enemies might eventually adapt through evolution to control all invaders, this is not a practical approach against threatening pests in Africa. Since adaptation has already occurred in the homelands of such invaders we can profit from natural enemies found during foreign exploration. Unfortunately, permits for such search and the necessary exportation have become increasingly difficult to obtain from countries like Brazil, India and others (Chapters 8 and 10).

Many new pests are difficult to identify, sometimes still unknown; the same applies for their parasitoids and predators (Chapter 4). Tests for accepting potential biological control agents for introduction generally ask about specificity and potential non-target insects, which in the African environment are often not yet known. Such studies demand a permanent in-house taxonomic capacity and worldwide links with other taxonomic institutions, which in a strictly project-based institute remains a problem. A parallel support and recognition of the Biodiversity Center in IITA-Cotonou for non-plant genetic resources as for the Genetic Resources Unit at IITA-Ibadan for plant resources is therefore the ultimate aim. Ideally, such institutions should also be created on other continents, and work together with CAB International, which has the greatest capacity in foreign exploration worldwide.

The initial impact of an invading exotic pest is often forgotten when, following the establishment of exotic natural enemies, population densities are kept below economic thresholds (mostly undefined) for years. Subsequent upsurges are therefore often dramatized, even if they are within the normal fluctuations. Or sometimes they are the consequence of insecticide abuse on other pests, which calls for integration in IPM programme. Sometimes, the new introduction concerns unrecognized sibling (of not yet described) species requiring further taxonomic studies (Chapter 4). Resurgences or incomplete successes can also be due to missing or neglected ecological links, for example missing alternative host plants during off-season on cowpea pests (Chapter 10). Which of these scenarios applies is often difficult to assess. Generally, larger resurgences, which are not attributable to insecticide misuse, should always be subjected to DNA-based identifications. Moreover, food web and competitive displacement studies as applied in many projects instead of linear predator-prey studies can give the necessary answers to stabilize the system.

None of the listed projects was reported as an absolute failure. As is usual in classical biological control though, many introduced parasitoids, coccinellid predators and phytoseiid predators could not establish. Detailed studies about these natural enemies sometimes revealed the reasons for their non-establishment or inefficiency, thereby contributing to the general science of biological control (Chapters 6, 7, 10 and 12). It must be stressed, however, that such non-established and in quarantine tested organisms do not cause ecological harm. Across all projects worldwide only about one in six organisms can be established as a natural enemy (Bellows, 2001). Such failures are part of biological control endeavours (much as crop varieties that are not accepted by farmers, or chemical compounds that eventually do not result in marketable pesticides). In each project, releasing several natural enemy species at the same time, though not in the same locations, is therefore an accepted practice and does not hinder establishment of the fittest, though this procedure is still discussed controversially (Osborne and Cuda, 2003).

Unfortunately, most classical and other biological control interventions have a relatively slow impact. An example is the desired transition from a pesticide to a biological control-dominated schedule, for instance when FAW is first subjected to chemical control before classical biological control options will be ripe for implementation (Prasanna et al., 2018; Chapters 4 and 7). Impact assessments of classical biological control introductions sometimes take years – too long for short project funding. A series of potential biological control successes mentioned above could therefore not be correctly evaluated as the problem seemingly faded from the scene and further demand by African clients was lacking. While donors will fund implementation and maybe economic impact assessment, they certainly do not give additional funds to carry out a final ecological assessment that might provide knowledge about reasons for success or failure: larger grain borer, water hyacinth, cowpea thrips and *Chromolaena* fall into this category. Unfortunately, these projects also do not provide good researchable (for PhD) topics because of the needed vast coverages and long durations.

3.1.3 The way ahead

Generally, judiciously implemented biological control can deliver ample 'hidden' environmental and human health benefits that are not captured by the prices of globally traded commodities (Wyckhuys et al., 2018b). Yet, full-scale quantitative studies considering socio-economic parameters are uncommon (about biological control successes see Neuenschwander, 2004; about pathogens Savary et al., 2012). To improve the effectiveness of research prioritization and impact assessment, the biophysical spatial framework as well as appropriate socio-economic attributes (Edreira et al., 2018) should be included in each project document, with final surveys to be executed several years after the completion of the project.

Despite improved quarantine measures, introductions of potential pests into Africa and across the globe continue and are likely to increase in the future (Seebens et al., 2018). Experience with cassava mealybug and other Hemiptera shows that, where new introductions lead to widespread serious damage, all financial support should go into biological control projects first. Since many pests for instance from South America (Bellotti et al., 2012; Bellotti, 2017) have not yet been accidentally introduced into Africa, an effective horizon scanning and early warning system for anticipated introductions is required together with improved quarantine services.

Judging when to attack invasive insects remains, however, tricky: on the one hand, it will be cheaper to fund proactive initiatives, even extermination of small foci, rather than use the firefighter approach as for FAW; on the other hand, some invasive species are later followed by their also exotic effective parasitoids, like spiralling whitefly, while others are controlled by indigenous parasitoids and predators, like leaf-mining flies as discussed above. Generally, a more efficient uptake of biological control practices is needed (Barratt et al., 2017).

3.2 The impact of IITA projects: resistant varieties

3.2.1 Comparing results

Resistant varieties are mentioned in 33 projects, and in 14 of them this was the most important (>50%), in six even the only, component. The mean importance and size of these 12 projects yielded a score of 2.2. Ten projects (31.3%) remained academic studies, 12 (37.5%) reached proof-of-concept stage or were already tested in the field and 10 (31.3%) were in use by farmers. Among the 14 projects with predominantly resistant variety input, three had size 4 or 5, that is reached the stage of *use*. Ten projects with size 1 to 3 did not go beyond *proof* of concept, again indicating that bigger projects generally brought higher rewards.

As a follow-up of the Green Revolution and the establishment of the CGIAR, the development of resistant varieties became the mainstay of plant protection for the first half of the indicated period (Chapter 2; review in Eriksson et al. 2018). Today, the use of host plant resistance is guided by the nature of crop–pest interactions, pest ecology and pathogen epidemiology, and the availability of novel resistant genes discovered with modern genomic tools. Generally, main successes in host plant resistance were achieved against bacterial, fungal and viral diseases. Against nematodes, impact was smaller because of the diversity of species involved. Resistance breeding against arthropod pests of IITA-mandated crops has proven more difficult and has only resulted, so far, in a number of resistant genotypes mostly tested under screen-house conditions (Chapters 7, 10 and 11). In addition, resistant varieties have been relevant in the broad context of IPM in combination with other compatible management options, but not as stand-alone interventions.

One of the first major breakthroughs in developing resistant varieties at IITA was achieved way back in 1972, when the cassava mosaic disease (CMD)-resistant genotype TMS 30572 was first cloned and later widely disseminated in Africa. Today, it remains one of the most popular cassava varieties still retaining the original level of resistance to CMD (Chapter 5). Other cassava pathogens such as cassava anthracnose disease can also be effectively managed by resistance breeding (Chapter 6).

Another breakthrough was achieved in the control of a downy mildew strain affecting maize, which appeared in the 1990s as an emerging disease in Nigeria. The successful approach included the development of a new screening methodology. The quantity of seeds of resistant varieties necessary to contain the epidemic could only be overcome by a concerted donor support for large-scale seed distribution schemes. As a result, the downy mildew epidemic was stopped in 2000, and the disease has not been reported anymore from Nigeria until today (Chapter 7). A similar success was achieved in controlling maize streak virus, which culminated in the 1990s with the release and farmer adoption of resistant varieties, often produced by national partners using sources of resistance discovered by IITA. However, these successes have recently been threatened by emerging epidemics caused by maize lethal necrosis. Efforts are currently underway to combine resistance genes against both viruses.

Other efforts to develop and deploy host plant resistance have not yet yielded the same scalable successes. Most of the available resistance is currently at the proof-of-concept level. Good examples for promising interventions are the development of aphid- and thrips-resistant cowpea varieties using modern and efficient phenotyping protocols, whereby hundreds of accessions from the IITA mini-core collection can be evaluated in a short time (Chapter 10). Coupled with new breeding approaches such as marker-assisted recurrent selection, there is hope for faster and more targeted progress to find durable resistance sources against these two pests.

Another interesting example is given by the lessons learned during the quest for resistance sources against the Asian soybean rust (ASR) which started to cause widespread epidemics in Africa in the mid-1990s. All varieties produced at that time in Africa were highly susceptible to the disease. Subsequently, some varieties with good levels of resistance were found, which are still being used today. However, ASR is capable of recombining into more virulent pathotypes, which needs monitoring using modern surveillance tools. At the same time, ASR-resistant genes are to be stacked for developing varieties with more durable resistance (Chapter 10).

Because of their intractable nature with conventional resistance breeding, banana *Xanthomonas* wilt (Chapter 9) and legume pod borer (Chapter 10) have been the object of intensified efforts to develop transgenic products. While IITA geneticists and breeders were the initiators of the transformation of bananas with two genes obtained from bell pepper, they were only marginally involved in the initiative to transform cowpea with the Cry1Ab gene construct to confer resistance against the legume pod borer. Both GM banana and cowpea varieties are currently tested under confined field trials in view of obtaining a deregulation under the prevailing national biosecurity laws.

3.2.2 Challenges

Perhaps the most disrupting challenge in host plant resistance breeding is the break down of good resistance levels already after a few years following their introduction due to mutations in the target pest organisms (Thrall et al., 2011). One of the classic examples was the loss of resistance in cowpea against the aphid *Aphis craccivora*, which had been mediated by a single dominant

Published by Burleigh Dodds Science Publishing Limited, 2019.

gene. Soon after the release of the resistant variety, new biotypes of the pest appeared, which were able to circumvent the resistant mechanism (Chapter 10). Other problematic organisms with high pathovar mutations are rust pathogens, as seen above for ASR.

Another technical obstacle in resistance breeding is the presence of co-infections with different types of plant pathogenic viruses. While improved screening methodologies yielded a number of cowpea lines highly resistant to potyviruses, this progress was halted by the presence of other viruses (cucumber mosaic virus, which belongs to the Bromoviridae) in the same plant tissue, which caused synergistic interactions. This problem has now been overcome by modifying the screening protocols for including co-infections in the selection process (Chapter 10).

One more recurring issue is the question of how to maintain high yields in existing varieties, while attempting to introgress resistant traits. Such resistance genes might have existed in the wild ancestors of the crop, but were inadvertently removed during the process of domestication and subsequent efforts to breed for high yields, as for example was the case during the Green Revolution. So the dilemma here is about the possible trade-offs between yield and resistance. As mentioned in the last section, there are some biotechnological approaches to this, but their application, so far, remains at the proof-of-concept level (Douglas, 2018).

As usual in plant breeding, only a small fraction of the genotypes with promising attributes is eventually developed into varieties that are given for free to national research programmes, for them to be named and/or adapted to their local conditions. During this process, the IITA origin is often lost, which in itself is what the CGIAR welcomes as part of the creation of international public goods. This approach can, however, pose problems of attribution when IITA's impact is being evaluated, for example by donor agencies funding that particular research.

With the advent of modern biotechnological tools leading to widespread deployment of transgenic crops, most conventional resistance breeding approaches have been abandoned in the developed world (Douglas, 2018). Also in African national programmes as well as at IITA, the modernization of breeding programmes has led to increased use of marker-assisted selection and other novel techniques. As mentioned above, two GMOs, transgenic bananas and cowpea, are now close to commercialization. However, they might face challenges in their implementation in most African countries, where the corresponding legal framework is still not ready. Compounding this problem is the weak seed (and tissue culture) production and distribution system, with low investments by national governments, and only marginal interest by the multinational private seed producing companies which prefer to invest in mainstream crops such as maize. However, with the further introduction of newer techniques such as CRISPR genome editing (Georges and Ray, 2017; Ryffel, 2017), which are posited to be more environmentally friendly and less controversial than, for example *Bt*-transformed crops, there might be new avenues for a true integration of GMOs into sustainably managed agro-ecosystems close to organic agriculture under a resistance management plan that would be easily implementable by low-literacy farmers (Ammann, 2007).

As already mentioned for biological control interventions, taxonomic problems – mostly misidentification of noxious organisms – can also hamper progress in host plant resistance (Chapter 4). Moreover, resistant traits are often species- and even strain-specific, so this needs to be taken into consideration when screening for resistant genotypes. To maintain this pool for the future development of new varieties, it is essential to maintain the genetic diversity of crop plants (Louwaars, 2018).

3.2.3 The way ahead

The CGIAR system is currently undergoing profound changes in its approach to breeding, particularly through its Excellence in Breeding Platform (http://excellenceinbreeding.org). It aims at more clearly defining breeding priorities by fixing client-oriented product profiles, an approach borrowed from the private sector. This new paradigm will also have implications for resistance breeding, whereby modern genotyping, phenotyping and bioinformatics tools will be applied throughout the product development pipeline.

The question remains whether these novel approaches will allow for the development of new varieties which can deliver metabolites and semiochemicals, for example those attracting biological control agents, or repelling herbivores (Wang et al., 2009; Pickett et al., 2014; Stenberg et al., 2015). In the past, breeders have been steadily responding to research managers, donors and possibly farmers by focussing mainly on 'genetic gains'. Today, there exist opportunities for re-visiting the way varieties should interact with their environment, including traits that can actively mediate plant–pest–natural enemies interactions. An even more unconventional approach concerns the modification of plants to deliver anti-microbial substances capable of suppressing endosymbionts in insects such as aphids, or to use microbial endosymbionts to deliver RNAi techniques; however, both approaches are still confined to the lab, with foreseeable challenges facing the fate of genetically modified microorganisms in the environment (Douglas, 2018).

Particularly since the wake-up call from the catastrophic invasion of FAW, IITA should be on the alert for new, potentially destructive invasive alien species. In fact, most staple crops in Africa are of exotic origin, and there are still a number of pests and diseases occurring in their respective area of origin, which are fortunately not present on our continent. These include, for example frog skin disease, superelongation disease and the witches' broom disease of cassava (Chapter 7). Being informed of such diseases and their causal agents, mode of spread as well as their impact is important in order to activate existing tools, for example sensitization, phytosanitary measures, efficient diagnosis and surveillance if an accidental introduction occurs. Perhaps more importantly, pathologists and breeders should be proactive in planning for a rapid response, which will mainly rely on resistance breeding in case of plant pathogens.

3.3 The impact of IITA projects: cultural and environmental control

3.3.1 Comparing results

Among most cultural practices developed at IITA in collaboration with extension services and farmers, relatively few address plant health protection or improvement directly. Cultural practices concerning phytosanitation and the use of virus-free planting material are mentioned in 38 projects, and in 20 of them it was the most important (>50%), in four even the only component. The mean importance and size of the 20 projects yielded a score of 2.4. Seven projects (35%) remained academic studies, six (30%) reached proof-of-concept stage or were already tested in the field and seven (35%) are in use by farmers. Among the 20 projects with predominantly cultural practices input, five had size 4 or 5, but only two of them reached the stage of *use*. The other five with *use* had only size 2 or 3. In other words, there was no relation between input as indicated by size of the project and its achievement.

Much of this type of control falls under the designation of 'good agricultural practices' and its success and adoption rate often depends on the local environment and is hence difficult to scale out without modifications. Moreover, all techniques require the active involvement of the farmers, their households and sometimes even entire villages, in order to assure impact. Farmers' education and gender-sensitive approaches are therefore at the heart of implementing these techniques.

One rather unconventional cultural control approach to clean plantain suckers of nematodes and weevils was developed for smallholder farmers in Central Africa (Chapter 9). A first version of this approach was contemplating the use of a high-volume tank of water, which was gradually heated to a constant temperature of 52°C. Suckers were submerged in the hot water for about 20 min, which would clean them of nematodes and weevils. However, this system was rather difficult to transport and assemble, required a substantial initial investment and proved difficult to operate by unskilled farmers. Hence, it was simplified by using locally available implements such as large cooking pans. Water would be boiling in these pans, and the suckers dipped in the boiling water for some 30 s. This approach delivered the same pest-free quality suckers as the first method, but with significantly less investment and labour.

Attempts at using cultural control to manage problem weeds showed good levels of technical performance, but remained largely non-adopted (Chapter 14). The introduction of cover or alley crops to control speargrass is such an example. In fact, planted fallows with *Mucuna* spp. and *Gliricidia* trees arranged as alleys were able to suppress *Imperata* over several years, but farmers did not find enough incentives to be able to perceive these technologies as profitable. Similarly, *Striga* management offered many promising technical approaches, but implementation was sporadic (Chapter 14).

3.3.2 Challenges

The main problem with cultural control can be described in simple terms as the lack of immediate, visible and direct benefit to the farmer. The most striking example is the non-adoption of the cover crop *Mucuna* for speargrass control in West Africa, which was related to negative attributes of its seeds (neither commercial nor nutritional value, lack of availability) on the one side, and perceived negative impacts on cropping systems (reduction of available cropping surface, competition with short duration crops and prone to dry season fires), on the other hand. The same fate occurred to alley cropping, where additional labour, water and nutrient competition, and interference with established cropping practices were perceived as key determinants for poor or non-adoption of the technique (Mutsaers et al., 2017).

This is not astonishing since traditional coping strategies by farmers mostly consider yield increases in the short term, often within a single cropping season. Long-term strategies and related investments are considered too risky in prevalently rain-fed African agriculture. This is also the case for approaches for managing the parasitic weed *Striga*, where several options were developed, but the adoption rate remained stagnant because of this lack of long-term perspective and difficulty of explaining in order to obtain commitment by low-literacy farmers.

It also needs to be emphasized that cultural control recommendations often are highly location-specific and sometimes difficult to transfer to other agro-ecological conditions. Stronger partnership with national programmes is particularly important here, because grass-roots staff in extension services, development projects and NGOs is best placed to implement and adapt such results.

3.3.3 The way ahead

As we have seen for the sucker-boiling technique, often simple but efficient methodologies can prevent losses. However, because of the lack of scaling pathways, mainly due to the inability to attract corresponding funding for such approaches, which are not as 'flashy' as breakthroughs in genetic gains, these efforts remain largely at local levels, often not well documented and even less widely appreciated. Yet, without good soil and weed management, the best improved varieties will not be able to express their full yield potential, and even biological control agents will perform poorly under such conditions. Similarly, a more sustainable and robust *Striga* reduction strategy should promote scientifically proven technologies co-developed with farming communities and provide tangible impacts on improving soil fertility, stopping *Striga* emergence, reducing the *Striga* seed bank and improving crop yields. Most soil-related biotic stresses such as *Striga* and nematodes can be considered symptoms of bad soil management requiring a holistic soil health management, that is 'improved' cultural practices. This implies that research results, which yield successes at the local level, are to be communicated and scaled out for wider adoption and impact. The generation and transfer of appropriate and gender-sensitive knowledge capable of transforming the way low-literacy farmers perceive agriculture is therefore critical.

Cultural practices were, however, not always tested in environments where they were most needed, that is where the farmers were with 'their backs to the wall' and had no other options anymore, as detailed in the appropriate GIS-based maps. Both mentioned practices fared better when introduced for instance in an area of low-fertility in the Mono Province of Benin: alley-cropping trees for farmers who owned their land and *Mucuna* for those who only rented it. Moreover, most tree species introduced in alley-cropping trials have now found their application as solid wood lots (Douthwaite et al., 2002; Carsky et al., 2002). It is concluded that considering cultural practices not as silver bullets, but as highly targeted options for adaptation by farmers will improve their success rate.

3.4 The impact of IITA projects: biorational control including use of botanicals and entomopathogens

3.4.1 Comparing results

Biorational control is mentioned in 14 projects; in five of them it was the most important (>50%) and in two even the only component. The mean importance and size of the five projects yielded a score of 3.2. Two projects (40%) reached only proof-of-concept stage and three (60%) are now in use by farmers. Among the five projects employing predominantly biorational methods, two had size

5, both reaching the stage of *use*. The other three had only size 2, but one of them still reached *use*.

The outstanding successes of two key products from the IITA 'kitchen', Green Muscle® and Aflasafe®, were only possible thanks to major investments by the donor community (Chapters 7 and 13). Private manufacturers were supposed to take up the relay in large-scale production and commercialization, but both approaches have still not reached cruising speed. Other biorational products of perhaps equal potential for impact have remained at the proof-of-concept level and continue to rely on the IITA-Benin pilot production unit for farmer demonstration trials. One of them is a diamond-back moth – active strain of the entomopathogenic fungus *Beauveria bassiana*, which has demonstrated excellent performance against this key pest of cabbage (Chapter 12), but has remained on the starting blocks owing to the lack of large-scale production.

3.4.2 Challenges

As opposed to East Africa, where export commodities such as cut flowers and green beans have driven the development of a competitive bio-pesticide industry to overcome excessive pesticide residue levels, there have been no similar developments in West Africa. Due to the lack of alternatives, horticultural farmers in particular are still using hazardous chemical pesticides, often of doubtful quality, in an attempt to save their yields. An exception to this scenario is currently happening in Benin, where at local levels, there has been a substantial increase in the production and supply of neem oil and derived products, with increasing demand also from neighbouring countries. Four key drivers are responsible for this recent development: (1) the efficacy of neem-derived biorationals against a range of pests in various cropping systems; (2) the competitive price as compared to chemical pesticides; (3) the growing perception of the health and environmental hazards posed by the indiscriminate use of pesticides; and (4) the development of a value chain around neem by-products, including oil cake, which is a good source of organic fertilizer with concurrent nematicidal properties (Chapters 10 and 11).

3.4.3 The way ahead

While IITA has been prominent in developing biorationals, we do not have a clear comparative advantage in their production and commercialization. This should be the prerogative of the private sector. Also, venturing into commercial production and the ensuing need for covering it with patenting rights might conflict with the CGIAR open access policy. All this calls for strengthening our collaboration with the private bio-pesticide industry, possibly through some innovative forms of partnership and financial co-investments.

Published by Burleigh Dodds Science Publishing Limited, 2019.

3.5 The impact of IITA projects: chemical control

3.5.1 Comparing results

The use of synthetic chemicals is mentioned in 15 out of the 58 projects (25.9%) only, mostly as a small component of up to 20%. Evidently, IITA's research develops environmentally friendly technologies in line with the IITA Refreshed Strategy 2012–20 (see below). As further developed in the next section about IPM, chemical control should be used as the last resort to rescue a crop from total failure, before other more sustainable control options are developed and implemented. This is best illustrated for the case of FAW.

In some projects, pesticides are used as seed coating (Chapter 14) or in bait sprays (Chapter 12), methods that do not much pollute the environment. Others are sprayed openly, among them many herbicides that in the past were considered harmless.

3.5.2 Challenges

IITA does not actively conduct research in developing or testing chemical pesticides; this is the mainstay of industry. However, IITA scientists, particularly in the early days of the cowpea programme, have been screening insecticides, mainly with the purpose of using the most efficient ones to protect their breeding plots. Moreover, recommendations for the use of pesticides usually come from national regulatory services and authorities, which in most cases adhere to standard rules such as the ones published by WHO (2009).

So, what are the main challenges for applying pesticides at the farm level? As noted in Chapters 10, 12 and 15, farmers are not always aware of the degree of danger they expose themselves to by applying chemical pesticides without any protecting equipment. This insufficient knowledge is compounded by the high costs related to the purchase of such implements, together with the discomfort of wearing them under a hot and humid tropical climate. The nature and quality of the pesticides themselves can have serious implications for human and environmental health. While there exist lists of recommended pesticides for different cropping systems, pesticide resellers – often untrained – offer to the farming community what sells 'well', that is pesticides that are cheaper than the recommended ones, mostly re-packaged imports from Asia and of doubtful quality. Often insecticides for example against cotton pests are deviated and used on cowpea, against store product pest and so on. Insecticides against agricultural pests, which are seeping into rivers and groundwater, also affect mosquitoes and lead to resistance against malaria (Yadouleton et al., 2011; Djouaka et al., 2016).

But why are the recommended pesticides more expensive? Only the tiniest fraction of synthetic chemical compounds eventually makes it to

Published by Burleigh Dodds Science Publishing Limited, 2019.

commercialization as a pesticide. The risks and costs are entirely carried by the chemical industry, which needs to recoup its expenses. Specificity is more and more required and also possible, but reduces the market and increases prices.

3.5.3 The way ahead

As highlighted in Chapter 15, there is still scope for integrating chemical pesticides into IPM schemes, particularly low-toxicity and specific ones, such as those used for seed coating against *Striga,* which certainly offer an environmentally friendly solution. However, the price of those products is most often beyond the means of subsistence farmers, or they are simply not available because of marginal interest for pesticide distributors and resellers.

Recently, special concerns were voiced regarding neonicotinoids and their effects, for instance, on bees (Godfray et al., 2014). Though these are insecticides that are mostly too expensive for our farmers, these warnings should be heeded in order not to repeat mistakes made elsewhere. Similarly, the herbicide glyphosate, which is effective against problematic weeds such as speargrass, will probably be phased out in Europe by 2020 because of its alleged potential for promoting cancer (IARC, 2017). It should therefore also be considered dangerous for Africans, and illiteracy and poverty not accepted as excuses for its indiscriminate use in Africa. Instead, other IPM interventions and products should be promoted.

The practical approach of our IITA plant health team is to try, wherever possible, to enforce the notion of not using pesticides unless there is an emergency situation, and surely not as the first and sole option. At the same time, we should continue the discourse with the industry, to make it aware that biorationals can be as efficient as chemical pesticides, and very often cheaper in production, while keeping in line with the need to make a profit.

3.6 The impact of IITA projects: combining techniques in IPM

3.6.1 Comparing results

In 7 out of 58 projects, no single technology predominated, that is, contributed over 50%. Three of them (42.9%) remained academic studies, four (57.1%) reached proof-of-concept stage or were tested, but none was employed. Among all 58 projects, 18 employed only one technology while 40 (69.0%) employed several techniques, thereby fulfilling the criteria for IPM. However, it needs to be emphasized that by implementing several options simultaneously without a proper concerted effort does not necessarily qualify as IPM. Currently, best examples of IPM implementation are given for cowpea and horticultural crops. In the past, these two systems have accounted for the highest occurrence of indiscriminate use of pesticides, but gradually they are on the path to

Published by Burleigh Dodds Science Publishing Limited, 2019.

applying novel approaches in IPM. For cowpea in particular, there exists a number of compatible insect management options ranging from biological control agents (against aphids, thrips and pod borers), resistant/tolerant varieties (mainly aphids and thrips), specific (e.g. baculoviruses) and broad-spectrum biopesticides (neem products), which can all be combined to obtain a synergistic effect in the field (Agunbiade et al., 2014; Chapter 10).

3.6.2 Challenges

To be functional, IPM needs to be understood, accepted and properly practised by farmers beyond the lifespan of yet another donor-funded IPM project. This is possibly the most critical aspect of IPM, which per se is a knowledge-intensive approach. IPM was 'invented' mainly to reduce inappropriate pesticide applications, but was often used as a legitimacy exercise to spray them in good conscience (Chapters 2 and 15). The old IPM paradigm was focussing on the notion of operational thresholds for applying pesticides – economic, intervention and damage thresholds – all without incorporating health and environmental considerations. This concept was difficult to understand by low-literacy farmers, even when inculcated through participatory approaches such as Farmer Field Schools (FFS). Later versions of FFS moved away from the overreliance on intervention thresholds to include agro-biodiversity notions in their curricula. Farmer 'zoos' were established to show the different ecosystem functions, from herbivory to predation and biological control by parasitoids. However, FFS had a major drawback: their price tag. Hence, they were only successful for the limited time frame of the project funding them. Also, in some socio-cultural settings, the participation of women in formal training sessions such as FFS was not acceptable, yet women were at the frontline in some of the plant health and post-harvest activities (Chapter 15).

The success of any IPM endeavour relies heavily on giving quality training to farmers, and this can be an additional obstacle because of the scarcity of skilled trainers in national extension systems who can carry out such functions. Alternatives exist, for example by piggybacking on development projects and grass-roots NGOs, but their interventions are often limited by financial and timeline constraints. Today, with the advent of cheap and stable internet connectivity even in rural areas in Africa, IPM is about to be re-branded into 'precision IPM' and relies on modern ICT tools, including smartphone-based decision making and information exchange, for its deployment and large-scale dissemination (Chapters 10 and 15).

One particular case, cocoa in West Africa (Chapter 12), highlights the need for better harmonization between the good intentions of the industry to preserve the rainforest (e.g. Rainforest Alliance), and the reality of implementing plant health measures on the ground. In fact, shaded cocoa as in a rainforest needs more pesticide applications (against black pod and mirids), while opening the canopies would lower the use of fungicides by small-scale farmers and make the system more sustainable. The question remains, what can be done to approach these kinds of trade-offs in IPM.

3.6.3 The way ahead

We argue that a paradigm shift is needed for IPM to succeed in the future. In our vision, IPM 2.0 is primarily a science-guided decision-making support platform that (1) can reach low-literacy farmers with a customized, gender-targeted message in their own language, empowering them to take the right decision about pest management at the right time by using modern and affordable ICT tools; (2) is based on ecological principles regulating pest populations by promoting biodiversity that is essential for ecosystem function, as for example against banana diseases, banana nematodes (Chapter 9), cowpea insect pests (Chapter 10), cassava root scale (Chapter 6); (3) is supported by next-generation genomics tools to understand pest (meta-)populations (Agunbiade et al., 2014), with their interacting populations of endosymbionts (Brinkmann et al., 2008), as well as populations of arthropod and microbial antagonists (Pierce et al., 2014); (4) promotes resistant/tolerant varieties and enhanced agro-ecosystem resilience (Willis et al., 2018); (5) advocates the use of biopesticides (based on plant extracts and/or entomopathogenic microorganisms) if the pest populations get out of ecological balance (Chapter 13); and (6) insists on using healthy seed and good planting materials (Chapter 15). As mentioned earlier, safe use of insecticides can be considered *as the last resort* (Jepson et al., 2014).

The IITA plant health team will implement this new 'precision IPM' paradigm to manage the recent invader, the FAW, in Africa (Chapters 4 and 7). One of the challenges will be to uncover the vast unknowns about this pest in its new territory, particularly with regard to its interactions with the natural vegetation, for example during the off-season when cultivated cereals are not available. The importance of trees for enhancing ecosystem services will also have to be considered (Kuyah et al., 2016), as suspected for *Striga* control (Chapter 14). Other ecosystem functions of natural vegetation are the hosting of (indigenous and introduced) biocontrol agents during the off-season when the main crop is not present, as in the case for maize and cowpea (Chapters 7 and 10), and which is expected to apply also for FAW. One option that is currently put forward for FAW control is the use of GM maize, which is much more compatible with biocontrol than synthetic pesticides (Bale et al., 2008). However, as already discussed in the section on resistant varieties, the challenges for deploying GMOs in Africa remain considerable at technical, institutional and policy levels.

One of the critical issues for efficient FAW management is the detection of early life stages - egg masses and young caterpillars - which are more vulnerable to the application of biorationals, and also better exposed. Our proposal to address this issue is to further improve the Farming Interface App, co-developed by IITA and partners at the Michigan State University (Chapter 10), for (1) guiding farmers scouting in their own field for detecting early infestations, (2) empowering them to take real-time decisions about pest management, (3) feeding back to the VIPS expert system, developed by NIBIO (2018), real-time data on crop phenology, performance, pest occurrence and severity for improving the accuracy of VIPS predictions at local and landscape level. This tool has been deployed in several countries across the world, is scale neutral, that is it can be successfully deployed at the smallholder household level in the village as well as by large commercial farms, but needs to be tailored to local circumstances. Communication and sensitization campaigns that take advantage of such modern ICT tools for better farmer involvement will go a long way towards the implementation of research results beyond the end of the project, which must already include their substantial costs. They are expected to reach and impact far more farmers than traditional FFS. At the moment, commercialization of IPM advice seems, however, out of reach in most African countries.

4 Overall assessment

Among the 58 projects assessed when writing this book, 40 projects were implemented in field situations and covered a large area (importance of project 2–5). Among all projects, 20.7% had a major contribution (>50%) from biological control, 24.1% from resistant varieties, 34.5% were based on cultural manipulations, 8.6% derived their success from biorational control and in 12.1% no technique predominated. Modest chemical control measures were taken in 25.9% of all projects, but were never a major component (contributing between 10% and 20% to the total success). While 18 projects employed only one technology, 40 (69.0%) used more than one. In predominantly biological control projects, 40% were tested in the field and adopted by farmers; for resistant varieties, the corresponding figure was 31%, and those for cultural control measures 35%. This means, across the institute, a good third of all projects delivered technologies that were adopted by farmers. One-third of all technologies, though tested, remained to be widely used, and one-third were academic studies, which formed the scientific basis for developing novel approaches in plant health. Bigger projects achieved higher impacts when biological control or resistant varieties were the main technologies; for cultural practices no relation between input and impact was found.

These numbers show that investments from donors in breeding and biological control offer good success rates, and hence should guide donors in their decision making on where and how to invest their limited resources for maximum outcome and impact. There is no doubt about IITA's impact in genetic improvement and biological control for better plant health (= increased yield for farmers); however, due to complicated and location-specific combinations of interventions, the impact of cultural control and IPM measures is quite difficult to assess and is therefore often neglected even by our collaborators (Ainembabazi et al., 2018). And often the origin of varieties is not known or successful biological control is not observed, as exemplified by the cover image (Fig. 2). It shows a female *Therophilus javanus* searching to oviposit into a *Maruca vitrata* larva on an infested cowpea pod. This parasitoid has been imported from Asia into Africa, mass reared at IITA and successfully established (Chapter 10). It is a perfect example of one of these contributions, in this case biological control, that go mostly unobserved by farmers and extension officers alike.

In general, IITA was promoting IPM that was based on research outputs beyond mere insecticide management. The projects demonstrate the practical implementation of sustainable IPM measures aiming at high productivity in the difficult African situation, though without fulfilling the stringent requirements of European 'organic' agriculture.

In Africa, agricultural production has so far largely been organic by default, due to farmers' lack of resources to buy or even have access to synthetic pesticides and other inputs. This, however, applies only to staple crops. In cash crops, overuse, misuse and abuse of synthetic pesticides are evident and likely increasing. Half-hearted attempts at introducing IPM in the form of pesticide management thereby only tend to perennially install and guarantee further insecticide use.

Environmental and health-related pesticide-induced problems are real, but not sufficiently addressed in most African countries. Likewise, many African

Figure 2 *Therophilus javanus* parasitizing *Maruca vitrata*. Photograph by Georg Goergen.

Published by Burleigh Dodds Science Publishing Limited, 2019.

countries have pesticide legislation in place, but there is no enforcement, therefore inappropriate sale and use of synthetic pesticides are not a concern. Only produce exported to Europe and abroad is analysed for pesticide residues, whereupon pesticide practices are adjusted accordingly to fulfil the requirements of the importing countries. There is still major work to be done in Africa in awareness creation and capacity building to address this problem (Chapters 3 and 15).

Globally, insecticide-free vegetable production relies on a series of biocontrol programmes (see European glasshouses: van Lenteren, 2000). At present this is, however, almost impossible to repeat under African field situations, though certainly a worthy long-term goal. Certified 'organic farming' is known only in some highly particular socio-economic environments in East and South Africa. Flower producers, for instance, could benefit from the well-established biocontrol programmes of European glasshouses. Presently, even producers of the so-called fair-trade products seem to ignore the opportunity to reduce risks to their workers and the environment, as documented by high residues of cocktails of insecticides in their produce (Roseth, 2010). Unfortunately, there is no incentive to change practice in Africa since importing countries do not ask for minimum levels of pesticide residues in non-food plants.

Particularly in weed control, eco-friendly IPM solutions often ask for more labour by the farmer, who – once the project is over – might not profit anymore from side benefits of the project. Unfortunately, IITA and its sister institutes have no means of pushing for implementation once financial support ceases, while sales people of chemical companies will continue to effectively peddle their wares, irrespective of the research results that usually aim at curbing pesticide abuse. Again, better education at all levels about long-term benefits of environment-friendly solutions is absolutely needed.

Further opportunities to reduce the use of synthetic pesticides as well as the burden of farmers in pest control are offered by exploiting insect–plant interactions. Surveillance and monitoring with pheromone traps, combined with prognosis and forecasting models and ICT tools, should be part of the future IPM strategy for Africa. The much-promoted push-pull technique exploits this avenue (Pickett et al., 2014), but is highly situation-specific. More research is needed on chemical ecology on both indigenous and alien pests, on crop varieties used in Africa and under African field conditions to further harvest the benefits of this technology.

5 IITA's plans for the future

While traditional farming systems with long fallow periods are sustainable, they need to become more productive. In most aspects, African agriculture still is organic farming, a benefit that should not be lost by repeating all mistakes (pesticide treadmill) of other continents (Chapter 2). For IITA, organic farming in the context of IPM is not a dogma; we rely on scientific evidence to provide a diversified, productive agro-ecological system (Frison et al., 2011) with the aim to make ecosystems resilient (Willis et al., 2018). This can only be done in close collaboration with other national and international organizations.

IT-based solutions to help farmers and extension services to improve plant protection practices in real time has become and will become more important in the future (Chapters 10 and 15). IITA is very much involved in what we could call the beginning of a plant protection revolution in Africa. New ICT tools, combined with the 'old' professional expertise such as taxonomy (morphological and DNA based), insect behaviour studies, phenology studies, agro-ecological knowledge and more, carve out a new area that can completely change future IPM strategies for many crops in Africa. IITA is presently in the forefront of developing ICT tools to help countries survey the spread of new pests and diseases, develop pest diagnostic tools and phenological models for prognosis and forecasting of pest outbreaks for broadcasting through different channels. Presently, we may not be able to provide country- or region-wide simulation modelling that is as precise as we would like, mainly due to the general lack of up-to-date and adequate networks of meteorological stations in most of Africa. Satellite data can partly compensate for lack of detailed data. These hindrances should not stop Africa from taking part in the possibilities that new ICT tools offer to farmers when and where they need them.

IITA is in agreement with much of the UN-commissioned IAASTD report (IAASTD, 2012), but not in its statements concerning pesticides and GMOs (as well as inorganic fertilizer) (see also Chapter 2). IITA supports the Malabo Declaration (2017) and will pragmatically develop solutions and demonstrate what works under local circumstances, by rallying African colleagues of universities, other research institutions and farming associations around a common goal for gradual change in legislation. IITA is an efficient actor situated between universities and implementing NGOs and government agencies. Unfortunately, national programmes, the original partners in research, who are supposed to pick up, adopt and adapt the results, often have no funding for scaling. This asks for better support to NARS partners, which in a project-based institute like IITA becomes difficult, but needs to be tackled.

Our direct and most successful clients often manage large farms with high inputs. They are well prepared to benefit from research results. The vast majority of farmers run, however, small subsistence farms. If IITA wants to increase its impact its projects have to cater to these clients too by facilitating access to cheap technology and education also to the young and to women, for instance by promoting youth entrepreneurs to address also production of biological control agents and biopesticides at local and national levels.

In Chapter 1, one of our partners recommended 'that IITA moves more decisively towards a holistic agro-ecology framework and away from a reductionist approach prioritizing short-term solutions such as GMOs, pesticides and synthetic fertilisers'. This book should provide concrete answers to this recommendation and demonstrate that we do not favour pesticides or GMOs. What we can certainly do better is communicating to the research community and the public in general about our strategy, which is solidly anchored in science-driven pest management, prioritizing biological control, resistant varieties and biorationals whenever possible. We suspect the above comment stems from the fact that we are certainly successful in publishing in indexed journals, but poor when it comes to 'making noise' about our successes, for example in the media, and more recently, necessarily, the social media. Our 2012–20 refreshed strategy clearly spells out the ecological base of our plant health approach, but evidently this is not the kind of information the general public is expecting to read. Otherwise our colleagues would possibly have noted that the IITA plant health team is far from prioritizing the mentioned short-term solutions.

In the refreshed IITA strategy 2012–20, three performance indicators are key, namely (1) lifting 7.2 million Africans out of poverty, (2) restoring 7 500 000 ha of degraded land and (3) increasing the yield of IITA's mandate crops by 60%. It is of course difficult to distinguish which factors (new varieties, plant health interventions, fertilizer and soil improvement techniques, inclusion of women, etc.) have contributed most towards the institute's overall achievements or impact, because they are all interdependent of each other. For instance, the potential genetic gain from a good variety can only be realized if seeds are healthy, the soil is not degraded, water and nutrition are optimal, but also if the social context allows its deployment. Whether this in turn leads to poverty reduction also relies on post-harvest technologies, access to market and whether increased profit benefits the family in need. As shown in this volume, improved plant health management interacts with most of these factors and is one of the key components contributing to the overall success in partly or fully achieving these targets by 2022. This was also recognized by the independent committee selecting IITA as the winner of the Africa Food Prize 2018.

How well then have IITA's responses to the triple challenge to agriculture across the world been met?

1 Feed more people on the same land: IITA's achievements clearly contribute to this goal.
2 Mitigate climate change: our classical biocontrol, smarter insecticides, resistant varieties (particularly if produced in Africa), less energy-hungry weed control, agroforestry measures clearly contribute.

3 Protect biodiversity: biocontrol implementation inherently supports this agenda. Recent biodiversity studies across all taxa by the biodiversity centre and its attached research forests broaden the approach.

6 Recommendations

- Raise awareness within the CGIAR, the broader donor community and African governments that single classical biological control projects against invasive alien species can produce benefits as high as complete multi-year breeding programmes, but require corresponding capacity building.
- Improve taxonomic capacities sustainably in collaboration with CAB International for foreign exploration and attribute a similar status to non-plant genetic resources as to plant genetic resources.
- Promote the biodiversity centre to teach citizens about the value of biodiversity, which can only survive if protected by the citizens.
- Support basic studies to understand the systems in collaboration with universities across the globe, favouring studies on competitive displacement, dispersal, food webs, alternative hosts and ecological niches, behaviour (particularly of ovipositing females) and tri-trophic interactions.
- Support horizon scanning and early warning systems for rapid response to emerging biological risks in collaboration with the Inter-African Phytosanitary Council of the African Union.
- Support holistic plant breeding approaches that also aim to breed for synergistic effects between varieties and natural enemies (instead of favouring some pests indirectly and unwillingly due to antagonistic effects).
- Develop and promote sustainable and cheap soil improvement practices that target weed and pest control without increasing labour demand, by developing sustainable IPM interventions and alternative products so that obsolete synthetic pesticides and herbicides like glyphosate become farmers' last choice.
- Strengthen collaboration with the private bio-pesticide industry, possibly through innovative forms of partnerships and financial co-investments. Make industry aware that biorationals can be as efficient as chemical pesticides, and often cheaper in production, while keeping profit margins.
- Develop IT-based solutions to help farmers and extension services to improve plant protection practices in real time by combining traditional IPM with modern ICT tools – including artificial intelligence – and climate modelling to develop 'precision IPM'.
- Evaluate the economic, environmental and health impacts of all projects 5 years after their closing in collaboration with major donors.

Published by Burleigh Dodds Science Publishing Limited, 2019.

- Publicize better sustainable and environment-friendly techniques, based on impact assessment of all plant protection measures in order to discourage farmers to use cheap and seemingly easy-to-use pesticides.
- Support NARS for scaling up IPM technologies, including production of biological control agents and biopesticides by youth entrepreneurs.

7 Acknowledgements

We thank all donors of the mentioned projects. Without their support our work would not have been possible (major donors are listed in the contents pages). We are also grateful to all contributors to this book, which was a huge communal enterprise and lists achievements that no single author could have assembled. We thank our colleagues in national and international organizations, without whose collaboration through the last 50 years IITA could not exist. Recently, this work has been partly coordinated in the CGIAR Research Projects listed separately. And finally we thank the staff of Burleigh Dodds Scientific Publishing for their untiring support and encouragement.

8 Where to look for further information

Neuenschwander, P. and Tamò, M. (Eds). 2019. *Critical Issues in Plant Health: 50 Years of Research in African Agriculture*. Burleigh Dodds Scientific Publishing, Cambridge, UK.

Chapter 1: Dashiell, K. Key challenges in plant health in sub-Saharan Africa: stakeholder priorities.
Chapter 2: Neuenschwander, P. and Tamò, M. Introduction to critical issues in plant health: a 50-year retrospective on African agriculture.
Chapter 3: Kumar, P. L., Legg, J. P., Ayodele M., Mahuku G., Ortega-Beltran A. and Bandyopadhyay, R. Disease surveillance, diagnostics and germplasm health in crop protection in sub-Saharan Africa.
Chapter 4: Goergen, G., Neuenschwander, P. and Coyne, D. Conserving and exploiting biodiversity in crop cultivation in sub-Saharan Africa.
Chapter 5: Legg, J., Kumar, P. L., Mahuku, G., Wosula, E., Stavolone, L., Terry, E. and Bosque-Pérez, N. Viruses affecting African crops and their vectors.
Chapter 6: Toko, M., Neuenschwander, P., Yaninek, J. S., Ortega-Betran, A., Fanou, A., Zinsou, V., Wydra, K., Hanna, R., Fotso, A. and Douro-Kpindou, O. K. Identifying and managing plant health risks for key African crops: cassava.
Chapter 7: Bandyopadhyay, R., Cardwell, K. F., Ortega-Beltran, A., Schulthess, F., Meikle, W., Sétamou, M. and Cotty, P. J. Identifying and managing plant health risks for key African crops: maize.

Chapter 8: Mignouna, B. D., Kumar, P. L., Coyne, D., Bandyopadhyay, R., Ortega-Beltran, A., Bhattacharjee, R. and De Koeyer, D. Identifying and managing plant health risks for key African crops: yams, taro and cocoyam.

Chapter 9: Hauser, S., Gold, C., Pasberg-Gauhl, C., Gauhl, F., Akello, J., Jacobsen, K., Norgrove, L., Coyne, D., Kumar, L., Mahuku, G., Kaushal, M., Nakato, V., Tripathi, L. and Tripathi, J. Identifying and managing plant health risks for key African crops: banana and plantain.

Chapter 10: Tamò, M., Afouda, L., Bandyopadhyay, R., Bottenberg, H., Cortada-Gonzales, L., Murithi, H., Ortega-Beltran, A., Pittendrigh, B., Sikirou, R., Togola, A. and Wydra, K. D. Identifying and managing plant health risks for key African crops: legumes.

Chapter 11: Godonou, I., Sæthre, M.-G., Tepa-Yotto, G., Gnanvossou, D., Douro-Kpindou, O. K. and Coyne, D. Identifying and managing plant health risks for key African crops: vegetables.

Chapter 12: Neuenschwander, P., Gnanvossou, D., Hauser, S., Goergen, G., Hanna, R., Norgrove, L., Negloh, K. and Agboton, C. Identifying and managing plant health risks for key African crops: fruit and other tree crops.

Chapter 13: Schreurs, F., Bandyopadhyay, R., Kooyman, C., Ortega-Beltran, A., Akande, A., Konlambigue, M. and Van den Bosch, N. Commercial products promoting plant health in African agriculture.

Chapter 14: Chikoye, D., Ekeleme, F., Hauser, S., Menkir, A., Kamara, A. Y., Neuenschwander, P., Ajuonu, O. and Ajeigbe, H. A. Weeds affecting field crops and water bodies in Africa.

Chapter 15: Coyne, D., Abberton, M., Adetonah, S., Ayodele, M., Cortada, L., Gbaguidi, B., Hauser, S., Kumar, L., Neuenschwander, P., Schut, M., Tamò, M. and Togola, A. What makes Integrated Pest Management (IPM) work in sub-Saharan Africa.

Chapter 16: Neuenschwander, P., Tamò, M. and Sæthre, M.-G. Improving plant health in sub-Saharan Africa: conclusions and future challenges.

9 References

Agunbiade, T. A., Coates, B. S., Datinon, B., Djouaka, R., Sun, W., Tamò, M. and Pittendrigh, B. R. 2014. Genetic differentiation among *Maruca vitrata* F. (Lepidoptera: Crambidae) populations on cultivated cowpea and wild host plants: Implications for insect resistance management and biological control strategies. *PLoS One* 9, e92072.

Ainembabazi, J. H., Abdoulaye, T., Feleke, S., Alene, A., Dontsop-Nguezet, P. M., Ndayisaba, P. C., Hicintuka, C., Mapatano, S. and Manyong, V. 2018. Who benefits from which agricultural research-for-development technologies? Evidence from farm household poverty analysis in Central Africa. *World Development* 108, 28–46.

Ammann, K. 2007. Why farming with high tech methods should integrate elements of organic agriculture. *New Biotechnology* 25, 378–88.

Bale, J. S., van Lenteren, J. C. and Bigler, F. 2008. Biological control and sustainable food production. *Philosophical Transactions of the Royal Society, London, B Biological Sciences* 363, 761–76.

Barratt, B. I. P., Moran, V. C., Bigler, F. and van Lenteren, J. 2017. The status of biological control and recommendations for improving uptake for the future. *BioControl* 63, 155–67.

Bellotti, A. 2017. Reliable region predictions for automated valuation models. *Annals of Mathematics and Artificial Intelligence* 81, 71–84.

Bellotti, A., Herrera Campo, B. V. and Hyman, G. 2012. Cassava production and pest management: Present and potential threats in a changing environment. *Tropical Plant Biology* 5, 39–72. doi:10.1007/s12042-011-9091-4.

Bellows, T. S. 2001. Restoring population balance through natural enemy introductions. *Biological Control* 21, 199–205.

Birdlife. 2018. State of the world's birds. *Birdlife* (The Magazine). July – September 2018, p. 17.

Brinkmann, N., Martens, R. and Tebbe, C. C. 2008. Origin and diversity of metabolically active gut bacteria from laboratory-bred larvae of *Manduca sexta* (Sphingidae, Lepidoptera, Insecta). *Applied and Environmental Microbiology* 74, 7189–96.

Carsky, R. J., Douthwaite, B., Schulz, S., Sanginga, N., Manyong, V. M., Diels, J. and Keatinge, D. J. D. H. 2002. Lessons for appropriate technology generation: Example of soil management technology at IITA. In: Jamin, J. Y., Seiny, B. L. and Floret, C. (Eds), *Savannes africaines: Des espaces en mutation, des acteurs face à de nouveaux défis: Actes du colloque, mai 2002, Garoua, Cameroun*. Centre de Coopération Internationale en Recherche Agronomique pour le Développement (CIRAD), Montpellier, FR, pp. 1–8.

Djouaka, R. F., Riveron, J., Yessoufou, A., Tchigossou, G., Akoton, R., Irving, H., Djegbe, I., Moutairou, K., Adeoti, R., Tamò, M., Manyong, V. and Wondji, C. S. 2016. Multiple insecticide resistance in an infected population of the malaria vector *Anopheles funestus* in Benin. *Parasites and Vectors* 9, 1–12. doi:10.1186/s13071-016-1723-y.

Douglas, A. E. 2018. Strategies for enhanced crop resistance to insect pests. *Annual Review of Plant Biology* 69, 637–60.

Douthwaite, B., Manyong, V. M., Keatinge, J. D. H. and Chianu, J. 2002. The adoption of alley farming and Mucuna: Lessons for research, development, and extension. *Agroforestry Systems* 56, 193–202.

Edreira, J. I. R., Cassman, K. G., Hochman, Z., van Ittersum, M. K., van Bussel, L., Claessens, L. and Grassini, P. 2018. Beyond the plot: technology extrapolation domains for scaling out agronomic science. *Environmental Research Letters* 13, 054027 (online). doi:10.1088/1748-9326/aac092.

Eriksson, D., Akoroda, M., Azmach, G., Labuschagne, M., Mahungu, N. and Ortiz, R. 2018. Measuring the impact of plant breeding on sub-Saharan African staple crops. *Outlook on Agriculture* 47, 163–80.

FAO (Food and Agriculture Organization of the United Nations). 1996. Code of conduct for the import and release of exotic biological control agents. International Standards for Phytosanitary Measures, Publication 3. FAO, Rome.

Frison, E. A., Cherfas, J. and Hodgkin, T. 2011. Agricultural biodiversity is essential for a sustainable improvement in food and nutrition security. *Sustainability* 3, 238–53. doi:10.3390/su3010238.

Georges, F. and Ray, H. 2017. Genome editing of crops: A renewed opportunity for food security. *GM Crops Food* 8, 1-12.

Godfray, H. C. J., Blacquière, T., Field, L. M., Hails, R. S., Petrokofsky, G., Potts, S. G., Raine, N. E., Vanbergen, A. J. and McLean, A. R. 2014. A restatement of the natural science evidence base concerning neonicotinoid insecticides and insect pollinators. *Proceedings of the Royal Society B* 281. pii:20140558. doi:10.1098/rspb.2014.0558.

Guerci, M. J., Norton, G. S., Ba, M. N., Baoua, J., Alwang, J., Amadou, L., Moumouni, O. Karimoune, L. and Muniappan, R. 2018. Economic feasibility of an augmentative biological control industry in Niger. *Crop Protection* 110, 34–40.

Guiaşu, R. C. 2016. *Non-native Species and Their Role in the Environment: The Need for a Broader Perspective*. Brill, Leiden, Boston, 916pp.

Howarth, F. G. 1991. Environmental impacts of classical biological control. *Annual Review of Entomology* 36, 485–509. doi:10.1146/annurev.en.36.010191.002413.

IAASTD. 2012. International assessment of agricultural knowledge, science and technology for development. Summary approved by governments attending the IAASTD Intergovernmental Plenary in Johannesburg, South Africa, 608pp.

IARC. 2017. http://www.iarc.fr/en/media-centre/iarcnews/pdf/IARC_Response_Reuters_October2017.pdf (last accessed 12 December 2018).

IPCC. 2014. Synthesis report: Summary for policymakers. Working Groups of the Intergovernmental Panel on Climate Change, Fifth Assessment Report, 31pp.

Jepson, P. C., Guzy, M., Blaustein, K., Sow, M., Sarr, M., Mineau, P. and Kegley, S. 2014. Measuring pesticide ecological and health risks in West African agriculture to establish an enabling environment for sustainable intensification. *Philosophical Transactions of the Royal Society B* 369, 20130491. http://dx.doi.org/10.1098/rstb.2013.0491

Kremen, C. and Merenlender, A. M. 2018. Landscapes that work for biodiversity and people. *Science* 362(6412), eaau6020. doi:10.1126/science.aau6020.

Kuyah, S., Öborn, I., Jonsson, M., Dahlin, A. S., Barrios, E., Muthuri, C., Malmer, A., Nyaga, J., Magaju, C., Namirembe, S., Nyberg, Y. and Sinclair, F. L. 2016. Trees in agricultural landscapes enhance provision of ecosystem services in Sub-Saharan Africa. *International Journal of Biodiversity Science, Ecosystem Services and Management*. doi:10.1080/21513732.2016.1214178.

Levine, J. M., Adler, P. B. and Yelenik, S. G. 2004. A meta-analysis of biotic resistance to exotic plant invasions. *Ecology Letters* 7, 975–89.

Louwaars, N. P. 2018. Plant breeding and diversity: A troubled relationship? *Euphytica* 214, 114–23. doi:10.1007/s10681-018-2192-5.

Lynch, L. D. and Thomas, M. B. 2000. Nontarget effects in the biocontrol of insects with insects, nematodes and microbial agents: The evidence. *Biocontrol News and Information* 21, 117N–30N.

Malabo Declaration. 2017. Inaugural biennial review report of the African Union Commission on the implementation of the Malabo Declaration on accelerated agricultural growth and transformation for shared prosperity and improved livelihoods. *Assembly Decision (Assembly/AU/2(XXIII)) of June 2014*. The 2017 progress report to the *Assembly Highlights on Intra-African Trade for Agriculture Commodities and Services: Risks and Opportunities*.

Masood, E. 2018. The battle for the soul of biodiversity: An ideological clash could undermine a crucial assessment of the world's disappearing plant and animal life. *Nature* 560, 423–5.

Published by Burleigh Dodds Science Publishing Limited, 2019.

McNeely, J. A. and Scherr, S. J. 2001. Common ground, common future. How ecoagriculture can help feed the world and save wild biodiversity. IUCN, Gland and Future Harvest, Washington DC, 24pp.

Mutsaers, H., Coyne, D., Hauser, S., Huising, J., Kamara, A., Nziguheba, G., Pipers, P., Taulya, G., van Asten, P. and Vanlauwe, B. 2017. *Soil and Fertility Management Research in sub-Saharan Africa: Fifty Years of Shifting Visions and Chequered Achievements*. Routledge and Taylor and Francis, London, 361pp.

Neuenschwander, P. 2004. Harnessing nature in Africa: Biological control can benefit the pocket, health and the environment. *Nature* 432, 801–2.

Neuenschwander, P., Murphy, S.T. and Coly, E. V. 1987. Introduction of exotic parasitic wasps for the control of *Liriomyza trifolii* (Dipt., Agromyzidae) in Senegal. *Tropical Pest Management* 33, 290–7.

NIBIO. 2018. https://www.vips-landbruk.no/information/1/ (accessed on 6 November 2018).

Osborne, L. S. and Cuda, J. P. 2003. Release of exotic natural enemies for biological control: A case of damned if we do and damned if we don't? *Journal of Land Use and Environmental Law* 18, 399–407.

Pickett, J. A., Woodcock, C. M., Midega, C. A. O. and Khan, Z. R. 2014. Push – pull farming systems. *Current Opinion in Biotechnology* 26, 125–32.

Pierce, A. A., de Roode, J. C., Altizer, S. and Bartel, R. A. 2014. Extreme heterogeneity in parasitism despite low population genetic structure among monarch butterflies inhabiting the Hawaiian Islands. *PLoS One* 9, e100061.

Prasanna, B. M., Huesing, J. E., Eddy, R. and Peschke, V. M. 2018 *Fall Armyworm in Africa: A Guide for Integrated Pest Management*. First Edition. CIMMYT, Mexico, CDMX, 106pp. http://repository.cimmyt.org/xmlui/handle/10883/19204

Pratt, C. F., Constantine, K. L. and Murphy, S. T. 2017. Economic impact of invasive alien species on African smallholder livelihoods. *Global Food Security* 14, 31–7.

Prinsloo, G. L. and Uys, M. 2015. *Insects of Cultivated Plants and Natural Pastures in Southern Africa*. Entomological Society of Southern Africa, Hatfield, South Africa, 785pp.

Roseth, R. 2010. Plant protection products (residues) in waste from production and imported flowers. Bioforsk Report 5, No. 67, pp. 33 (in Norwegian).

Rousse, P. and Quilici, S. 2009. Chapter 5: Recent advances in biological control of fruit fly pests. In: Bandeira, R. R. (Ed.), *Current Trends in Fruit Fly Control on Perennial Crops and Research Prospects 2009*. Transworld Research Network, Trivandrum, Kerala, India, pp. 91–115. ISBN: 978 81 7895 411 0.

Roy, H. E., Brown, P. M., Adriaens, T., Berkvens, N., Borges, I., Clusella-Trullas, S., Comont, R. F., De Clercq, P., Eschen, R., Estoup, A., Evans, E. W., Facon, B., Gardiner, M. M., Gil, A., Grez, A. A., Guillemaud, T., Haelewaters, D., Herz, A., Honek, A., Howe, A. G., Hui, C., Hutchison, W. D., Kenis, M., Koch, R. L., Kulfan, J., Lawson Handley, L., Lombaert, E., Loomans, A., Losey, J., Lukashuk, A. O., Maes, D., Magro, A., Murray, K. M., Martin, G. S., Martinkova, Z., Minnaar, I. A., Nedved, O., Orlova-Bienkowskaja, M. J., Osawa, N., Rabitsch, W., Ravn, H. P., Rondoni, G., Rorke, S. L., Ryndevich, S. K., Sæthre, M.-G., Sloggett, J. J., Soares, A. O., Stals, R., Tinsley, M. C., Vandereycken, A., Wielink, P., Viglášová, S., Zach, P., Zakharov, I. A., Zaviezo, T. and Zhao, Z. 2016. The harlequin ladybird, *Harmonia axyridis*: Global perspectives on invasion history and ecology. *Biological Invasions* 18, 997–1044.

Ryffel, G. U. 2017. Editorial. I have a dream: Organic movements include gene manipulation to improve sustainable farming. *Sustainability* 392, 1–9. doi:10.3390/su9030392.

Savary, S., Ficke, A., Aubertot, J. N. and Hollier, C. 2012. Crop losses due to diseases and their implications for global food production losses and food security. *Food Security* 4, 519. doi:10.1007/s12571-012-0200-5.

Seebens, H., Blackburn, T. M., Dyer, E. E., Genovesi, P., Hulme, P. E, Jeschke, J. M., Pagad, S., Pyšek, P., van Kleunen, M., Winter, M., Ansong, M., Arianoutsou, M., Bacher, S., Blasius, B., Brockerhoff, E. G., Brundu, G., Capinha, C., Causton, C. E., Celesti-Grapow, L., Dawson, W., Dullinger, S., Economo, E. P., Fuentes, N., Guénard, B., Jäger, H., Kartesz, J., Kenis, M., Kühn, I., Lenzner, B., Liebholdj, A. M., Mosena, A., Moser, D., Nentwig, W., Nishino, M., Pearmann, D., Perg, I., Rabitsch, W., Rojas-Sandoval, J., Roques, A., Rorke, S., Rossinelli, S., Roy, H. E., Scalera, R., Schindle, S., Stajerová, K., Tokarska-Guzik, B., Walker, K., Ward, D. F., Yamanaka, T. and Essl, F. 2018. Global rise in emerging alien species results from increased accessibility of new source pools. *PNAS*. February 2018. doi:10.1073/pnas.1719429115.

Stenberg, J. A., Heil, M., Åhman, I. and Björkman, C. 2015. Optimizing crops for biocontrol of pests and disease. *Trends in Plant Science* 20, 698–712. doi:10.1016/j.tplants.2015.08.007.

Stokes, D. E. 1997. *Pasteur's Quadrant – Basic Science and Technological Innovation*. Brookings Institution Press, Washington DC, pp. 1–196. ISBN: 9780815781776.

Thomas, M. and Waage, J. 1996. *Integration of Biological Control and Host-plant Resistance Breeding*. CTA, Wageningen, The Netherlands, 99pp.

Thrall, P. H., Oakeshott, J. G., Fitt, G., Southerton, S., Burdon, J. J., Sheppard, A., Russell, P. J., Zalucki, M., Heino, M. and Denison, R. F. 2011. Evolution in agriculture: The application of evolutionary approaches to the management of biotic interactions in agro-ecosystems. *Evolutionary Applications* 4, 200–15.

van den Bosch, R., Bess, H. A. and Haramoto, F. H. 1951. Status of oriental fruit fly parasites in Hawaii. *Journal of Economic Entomology* 44, 753–9.

van Driesche, R. G., Carruthers, R. I., Center, T., Hoddle, M. S., Hough-Goldstein, J., Morin, L., Smith, L., Wagner, D. L., Blossey, B., Brancatini, V., Casagrande, R., Causton, C. E., Coetzee, J. A., Cuda, J., Ding, J., Fowler, S. V., Frank, J. H., Fuester, R., Goolsby, J., Grodowitz, M., Heard, T. A., Hill, M. P., Hoffmann, J. H., Huber, J., Julien, M., Kairo, M. T. K., Kenis, M., Mason, P., Medal, J., Messing, R., Miller, R., Moore, A., Neuenschwander, P., Newman, R., Norambuena, H., Palmer, W. A., Pemberton, R., Perez Panduro, A., Pratt, P. D., Rayamajhi, M., Salom, S., Sands, D., Schooler, S., Sheppard, A., Shaw, R., Schwarzländer, M., Tipping, P. W. and van Klinken, R. D. 2010. Review: Classical biological control for the protection of natural ecosystems. *Biological Control* 54, S2–S33.

van Lenteren, J. C. 2000. A greenhouse without pesticides: Fact or fantasy? *Crop Protection* 19, 375–84.

van Wilgen, B. W. and de Lange, W. J. 2011. The costs and benefits of biological control of invasive alien plants in South Africa. *African Entomology* 19, 504–14.

Wang, X. G., Nadel, H., Johnson, M. W., Daane, K. M., Hoelmer, K., Walton, V. M., Pickett, C. H. and Sime, K. R. 2009. Crop domestication relaxes both top-down and bottom-up effects on a specialist herbivore. *Basic and Applied Ecology* 10, 216–27.

WHO. 2009. The WHO recommended classification of pesticides by hazard and guidelines to classification 2009. https://www.who.int/ipcs/publications/pesticides_hazard_2009.pdf?ua=1.

Willis, K. J., Jeffers, E. S. and Tovar, C. 2018. What makes a terrestrial ecosystem resilient? *Science* 359, 988–9.

Wyckhuys, K. A. G., Wongtiem, P., Rauf, A., Thancharoen, A., Heimpel, G. E., Le, N. T. T., Fanani, M. Z., Gurr, G. M., Lundgren, J. G., Burra, D. D., Palao, L. K., Hyman, G., Graziosi, I., Le, V. X., Cock, M. J. W., Tscharntke, T., Wratten, S. D., Nguyen, L. V., You, M., Lu, Y., Ketelaar, J. W., Goergen, G. and Neuenschwander, P. 2018a. Continental-scale suppression of an invasive pest by a host-specific parasitoid underlines both environmental and economic benefits of arthropod biological control. *Peer Journal* 6, e5796. doi:10.7717/peerj.5796.

Wyckhuys, K. A. G., Zhang, W., Prager, S. D., Kramer, D. B., Delaquis, E., Gonzalez, C. E. and van der Werf, W. 2018b. Biological control of an invasive pest eases pressures on global commodity markets. *Environmental Research Letters* 13, 094005.

Yadouleton, A., Martin, T., Padonou, G., Chandre, F., Asidi, A., Djogbenou, L., Dabiré, R., Aïkpon, R., Boko, M., Glitho, I. and Akogbeto, M. 2011. Cotton pest management practices and the selection of pyrethroid resistance in *Anopheles gambiae* population in northern Benin. *Parasites and Vectors* 4, 60. doi:10.1186/1756-3305-4-60.

Further reading

Hershey, C. (Ed.). 2017a. *Achieving Sustainable Cultivation of Cassava Volume 1: Cultivation Techniques*. Burleigh Dodds Science Publishing, Cambridge, UK. ISBN: 978 1 78676 000 5.

Hershey, C. (Ed.). 2017b. *Achieving Sustainable Cultivation of Cassava Volume 2: Genetics, Breeding, Pests and Diseases*. Burleigh Dodds Science Publishing, Cambridge, UK. ISBN: 978 1 78676 004 3.

Kema, G. and Drenth, A. (Eds). 2018. *Achieving Sustainable Cultivation of Bananas Volume 1: Cultivation Techniques*. Burleigh Dodds Science Publishing, Cambridge, UK. ISBN: 978 1 78676 156 9.

Kogan, M. and Higley, L. (Eds). 2019. *Integrated Management of Insect Pests: Current and Future Developments*. Burleigh Dodds Science Publishing, Cambridge, UK. ISBN: 978 1 78676 260 3.

Maredia, K., Dakouo, D. and Mota-Sanchez, D. (Eds). 2003. *Integrated Pest Management in the Global Arena*. CAB International, Wallingford, UK, 560pp.

Neuenschwander, P., Borgemeister, C. and Langewald, J. (Eds). 2003. *Biological Control in IPM Systems in Africa*. CAB International, Wallingford, UK, 448pp.

PestSmart CABI. https://www.cabi.org/elearning/pestsmart/ (last accessed 12 December 2018).

Sivasankar, S., Bergvinson, D., Gaur, P., Agrawal, S. K., Beebe, S. and Tamò, M. (Eds). 2018a. *Achieving Sustainable Cultivation of Grain Legumes Volume 1: Advances in Breeding and Cultivation Techniques*. Burleigh Dodds Science Publishing, Cambridge, UK. ISBN: 978 1 78676 13 6 1.

Sivasankar, S., Bergvinson, D., Gaur, P., Agrawal, S. K., Beebe, S., Tamò, M. and Kelly, J. D. (Eds). 2018b. *Achieving Sustainable Cultivation of Grain Legumes Volume 2:*

Improving Cultivation of Particular Grain Legumes. Burleigh Dodds Science Publishing, Cambridge, UK. ISBN: 978 1 78676 140 8.

Watson, D. (Ed.). 2017a. *Achieving Sustainable Cultivation of Maize Volume 1: From Improved Varieties to Local Applications*. Burleigh Dodds Science Publishing, Cambridge, UK. ISBN: 978 1 78676 008 1.

Watson, D. (Ed.). 2017b. *Achieving Sustainable Cultivation of Maize Volume 2: Cultivation Techniques, Pest and Disease Control*. Burleigh Dodds Science Publishing, Cambridge, UK. ISBN: 978 1 78676 012 8.

Zimdahl, R. (Ed.). 2017. *Integrated Weed Management for Sustainable Agriculture*. Burleigh Dodds Science Publishing, Cambridge, UK. ISBN: 978 1 78676 164 4.

Index

Published by Burleigh Dodds Science Publishing Limited, 2019.

Published by Burleigh Dodds Science Publishing Limited, 2019.

Published by Burleigh Dodds Science Publishing Limited, 2019.

Published by Burleigh Dodds Science Publishing Limited, 2019.

Published by Burleigh Dodds Science Publishing Limited, 2019.

Published by Burleigh Dodds Science Publishing Limited, 2019.